Antibiotic Materials in Healthcare

Antibiotic Materials in Healthcare

Edited by

VARAPRASAD KOKKARACHEDU, MSC, PHD, MRSC
Centro de Investigacion de Polimeros Avanzados
Concepcion, Chile

VIMALA KANIKIREDDY, MSC, PHD
Department of Chemistry
Osmania University
Hyderabad, India

EMMANUEL ROTIMI SADIKU, PHD
Department of Chemical
Metallurgical and Materials Engineering
Tshwane University of Technology, South Africa

ACADEMIC PRESS
An imprint of Elsevier

ELSEVIER

Academic Press is an imprint of Elsevier
125 London Wall, London EC2Y 5AS, United Kingdom
525 B Street, Suite 1650, San Diego, CA 92101, United States
50 Hampshire Street, 5th Floor, Cambridge, MA 02139, United States
The Boulevard, Langford Lane, Kidlington, Oxford OX5 1GB, United Kingdom

Notices
Knowledge and best practice in this field are constantly changing. As new research and experience broaden our understanding, changes in research methods, professional practices, or medical treatment may become necessary.

Practitioners and researchers must always rely on their own experience and knowledge in evaluating and using any information, methods, compounds, or experiments described herein. In using such information or methods they should be mindful of their own safety and the safety of others, including parties for whom they have a professional responsibility.

To the fullest extent of the law, neither the Publisher nor the authors, contributors, or editors, assume any liability for any injury and/or damage to persons or property as a matter of products liability, negligence or otherwise, or from any use or operation of any methods, products, instructions, or ideas contained in the material herein.

Library of Congress Cataloging-in-Publication Data
A catalog record for this book is available from the Library of Congress

British Library Cataloguing-in-Publication Data
A catalogue record for this book is available from the British Library

ISBN: 978-0-12-820054-4

For information on all Academic Press publications visit our website at
https://www.elsevier.com/books-and-journals

Publisher: Andre Gerhard Wolff
Acquisitions Editor: Linda Versteeg-buschman
Editorial Project Manager: Pat Gonzalez
Production Project Manager: Sreejith Viswanathan
Cover Designer: Matthew Limbert

Typeset by TNQ Technologies

Working together to grow libraries in developing countries

www.elsevier.com • www.bookaid.org

List of Contributors

A.A. Adeboje, BTech, MSc, DEng
Institute for NanoEngineering Research (INER) and
    Department of Civil Engineering
Tshwane University of Technology
Pretoria, Gauteng, South Africa

T.A. Adegbola, BTech, MTech, MSc, DEmg
Department of Mechanical and Automation
    Engineering
Tshwane University of Technology
Pretoria, Gauteng, South Africa

Gbolahan Joseph Adekoya, BEng, MSc
Institute of NanoEnginieering Research (INER)
Department of Chemical
Metallurgical and Materials Engineering
Faculty of Engineering and the Built Environment
Tshwane University of Technology
Pretoria, Gauteng, South Africa

Department of Polymer and Textile Technology
Yaba College of Technology
Lagos, Lagos, Nigeria

Oluwasegun Chijioke Adekoya, ND (SLT),
    BSc (Hons), MSc (In view)
Department of Medical Laboratory Science
Faculty of Health Sciences
College of Medicine
University of Nigeria
Enugu, Enugu, Nigeria

Blessing A. Aderibigbe, PhD
Department of Chemistry
University of Fort Hare
Alice Campus
Alice, Eastern Cape, South Africa

Victor Chike Agbakoba, BSc, BSc Hons, MSc
Department of Chemistry
Faculty of Science
Nelson Mandela University
Port Elizabeth, Eastern Cape, South Africa

CSIR Material Science and Manufacturing
Polymers and Competence Area
Port Elizabeth, Eastern Cape, South Africa

O. Agboola, BEng, MTech, DTech
Department of Chemical Engineering
Covenant University
Ota, Ogun, Nigeria

J.O. Ajibola
Nursing and Allied Health Division
South Louisiana Community College
Lafayette, LA, United States

K.K. Alaneme
Department of Metallurgical and Materials Engineering
Federal University of Technology
Akure, Ondo, Nigeria

S. Alven, MSc
Department of Chemistry
University of Fort Hare
Alice Campus
Alice, Eastern Cape, South Africa

Ahamdu George Apeh, HND, BSc (in view)
Department of Polymer and Textile Technology
Yaba College of Technology
Lagos, Lagos, Nigeria

Abayomi Awosanya, ND, HND, BSc, MSc, PhD (in view)
Department of Polymer and Textile Technology
Yaba College of Technology
Lagos, Lagos, Nigeria

O.O. Ayeleru, BEng, MTech
Centre for Nanoengineering and Tribocorrosion (CNT)
Department of Chemical Engineering
School of Mines
Metallurgy and Chemical Engineering
Faculty of Engineering and the Built Environment
University of Johannesburg
Johannesburg, Gauteng, South Africa

A.M. Berhe, BSc
Department of Civil Engineering
Tshwane University of Technology
Pretoria, Gauteng, South Africa

W. Bezuidenhout, BTech
Department of Civil Engineering
Tshwane University of Technology
Pretoria, Gauteng, South Africa

Olusesan Frank Biotidara, ND, HND, MTech, PhD
Department of Polymer and Textile Technology
Yaba College of Technology
Lagos, Lagos, Nigeria

Babatunde Bolasodun, PhD
Department of Metallurgical and Materials Engineering
Faculty of Engineering
University of Lagos
Lagos, Lagos, Nigeria

D.A. Branga-Peicu, BSc, MSc
Department of Civil Engineering
Tshwane University of Technology
Pretoria, Gauteng, South Africa

Rodrigo Briones, PhD
Centre for Advanced Polymer Research (CIPA)
Concepción, Biobío Region, Chile

A.A. Busari, BTech, MEng, PhD
Institute for NanoEngineering Research (INER) and Department of Civil Engineering
Tshwane University of Technology
Pretoria, Gauteng, South Africa

B. Buyana, MSc
Department of Chemistry
University of Fort Hare
Alice Campus
Alice, Eastern Cape, South Africa

Oluyemi Ojo Daramola, BEng, MEng, PhD
Institute of NanoEnginieering Research (INER)
Department of Chemical
Metallurgical and Materials Engineering
Faculty of Engineering and the Built Environment
Tshwane University of Technology
Pretoria, Gauteng, South Africa
Department of Metallurgical and Materials Engineering
Federal University of Technology
Akure, Ondo, Nigeria

L. De Villiers, BTech
Department of Civil Engineering
Tshwane University of Technology
Pretoria, Gauteng, South Africa

D.A. Desai
Department of Mechanical and Automation Engineering
Tshwane University of Technology
Pretoria, Gauteng, South Africa

Víctor Díaz-García, Bs, PhD
Facultad de Ingeniería y Tecnología
Universidad San Sebastián
Concepción, Concepción, Chile

Ehigie David Esezobor, MSc
Department of Metallurgical and Materials Engineering
Faculty of Engineering
University of Lagos
Lagos, Lagos, Nigeria

A.A. Eze, HND, MTech
Department of Mechanical and Automation Engineering
Tshwane University of Technology
Pretoria, Gauteng, South Africa

Victoria Oluwaseun Fasiku, BTech, MSc
Biological Sciences
North West University
Mahikeng, North West, South Africa

Department of Biochemistry
North West University
Mafikeng Campus
Mahikeng, North West, South Africa

Department of Pharmaceutical Sciences
University of Kwazulu-Natal
Durban, KwaZulu-Natal, South Africa

Oladipo Folorunso, BEng, MEng, PhD
Department of Electrical Engineering
Faculty of Engineering and the Built Environment
Tshwane University of Technology
Pretoria, Gauteng, South Africa

A. Frattari, MSc
Laboratory of Building Design (LBD)
University Centre for Smart Building (CUNEDI) and
    Department of Civil
Environmental and Mechanical Engineering
Trento, Italy

Mariel Godoy, MSc
Departamento de Ciencias del Ambiente
Facultad de Química y Biología
Universidad de Santiago de Chile
USACH
Santiago, Santiago, Chile

Yskander Hamam, BEE, MSc, PhD, HDR (DSc)
French South African Institute of Technology
    (F'SATI)/Department of Electrical Engineering
Tshwane University of Technology
Pretoria, Gauteng, South Africa
École Supérieure d'Ingénieurs en Électrotechnique et
    Électronique
Cité Descartes
Paris, France

Daniel Hassan, BSc, MSc, MBA, PhD
Department of Pharmaceutical Sciences
University of Kwazulu-Natal
Durban, KwaZulu-Natal, South Africa

Shanganyane Percy Hlangothi, BSc, MSc, PhD
Department of Chemistry
Faculty of Science
Nelson Mandela University
Port Elizabeth, Eastern Cape, South Africa

Idowu David Ibrahim, BEng, MTech
Department of Mechanical Engineering
Mechatronics and Industrial Design
Tshwane University of Technology
Pretoria, Gauteng, South Africa

M.J. John
Department of Chemistry
Nelson Mandela University
Port Elizabeth, Eastern Cape, South Africa

CSIR Materials Science and Manufacturing
Polymers and Composites
Port Elizabeth, Eastern Cape, South Africa

Organisational Unit
School of Mechanical
Industrial & Aeronautical Engineering
University of the Witwatersrand
Johannesburg, South Africa

C. Kambole, BEng, MSc, DEng
Institute for NanoEngineering Research (INER) and
    Department of Civil Engineering
Tshwane University of Technology
Pretoria, Gauteng, South Africa

Vimala Kanikireddy, MSc, PhD
Department of Chemistry
Osmania University
Hyderabad, Telangana, India

Chandrasekaran Karthikeyan, MSc, PhD
Centre for Advanced Polymer Research (CIPA)
Concepción, Biobío Region, Chile

Vuyolwethu Khwaza, MSc
Department of Chemistry
University of Fort Hare
Alice Campus
Alice, Eastern Cape, South Africa

Kehinde Williams Kupolati, BSc, MSc, PhD
Department of Civil Engineering
Faculty of Engineering and the Built Environment
Tshwane University of Technology
Pretoria, Gauteng, South Africa

Jimmy Lolu Olajide, BEng, MEng
Department of Mechanical and Automation
    Engineering
Tshwane University of Technology
Pretoria, Gauteng, South Africa

Shadrack Joel Madu, BPharm, MSc
Faculty of Pharmacy
Department of Pharmaceutics and Microbiology
University of Maiduguri
Maiduguri, Borno, Nigeria

M.R. Maite, BTech
Institute for NanoEngineering Research (INER) and
    Department of Civil Engineering
Tshwane University of Technology
Pretoria, Gauteng, South Africa

Nyemaga Masanje Malima, PhD Candidate
Department of Chemistry
University of Zululand
KwaDlangezwa, KwaZulu-Natal, South Africa

Zintle Mbese, BSc Hons, MSc
Department of Chemistry
University of Fort Hare
Alice Campus
Alice, Eastern Cape, South Africa

M.J. Mochane, BSc, BSc Hons, MSc, PhD
Department of Life Sciences
Department of Life Sciences
Central University of Technology
Bloemfontein, Free State, South Africa

K.S. Mojapelo, BTech, MTech
Institute for NanoEngineering Research (INER) and
    Department of Civil Engineering
Tshwane University of Technology
Pretoria, Gauteng, South Africa

M.M. Mokae, BTech
Department of Civil Engineering
Tshwane University of Technology
Pretoria, Gauteng, South Africa

T.C. Mokhena, BSc, BSc Hons, MSc, PhD
Senior Researcher
Department of Chemistry
Nelson Mandela University
Port Elizabeth, Eastern Cape, South Africa

Senior Researcher
CSIR Materials Science and Manufacturing
Polymers and Composites
Port Elizabeth, Eastern Cape, South Africa

R.J. Moloisane, BSc, MSc
Department of Civil Engineering
Tshwane University of Technology
Pretoria, Gauteng, South Africa

N. Motsilanyane, BTech
Institute for NanoEngineering Research (INER) and
    Department of Civil Engineering
Tshwane University of Technology
Pretoria, Gauteng, South Africa

T.S. Motsoeneng, BSc, BSc Hons, MSc
University of South Africa
Department of Chemistry
Florida, Gauteng, South Africa

Mtibe
CSIR Materials Science and Manufacturing
Polymers and Composites
Port Elizabeth, Eastern Cape, South Africa

Jamilu Muazu, BPharm, MSc, PhD
Faculty of Pharmacy
Department of Pharmaceutics and Microbiology
University of Maiduguri
Maiduguri, Borno, Nigeria

Emmanuel Mukwevho, MSc, PhD
Department of Biochemistry
North West University
Mafikeng Campus
Mahikeng, North West, South Africa

**Wakufwa Bonex Mwakikunga, B.Ed (Hons),
    B.Ed (Science), MSc, PhD**
National Centre for Nano-Structured Materials
Council for Scientific and Industrial Research
CSIR
Pretoria, Gauteng, South Africa

J.M. Ndambuki, BSc, MSc, PhD
Department of Civil Engineering
Tshwane University of Technology
Pretoria, Gauteng, South Africa

M. Ndlovu, BTech
Department of Civil Engineering
Tshwane University of Technology
Pretoria, Gauteng, South Africa

Jethro Nkomo, BPharm, MPharm
Pharmacy Department
Faculty of Health Sciences
Nelson Mandela University
Port Elizabeth, Eastern Cape, South Africa

X. Nqoro, MSc
Department of Chemistry
University of Fort Hare
Alice Campus
Alice, Eastern Cape, South Africa

Dariela Núñez, PhD
Centre for Advanced Polymer Research (CIPA)
Concepción, Concepción, Chile

Omonefe Joy Odubunmi, BSc, MSc
Department of Polymer and Textile Technology
Yaba College of Technology
Lagos, Lagos, Nigeria

Omondi Vincent Ojijo, DTech, MTech
National Centre for Nano-Structured Materials
Council for Scientific and Industrial Research
CSIR
Pretoria, Gauteng, South Africa

P.A. Olubambi, BEng, MEng, PhD
Centre for Nanoengineering and Tribocorrosion (CNT)
Department of Chemical Engineering
School of Mines
Metallurgy and Chemical Engineering
Faculty of Engineering and the Built Environment
University of Johannesburg
Johannesburg, Gauteng, South Africa

M.S. Onyango, BSc, MSc, DEng
Department of Chemical
Metallurgical and Materials Engineering
Polymer Technology Unit
Tshwane University of Technology
Pretoria, Gauteng, South Africa

Shesan John Owonubi, PhD
Postdoctoral Researcher
Polymer Technology
Department of Chemistry
University of Zululand
KwaDlangezwa, KwaZulu-Natal, South Africa

Chemistry
Nelson Mandela University
Port Elizabeth, Eastern Cape, South Africa

Opeoluwa O. Oyedeji, BSc Hons, MSc, PhD
Department of Chemistry
University of Fort Hare
Alice Campus
Alice, Eastern Cape, South Africa

Suprakas Sinha Ray, PhD
National Centre for Nano-Structured Materials
Council for Scientific and Industrial Research
CSIR
Pretoria, Gauteng, South Africa

Gerardo Retamal-Morales, PhD
Santiago, RM
Chile

Neerish Revaprasadu, PhD
Department of Chemistry
University of Zululand
KwaDlangezwa, KwaZulu-Natal, South Africa

Emmanuel Rotimi Sadiku, PhD
Institute of NanoEngineering Research (INER)
Department of Chemical
Metallurgical & Materials Engineering
Tshwane University of Technology
Pretoria, Gauteng, South Africa

R.W. Salim, BSc, MSc, PhD
Department of Civil Engineering
Tshwane University of Technology
Pretoria, Gauteng, South Africa

Julio Sánchez, PhD
Departamento de Ciencias del Ambiente
Facultad de Química y Biología
Universidad de Santiago de Chile
USACH
Santiago, Chile

Associate Professor
Environmental Sciences
University of Santiago
Santiago, Chile

Motshabi Alinah Sibeko, BSc, BSc Hons, MSc, PhD
Doctor
Department of Chemistry
Faculty of Science
Nelson Mandela University
Port Elizabeth, Eastern Cape, South Africa

J. Snyman, BSc, MTech, DTech
Department of Civil Engineering
Tshwane University of Technology
Pretoria, Gauteng, South Africa

P.C. Tsipa
Department of Chemistry
Nelson Mandela University
Port Elizabeth, Eastern Cape, South Africa

Ugonna Kingsley Ugo, BEng, MSc
Department of Polymer and Textile Technology
Yaba College of Technology
Lagos, Lagos, Nigeria

Kokkarachedu Varaprasad, MSc, PhD, MRSC
Doctor
Centre for Advanced Polymer Research (CIPA)
Concepción, Concepción, Chile

Walther Ide, BSc
Centro de Investigación de Polímeros Avanzados
 (CIPA)
Edificio de Laboratorios CIPA
Concepción, Chile

Ebiowei Moses Yibowei, OND, BEng, MSc
Department of Polymer and Textile Technology
Yaba College of Technology
Lagos, Lagos, Nigeria

Yousof Farrag, PhD
Universidade da Coruña
Grupo de Polímeros
Departamento de Física
Escuela Universitaria Politécnica
Serantes, Ferrol, Spain

Contents

CHAPTER 1

Antibiotic Nanomaterials

KOKKARACHEDU VARAPRASAD • CHANDRASEKARAN KARTHIKEYAN •
VIMALA KANIKIREDDY • DARIELA NÚÑEZ • EMMANUEL ROTIMI SADIKU •
RODRIGO BRIONES

1 INTRODUCTION

Antibiotics have been widely used in healthcare applications for the control of bacterial infectious diseases [1–3]. In a more detailed situation, they can kill or inhibit the growth of bacteria, fungi, viruses, archaea, protozoa, microalgae, and other microorganisms [3]. According to a recent report, the abuse of traditional antibiotics (anti = against, biotic = life) has led to the rapid assembly of multidrug-resistant bacteria and they are killing several people worldwide [4]. The World Health Organization (WHO) published the list of new multidrug-resistant bacteria information, which is harmful to the living systems [5,6]. However, the drug-resistant bacteria cell envelope is strong due to poor antibiotic internalization, which can cause the resistance of antibiotics. In addition, bacterial resistance to antibiotics can ascend the expression of particular genes of resistance [7]. Generally, when there is increasing bacterial resistance in the living system, the dosages of antibiotics also increase to control bacteria, which causes the side effects to the human body [8]. To solve these issues, several researchers have been working on the generation of new antibiotic materials, by employing several methods. Fenati and his coresearchers synthesized inventive biofilm from Oxacillin, G-quadruplexes, β-lactam antibiotic by coupling [9]. Owing to its oxidizing behavior (peroxidase mimicking), it showed significant action toward *Staphylococcus aureus* bacteria. They explained the fact that the synthesis process is more economical, with wider operating conditions and have the ability to react with various substrates. Finally, they concluded that this novel system can provide a better candidate for peroxidase-like antimicrobial systems in the future. However, by using nanotechnology, several researchers have invested in new antibiotic materials, toward multidrug-resistant bacteria in healthcare applications. Nanocomposite materials, due to their advanced physical, chemical, and biological properties (size, solubility), can easily interact with bacterial envelopes and inhibit bacterial growth. In addition, they can easily carry drugs and distort the infection of bacteria.

Antibiotics are often composed of inorganic or organic materials. Lately, to enhance antibacterial properties, they are prepared with organic and inorganic nanomaterials several methods. This is because metal-based antibiotic nanoparticles are often more effective in inhibiting multidrug-resistant bacteria and specifically, they are excellent alternatives to the conventional small-molecule antibiotics [10]. These engineered antibiotic materials can have certain levels of toxicity with less degradation. However, up to today, several researchers have been working on the generation of new antibiotics nanomaterials with little or no side effects to the living systems in healthcare. Overall, antibiotics are used for curing human and animal diseases, with less said effects in clinical applications.

2 NANOPARTICLES

Nanotechnology can offer new futures to organic and inorganic nanomaterials when compared to bulk organic/inorganic materials for use in healthcare [10–12]. In addition, complex antibiotic nanomaterials can reduce the severe toxicity of the materials, hence, overcoming the anticipated resistance and lowering cost, thereby enhancing their applicability in clinical applications [13]. Through nanotechnology, researchers have synthesized small size nanomaterials, and their characteristics greatly enhanced their applicability in biomedical applications, especially, toward microbe's infections. Stauber et al. [14] have clearly enunciated the strong interactions between nanomaterials and human cells (Fig. 1.1). They stated the fact that the many nanomaterials assets and attributes define their technological applications. Therefore it is obvious that the mainly advanced physicochemical properties of the

Antibiotic Materials in Healthcare. https://doi.org/10.1016/B978-0-12-820054-4.00001-X

(A)

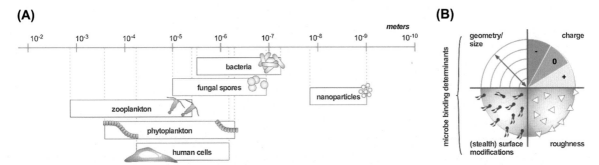

FIG. 1.1 **(A)** Relative size range of microbes, cells, and nanomaterials. **(B)** Physicochemical characteristics of nanomaterials potentially determine their interactions with microbes [14].

nanoantibiotic materials can lead to the alteration of the pharmacokinetics materials [15]. According to the literature, several nanomaterials have themselves exhibited a very high degree of antimicrobial activity [16]. This antibacterial property mainly depends on size, shape, chemical composition, chemical modification, and coating of nanoparticles, as well as the solvent, used [17]. They are stable (pH, temperature), storable for a long period, and can control infection by using in vitro and in vivo methods [13]. Nanomaterials can be used to carry nanodrugs, for sustained and controlled delivery of antibiotics, improved solubility, controllable and uniform distribution to the target places, reduce the side effects of antibiotics, and achieve superior cellular internalization [13,18,19]. However, by using nanotechnology, several researchers have developed effective nanomedicine against drug recentness bacteria. In addition, antibiotics are often encapsulated into biodegradable polymeric nanoparticles, which can provide protection to antibiotics against environmental deactivation and, hence, alter antibiotic drug movement (pharmacokinetics) and biodistribution [20]. Of importance is the advantage of polymeric particles, as they can easily be modified according to the target site, tissue, cells, and delivery of the drug. They are biocompatible, stable, and have good multifunctional properties to be used for in vitro and in vivo applications. By using polymer particles, drug dosage and dosing frequency can also minimize the drug side effects and improve the quality of life [21]. Overall, nanomaterials improve antibacterial efficacy against biofilm-related infectious diseases.

3 PHYSICAL AND CHEMICAL PROPERTIES NANOMATERIAL INFLUENCES ON BACTERIAL GROWTH

The nanoparticles' physical and chemical properties, such as size, high surface-to-volume ratio, and charges, are important parameters [22–25] for efficient antibacterial activity (Fig. 1.2). According to the literature, it is a well-known fact that smaller size nanoparticles have higher antibacterial activities than their larger counterparts [26]. This is because the size of nanoparticles is an important factor in their toxicity. In fact, small nanoparticles have relatively large surface-to-volume ratios when compared to the bulk molecules. Therefore small nanoparticles effectively have better interaction with bacterial and can easily pass the bacterial cell wall than with larger nanomaterials [27]. According to Wang et al. stories, few types of antibacterial nanomaterials do not depend on the size of the nanoparticle. However, antibacterial capacity depends on the production of reactive free radicals and nonradicals (ROS, RNS, RSS), which can destroy and inactivate the essential macromolecules, such as DNA, proteins, and lipids [27]. In addition, the morphology of nanoparticles with rough surfaces and rough edges have been found to adhere to the bacterial cell wall and cause damage to the cell membrane. Recently, Tong et al. explained on polymorphous ZnO nanocomplexes that exhibit spherical aggregates, fusiform-shaped microrods, nanosheet-based flowers, microrod-composed flowers, and nanopetal-built flower structures [28]. However, they specified that better antibacterial activity is derived from more effective antibacterial surfaces. The antibacterial surfaces depend mainly on the surface area, nanomaterials, and charges.

In clinical applications, the charge on nanomaterials also plays a key role in the control of bacterial infection. The charge of nanomaterials is calculated from the zeta potential and electron paramagnetic resonance spectroscopy. The nanoparticles are positively charged (at low pH), and they can electrostatically attract the negative charge of the bacterial cell wall. Owing to this phenomenon, active ions are released for antibiotic nanomaterials, which can penetrate into the outside cell membrane, react with biomolecules (DNA, proteins, lipids) inside the cell wall, and disrupt the cell integrity [29]. According to several reports, positively charged

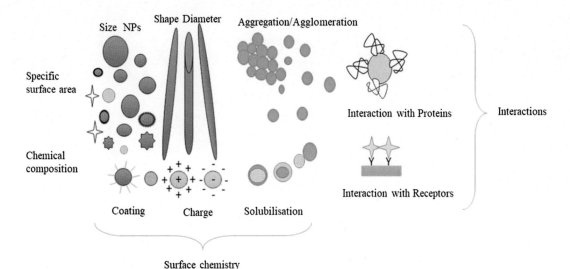

FIG. 1.2 Physicochemical properties of nanoparticles involved in biological activity [25].

nanoparticles generated a significant amount of active free radicals than the negatively charged and neutral nanoparticles. However, the negatively charged nanoparticles also control the bacterial growth at high concentration levels. On the other hand, the antibacterial activity of nanoparticles depends [24,30] on the environmental conditions, such as temperature and pH (dissolution of nanoparticles depends on pH). This is because of an active free radical generation that depends on their environmental conditions.

4 MECHANISM OF NANOMATERIALS

Multidrug-resistant bacteria are greater challenges in public healthcare [31,32]. Most infections produced by such resistant strains are on the increase, globally. Drug-resistant pathogens are a potential challenge for many antimicrobial drugs. Nanoscience and nanotechnology are the new routes to developing novel functional materials, based on distinct types of organic and inorganic antibiotic nanoparticles with different sizes and shapes and flexible antimicrobial properties [26,33]. In the more hybrid nanomaterials, most of the inorganic materials play vital roles in enhancing the stability, robustness, and shelf-life of the antibiotic nanomaterials [22].

The mechanism of antibiotics can involve in antibacterial activity in different ways, such as (a) direct interaction with the bacterial cell wall, (b) generation of free radicals (ROS, RNS, and RSS), and (c) induction of intracellular effects (e.g., interactions with macromolecules, such as DNA and/or proteins). A schematic diagram of organic, inorganic, and antibiotic mechanisms of nanomaterials is shown in Fig. 1.3.

Active free radicals of inorganic or organic antibiotics with one or more unpaired electron(s), denoted by ($^{\bullet}$), must be highly reactive but with a short time. The free radical formation may be due to chemical cleavage of covalent bonds when one electron each, of a common electron pair, remains with the split-off piece. The hydroxyl radical ($^{\bullet}OH$) or superoxide radical ($^{\bullet}O_2^-$) and hydroxyl radicals (from H_2O_2) and the reactive oxygen species (ROS) are of particular biological importance [34]. Singlet oxygen ($^1O_2$) is also a highly reactive agent, generated in phagocytic processes as well as in various reactions of photosensitization; it has both a radical and nonradical forms. Hydroxyl radicals ($^{\bullet}OH$) are generated from hydrogen peroxide (H_2O_2) and peroxyl radicals in enzymatic pathways (oxidoreductases, xanthine oxidase, nicotinamide adenine dinucleotide phosphate [NADPH] oxidase) and electron transmission from flavins and quinolones. Such reactions occur mainly in the context of inflammation processes and/or phagocytosis by macrophages. The $^{\bullet}OH$ radical is one of the most reactive antibiotic agents. The generation of $^{\bullet}OH$ radicals is initiated by reactions of metal ions with hydrogen peroxide (Fenton reaction), ionizing radiation, or by exposition to ozone. The most important sources of the superoxide radicals ($^{\bullet}O_2^-$) are from the mitochondrial respiratory chain and the endoplasmatic reticulum. In the mitochondria, the generation of $^{\bullet}O_2^-$ is based on electron transport chains that function imprecisely and leak; that may cause a single electron transfer to oxygen. In the respiratory chain, this leaky

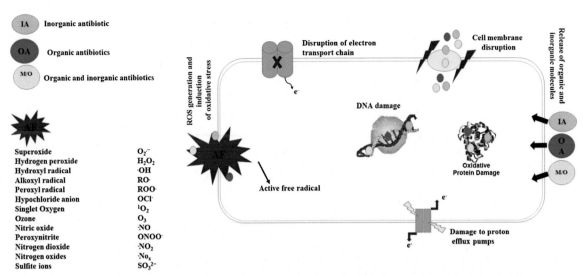

IA   Inorganic antibiotic

OA   Organic antibiotics

M/O   Organic and inorganic antibiotics

Superoxide	$O_2^{\cdot-}$
Hydrogen peroxide	H_2O_2
Hydroxyl radical	$\cdot OH$
Alkoxyl radical	$RO\cdot$
Peroxyl radical	$ROO\cdot$
Hypochloride anion	OCl^-
Singlet Oxygen	$^1O_2$
Ozone	O_3
Nitric oxide	$\cdot NO$
Peroxynitrite	$ONOO^-$
Nitrogen dioxide	$\cdot NO_2$
Nitrogen oxides	$\cdot No_x$
Sulfite ions	SO_3^{2-}

FIG. 1.3 Schematic diagram of an organic and inorganic and antibiotic mechanism of nanomaterials.

location is related, mainly, to the NADH-coenzyme-Q-reductase complex. However, the generation of hydroxyl radicals by the reaction of H_2O_2 with $\cdot O_2^-$ and catalyzed by transition metal ions is known as the *Haber–Weiss* reaction, which provokes the harmful effect of superoxide. By the regeneration of bivalent iron, a precondition for the Fenton reaction, $\cdot O_2^-$ radicals also take part in the generation of alkoxy radicals ($RO^\bullet$) from hydrogen peroxide. In the organism, this is particularly important when lipid peroxidation causes a chain reaction that can lead to cell death.

The generation of reactive nitrogen species (RNS), the nitrogen dioxide ($\cdot NO_2$) radicals, is a common indoor and outdoor gaseous pollutant (caused by combustion/traffic) that has been shown experimentally, to cause epithelial damage and to induce inflammation in the airways. Nitrogen monoxide ($\cdot NO$) and nitrogen dioxide ($\cdot NO_2$) are also highly reactive free radicals and the main constituents of urban smog. Moreover, sulfur dioxide (SO_2) itself is not a free radical, but it may react to produce sulfite ions (SO_3^{2-}) and upon reaction, superoxide generates other radicals, such as $SO_3^{\cdot-}$, $SO_4^{\cdot-}$, and $SO_5^{\cdot-}$ that contribute to urban smog [35]. The effects of organic and inorganic antibiotics on an antibacterial response are given in Table 1.1.

5 ENGINEERED NANOCOMPOSITES AND THEIR APPLICATIONS

Recently, microbial engineered nanomaterials have received considerable attention and interest from several researchers, and studies over the past few decades owe developments to the key role they can perform in enhancing humans and animal health. Lately, Elodous et al. reported on the significance of nanomaterials, which are classified as organic nanoparticles (liposomes, ferritin, dendrimers and micelles, polymer nanoparticles), inorganic nanoparticles (metal, metal oxides), and carbon-based nanomaterials (made of carbon, such as carbon nanotubes, fullerenes, activated black carbon, carbon nanofibers, and graphene) and their biomedical applications [50]. According to the literature based on nanoparticle toxicity, biodegradability, and sensitivity, they can be used for respective different applications [50]. Owing to their antibacterial properties (cell-damaging metabolites by the generation of free radicals and nonradical species) and small size (then the few living system and biomolecules), they are, often, used as multidrug-resistant microbes and other biomedical applications. In addition, due to their electronic conductivity, magnetic, and physicochemical properties, they are used for the detection of antibiotics [51]. Xia et al. reported on graphene-based nanomaterials and their antibacterial activities to protect human health [52]. Generally, graphene nanomaterials (graphene oxide, reduced graphene oxide) have large surface areas (two-dimensional crystal layers), good photothermal effect, and relatively low price [52]. In addition, its edge-cutting effect, oxidative stress effect, and cell entrapment ability make them popular in antibacterial investigation [52]. However, graphene can easily band with other bioactive agents, which can improve the functional property as well as antibacterial activity. Lin et al. developed photosensitizer Chlorin e6-doped silica nanoparticles

TABLE 1.1
Effect of Organic and Inorganic Antibiotics on Antibacterial Response.

S. No	Materials	Bacteria Name	Mechanism
1	ZnO	E. coli and E. faecalis	The high amount of active surface nanoparticles affects the bacterial biochemical processes by causing damage to DNA and protein denaturation; due to the effect, it triggers apoptosis or cell death as the bacteria fail to replicate normally [36].
2	ZnO—Ag	E. coli and S. aureus	The strong interaction between semiconductor ZnO and metallic Ag for bacteria [37].
3	ZnO	K. pneumoniae, E. coli, P. aeruginosa, and E. aerogenes	The nanoparticles lead to changes in antibacterial activities including ROS, and thereby cause membrane damage leading to cell death [38].
4	ZnO	E. coli and S. aureus	The antibacterial mechanism of ZnO particles can be attributed to both Zn^{2+} release and oxidative stress [39].
5	ZnO	E. coli, S. aureus, and S. enterica Typhimurium	The release of Zn^{2+} ions; these ions will penetrate the cell wall of bacteria and kill the microorganism [40].
6	Ag-saponite composites	S. aureus, B. subtilis, E. coli and P. aeruginosa, P. mirabilis, and S. epidermidis	The antibacterial effect of silver NPs is associated with the amount of released silver ions [41]
7	Ag/cellulose fibers	E. coli	Ag NPs have larger surface areas due to the effects of higher antibacterial and antifungal activities [42]
8	Polypyrrole@CuO	S. aureus and E. coli	Polypyrrole@CuO nanocomposites have increased its cationic behavior, inducing electrostatic/ionic interactions with bacteria. This interaction at the microorganism wall can lead to the leakage of intracellular electrolytes, causing cell death [43].
9	CuO	B. subtilis, S. aureus, E. coli, K. pneumonia, A. hydrophila, P. fluorescence, F. branchiophilum, E. tarda, and Y. ruckeri	Large surface areas are provided better contact microorganism and release of Cu^{2+} ions may be subsequently bound with DNA molecules that lead to helical structure disorder by inter- and intralinking nucleic acid strands [44]
10	CuO—ZnO composites	E. coli and methicillin-resistant Staphylococcus aureus (MRSA)	Ion release of Cu^{2+} and Zn^{2+}, and electrostatic interaction of NPs, which can change membrane permeability in virtue depolarization [45]
11	ZnO—CuO	MRSA (S. aureus)	The generation oxidative stress by electron transfer pathway and reactive oxygen species (ROS) generation effects killing the bacteria [46]

Continued

TABLE 1.1
Effect of Organic and Inorganic Antibiotics on Antibacterial Response.—cont'd

S. No	Materials	Bacteria Name	Mechanism
12	β-lactam antibiotics: amoxicillin, ampicillin, penicillin V, and piperacillin	*B. subtilis*	The loss of bactericide activity may be a transformation of the β-lactam (amoxicillin, ampicillin, and piperacillin), which is responsible for the antibiotic effects [47]
13	C_3N_4/perylene-3,4,9,10-tetracarboxylic diimide (PDINH)	*E. coli* and *S. aureus*	From the photocatalytic effects, the C_3N_4/ PDINH heterostructure can produce more ROS to have a better bactericidal effect [48]
14	Streptomycin functionalization on Ag NPs	*E. coli* and *S. aureus*	The toxicity toward bacteria is dependent on the surface functionalization of the AgNPs [49]

to overcome methicillin-resistant *Staphylococcus aureus* [53]. This system improves the photostability of Chlorin e6, which can generate (singlet oxygen) active free radicals under light illumination. These biocompatible nanoparticles were tested in vivo, and they improved wound healing efficiency when compared to pure Chlorin e6.

Principally, antibiotics are often composed of organic and inorganic materials, which can be used for the treatment of bacterial infection. In addition, decreasing the dose of antibiotics can enhance their lifetime. Fenouillet et al. developed hybrid nanosystems from ampicillin and gold nanoparticles, which have strong intermolecular interactions and with high stability [5]. They designed these nanosystems to reduce the dosage and increase the stability of ampicillin antibiotics. Cha et al. developed new antibiotic gold nanoparticles to cure a bacterial infection, with obvious benefit to intestinal microflora. They studied the microflora, distribution of gold, and biomarkers in mice. The result showed that 4,6-diamino-2-pyrimidinethiol-coated gold nanoparticles cured bacterial infection more effectively than levofloxacin, without harming intestinal microflora [54]. These nanoparticles are alternatives to antibiotics for the therapy of bacterial infections. However, to improve the antibacterial stability, Wu et al. employed a self-assembly method for the synthesis of nisin-encapsulated chitosan nanoparticles [55]. The results showed that the nanoscale antibiotics were spherical in shape with ~150 nm mean size. They concluded that the formation of nanoscale antibiotics depended on the interactions (hydrogen bonds and electrostatic) among the raw material functional species. However, the nanomaterials have good

stability with stable antibacterial activity (inhibiting the growth of microorganisms), and they exemplified a broad potential as a preservative material in the seafood industry.

Varaprasad et al. reported on the development of new engineered core—shell antibiotic nanoparticles by using bioactive curcumin and zinc oxide nanoparticles. They obtained ~45 nm ZnO core and ~12 nm curcumin shell layer nanoparticles. The complex nanoantibiotic nanoparticles have good solubility in distilled water due to the high surface area of the nanomaterials. However, bioactive curcumin has poor solubility due to its hydrophobic chemical nature and surface area. Owing to these factors, core—shell nanoparticles showed better antibacterial activity than pure curcumin and the commercial antibiotic amoxicillin. Finally, they observed that the antibacterial property of curcumin-zinc oxide nanoparticles depends mainly on the core/shell nanoparticle composition and the production of active free radicals [56]. Li et al. developed core—shell supramolecular gelatin nanoparticles for adaptive and one demand antibiotic delivery with a low (minimum) dose for the treatment of bacterial infections [57]. They explained that antibiotic delivery systems (nanobiomaterials) can improve the biodistribution and bioavailability of antibiotics is a practical strategy that can reduce the generation of antibiotic resistance and increase the lifespan of the newly developed antibiotics. However, the killing of bacteria with high efficacy depends on the core—shell nanomaterials compositions.

Furthermore, to kill the drug-resistant bacteria, Wu et al. synthesized layer-by-layer self-assembled biohybrid nanomaterials from antibiotics, enzymes, polymers, hyaluronic acid, and silica nanoparticles [58]. In here, cation polymer, lysozyme, and hyaluronic acid (showed bacteriostatic effects) improved the antibacterial activities of the nanomaterials. However, the biohybrid nanomaterials showed significant inhibition capacity on pathogens in bacteria-infected wounds in vivo. Rivas et al. developed the high antitumor activity of nanoparticles by using hydroxyapatite, antibiotic chloramphenicol, calcium phosphate, and pyrophosphate [59]. In their research, they produced a nanoplatform for anticancer therapy, which is based, mainly, on the combination of three materials: 1) **Antibiotics**: antibiotic that can target selectively the mitochondria of cancer cells, thereby inhibiting their functions, 2) **Nanoparticles**: mineral nanoparticles that are able to encapsulate the antibiotic and to enter into the cells across the cell membrane, and 3) **Coating**: a biocoating process that is needed to protect the antibiotic and regulate the release of the antibiotic, thereby increasing its therapeutic efficacy.

Multidrug-resistant bacteria have increased due to several factors. To control bacteria, multidrug-encapsulated nanomaterials have been widely synthesized, in recent years, by employing nanotechnology; this is because of the nanomaterials size, size-dependent plasmonic optical properties, and their drug-delivery abilities. Dig et al. developed three different sizes of silver nanoparticles, functionalized with 11-amino-1-undecanethiol and covalently conjugated them with two different antibiotics (ofloxacin, oflx) to inhibit *Pseudomonas aeruginosa* [14]. They, however, concluded that the smaller nanoparticles (smallest nanocarriers) showed a lower inhibitory effect than free oflx. Recently, photothermal therapy was employed for the treatment of bacterial infections by relying on multifunctional antibiotic nanomaterials that can exhibit magnetic and heat transfer characteristics. These special properties can destroy the bacterial infection by applying irradiation with near-infrared light. Huang et al. developed $Fe_3O_4@Au$ nanoeggs as photothermal agents for the selective killing of nosocomial and antibiotic-resistant bacteria [60]. They proposed that these materials have reasonably promising applicability, conceivably as an adjuvant therapeutic method, for the treatment of patients suffering from serious bacterial infections.

6 FUTURE PERSPECTIVES OF ANTIBIOTIC NANOMATERIALS

During the last few decades, there have been reports of active bacteria that have high resistance to the commonly available antibiotic drugs [61]. In addition, bulk antibiotics dosage forms create several health problems (delivery drug, side effects, frequency of administration) in healthcare applications. To solve these problems, several researchers have employed nanotechnology to develop innovative nanomaterials with advanced physicochemical and biological features [60]. They are a potent candidate for the treatment of bacterial infection without any side effects and with increased patient compliance in healthcare applications. Recently, Sonawane et al. developed a new lipid dendrimer hybrid nanoparticle to effectively deliver vancomycin to methicillin-resistant *Staphylococcus aureus* infections [62]. They reported that the lipid dendrimer hybrid nanoparticle system developed, improved encapsulation efficiency, provided sustained drug release, and enhanced antibacterial activity. They suggested that there is the need to study, with other classes of drugs, the effective management of various disease conditions. Agreeing to the Hossain et al. report, nanomaterials (especially carbon-based nanomaterials) have relatively few adverse effects on human health and the environment and certainly, organized research can increase its benefit and decrease its unfavorable effects [63]. They suggested that further investigation is required to overcome the adverse impacts of nanomaterials and hence, convert those impacts as benefits for healthcare applications.

However, in the future, more economically feasible antibiotic nanomaterials should be developed with a simple process, by employing suitable organic and inorganic materials. In addition, more stable antibiotic nanomaterials need to be developed to further control multidrug-resistant bacteria in healthcare.

7 CONCLUSIONS

Many researchers have developed nanoantibiotics to control bacterial infectious diseases by using a low dosage without any side effects. These nanoantibiotic materials were developed from the organic and inorganic nanomaterials to improve drug delivery, dissolution property, and the stability for a long period of the antibiotic nanomaterials. The mechanism of nanoantibiotic materials' activity depends mainly on direct interaction with the bacterial cell wall, the generation of free radicals, and the induction of intracellular effects. In addition, new and economically viable antibiotic nanomaterials are expected from organic and inorganic materials to control the associated bacterial infectious diseases that are common in the healthcare fields.

ACKNOWLEDGMENTS

The authors wish to acknowledge the Fondecyt Incioacion 11160073 (KVP), Fondecyt Postdoctoral Project 3190029 (KC and KVP), Programa de Cooperación Internacional/REDES180165 (KVP and RB), Centro de Investigación de Polímeros Avanzados (CIPA), CONICYT Regional and GORE BIO-BIO R17A10003.

REFERENCES

[1] K. Varaprasad, G.M. Raghavendra, T. Jayaramudu, J. Seo, Nano zinc oxide—sodium alginate antibacterial cellulose fibres, Carbohydrate Polymers 135 (2016) 349—355, https://doi.org/10.1016/j.carbpol.2015.08.078.

[2] K. Varaprasad, Co-assembled ZnO (shell)—CuO (core) nano-oxide materials for microbial protection, Phosphorus, Phosphorus, Sulfur and Silicon and the Related Elements 193 (2018) 74—80, https://doi.org/10.1080/10426507.2017.1417301.

[3] S. Liu, L. Mei, X. Liang, L. Liao, G. Lv, S. Ma, S. Lu, A. Abdelkader, K. Xi, Anchoring Fe 3 O 4 nanoparticles on carbon nanotubes for microwave-induced catalytic degradation of antibiotics, ACS Applied Materials and Interfaces 10 (2018) 29467—29475, https://doi.org/10.1021/acsami.8b08280.

[4] X. Yang, Q. Wei, H. Shao, X. Jiang, Multivalent aminosaccharide-based gold nanoparticles as narrow-spectrum antibiotics in vivo, ACS Applied Materials and Interfaces 11 (2019) 7725—7730, https://doi.org/10.1021/acsami.8b19658.

[5] X. Fenouillet, M. Benoit, N. Tarrat, On the role of inter-molecular interactions in stabilizing AuNP@Ampicillin nano-antibiotics, Materialia 4 (2018) 297—309, https://doi.org/10.1016/j.mtla.2018.09.036.

[6] B. Aslam, W. Wang, M.I. Arshad, M. Khurshid, S. Muzammil, M.H. Rasool, M.A. Nisar, R.F. Alvi, M.A. Aslam, M.U. Qamar, M.K.F. Salamat, Z. Baloch, Antibiotic resistance: a rundown of a global crisis, Infection and Drug Resistance 11 (2018) 1645—1658, https://doi.org/10.2147/IDR.S173867.

[7] R.S. Santos, C. Figueiredo, N.F. Azevedo, K. Braeckmans, S.C. De Smedt, Nanomaterials and molecular transporters to overcome the bacterial envelope barrier: towards advanced delivery of antibiotics, Advanced Drug Delivery Reviews 136—137 (2018) 28—48, https://doi.org/10.1016/j.addr.2017.12.010.

[8] J. Tian, J. Zhang, J. Yang, L. Du, H. Geng, Y. Cheng, Conjugated polymers act synergistically with antibiotics to combat bacterial drug resistance, ACS Applied Materials and Interfaces 9 (2017) 18512—18520, https://doi.org/10.1021/acsami.7b03906.

[9] R.A. Fenati, K. Locock, Y. Qu, A.V. Ellis, Oxacillin coupled G-quadruplexes as a novel biofilm-specific antibiotic for *Staphylococcus aureus* biofilms, ACS Applied Bio Materials 2 (2019) 1—7, https://doi.org/10.1021/acsabm.9b00336.

[10] D. Bradley, Moth helps develop nano-antibiotics, Materials Today 24 (2019) 3—4, https://doi.org/10.1016/j.mattod.2019.01.008.

[11] N. Khatoon, H. Alam, A. Khan, K. Raza, M. Sardar, Ampicillin silver nanoformulations against multidrug resistant bacteria, Scientific Reports 9 (2019) 1—10, https://doi.org/10.1038/s41598-019-43309-0.

[12] G.M. Raghavendra, T. Jayaramudu, K. Varaprasad, G.S. Mohan Reddy, K.M. Raju, Antibacterial nanocomposite hydrogels for superior biomedical applications: a Facile eco-friendly approach, RSC Advances 5 (2015), https://doi.org/10.1039/c4ra15995k.

[13] A.J. Huh, Y.J. Kwon, "Nanoantibiotics": a new paradigm for treating infectious diseases using nanomaterials in the antibiotics resistant era, Journal of Controlled Release 156 (2011) 128—145, https://doi.org/10.1016/j.jconrel.2011.07.002.

[14] R.H. Stauber, S. Siemer, S. Becker, G. Bin Ding, S. Strieth, S.K. Knauer, Small meets smaller: effects of nanomaterials on microbial biology, pathology, and ecology, ACS Nano 12 (2018) 6351—6359, https://doi.org/10.1021/acsnano.8b03241.

[15] S. Soares, J. Sousa, A. Pais, C. Vitorino, Nanomedicine: principles, properties, and regulatory issues, Frontiers in Chemistry 6 (2018) 1—15, https://doi.org/10.3389/fchem.2018.00360.

[16] Q. Li, S. Mahendra, D.Y. Lyon, L. Brunet, M. V Liga, D. Li, P.J.J. Alvarez, Antimicrobial nanomaterials for water disinfection and microbial control: potential applications and implications, Water Research 42 (2008) 4591—4602, https://doi.org/10.1016/j.watres.2008.08.015.

[17] N. Beyth, Y. Houri-Haddad, A. Domb, W. Khan, R. Hazan, Alternative antimicrobial approach: nano-antimicrobial materials, Evidence-Based Complementary and Alternative Medicine 2015 (2015), https://doi.org/10.1155/2015/246012.

[18] N. Abed, P. Couvreur, Nanocarriers for antibiotics: a promising solution to treat intracellular bacterial infections, International Journal of Antimicrobial Agents 43 (2014) 485—496, https://doi.org/10.1016/j.ijantimicag.2014.02.009.

[19] A. Sosnik, Á.M. Carcaboso, R.J. Glisoni, M.A. Moretton, D.A. Chiappetta, New old challenges in tuberculosis: potentially effective nanotechnologies in drug delivery, Advanced Drug Delivery Reviews 62 (2010) 547—559, https://doi.org/10.1016/j.addr.2009.11.023.

[20] M.H. Xiong, Y. Bao, X.Z. Yang, Y.H. Zhu, J. Wang, Delivery of antibiotics with polymeric particles, Advanced Drug Delivery Reviews 78 (2014) 63—76, https://doi.org/10.1016/j.addr.2014.02.002.

[21] K. Hadinoto, W.S. Cheow, Nano-antibiotics in chronic lung infection therapy against *Pseudomonas aeruginosa*, Colloids and Surfaces B: Biointerfaces 116 (2014) 772—785, https://doi.org/10.1016/j.colsurfb.2014.02.032.

[22] Y.N. Slavin, J. Asnis, U.O. Häfeli, H. Bach, Metal nanoparticles: understanding the mechanisms behind antibacterial activity, Journal of Nanobiotechnology 15 (2017) 1—20, https://doi.org/10.1186/s12951-017-0308-z.

[23] O. Choi, Z. Hu, Size dependent and reactive oxygen species related nanosilver toxicity to nitrifying bacteria, Environmental Science and Technology 42 (2008) 4583−4588, https://doi.org/10.1021/es703238h.

[24] L. Wang, C. Hu, L. Shoa, The antimicrobial activity of nanoparticles: present situation and prospects for the future, International Journal of Nanomedicine (2017) 1227−1249, https://doi.org/10.2147/IJN.S121956.

[25] A. Khezerlou, M. Alizadeh-Sani, M. Azizi-Lalabadi, A. Ehsani, Nanoparticles and their antimicrobial properties against pathogens including bacteria, fungi, parasites and viruses, Microbial Pathogenesis 123 (2018) 505−526, https://doi.org/10.1016/j.micpath.2018.08.008.

[26] K.R. Raghupathi, R.T. Koodali, A.C. Manna, Size-dependent bacterial growth inhibition and mechanism of antibacterial activity of zinc oxide nanoparticles, Langmuir 27 (2011) 4020−4028, https://doi.org/10.1021/la104825u.

[27] N.H. Zaki, Z. Husain, Enhanced antibacterial and anti-biofilm activities of biosynthesized silver nanoparticles against pathogenic bacteria, Journal of Genetic and Environmental Resources Conservation 4 (2016) 197−203.

[28] G.X. Tong, F.F. Du, Y. Liang, Q. Hu, R.N. Wu, J.G. Guan, X. Hu, Polymorphous ZnO complex architectures: selective synthesis, mechanism, surface area and Zn-polar plane-codetermining antibacterial activity, Journal of Materials Chemistry B 1 (2013) 454−463, https://doi.org/10.1039/c2tb00132b.

[29] A.S. Haja Hameed, C. Karthikeyan, V. Senthil Kumar, S. Kumaresan, S. Sasikumar, Effect of Mg^{2+}, Ca^{2+}, Sr^{2+} and Ba^{2+} metal ions on the antifungal activity of ZnO nanoparticles tested against Candida albicans, Materials Science and Engineering: C 52 (2015) 171−177, https://doi.org/10.1016/j.msec.2015.03.030.

[30] A. Ivask, A. Elbadawy, C. Kaweeteerawat, D. Boren, H. Fischer, Z. Ji, C.H. Chang, R. Liu, T. Tolaymat, D. Telesca, J.I. Zink, Y. Cohen, P.A. Holden, H.A. Godwin, Toxicity mechanisms in Escherichia coli vary for silver nanoparticles and differ from ionic silver, ACS Nano 8 (2014) 374−386, https://doi.org/10.1021/nn4044047.

[31] J.P. Horcajada, F. Belvis, X. Castells, S. Grau, M. Riu, E. Morales, M. Montero, F. Cots, M. Sala, M. Salvadó, M. Comas, Hospital costs of nosocomial multi-drug resistant Pseudomonas aeruginosa acquisition, BMC Health Services Research 12 (2012) 122, https://doi.org/10.1186/1472-6963-12-122.

[32] J.E. McGowan, Resistance in nonfermenting gram-negative bacteria: multidrug resistance to the maximum, American Journal of Infection Control 34 (2006), https://doi.org/10.1016/j.ajic.2006.05.226.

[33] V.B. Schwartz, F. Thétiot, S. Ritz, S. Pütz, L. Choritz, A. Lappas, R. Förch, K. Landfester, U. Jonas, Antibacterial surface coatings from zinc oxide nanoparticles embedded in poly(N-isopropylacrylamide) hydrogel surface layers, Advanced functional materials 22 (2012) 2376−2386, https://doi.org/10.1002/adfm.201102980.

[34] M.P. Murphy, A. Holmgren, N.G. Larsson, B. Halliwell, C.J. Chang, B. Kalyanaraman, S.G. Rhee, P.J. Thornalley, L. Partridge, D. Gems, T. Nyström, V. Belousov, P.T. Schumacker, C.C. Winterbourn, Unraveling the biological roles of reactive oxygen species, Cell Metabolism 13 (2011) 361−366, https://doi.org/10.1016/j.cmet.2011.03.010.

[35] N. Carolina, Ll, 244 (1969).

[36] S.N.A. Mohamad Sukri, K. Shameli, M.Mei-T. Wong, S.Y. Teow, J. Chew, N.A. Ismail, Cytotoxicity and antibacterial activities of plant-mediated synthesized zinc oxide (ZnO) nanoparticles using Punica granatum (pomegranate) fruit peels extract, Journal of Molecular Structure 1189 (2019) 57−65, https://doi.org/10.1016/j.molstruc.2019.04.026.

[37] M. Zare, K. Namratha, S. Alghamdi, Y.H.E. Mohammad, A. Hezam, M. Zare, Q.A. Drmosh, K. Byrappa, B.N. Chandrashekar, S. Ramakrishna, X. Zhang, Novel green biomimetic approach for synthesis of ZnO-Ag nanocomposite; antimicrobial activity against food-borne pathogen, biocompatibility and solar photocatalysis, Scientific Reports 9 (2019) 1−15, https://doi.org/10.1038/s41598-019-44309-w.

[38] Y. Gao, M.A.V. Anand, V. Ramachandran, V. Karthikkumar, V. Shalini, S. Vijayalakshmi, D. Ernest, Biofabrication of zinc oxide nanoparticles from Aspergillus Niger, their antioxidant, antimicrobial and anticancer activity, Journal of Cluster Science 30 (2019) 937−946, https://doi.org/10.1007/s10876-019-01551-6.

[39] T. Yang, S. Oliver, Y. Chen, C. Boyer, R. Chandrawati, Tuning crystallization and morphology of zinc oxide with polyvinylpyrrolidone: formation mechanisms and antimicrobial activity, Journal of Colloid and Interface Science 546 (2019) 43−52, https://doi.org/10.1016/j.jcis.2019.03.051.

[40] M. Kaushik, R. Niranjan, R. Thangam, B. Madhan, V. Pandiyarasan, C. Ramachandran, D.H. Oh, G.D. Venkatasubbu, Investigations on the antimicrobial activity and wound healing potential of ZnO nanoparticles, Applied Surface Science 479 (2019) 1169−1177, https://doi.org/10.1016/j.apsusc.2019.02.189.

[41] M. Sprynskyy, H. Sokol, K. Rafińska, W. Brzozowska, V. Railean-Plugaru, P. Pomastowski, B. Buszewski, Preparation of AgNPs/saponite nanocomposites without reduction agents and study of its antibacterial activity, Colloids and Surfaces B: Biointerfaces 180 (2019) 457−465, https://doi.org/10.1016/j.colsurfb.2019.04.066.

[42] K. Matsuyama, K. Morotomi, S. Inoue, M. Nakashima, H. Nakashima, T. Okuyama, T. Kato, H. Muto, H. Sugiyama, Antibacterial and antifungal properties of Ag nanoparticle-loaded cellulose nanofiber aerogels prepared by supercritical CO_2 drying, The Journal of Supercritical Fluids 143 (2019) 1−7, https://doi.org/10.1016/j.supflu.2018.08.008.

[43] M. Maruthapandi, A.P. Nagvenkar, I. Perelshtein, A. Gedanken, Carbon-dot initiated synthesis of polypyrrole and polypyrrole@CuO micro/nanoparticles with

enhanced antibacterial activity, ACS Applied Polymer Materials 1 (2019) 1181–1186, https://doi.org/10.1021/acsapm.9b00194.

[44] P.N.V.K. Pallela, S. Ummey, L.K. Ruddaraju, P. Kollu, S. Khan, S.V.N. Pammi, Antibacterial activity assessment and characterization of green synthesized CuO nano rods using *Asparagus racemosus* roots extract, SN Applied Sciences 1 (2019), https://doi.org/10.1007/s42452-019-0449-9.

[45] A.S. Lozhkomoev, O.V. Bakina, A.V. Pervikov, S.O. Kazantsev, E.A. Glazkova, Synthesis of CuO–ZnO composite nanoparticles by electrical explosion of wires and their antibacterial activities, Journal of Materials Science: Materials in Electronics 30 (2019) 13209–13216, https://doi.org/10.1007/s10854-019-01684-4.

[46] T. Jan, S. Azmat, Q. Mansoor, H.M. Waqas, M. Adil, S.Z. Ilyas, I. Ahmad, M. Ismail, Superior antibacterial activity of ZnO-CuO nanocomposite synthesized by a chemical Co-precipitation approach, Microbial Pathogenesis 134 (2019), 103579, https://doi.org/10.1016/j.micpath.2019.103579.

[47] A. Timm, E. Borowska, M. Majewsky, S. Merel, C. Zwiener, S. Bräse, H. Horn, Photolysis of four β-lactam antibiotics under simulated environmental conditions: degradation, transformation products and antibacterial activity, The Science of the Total Environment 651 (2019) 1605–1612, https://doi.org/10.1016/j.scitotenv.2018.09.248.

[48] L. Wang, X. Zhang, X. Yu, F. Gao, Z. Shen, X. Zhang, S. Ge, J. Liu, Z. Gu, C. Chen, An all-organic semiconductor C3N4/PDINH heterostructure with advanced antibacterial photocatalytic therapy activity, Advanced Materials (2019) 1–9, https://doi.org/10.1002/adma.201901965.

[49] T.S. Shruthi, M.R. Meghana, M.U. Medha, S. Sanjana, P.N. Navya, H. Kumar Daima, Streptomycin functionalization on silver nanoparticles for improved antibacterial activity, Materials Today: Proceedings 10 (2019) 8–15, https://doi.org/10.1016/j.matpr.2019.02.181.

[50] M. Abd Elkodous, G.S. El-Sayyad, I.Y. Abdelrahman, H.S. El-Bastawisy, A.E. Mohamed, F.M. Mosallam, H.A. Nasser, M. Gobara, A. Baraka, M.A. Elsayed, A.I. El-Batal, Therapeutic and diagnostic potential of nanomaterials for enhanced biomedical applications, Colloids and Surfaces B: Biointerfaces 180 (2019) 411–428, https://doi.org/10.1016/j.colsurfb.2019.05.008.

[51] L. Lan, Y. Yao, J. Ping, Y. Ying, Recent advances in nanomaterial-based biosensors for antibiotics detection, Biosensors and Bioelectronics 91 (2017) 504–514, https://doi.org/10.1016/j.bios.2017.01.007.

[52] M.Y. Xia, Y. Xie, C.H. Yu, G.Y. Chen, Y.H. Li, T. Zhang, Q. Peng, Graphene-based nanomaterials: the promising active agents for antibiotics-independent antibacterial applications, Journal of Controlled Release 307 (2019) 16–31, https://doi.org/10.1016/j.jconrel.2019.06.011.

[53] J.F. Lin, J. Li, A. Gopal, T. Munshi, Y.W. Chu, J.X. Wang, T.T. Liu, B. Shi, X. Chen, L. Yan, Synthesis of photo-excited Chlorin e6 conjugated silica nanoparticles for enhanced anti-bacterial efficiency to overcome methicillin-resistant: *Staphylococcus aureus*, Chemical Communications 55 (2019) 2656–2659, https://doi.org/10.1039/c9cc00166b.

[54] J. Li, R. Cha, X. Zhao, H. Guo, H. Luo, M. Wang, F. Zhou, X. Jiang, Gold nanoparticles cure bacterial infection with benefit to intestinal microflora, ACS Nano 13 (2019) 5002–5014, https://doi.org/10.1021/acsnano.9b01002.

[55] C. Wu, Y. Hu, S. Chen, J. Chen, D. Liu, X. Ye, Formation mechanism of nano-scale antibiotic and its preservation performance for silvery pomfret, Food Control 69 (2016) 331–338, https://doi.org/10.1016/j.foodcont.2016.05.020.

[56] K. Varaprasad, M.M. Yallapu, D. Núñez, P. Oyarzún, M. López, T. Jayaramudu, C. Karthikeyan, Generation of engineered core-shell antibiotic nanoparticles, RSC Advances 9 (2019) 8326–8332, https://doi.org/10.1039/c9ra00536f.

[57] L.L. Li, J.H. Xu, G. Bin Qi, X. Zhao, F. Yu, H. Wang, Core-shell supramolecular gelatin nanoparticles for adaptive and "on-demand" antibiotic delivery, ACS Nano 8 (2014) 4975–4983, https://doi.org/10.1021/nn501040h.

[58] Y. Wu, Y. Long, Q.L. Li, S. Han, J. Ma, Y.W. Yang, H. Gao, Layer-by-Layer (LBL) self-assembled biohybrid nanomaterials for efficient antibacterial applications, ACS Applied Materials and Interfaces 7 (2015) 17255–17263, https://doi.org/10.1021/acsami.5b04216.

[59] M. Rivas, L.J. Del Valle, A.M. Rodríguez-Rivero, P. Turon, J. Puiggalí, C. Alemán, Loading of antibiotic into biocoated hydroxyapatite nanoparticles: smart antitumor platforms with regulated release, ACS Biomaterials Science and Engineering 4 (2018) 3234–3245, https://doi.org/10.1021/acsbiomaterials.8b00353.

[60] W.C. Huang, P.J. Tsai, Y.C. Chen, Multifunctional Fe$_3$O$_4$@Au nanoeggs as photothermal agents for selective killing of nosocomial and antibiotic-resistant bacteria, Small 5 (2009) 51–56, https://doi.org/10.1002/smll.200801042.

[61] K. Niemirowicz, R. Bucki, Enhancing the fungicidal activity of antibiotics: are magnetic nanoparticles the key? Nanomedicine 12 (2017) 1747–1749, https://doi.org/10.2217/nnm-2017-0051.

[62] S.J. Sonawane, R.S. Kalhapure, S. Rambharose, C. Mocktar, S.B. Vepuri, M. Soliman, T. Govender, Ultra-small lipid-dendrimer hybrid nanoparticles as a promising strategy for antibiotic delivery: in vitro and in silico studies, International Journal of Pharmaceutics 504 (2016) 1–10, https://doi.org/10.1016/j.ijpharm.2016.03.021.

[63] F. Hossain, O.J. Perales-Perez, S. Hwang, F. Román, Antimicrobial nanomaterials as water disinfectant: applications, limitations and future perspectives, The Science of the Total Environment 466–467 (2014) 1047–1059, https://doi.org/10.1016/j.scitotenv.2013.08.009.

CHAPTER 2

Therapeutic Efficacy of Antibiotics in the Treatment of Chronic Diseases

VUYOLWETHU KHWAZA • ZINTLE MBESE • BLESSING A. ADERIBIGBE • OPEOLUWA O. OYEDEJI

1 INTRODUCTION

Chronic diseases such as cancer, human immune deficiency virus (HIV), and malaria seriously affect peoples' lives. The GLOBOCAN 2018, recently, reported that in 2018 about 18.1 million people were diagnosed with cancer and more than 9.6 million deaths were reported to be caused by cancer [1]. It is predicted that by 2025, over 20 million people will be living with cancer [2]. According to the World Health Organization (WHO), around 37 million people worldwide were diagnosed with HIV [3]. In addition, in 2016, malaria affected about 216 million people and caused 445,000 deaths worldwide [4]. This indicates the urgent need for the development of new therapeutic strategies to combat these diseases. Currently, Dave et al. 2018 reported that there is a shortage of drugs caused by the increase in drug prices [5]. The issue of drug productivity along with increasing drug prices have shifted the focus of drug developers to discover alternative strategies such as drug repurposing, combination therapy, and synthesis. The emergence and reemergence of various infections caused by cancer, malaria, and HIV have put a lot of pressure on the development of new therapeutic agents. The rapid occurrence of many drug-resistant pathogens such as drug-resistant parasites, viruses, and bacteria has been wildly reported [6–8]. Effective drugs are still not available for numerous infections such as cancer, malaria, and HIV. The development of novel broad-spectrum therapeutic drugs is increasingly difficult. Thus alternative strategies of using the existing drugs, such as drug repurposing, combination therapy, and synthesis, are needed to fight the emergence of drug-resistant infectious diseases. This chapter reports the therapeutic efficacy of various antibiotics in the treatment of chronic diseases such as cancer, HIV, and malaria. This chapter also gives evidence supporting the role of antibiotics and their potential clinical benefits in the management of chronic conditions.

Drug repurposing (also known as drug repositioning, redirecting, retasking, or profiling) is the strategy for finding new therapeutic effects of the existing drugs apart from their original medication indication [9,10]. Many drugs are used to treat diseases for which they were not initially designed for [11]. Drug repurposing has been reported to be useful when traditional anticancer monotherapy failed to give a safe and tolerable treatment for cancer patients. For instance, in cancer, the combination of drugs may consist of a repurposed neoprotector agent, such as a cytostatic agent that protects normal cells by arresting cell growth and a secondary or tertiary agent that kills cancer cells [12]. Thus many antibiotics have been repurposed for the treatment of various cancers (Table 2.1). Drug repurposing is cheap and the faster approach that offers many advantages over the lengthy process of traditional drug development. The development of new therapeutic drugs takes time and is resource consuming [13]. The complete process of developing a new drug and its approval to be used in humans take about 12–16 years [14]. Many pharmaceutical companies have adopted drug repurposing programs into their drug-development agenda [15]. Repurposing the existing drugs provides an alternative approach for the rapid identification of new therapeutic agents to treat infectious diseases with drug-resistant pathogens and other emerging infectious diseases. The data for the human pharmacokinetic profile, drug safety, and the preclinical results are already available for the approved drugs. From the traditional drug-development process, about one-third of the investigated drugs fail in clinical trials due to human toxicity and lack of efficacy [16,17].

Antibiotic Materials in Healthcare. https://doi.org/10.1016/B978-0-12-820054-4.00002-1

TABLE 2.1
The Mechanism of Action of Antibiotics in Various Types of Cancers.

Types of Antibiotics	Types of Cancer	Mechanism of Action	References
		CANCER	
Doxycycline	Prostate cancer	Inhibit the proliferation of castration-resistant prostate cancer cells	[89]
		Upregulates urokinase receptor activities including resistance to anoikis in human prostate cancer cell lines	[90]
	Breast cancer	Inhibit the proliferation of breast cancer cells and self-renewal of breast cancer stem cells	[91]
		Decreases plasma lysophosphatidate concentrations and inhibit NF-κB activation	[56]
	Lung cancer	Suppresses the proliferation and colony formulation, invasion, and migration of small-cell lung cancer cells and induces apoptosis	[92]
	Ovarian cancer	Sensitizes the ovarian cancer cells to cisplatin and inhibit the ability of invasion of epithelial ovarian cancer cells	[24]
	Colorectal cancer	Induces apoptosis of colorectal cancer cells via caspase 3 activation	[93]
Levofloxacin	Breast cancer	Inhibits the mitochondrial biogenesis by deactivating the PI3K/Akt/mTOR and MAPK/ERK pathways	[46]
	Lung cancer	Inhibits the mitochondrial respiration and reduces the ATP production	[49]
Salinomycin	Breast cancer	Reduces growth, proliferation, and metastasis of cisplatin-resistant breast cancer cells via NF-kB deregulation	[65]
		Inhibits progression of breast cancer by targeting HIF-1α/VEGF-mediated tumor angiogenesis in vitro and in vivo	[1]
	Colorectal cancer	Induces apoptosis in cisplatin-resistant colorectal cancer cells by the accumulation of reactive oxygen species	[64]
		Targets Wnt signaling in human colon adenocarcinoma cell line LoVo	[58]
	Prostate cancer	Decreases NF-κB-mediated CD44+ CSCs population and inhibition of growth and migration	[94]
		Cell death occurs through autophagy and mitophagy responses in PC-3 cells	[95]
	Lung cancer	Caspase-mediated cell death and inhibition of cell migration and invasion in LNM35 and A549 human lung cancer cells	[96]
		ER stress-mediated autophagic response in A549 cells	[97]
	Ovarian cancer	Apoptosis induction in human ovarian epithelial carcinoma cell line A2780 through inhibition of multidrug resistance protein gp170	[98]
		Caspase-mediated cell death in OVCAR-8 cell line in vitro	[99]
		Stat3 inactivation downregulates Skp-2 in OVCAR-8 ovarian cancer cell line and its multidrug-resistant derivative cells	[100]

Chronic lymphocytic leukemia	Targets Wnt signaling in cells isolated from patients of chronic lymphocytic leukemia	[101]
Head and neck cancer	Induces cell death and decreases CD44+ population in JLO-1 and UMSSC-10B in head and neck squamous cell carcinoma in vitro	[102]
Burkitt's lymphoma	Induces apoptosis in human CD4+ T-cell leukemia cells from the human patient	[103]
Hepatocellular carcinoma	Induces apoptosis by modulating intracellular Ca^{2+} and Wnt/β-catenin pathway in HepG2, SMMC-7721, and BEL-7402 cells in vitro and HepG2 orthotopic tumor in nude mice	[104]
Osteosarcoma	Inhibits osteosarcoma stem cells of U2OS, MG63, and SAOS2 in vitro and tumor regression of human osteosarcoma cells ZOS in nude mice	[105]
Uterine sarcoma	Decrease P-gp expression in MDR human uterine sarcoma cells MESSA/Dx5 expressing P-gp in vitro	[103]
Ciprofloxacin		
Prostate cancer	Sensitizes the hormone-refractory prostate cancer (HRPC) cells	[77]
Lung cancer	Mediates cancer stem cell phenotypes in lung cancer cells via caveolin-1-dependent mechanism	[106]
	Enhances TRAIL-induced apoptosis in lung cancer cells by upregulating the expression and protein stability of death receptors through CHOP expression	[88]
Ovarian cancer	Inhibits the topoisomerase II, which leads to cell death by apoptosis in malignant. After oral administration, it accumulates in ovarian	[81]
Colon carcinoma	Decreases the proliferation and induces apoptosis of human colon carcinoma cells, by blocking mitochondrial DNA synthesis	[79]
Osteosarcoma	Induces the proliferation of osteoblast-like MG-63 on human osteosarcoma cells in vitro	[107,108]
Monensin		
Prostate cancer	Induces oxidative stress and inhibit androgen signaling leading to apoptosis in prostate cancer cells	[71]
Lung cancer	Enhances cell cycle arrest and apoptosis induced by epidermal growth factor receptor inhibitors in lung cancer cells	[109]
Ovarian cancer	Inhibits the proliferation of cells, migration, and progression of cell cycle and induction of apoptosis in human ovarian cancer cells	[70]
Colorectal cancer	Inhibits canonical Wnt signaling in human colorectal cancer cells	[110]

The new therapeutic effect of a drug candidate can be found serendipitously or can be hypothesis-driven through rational approaches. The hypothesis-driven approaches for drug repurposing include experimental and computational approaches with enormous potential to provide a better understanding of mechanisms or pathways involved in disease pathogenesis [18]. The experimental drug-repurposing strategy includes the binding assays and the phenotypic screening methods used to identify binding interactions of ligands to assay components and also to identify lead compounds from large compound libraries, respectively [19]. The computational strategies are categorized into knowledge-based, target-based, signature-based, pathway-based, and target-mechanism-based approaches [20,21].

According to the WHO, the combination therapy of two or more drugs with different modes of action can delay or prevent the development of resistance [22]. The combination of antibiotics with anticancer drugs can result in a greater anticancer effect and less toxicity. Norouzi and colleagues revealed the efficacy of nanofibres loaded with an antibiotic, salinomycin which resulted in reduced glioblastoma cells in vitro. Over 50% of the cells showed apoptosis due to the capability of the drug-loaded fibres to induce intracellular reactive oxygen species over a period of 48 h. The action of salinomycin reduced the activity of the Wnt signaling pathway that is indispensable for the survival of stem cells (CSCs). Additionally, salinomycin increased the activity of tumor suppressors and the activity of caspase 3 the tumor death enzyme [23]. Wu and colleagues demonstrated that antibiotic doxycycline with an inhibitory effect on cancer can also enhance the chemosensitivity of cisplatin [24].

2 ANTIBIOTICS

The term antibiotics refers to any compounds that are used in the treatment of bacterial infections. Apart from their antibacterial activity, many other implications have been reported for antibiotics such as anticancer [25] and antimalarial [26]. Some are still under investigational and clinical trials [10,27]. In the treatment of cancer, DNA is the most common molecular target of many antibiotics. They affect synthesis and replication of DNA by interfering with DNA sequence, interacting with DNA and inhibit the topoisomerase and ultimately prevent the cancer cells from further division [28,29]. Antibiotics like tetracycline target the ribosome and synthesis of protein within the cell causing the prevention of the binding of aminoacyl-t-RNA to the 30S ribosomal subunit [28,30]. Additionally, some clinical trials performed with antibacterial drugs revealed that antibiotics alone are not suitable

to cure chronic diseases like malaria but they are highly effective in combination with antimalarial agents even against cases of drug-resistant malaria [31,32].

3 CANCER

Cancer is characterized by uncontrollable cell proliferation that interferes with normal organ function due to the developing number of tumor cells. It spreads to other organs and become very dangerous after metastasis. The cancer pathogenicity is based on cell proliferation, and the anticancer drugs rapidly damage the multiplying tumor cells [33]. Anticancer drugs inhibit the DNA metabolism and mitosis progression leading to apoptosis. Bacterial cells and cancer cells share many similar properties, like high replication rates, the way they spread within the host, the rapid development of resistance mechanisms against chemotherapeutic agents, and their tendency to become more aggressive during disease progression [34]. In addition, it has been hypothesized that cancer cells utilize cell—cell communication systems analogous to those of bacterial cells that allow them to successfully coordinate their attacks against the host [35,36]. The antibiotics discussed in this review have been demonstrated to target one or more cancer hallmarks (Table 2.1).

4 CHALLENGES IN CANCER TREATMENT

Drug resistance is a well-known concept that occurs when pharmaceutical drugs become ineffective. This phenomenon was first discovered when bacterial pathogens became resistant to a certain number of antibiotics; after that, similar mechanisms were also observed in other diseases such as cancer. Other methods of drug resistance are disease-specific, while some, such as drug efflux, which occurs in microbes and human drug-resistant cancers have also been reported [37]. Resistance in cancer cells is a major problem and sensitizing them toward chemotherapeutic drugs is a challenging task [38]. Drug resistance is a technique exhibited by various types of cancers that involve cellular and noncellular mechanisms for their resistance to survive. The main mechanism of drug resistance in cancer cells is the energy-dependent efflux of drugs, driven by P-gp pump [38]. Many treatment strategies, such as chemotherapy, targeted therapy, and newer immunotherapy, have been developed and modified to increase the therapeutic effect. The use of platinum-based chemotherapy was among the promising strategies in the treatment of cancer. Cisplatin and its two derivatives, oxaliplatin and carboplatin, have been used to successfully treat a number of solid

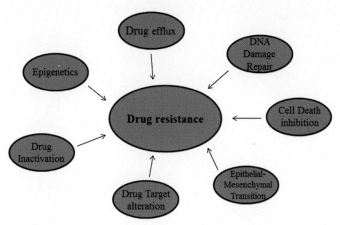

FIG. 2.1 Groups of mechanisms that can promote drug resistance in human cancer cells. These mechanisms can act independently or in combination and through many signal transduction pathway.

tumors, such as ovarian cancer, lung cancer, testicular cancer, and head and neck cancers [39]. The mechanism of action of these platinum-based drugs is based on intracellular hydrolysis (aquation) resulting in the formation of the positively charged aqua/hydroxido-species, capable to covalently attack the target DNA molecule, while the N-donor carrier ligands (NH_3 or RRdach) remain intact during the described mechanistic processes [40]. "In spite of all the advantages of platinum-based drugs, severe side effects such as neurotoxicity and nephrotoxicity limit their clinical application [41,42].

Although various successful strategies for cancer treatment have been proposed, drug resistance is still a limitation associated with various chemotherapeutic drugs. Unfortunately, mechanistic links between various anticancer drugs and the emergence of drug resistance have not been fully understood. Identifying the mechanisms of drug resistance is a potential approach for innovative drug development and improving the outcome of cancer patients [43]. Housman and colleagues stated that drug resistance is influenced by drug inactivation, drug efflux, drug target alteration, DNA damage repair, epithelial-mesenchymal transition, cell death inhibition, epigenetic effects, inherent cell heterogeneity, or any combination of these mechanisms (Fig. 2.1) [37].

Elshimali et al. also reported two different categories of mechanisms of resistance to cancer drugs: cellular mechanism and noncellular mechanism. The cellular mechanism addresses tumor cell-autonomous signaling pathways, insensitive to natural growth arrest signals, abolishes the cell contact inhibition, the ability to evade apoptosis, and the role of the tumor microenvironment. The noncellular mechanism is related to the pharmacological response [44]. Many factors are associated with the tumor cell-intrinsic modifications such as genetic alteration, chromatin modification, enrichment of cancer stem cells, loss of cell polarity, alteration in cell–cell and cell–matrix adhesion, and the regulation of receptor kinase signaling. All these factors together support detachment, the migration, and invasion of tumor cells that negatively influence the response to chemotherapy [45].

5 ANTIBIOTICS EFFECTIVE FOR CANCER TREATMENT

5.1 Levofloxacin

Levofloxacin (1) is the fluoroquinolone derivative used as an antibiotic to kill bacterial pathogens through inhibition of topoisomerase enzymes and prevent DNA replication (Fig. 2.2) [46]. Most antibiotics targeting DNA replication in bacteria have been proven to be effective against tumor cells by inhibiting mitochondrial biogenesis [47,48]. Song et al. (2016) revealed the inhibitory effects of levofloxacin on lung cancer cells in vitro and in vivo. Its mode of action consists of the inhibition of mitochondrial respiration, suppression of mitochondrial complex activities and the induction of oxidative damage [49]. Yu et al. (2016) also studied the effect of levofloxacin in breast cancer cells and concluded that its effect on breast cancer is dependent on its ability to suppress mitochondrial biogenesis [46]. Both studies recommended the repurpose of levofloxacin for lung and breast cancer treatment.

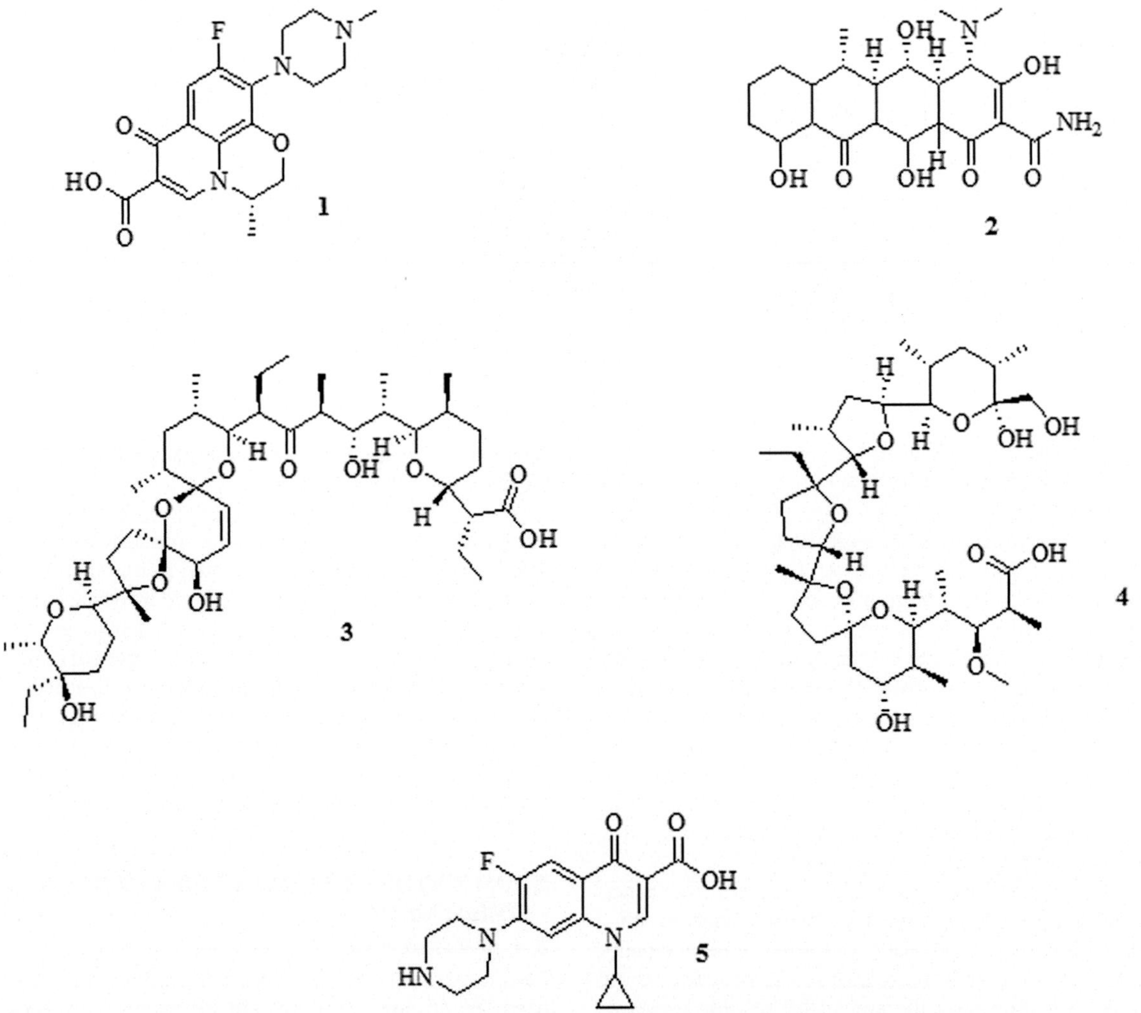

FIG. 2.2 Antibiotics with the therapeutic effect in the treatment of cancer.

5.2 Doxycycline

Antibiotic doxycycline (2) is the semisynthetic derivative of oxytetracycline, effective against a number of bacterial infectious diseases (Fig. 2.2). Tetracycline antibiotics entered the clinical usage in the 1940s and were authorized by the Food and Drug Administration in 1948 [50]. The majority of tetracyclines are known to be effective inhibitors of angiogenesis [51]. The antibacterial activity of doxycycline has antiproliferating and cytotoxic properties in different cancer cells [52,53]. The anticancer effect of doxycycline was further demonstrated in an experimental system including its inhibition of antiangiogenesis, matrix metalloproteinase, and cytostatic effects on cancer cells [54]. Fife and colleagues presented data showing that doxycycline

kills human prostate cancer cells in vitro and the observed cell death was due to enhanced apoptosis [55]. Tang et al. (2017) demonstrated a novel effect of doxycycline in decreasing extracellular lysophosphatidate (LPA) concentrations by increasing the "ectoactivity" of the lipid phosphate phosphatases. Doxycycline further showed action in decreasing NF-κB activation without the LPA signal. These actions make it a potential candidate for adjuvant treatment for cancer and some inflammatory diseases [56].

5.3 Salinomycin

Salinomycin (3) is a polyether antibiotic isolated from -*Streptomyces albus*, which shows a broad spectrum of various biological activities such as antifungal, antiviral, antiparasitic, and anticancer activities [57,58] (Fig. 2.2). Salinomycin is specifically known to induce cell apoptosis disturbing the Na^+/K^+ ion equilibrium in cellular membranes including cytoplasm and mitochondria [59]. It has been proved to have antitumor activities in various types of cancer cells [60]. It reduces the proportion of breast cancer stem cells by at least 100 times more than the commonly used anticancer drug, paclitaxel [61]. It is very effective even in combination therapy with plant-derived natural compounds and enhances its antiproliferative efficacy by different mechanisms in many cancers [62]. It also had a potent synergistic effect when combined with histone deacetylase inhibitor (LBH589) for triple-negative breast cancer [63]. Salinomycin alone has the ability to sensitize the *cis*-diamminedichloro-platinum (CDDP)-resistant colorectal cancer cells [64]. Its efficacy or mechanism of action against many cancers has been discovered as summarized in Table 2.1; the efficacy and its mode of action on CDDP-resistant cancer are not well known [65].

5.4 Monensin

Monensin (4) is an ionophoric antibiotic originally isolated from *Streptomyces cinnamonensis* (Streptomycetaceae family) [66,67] (Fig. 2.2). Recent studies have demonstrated that monensin may be repurposed to treat chemoresistant pancreatic cancer. It exhibits a synergistic effect when combined with other anticancer drugs, including erlotinib or gemcitabine, for the treatment of drug-resistant cancer [68]. Monensin inhibits the signaling pathways related to the growth of cancer such as NF-κB and STAT and also reduces the expression of epidermal growth factor receptors [69,70]. The main inhibitory effect of this antibiotic includes the inhibition of cell proliferation and migration of ovarian cancer cells as well as the induction of apoptosis [70]. Monensin decreases the expression of cyclin A, cyclin D1, and CDK6 by inducing programmed cell death-related gene activity, such as caspase-3, caspase-8, and Bax, as well as stimulating mitochondria transmembrane potential in different types of human cancer cell lines [25]. Ketol et al. demonstrated that monensin effects at nanomolar concentrations (nM) are linked to the apoptosis induction and an effective reduction of androgen receptor (mRNA) and protein in prostate cancer cells [71].

5.5 Ciprofloxacin

Ciprofloxacin (5) is an antibiotic drug from the fluoroquinolone family and has been commonly used in the treatment of conditions such as inflammatory bowel disease and urinary tract infections (Fig. 2.2) [72,73].

Other studies reported that ciprofloxacin has an antitumor effect in different human cancer cell lines [74−78]. The efficacy of ciprofloxacin in cancer was confirmed by in vitro studies on animal and human tumor cell lines including rat (CC-531) colorectal cell lines [79], human bladder cancer cell line (HT B9) [80], human and hamster ovarian cancer cell lines (HT-29, SW-403, and CHO AA8) [81], human nonsmall cell lung cancer (A549), human hepatocellular carcinoma (HepG2), mouse melanoma cell line (B16) [76,82], rat glioblastoma cell line (C6) [76], or pancreatic cancer cell line (MIAPaCa-2, Panc-1) [78]. It can also manage the chemotherapy-induced febrile neutropenia in patients with cancer [83]. In addition, ciprofloxacin inhibits the proliferation of human colonic and nonsmall cell lung cancers [74−76], and the induction abilities, cytotoxicity, and apoptosis were observed in human pancreatic and prostate cancers [77,78]. These antibiotics alone or when combined with other chemotherapeutic drugs induce apoptosis and cell cycle arrest in many cancer cell lines [75,79,84−86]. Ciprofloxacin enhances the chemosensitivity of various cancer cells to ABCB1 substrates [87]. Recently, Lim et al. reported that it can enhance the tumor necrosis factor-related apoptosis-inducing ligand (TRAIL) apoptosis in lung cancer cells [88].

6 MALARIA

Malaria is caused by protozoan parasites that are transmitted by blood-feeding Anopheline mosquitoes. Five different types of human malaria parasites such as *Plasmodium falciparum*, *P. vivax*, *P. malariae*, *P. knowlesi*, and *P. ovale* are responsible for malaria cases around the world. When infecting a human host, the parasites first target liver hepatocytes where they cause an initial-asymptomatic infection that ensures multiplication before continuing their life cycle inside human erythrocytes. The intraerythrocytic infection cycle would cause the symptoms of the disease such as fever and anemia, which in many cases is followed by respiratory distress and organ failure [111]. In the literature, the complete life cycle of the malaria parasite is well described [112].

Currently, the reported mortality rate caused by malaria has decreased in the past 5 years. Despite this, malaria is still among the leading causes of death, with 445,000 deaths and 212 million cases estimated in 2016 [4,113]. Drug-resistant malaria remains a great challenge for human health. Based on the full malaria parasite genome [18], new antimalarial targets for drug development can be identified and validated. The antimalarial targets can be generally categorized into three different groups: (1) biosynthesis, which targets malarial enzymes, involved in numerous biosynthesis pathways, including the folate biosynthesis pathway [114,115]. (2) Membrane transport and signaling: targets several pathways such as that of (1), transport of nutrients into the cell, and generation and maintenance of transmembrane electrochemical gradients (i.e., the plasmodial surface anion channel) [116–118]. (3) Hemoglobin catabolism: participates in hemoglobin digestion and plays an important role in the development of intraerythrocytic malaria parasites [119].

A lot of new therapeutic strategies have been developed to overcome the problem of drug resistance such as drug repurposing and the combination of two or more molecules (i.e., artemisinin-based combination for malaria) and antibiotics have been proven to have a therapeutic effect against malaria parasites [26]. A number of antibiotics such as clindamycin [120], tetracycline [121], and rifampin [122] showed antimalarial activities in vivo either alone or in combination with commonly used antimalarial drugs. Various studies have elucidated the mechanisms of action of antimalarials. The malaria parasite's mitochondria and apicoplast are of prokaryotic origin and are susceptible to certain antibiotics [123]. The antibacterial targets between the two organelles but the stage of malaria parasite life cycle are not well known. The apicoplast is a reduced cyanobacterium found inside malaria parasites that have the bacterial housekeeping machinery such as DNA replication, translation, transcription, and anabolic pathways for the synthesis of fatty acids, haem, isoprenoid precursors, and iron–sulfur complexes. Selected antibiotics targeting these pathways have previously been shown to kill malaria parasites [124,125].

6.1 Challenges in the Malarial Treatment

The development and increase of parasite resistance to a number of antimalarial drugs have limited the disease management in malaria-endemic areas and contributed to the increase in malaria-related deaths in recent years. The intensive use of antimalarial drugs has dramatically led to an increase in drug-resistant parasites and failure to control or eradicate malaria worldwide [126]. The majority of the existing antimalarial drugs have been used for decades and their use is limited by the emergence of drug resistance. Chloroquine, mefloquine, and quinine are typically fast-acting schizonticidal drugs. Pyrimethamine, sulfonamides, and sulfone also possess schizonticidal properties, and their action is slow. The novel kinase inhibitors such as primaquine and tafenoquine have gametocidal activities. Primaquine and proguanil are the main

sporontocidal drugs [127]. All these antimalarial agents were designed based on their major metabolic differences of malaria parasite with its host. Heme toxification, nucleic acid metabolism, oxidative stress, and fatty-acid biosynthesis are the major pathways that are targeted for antimalarial drug design [128]. The threat posed by failing antimalarial monotherapies has led to an effort to replace these antimalarial agents with more effective artemisinin-based combination therapies (ACTs) for the management of malaria [129]. The introduction of ACT with promising efficacy largely revolved around the issue of costs and sustainability.

7 ANTIBIOTICS EFFECTIVE FOR MALARIA TREATMENT

7.1 Tigecycline

Tigecycline (6) is the minocycline derivative used as an antibiotic for the treatment of difficult-to-treat bacterial pathogens like vancomycin-resistant *Acinetobacter baumannii*, *Enterococcus* spp., methicillin-resistant *Staphylococcus aureus*, and the Gram-negative bacterial strains that yield extended-spectrum β-lactamases (Fig. 2.3) [130]. Additionally, tigecycline causes low organ toxicity, and its dosage adjustment can be ignored in most patient populations [131]. Tetracycline interferes with mitochondrial function [123]. Other studies have reported the antimalarial efficacy of tigecycline [132–134]. Tigecycline has been reported to be the most effective drug against *Plasmodium* when compared to other antibiotics [134]. Tigecycline indirectly targets the *Plasmodium falciparum* mitochondria via Dihydroorotate dehydrogenase (DHOD) activity and apicoplast transcription [135,136]. Despite the improved biological activities of tigecycline, its clinical use in the treatment of malaria is limited due to its pharmacokinetic profile. Its poor absorption and rapid metabolism [137] might expose the parasite populations to a prolonged period of subtherapeutic concentrations and lead to a high risk of resistance [132].

7.2 Rifampicin

Rifampicin (7) is an antibiotic commonly used for the treatment of tuberculosis (Fig. 2.3) [138]. It has a potent antimalarial activity against *Plasmodium chabaudi* in rodents, *Plasmodium vivax* in humans, and chloroquine-resistant *P. falciparum* in vitro [139]. Rifampicin blocks bacterial-type transcription and abrogates mRNA production by the apicoplast-encoded RpoB gene [135,140]. Rifampicin induces hepatic metabolism and influences the pharmacokinetic properties of other drugs [141]. This compound has been shown to have

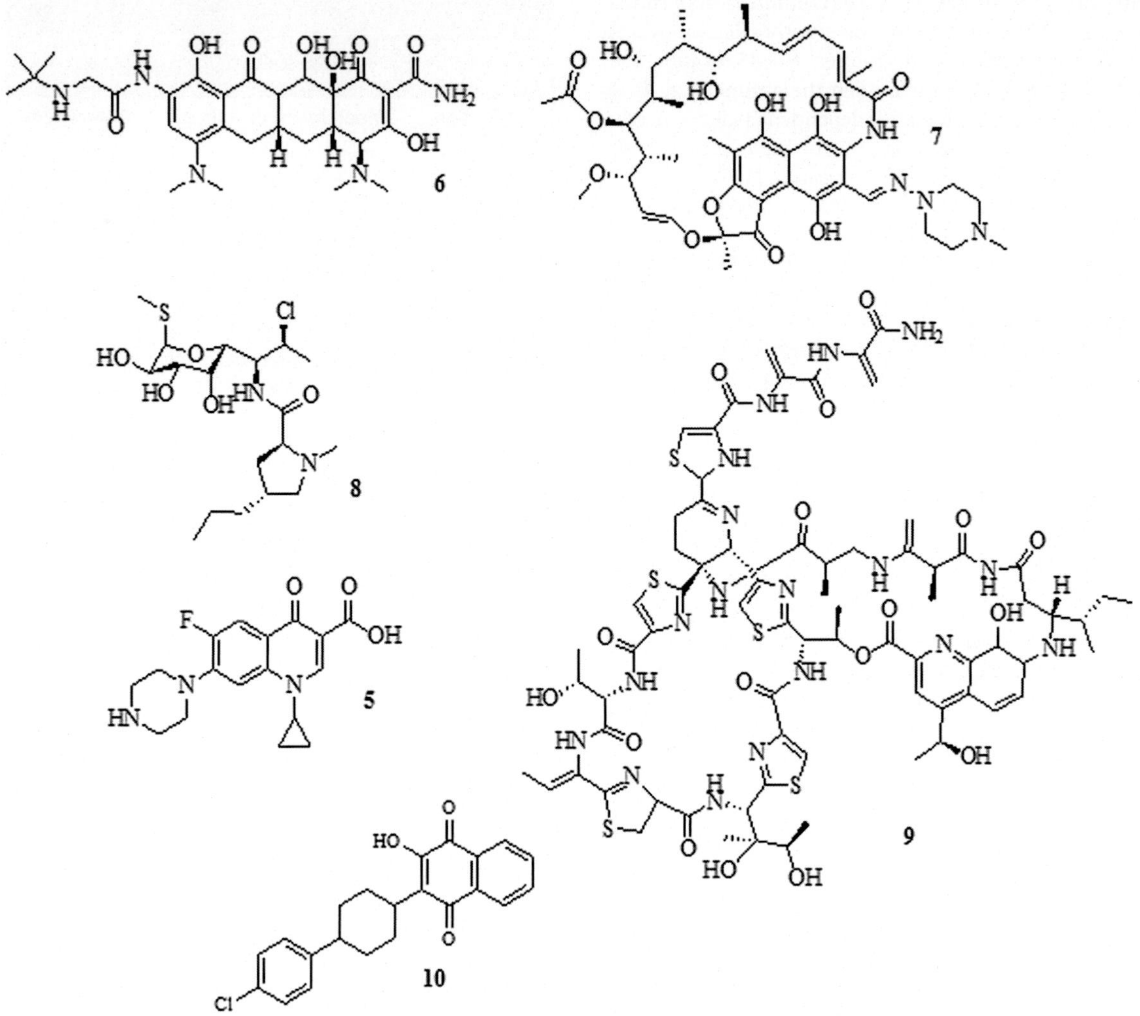

FIG. 2.3 Antibiotics with the therapeutic effect against malaria.

slow and relatively weak antimalarial effects, but it could cause a synergistic effect when combined with other potent antimalarial drugs.

The combination of rifampicin and primaquine showed a synergistic effect against *Plasmodium vivax* infection in humans [142]. In that experimental study, rifampicin reduced the parasitemia and was effective against blood-stage infection of *Plasmodium vivax*. Badejo et al. showed the beneficial interaction between rifampicin and amodiaquine in mice infected with *Plasmodium berghei*. Their results demonstrated the significant reduction of parasitemia and improved survival rate in animals treated with the combination of rifampicin and amodiaquine [139].

7.3 Clindamycin

Clindamycin (8) is the semisynthetic derivative of lincomycin with a broad spectrum of antimicrobial activities against a number of Gram-positive, and anaerobic bacteria (Fig. 2.3) [143]. Clindamycin has a good tissue penetration including bone tissues and could be used in the treatment of osteoarthritis [144]. Clindamycin's safety profile and its spectrum coverage make it a good choice for the treatment of *Plasmodium falciparum* malaria. Clindamycin targets the prokaryote-like ribosomes of the apicoplast by inhibiting self-replication of the organelle and kills the parasite in the second-replication cycle [145]. Its use as an antimalarial drug is limited by its unacceptably slow and weak initial clinical response rates [120]. The combination of clinda-mycin with fosmidomycin was poorly effective against *P. falciparum* malaria in children (6−35 months age) [146]. Fosmidomycin is among the antimalarial drugs that block the mevalonate-independent 1-deoxy-D-xylu-lose-5-phosphate pathway in the apicoplast of the malaria parasite *Plasmodium falciparum* [147].

7.4 Thiostrepton

Thiostrepton (**9**) is a well-known antibiotic belonging to the thiopeptide family of highly modified macrocyclic peptides produced as secondary metabolites by actinomycetes from the genus *Streptomyces* (Fig. 2.3) [148]. Thiostrepton inhibits the mitochondrion and the apicoplast protein synthesis by binding the apico-plast ribosomal RNA [149]. McConkey et al. showed that the sensitivity of *Plasmodium falciparum* to thiostrep-ton confirms that the plastid-like genome is essential for the erythrocytic cycle [140]. Thiostrepton further exhibits the gametocytocidal activity through the elimination of gametocytes (the sexual-precursor cells essential for para-site transmission to the mosquito). Thiostrepton causes a delayed death phenotype found in parasite killing during the second-replication cycle following administration. Thiostrepton-based derivatives interfere with the eukary-otic proteasome, an important multimeric protease for the degradation of ubiquitinated proteins. These deriva-tives have potent activity against sensitive and resistant chloroquine *P. falciparum*, by eliminating the parasites before DNA replication [150].

7.5 Ciprofloxacin

Ciprofloxacin is a broad-spectrum antibacterial agent, which is effective against many strains of *Plasmodium falciparum* malaria [151] and is known to target the api-coplast genome by inhibiting the DNA replication [152]. Although ciprofloxacin has a good pharmacoki-netic profile and efficiently enters the erythrocytes, its monotherapy as an antimalarial agent has been discouraged because it is only effective only at unaccept-able high dose where the recommended serum concen-tration (C_{max}) can be achieved [153].

Combination therapy of ciprofloxacin with an antimalarial drug, amodiaquine, against *Plasmodium berghei* showed a synergistic antiplasmodial effect at their recommended dosage levels [153,154]. Gbotosho et al. (2012) and Andrade et al. (2007) also reported the potential clinical value of ciprofloxacin in combination with other good acting antimalarial drugs such as chloroquine and mefloquine against drug-resistant plasmodium infection. Their results showed that the pharmacokinetic interaction between ciprofloxacin and chloroquine limits the efficacy of the combination

TABLE 2.2
Antibiotics and Their Mechanism of Action in the Treatment of Malaria Parasite.

Antibiotics	Mechanism of Action	References
Tigecycline	Interferes with mitochondrial function	[123]
Rifampicin	Induces the hepatic metabolism and influences the pharmacokinetic properties of other drugs	[141]
Clindamycin	Inhibit self-replication of the organelle and kills the parasite in the second-replication cycle	[145]
Thiostrepton	Inhibit the mitochondrial activity	[149]
Ciprofloxacin	Interfere with erythrocyte	[153]

at lower doses of ciprofloxacin with chloroquine (Table 2.2)[155,156].

7.6 Atovaquone

Atovaquone (**10**) is known as a unique naphthoqui-none antibiotic with a broad-spectrum antiprotozoal action. Atovaquone has found to have a broad-spectrum action against *P. carinii*, *Plasmodium* spp., *Toxoplasma gondii*, and *Babesia* spp. [157]. Atovaquone was found to be active in the treatment and inhibition of *Pneumocystis carinii* pneumonia (PCP). Atovaquone is normally used in combination with other agents as the treatment failed in clinical trials approximately 30% when tested against *Plasmodium falciparum* malaria and resistant parasites appeared quick [157,158]. Atovaquone-proguanil combination has shown posi-tive results in clinical trials with treatment rates of about 100% and slight proof of the emergence of resistance [158]. The combination of atovaquone and proguanil was found to be effective for the treatment of children and adults with uncomplicated malaria or as a chemo-prophylactic agent for inhibiting malaria in travelers [129] and was effective for the treatment of babesiosis in combination with azithromycin [157].

8 HUMAN IMMUNODEFICIENCY VIRUS

HIV is a chronic disease that belongs to the family of Retroviridae; moreover, in an advanced stage of HIV infection, it causes acquired immunodeficiency syn-drome (AIDS) [159]. Viruses are a major threat to human health. Many viruses can cause various diseases in human but most of them are still incurable. Emerging or reemerging viruses are a concern of public health [160]. HIV is the pandemic disease that emerged in the 1980s and its cure is still unknown although antiretroviral therapy can control virus replication.

In 2016, the WHO stated that HIV is one of the most leading causes of deaths, and it continues to be the main public health issue globally. In 2016, more than 35 million people were living with HIV and one million died from HIV-related causes. In 2016, most people who were affected by HIV infection were in the Africa region [159]. WHO suggested that, in 2013, antiretroviral therapy (ART) should be given to people infected with HIV whose CD4+ T-cell (CD4) counts are less than 500 cells per microliter (μL), to pregnant women who are infected by this disease and also to individuals with certain medical conditions like active hepatitis B, active tuberculosis, and to person with advanced HIV clinical disease [161]. ART in HIV patients is well recognized to increase CD4 count and minimize plasma HIV ribonucleic acid (RNA) load. The immunity may be restored by the ART in the patient, and the risk of acquiring new infections can be reduced. In an acute disease, the ART can inhibit the progression of HIV to AIDS-related infection, enhance the health and value of life [162]. Even though ART plays a major role in a patient who is infected with HIV, drug–drug interactions, multidrug resistance, and toxicities are still the biggest problems in HIV treatment [163]. HIV infection has a very complex pathogenesis and varies substantially in different patients. Therefore it can easily be considered as a very host-specific infection. The specificity of pathogenesis often complicates treatment options that are currently available for HIV infection [164].

AIDS is a medical condition caused by HIV infection, an RNA virus that attacks the immune system. HIV infection targets the CD4+ lymphocytes (T helper/H), which are the one that plays an important role in the immune system. During the early stage of the illness, it is easy for the immune system to control the HIV infection, but in the course of the illness, HIV causes a minimum number of CD4+ lymphocytes, disturbance of homeostasis, and operate other cells in the immune system [165]. Owing to the aforementioned factors, many symptoms occur because of a disturbance in the cellular and humoral immunity including hypersensitivity reactions, an immune deficiency that causes opportunistic viruses and neurological disorders, or unusual malignancies [165]. The probability of acquitting other diseases in an HIV/AIDS patient is high: diseases such as tuberculosis (TB), pneumonia, and/or other mycobacterial opportunistic infections. Specific

treatments have been developed for the treatment of TB disease in HIV patients [166]. Moreover, the effect of anti-TB medications is higher among HIV patients because of the interactions with ART resulting in higher mortality [167,168].

The high rate of bacterial infections in HIV/AIDS persons raises the chances of consumption of high levels of antibiotics. Misusing and overusing antibiotics promote many resistance mechanisms among bacteria. HIV-infected persons are very vulnerable to many opportunistic infections (OIs) due to their weak immune systems. Patients with HIV are normally exposed to a high level of antimicrobial agents, which leads to multidrug-resistant bacteria (MDR) [159]. MDR bacteria are difficult to treat with the current antibiotics and MDR bacteria are growing in HIV persons and it becomes the main challenge in the clinical management of HIV infection, thus the HIV treatment options remain limited [159]. Among the most overwhelming effects of HIV infection is the compromising immune system which makes the infected individuals prone to opportunistic infections (OIs) which in turn makes HIV person to need prolonged antibiotic use [169]. PCP followed by bacterial pneumonia and HIV/AIDS can lead to important mortality and morbidity. As the first HIV/AIDS cases were related to OIs like PCP, the significance of the bacterial disease, with bacterial pneumonia, was not known at the commencement of the epidemic. However, researchers have revealed that bacterial infections take place more often than other OIs in patients with HIV/AIDS [170].

9 CHALLENGES OF ANTIRETROVIRALS

The morbidity and mortality associated with HIV/AIDS infection have been reduced using antiretroviral (ARV) treatment. Researchers have reported that the utilization of combination therapies can improve HIV individuals' chances to survive for a very long time. Unfortunately, the effectiveness of ARVs treatment is hampered by the emergence of drug resistance [171]. The significant cause of treatment failure is due to the drug resistance in some HIV patients [172]. Drug-resistant infections are frequently resistant to multiple ARV remedies [171]. ARV drug resistance is rapidly increasing in HIV patients at a high frequency, thereby making the management of HIV infections complex [172].

Highly active antiretroviral therapy used to treat persons with HIV infection caused serious side effects, such as peripheral neuropathy, treatment-induced diarrhea, lipodystrophy, mitochondrial toxicity, and osteoporosis [173].

10 ANTIBIOTICS EFFECTIVE FOR HIV TREATMENT

10.1 Azithromycin

Azithromycin (**11**) is known as a nontoxic antibiotic that is used to treat bacterial infections namely, pneumonia; bronchitis and sexually transmitted diseases; and infections of the lungs, skin, ears, sinuses, throat, and reproductive organs (Fig. 2.4) [174]. Azithromycin is in a class of medications called macrolide antibiotics with bacteriostatic action against the highest common respiratory bacterial infections [175]. Azithromycin is used for the treatment of ocular toxoplasmosis because it is nontoxic and exhibits good bioavailability, can cross the blood–brain barrier and has been reported to be broadly dispersed to the brain tissue [174]. Furthermore, it also has a strong immune-modulatory effect with reduced production of inflammatory cytokines in the critical stage and at a later stage with a good resolution of chronic inflammation. The aforementioned characteristics have also resulted in its usage for the management of several chronic lung diseases, such as cystic fibrosis, noncystic fibrosis, bronchiolitis, bronchiectasis, obliterans syndrome, and chronic obstructive pulmonary disease [175]. Azithromycin is recommended as a strong candidate in the treatment of HIV infections associated with chronic diseases, given its antiinflammatory, immune-modulatory and broad-spectrum antibiotic activity, as well as displayed action in similar chronic lung diseases [175].

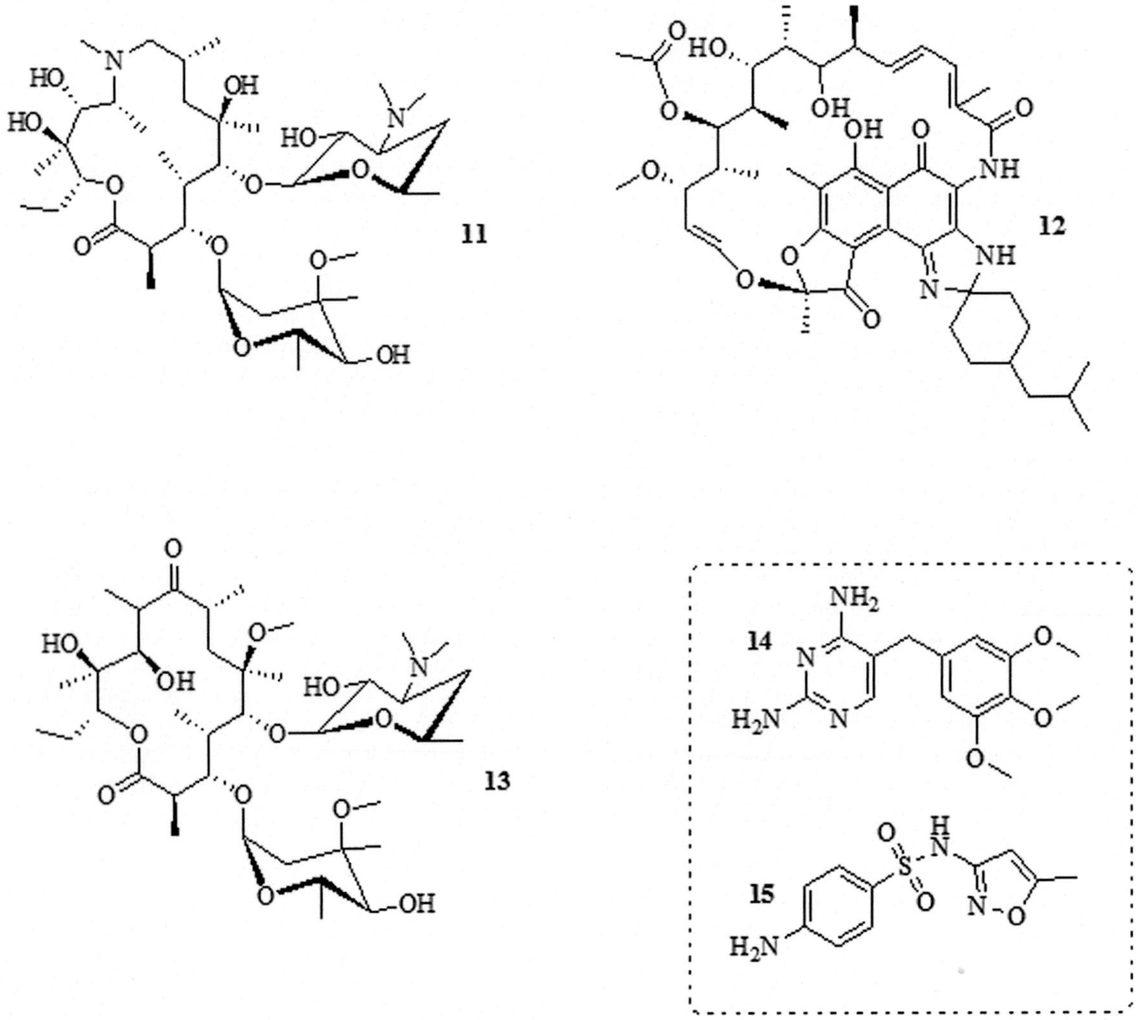

FIG. 2.4 Antibiotics with the therapeutic effect against malaria.

It is also used to treat HIV patients affected with disseminated *Mycobacterium avium complex* (MAC) infection [176]. In 2014, new cases of HIV patients affected with TB were about 1.2 million in the world, suggesting that HIV infection is responsible for the increase in TB infection in individuals with HIV infection. Antibiotics like Azithromycin and Trimethoprim-Sulfamethoxazole with antimicrobial prophylaxis have been used for the treatment and prevention of TB and other infectious diseases in people with HIV [177]. The clinical efficiency of combining metronidazole and amoxicillin has been extensively studied for the treatment of infectious diseases. The studies in vitro and in vivo on the combination of these antibiotics did not show any efficacy [178]. However, the macrolide antibiotic azithromycin has been commended as a possible alternative way for the treatment of the disease. Owing to many potential benefits of this antibiotic such as antiinflammatory activity, broad antimicrobial spectrum, frequency of intake and lower intake dosage, it is useful for HIV and or/TB therapy [178].

Macrolide antibiotic azithromycin prevents microbial growth and inhibits protein synthesis [179]. It is employed in the treatment of chronic inflammatory lung conditions in HIV infections [179].

10.2 Rifabutin

Rifabutin (**12**) is an antibiotic that is used to treat TB and treat/inhibit the MAC [180,181]. MAC opportunistic infection is common in HIV-positive patients. The rate of MAC infection rises as the CD4 cell count decreases below 50 cells mlx1 [182]. Primary prophylaxis for MAC includes azithromycin, clarithromycin, or rifabutin [182]. Rifabutin is recommended for people living with active TB [183]. Rifabutin and rifampin have similar effects on the management of TB. However, the half-life of rifabutin is much longer when compared to rifampin (i.e., rifabutin is 35 h and rifampin is 3.5 h). These difference in pharmacokinetics is responsible for acquired rifamycin resistance in HIV-positive patients receiving intermittent medication for TB with a rifabutin-containing regimen [180]. Its high cost renders it inaccessible for TB management. It is used as an alternative to rifampicin and it acts by inhibiting RNA polymerase of the bacteria responsible for TB infection thereby preventing the bacteria from synthesizing important proteins. Rifabutin suffers from side effects such as neutropenia and diarrhea. It interacts with antiretroviral therapy by affecting the concentrations of ARVs in the body and it loses its effectiveness when combined with selected ARVs.

10.3 Clarithromycin

Clarithromycin (**13**) is an antimicrobial drug used in the treatment of diseases namely leprosy and TB and is used to treat the infections caused by MAC in AIDS individuals. Clarithromycin and ethambutol can treat the *Mycobacterium avium* infections, except if the isolate exhibits resistance to macrolides antibiotics [184]. Clarithromycin displays a broad antibacterial spectrum in vitro. In the treatment of MAC infections, a combination of antimicrobial agents has been used among which clarithromycin has shown to be an effective agent for both in vivo and in vitro [185]. Clarithromycin is acid-stable with a half-life of about 5 h compatible with a twice per day when administered and has good bioavailability [186]. MAC infections are very common in people living with AIDS. The clinical treatment of *Mycobacterium avium* infections in a person with AIDS is complex because of the development of drug resistance [187]. The treatment of HIV infection usually involves the simultaneous administration of multiple antimycobacterial drugs that frequently have side effects. Thus macrolide antibiotics have constantly shown to be the most effective against *M. avium* infection, whereas they are combined with other therapeutics to enhance their therapeutic efficacy and to prevent the development of resistance by the bacteria [187].

10.4 Cotrimoxazole

Cotrimoxazole is a combination of two antibiotics namely trimethoprim (**14**) and sulfamethoxazole (**15**), which has a broad spectrum of antimicrobial agents and is active against gram-positive and gram-negative bacteria (Fig. 2.4). The first-used clinically antibiotic agent was sulfonamides, which were effective against a wide range of bacterial infections. In 1968, sulfonamides and Trimethoprim (TMP) combinations were registered for clinical use [188]. Cotrimoxazole has been used for therapy and prophylaxis of *Pneumocystis carinii* pneumonia in people living with AIDS. It is affordable and has been used for many years globally in the management of community-acquired infections [189–192]. HIV destroys the body's immune system, making an OI more common. Therapy and inhibition of these infections are integral to the treatment of individuals with HIV infection. Cotrimoxazole exhibit a wide range of activity against similar bacteria, fungi, yeast, and parasites [193]. Cotrimoxazole is the key antibiotic used for the treatment of pneumonia in HIV/AIDS-infected individuals [194]. Cotrimoxazole antibiotic therapy is recommended by WHO to inhibit OI in people who have been living with HIV/AIDS and is also recommended for the treatment of infections in children who are infected with HIV [195–197]. When the immune system is being attacked with HIV, it leaves patients vulnerable to infections; the infections frequently cause death or critical complications. Cotrimoxazole is used to inhibit infections [198]. It is utilized as long-term prophylaxis in persons with HIV infection, and its therapeutic outcome has made it a perfect first-line medication for the treatment and prophylaxis of pneumocystis infection [199].

Trimethoprim and sulfamethoxazole combination synergistically hinder the microbial synthesis of folic acid, an essential cofactor in the making of thymidine and purines [200]. A good inhibitor of dihydrofolic acid synthesis is sulfamethoxazole, and on the other side, trimethoprim inhibits the physiological action of tetrahydrofolic acid. Even though each antibiotic has bacteriostatic agents, the combinations of these antibiotics can block the two sequential enzymes involved in bactericidal action [201]. Moreover, the combination of antimicrobials was suggested to reduce the emergence of resistance; thus, cotrimoxazole has essential action against a broad spectrum of fungal, bacterial, and protozoal pathogens [200].

Cotrimoxazole is a major treatment for prophylaxis against *Pneumocystis carinii* pneumonia, which is the main common OI in individuals living with HIV/AID [202]. Clinical trials reported cotrimoxazole as a safe and effective combination with rare side effects in immune-competent persons with infections [188]. *Pneumocystis jirovecii* pneumonia is still one of the major vital reasons for pulmonary OIs in patients with HIV infection in the cART. The rate mortality of *P. jirovecii* pneumonia is about 10%–12% in patients affected with HIV infection. On the other side, cotrimoxazole is a recommended therapy for *P. jirovecii* pneumonia in patients with HIV infection [203]. However, it can cause numerous side effects namely hepatotoxicity, allergy, hyperkalemia, bone marrow suppression, and nephrotoxicity. The utilization of Cotrimoxazole in the treatment of *P. jirovecii* pneumonia has been revealed to cause severe psychosis in patients with HIV infection. In a clinical trial, hepatotoxicity was reported to occur in approximately less than 10% in patients administered cotrimoxazole for the treatment of pneumocystosis [203].

Another side effect of Cotrimoxazole is drug hypersensitivity [204]. **Cotrimoxazole** drugs mode of action is by sulfamethoxazole inhibition of bacteria production of dihydropteroate from the two folate predecessors, 6-hydroxymethyl 7,8-dihydropterin pyrophosphate and *p*-aminobenzoic acid and trimethoprim inhibition of the conversion of dihydrofolate to Tetrahydrofolate (THF) [188].

11 CONCLUSION AND FUTURE PERSPECTIVES

In summary, the therapeutic efficacy of antibiotics in the treatment of chronic diseases was reviewed. The antibiotics mentioned in this review have good therapeutic efficacy in the treatment of chronic diseases such as cancer and malaria. However, there is still a great need to discover other antibiotics. Many discoveries can arise from the chemical modifications of the currently used antibiotics. One advantage of using antibiotics already approved for the treatment of infectious and chronic diseases such as doxycycline and tigecycline is that they are affordable when compared to the clinical development of new drugs. Antibiotics that act at new targets via distinct mechanisms have the greatest potential to overcome resistance [205]. The differences in modes of action of antibiotics imply that they are good precursors for combination therapy. The combination therapy of two or more drugs with different modes of action can also delay or prevent the development of resistance. Despite this advantage of combination therapy, some reports have proven that it has failed in several patients [206,207]. The clinical development of good combination therapy is expensive and drug–drug interactions may cause additive toxic side effects [208]. Another challenging problem is the selection of drugs and dosages for combination therapy due to the differences in stability, solubility, and pharmacokinetic profile of the combined drugs. The modification of the existing drugs through the synthesis of hybrid compounds could be a noble solution. Combining two or more active pharmacophores into a single molecule (hybrid compound) can overcome combination therapy while enhancing the overall potency [209,210].

ACKNOWLEDGMENTS

The financial assistance of the Medical Research Council (Self-Initiated Research) (MRC) and National Research Foundation (NRF), South Africa toward this research is hereby acknowledged. The views and opinions expressed in this manuscript are those of the authors and not of MRC or NRF.

REFERENCES

[1] J. Dewangan, S. Srivastava, S. Mishra, A. Divakar, S. Kumar, S.K. Rath, Salinomycin inhibits breast cancer progression via targeting HIF-1α/VEGF mediated tumor angiogenesis in vitro and in vivo, Biochemical Pharmacology 164 (2019) 326–335.

[2] F. Made, et al., Distribution of cancer mortality rates by province in South Africa, Cancer Epidemiol 51 (2017) 56–61.

[3] World Health Organisation. https://www.who.int/hiv-aids/latest-news-and-events/why-the-hiv-epidemic-is-not-over. (Accessed on 27 May 2019).

[4] K. Plewes, S.J. Leopold, H.W.F. Kingston, A.M. Dondorp, Malaria: what's new in the management of malaria? Infectious Disease Clinics of North America 33 (1) (2019) 39–60.

[5] C.V. Dave, A. Pawar, E.R. Fox, G. Brill, A.S. Kesselheim, Predictors of drug shortages and association with generic drug prices: a retrospective cohort study, Value in Health 21 (11) (2018) 1286–1290.

[6] L.S. Ross, D.A. Fidock, Elucidating mechanisms of drug-resistant plasmodium falciparum, Cell Host and Microbe 26 (1) (2019) 35–47.

[7] M. Sylla, et al., Second-line antiretroviral therapy failure and characterization of HIV-1 drug resistance patterns in children in Mali, Archives of Pediatrics 26 (5) (2019) 254–258.

[8] V. Bonnet, et al., Influence of bacterial resistance on mortality in intensive care units: a registry study from 2000 to 2013 (IICU Study), Journal of Hospital Infection 102 (3) (2019) 317–324.

[9] S. Pushpakom, F. Iorio, P.A. Eyers, K.J. Escott, Drug repurposing: progress, challenges and recommendations, Nature Publishing Group 18 (1) (2018) 41–58.

[10] V. Yadav, P. Talwar, Biomedicine & pharmacotherapy repositioning of fluoroquinolones from antibiotic to anti-cancer agents : an underestimated truth, Biomedicine and Pharmacotherapy 111 (2019) 934–946.

[11] H. Matthews, M. Usman-Idris, F. Khan, M. Read, N. Nirmalan, Drug repositioning as a route to anti – malarial drug discovery: preliminary investigation of the in vitro anti - malarial efficacy of emetine dihydrochloride hydrate, Malaria Journal 12 (1) (2016) 359.

[12] R.B. Mokhtari, et al., Combination therapy in combating cancer, Oncotarget 8 (23) (2017) 38022–38043.

[13] J.A. Dimasi, H.G. Grabowski, R.W. Hansen, Innovation in the pharmaceutical industry: new estimates of R & D costs, Journal of Health Economics 47 (2016) 20–33.

[14] M. García-Serradilla, C. Risco, B. Pacheco, Drug repurposing for new, efficient, broad spectrum antivirals, Virus Research 264 (2019) 22–31.

[15] S. Murteira, Z. Ghezaiel, S. Karray, M. Lamure, Drug reformulations and repositioning in pharmaceutical industry and its impact on market access: reassessment of nomenclature, Journal of Market Access and Health Policy 1 (1) (2013) 21131.

[16] E. Petrova, Innovation and Marketing in the Pharmaceutical Industry: The Process of Drug Discovery and Development, Springer, New York, NY, 2014, pp. 19–81.

[17] W. Zheng, W. Sun, A. Simeonov, Drug repurposing screens and synergistic drug-combinations for infectious diseases, British Journal of Pharmacology 175 (2) (2018) 181–191.

[18] S. Murteira, A. Millier, Z. Ghezaiel, M. Lamure, Drug reformulations and repositioning in the pharmaceutical industry and their impact on market access: regulatory implications, Journal of Market Access and Health Policy 2 (1) (2014) 22813.

[19] W. Zheng, N. Thorne, J.C. McKew, Phenotypic screens as a renewed approach for drug discovery, Drug Discovery Today 18 (21–22) (2013) 1067–1073.

[20] G. Jin, S.T.C. Wong, Toward better drug repositioning: prioritizing and integrating existing methods into efficient pipelines, Drug Discovery Today 19 (5) (2014) 637–644.

[21] B. Delavan, R. Roberts, R. Huang, W. Bao, W. Tong, Z. Liu, Computational drug repositioning for rare diseases in the era of precision medicine, Drug Discovery Today 23 (2) (2018) 382–394.

[22] A.Ç. Karagöz, et al., Synthesis of new betulinic acid/betulin-derived dimers and hybrids with potent antimalarial and antiviral activities, Bioorganic & Medicinal Chemistry 27 (1) (2019) 110–115.

[23] M. Norouzi, Z. Abdali, S. Liu, D.W. Miller, Salinomycin-loaded nanofibers for glioblastoma therapy, Scientific Reports 8 (2018) 9377.

[24] W. Wu, L.H. Yu, B. Ma, M.J. Xu, The inhibitory effect of doxycycline on cisplatin-sensitive and -resistant epithelial ovarian cancer, PLoS One 9 (3) (2014) e89841.

[25] A. Markowska, J. Kaysiewicz, J. Markowska, A. Huczyński, Doxycycline, salinomycin, monensin and ivermectin repositioned as cancer drugs, Bioorganic and Medicinal Chemistry Letters 29 (13) (2019) 1549–1554.

[26] T. Gaillard, M. Madamet, F.F. Tsombeng, J. Dormoi, Antibiotics in malaria therapy: which antibiotics except tetracyclines and macrolides may be used against malaria? Malaria Journal 15 (556) (2016) 1–10.

[27] K. Kochanowski, L. Morinishi, S.J. Altschuler, L.F. Wu, Drug persistence – from antibiotics to cancer therapies, Current Opinion in Structural Biology 10 (2018) 1–8.

[28] M. Lee, S. Hung, M. Huang, T. Tsai, C. Chen, Doxycycline potentiates antitumor effect of 5- aminolevulinic acid-mediated photodynamic therapy in malignant peripheral nerve sheath tumor cells, PLoS One 12 (5) (2017) e0178493.

[29] M. Juarez, A. Schcolnik-Cabrera, A. Dueñas-Gonzalez, The multitargeted drug ivermectin: from an antiparasitic agent to a repositioned cancer drug, American Journal of Cancer Research 8 (2) (2018) 317–331.

[30] F. Nguyen, A.L. Starosta, S. Arenz, D. Sohmen, A. Dönhöfer, Tetracycline antibiotics and resistance mechanisms, Biological Chemistry 395 (5) (2014) 559–575.

[31] S.K. Dixit, et al., Synthesis and in vitro antiplasmodial activities of fluoroquinolone analogs, European Journal of Medicinal Chemistry 51 (2012) 52–59.

[32] G. Watt, G.D. Shanks, M.D. Edstein, K. Pavanand, H.K. Webster, Ciprofloxacin treatment of drug-resistant falciparum malaria, The Journal of Infectious Diseases 164 (3) (1991) 602–604.

[33] Y.A. Luqmani, Mechanisms of drug resistance in cancer chemotherapy, Medical Principles and Practice 14 (Suppl. 1) (2005) 35–48.

[34] D. Benharroch, L. Osyntsov, Infectious diseases are analogous with cancer. Hypothesis and implications, Journal of Cancer 3 (2012) 117—121.

[35] I. Castillo-juárez, et al., Role of quorum sensing in bacterial infections, World Journal of Clinical Cases 3 (7) (2015) 575—599.

[36] J. Hickson, S.D. Yamada, J. Berger, J. Alverdy, J.O. Keefe, B. Bassler, Societal interactions in ovarian cancer metastasis : a quorum-sensing hypothesis, Clinical and Experimental Metastasis 26 (2009) 67—76.

[37] G. Housman, et al., Drug resistance in cancer: an overview, Cancers 6 (3) (2014) 1769—1792.

[38] J. Dewangan, S. Srivastava, S.K. Rath, Salinomycin: a new paradigm in cancer therapy, Tumor Biology 39 (3) (2017).

[39] J.T. Sloop, J.A. Carter, U. Bierbach, B.T. Jones, G.L. Donati, Effects of platinum-based anticancer drugs on the trace element profile of liver and kidney tissue from mice, Journal of Trace Elements in Medicine and Biology 54 (2019) 62—68.

[40] P. Štarha, J. Vančo, Z. Trávníček, Platinum iodido complexes: a comprehensive overview of anticancer activity and mechanisms of action, Coordination Chemistry Reviews 380 (2019) 103—135.

[41] F.Y. Wang, K. Bin Huang, H.W. Feng, Z.F. Chen, Y.N. Liu, H. Liang, New Platinum(II) agent induces bimodal death of apoptosis and autophagy against A549 cancer cell, Free Radical Biology and Medicine 129 (2018) 418—429.

[42] F.Y. Wang, R. Liu, K. Bin Huang, H.W. Feng, Y.N. Liu, H. Liang, New platinum(II)-based DNA intercalator: synthesis, characterization and anticancer activity, Inorganic Chemistry Communications 105 (2019) 182—187.

[43] D. Liu, et al., Neural regulation of drug resistance in cancer treatment, Biochimica et Biophysica Acta (BBA) — Reviews on Cancer 1871 (1) (2019) 20—28.

[44] T. Fiaschi, P. Chiarugi, Oxidative stress, tumor microenvironment, and metabolic reprogramming: a diabolic liaison, International Journal of Cell Biology 2012 (2012).

[45] Y.I. Elshimali, et al., Optimization of cancer treatment through overcoming drug resistance, Journal of Cancer Research and Oncobiology 1 (2) (2018).

[46] M. Yu, R. Li, J. Zhang, Repositioning of antibiotic levofloxacin as a mitochondrial biogenesis inhibitor to target breast cancer, Biochemical and Biophysical Research Communications 471 (4) (2016) 639—645.

[47] S. Kalghatgi, et al., Bactericidal antibiotics induce mitochondrial dysfunction and oxidative damage in mammalian cells, Science Translational Medicine 5 (192) (2013) 192ra85.

[48] R. Lamb, et al., Antibiotics that target mitochondria effectively eradicate cancer stem cells, across multiple tumor types: treating cancer like an infectious disease, Acta Poloniae pharmaceutica drug research 44 (1) (2015) 85—86.

[49] M. Song, et al., Antibiotic drug levofloxacin inhibits proliferation and induces apoptosis of lung cancer cells through inducing mitochondrial dysfunction and oxidative damage, Biomedicine and Pharmacotherapy 84 (2016) 1137—1143.

[50] G.P. Wormser, R.P. Wormser, F. Strle, R. Myers, B.A. Cunha, How safe is doxycycline for young children or for pregnant or breastfeeding women? Diagnostic Microbiology and Infectious Disease 93 (3) (2019) 238—242.

[51] W.C.M. Duivenvoorden, et al., Doxycycline decreases tumor burden in a bone metastasis model of human breast cancer, Cancer Research 62 (6) (2002) 1588—1591.

[52] T. Onoda, T. Ono, D.K. Dhar, A. Yamanoi, T. Fujii, N. Nagasue, Doxycycline inhibits cell proliferation and invasive potential: combination therapy with cyclooxygenase-2 inhibitor in human colorectal cancer cells, The Journal of Laboratory and Clinical Medicine 143 (4) (2004) 207—216.

[53] L.C. Shen, Y.K. Chen, L.M. Lin, S.Y. Shaw, Anti-invasion and anti-tumor growth effect of doxycycline treatment for human oral squamous-cell carcinoma - in vitro and in vivo studies, Oral Oncology 46 (3) (2010) 178—184.

[54] C.L. Addison, et al., A phase 2 trial exploring the clinical and correlative effects of combining doxycycline with bone-targeted therapy in patients with metastatic breast cancer, Journal of Bone Oncology 5 (4) (2016) 173—179.

[55] R.S. Fife, G.W. Sledge, B.J. Roth, C. Proctor, Effects of doxycycline on human prostate cancer cells in vitro, Cancer Letters 127 (1—2) (1998) 37—41.

[56] X. Tang, X. Wang, Y.Y. Zhao, J.M. Curtis, D.N. Brindley, Doxycycline attenuates breast cancer related inflammation by decreasing plasma lysophosphatidate concentrations and inhibiting NF-KB activation, Molecular Cancer 16 (1) (2017) 1—13.

[57] W. Zhang, et al., Design and synthesis of conformationally constrained salinomycin derivatives, European Journal of Medicinal Chemistry 138 (2017) 353—356.

[58] A. Huczyński, J. Janczak, M. Antoszczak, J. Wietrzyk, E. Maj, B. Brzezinski, Antiproliferative activity of salinomycin and its derivatives, Bioorganic and Medicinal Chemistry Letters 22 (23) (2012) 7146—7150.

[59] A. Michał, R. Jacek, H. Adam, Structure and Biological Activity of Polyether Ionophores and Their Semisynthetic Derivatives, 2015.

[60] L. Liu, et al., Chemico-Biological Interactions Salinomycin suppresses cancer cell stemness and attenuates TGF-β-induced epithelial-mesenchymal transition of renal cell carcinoma cells, Chemico-Biological Interactions 296 (2018) 145—153.

[61] P.B. Gupta, et al., Identification of selective inhibitors of cancer stem cells by high-throughput screening, Cell 138 (4) (2009) 645—659.

[62] J. Dewangan, D. Tandon, S. Srivastava, A. Kumar, Novel combination of salinomycin and resveratrol synergistically enhances the anti-proliferative and pro-apoptotic effects on human breast cancer cells, Apoptosis 22 (10) (2017) 1246–1259.

[63] M. Kai, et al., Targeting breast cancer stem cells in triple-negative breast cancer using a combination of LBH589 and salinomycin, Breast Cancer Research and Treatment 151 (2) (2015) 281–294.

[64] J. Zhou, et al., Salinomycin induces apoptosis in cisplatin-resistant colorectal cancer cells by accumulation of reactive oxygen species, Toxicology Letters 222 (2) (2013) 139–145.

[65] M. Tyagi, B.S. Patro, Salinomycin reduces growth, proliferation and metastasis of cisplatin resistant breast cancer cells via NF-kB deregulation, Toxicology in Vitro 60 (May) (2019) 125–133.

[66] D. Łowicki, A. Huczyński, Structure and antimicrobial properties of monensin A and its derivatives: summary of the achievements, Biomed Research International 2013 (2013) 1–14.

[67] D.A. Kevin, D.A. Meujo, T.H. Mark, Polyether ionophores: broad-spectrum and promising biologically active molecules for the control of drug-resistant bacteria and parasites, Expert Opinion on Drug Discovery 4 (2) (2016) 109–146.

[68] X. Wang, et al., Monensin inhibits cell proliferation and tumor growth of chemo-resistant pancreatic cancer cells by targeting the EGFR signaling pathway, Scientific Reports 8 (1) (2018) 1–15.

[69] A. Huczyński, M. Ratajczak-Sitarz, A. Katrusiak, B. Brzezinski, Molecular structure of the 1:1 inclusion complex of monensin A lithium salt with acetonitrile, Journal of Molecular Structure 871 (1–3) (2007) 92–97.

[70] Y. Deng, et al., Antibiotic monensin synergizes with EGFR inhibitors and oxaliplatin to suppress the proliferation of human ovarian cancer cells, Scientific Reports 5 (2015) 1–16.

[71] K. Ketola, P. Vainio, V. Fey, O. Kallioniemi, K. Iljin, Monensin is a potent inducer of oxidative stress and inhibitor of androgen signaling leading to apoptosis in prostate cancer cells, Molecular Cancer Therapeutics 9 (12) (2010) 3175–3185.

[72] I. Nordgaard-Lassen, K.A. Krogfelt, H. Mirsepasi, S. Halkjaer, A.M. Petersen, E.M. Mortensen, Ciprofloxacin and probiotic Escherichia coli nissle as add-on treatment in active ulcerative colitis; a double-blinded randomized placebo controlled clinical trial, Journal of Crohn's and Colitis 8 (5) (2014) 1498–1505.

[73] H. Saini, S. Chhibber, K. Harjai, Azithromycin and ciprofloxacin: a possible synergistic combination against Pseudomonas aeruginosa biofilm-associated urinary tract infections, International Journal of Antimicrobial Agents 45 (4) (2015) 359–367.

[74] L.A. Bourikas, et al., Ciprofloxacin decreases survival in HT-29 cells via the induction of TGF-β1 secretion and enhances the anti-proliferative effect of 5-fluorouracil, British Journal of Pharmacology 157 (3) (2009) 362–370.

[75] E.R. Mondal, S.K. Das, P. Mukherjee, Comparative evaluation of antiproliferative activity and induction of apoptosis by some fluoroquinolones with a human non-small cell lung cancer cell line in culture, Asian Pacific Journal of Cancer Prevention 5 (2) (2004) 196–204.

[76] T. Kloskowski, et al., Ciprofloxacin is a potential topoisomerase II inhibitor for the treatment of NSCLC, International Journal of Oncology 41 (6) (2012) 1943–1949.

[77] A.C. Pinto, J.N. Moreira, S. Simões, Ciprofloxacin sensitizes hormone-refractory prostate cancer cell lines to doxorubicin and docetaxel treatment on a schedule-dependent manner, Cancer Chemotherapy and Pharmacology 64 (3) (2009) 445–454.

[78] V. Yadav, P. Varshney, S. Sultana, J. Yadav, N. Saini, Moxifloxacin and ciprofloxacin induces S-phase arrest and augments apoptotic effects of cisplatin in human pancreatic cancer cells via ERK activation, BMC Cancer 15 (1) (2015) 1–15.

[79] C. Herold, M. Ocker, M. Granslmayer, E.G. Hahn, D. Schuppan, Ciprofloxacin induces apoptosis and inhibits proliferation of colorectal carcinoma cell in vitro, British Journal of Cancer 86 (2002) 443–448.

[80] O. Aranha, D.P. Wood, F.H. Sarkar, Ciprofloxacin mediated cell growth inhibition, S/G2-M cell cycle arrest, and apoptosis in a human transitional cell carcinoma of the bladder cell line, Clinical Cancer Research 6 (3) (2000) 891–900.

[81] T. Kloskowski, J. Olkowska, A. Nazlica, T. Drewa, The influence of ciprofloxacin on hamster ovarian cancer cell line CHO AA8, Acta Poloniae pharmaceutica drug research 67 (4) (2010) 345–349.

[82] T. Kloskowski, et al., Influence of ciprofloxacin on viability of A549 , HepG2 ,, Acta Poloniae pharmaceutica drug research 68 (6) (2011) 859–865.

[83] N. Pherwani, J.M. Ghayad, L.M. Holle, E.L. Karpiuk, Outpatient management of febrile neutropenia associated with cancer chemotherapy: risk stratification and treatment review, American Journal of Health-System Pharmacy 72 (8) (2015) 619–631.

[84] A. Gürbay, M. Osman, A. Favier, F. Hincal, Ciprofloxacin-induced cytotoxicity and apoptosis in HeLa cells, Toxicology Mechanisms and Methods 15 (5) (2005) 339–342.

[85] D. Reuveni, D. Halperin, I. Fabian, G. Tsarfaty, N. Askenasy, I. Shalit, Moxifloxacin increases antitumor and anti-angiogenic activity of irinotecan in human xenograft tumors, Biochemical Pharmacology 79 (8) (2010) 1100–1107.

[86] D. Reuveni, D. Halperin, I. Shalit, E. Priel, I. Fabian, Quinolones as enhancers of camptothecin-induced cytotoxic and anti-topoisomerase I effects, Biochemical Pharmacology 75 (6) (2008) 1272–1281.

[87] P. Gupta, H.L. Gao, Y.V. Ashar, N.M. Karadkhelkar, S. Yoganathan, Z.S. Chen, Ciprofloxacin enhances the chemosensitivity of cancer cells to ABCB1 substrates, International Journal of Molecular Sciences 20 (2) (2019) 1–12.

[88] E.J. Lim, Y.J. Yoon, J. Heo, T.H. Lee, Y. Kim, Ciprofloxacin enhances TRAIL-induced apoptosis in lung cancer cells by upregulating the expression and protein stability of death receptors through CHOP expression, International Journal of Molecular Sciences 3187 (19) (2018) 1−12.

[89] C. Zhu, X. Yan, A. Yu, Y. Wang, Doxycycline synergizes with doxorubicin to inhibit the proliferation of castration-resistant prostate cancer cells, Acta Biochimica et Biophysica Sinica 49 (11) (2017) 999−1007.

[90] M. Hasanuzzaman, R. Kutner, S. Agha-Mohammadi, J. Reiser, I. Sehgal, A doxycycline-inducible urokinase receptor (uPAR) upregulates uPAR activities including resistance to anoikis in human prostate cancer cell lines, Molecular Cancer 6 (2007) 1−8.

[91] L. Zhang, L. Xu, F. Zhang, E. Vlashi, Doxycycline inhibits the cancer stem cell phenotype and epithelial-to-mesenchymal transition in breast cancer, Cell Cycle 16 (8) (2017) 737−745.

[92] S.Q. Wang, et al., New application of an old drug: antitumor activity and mechanisms of doxycycline in small cell lung cancer, International Journal of Oncology 48 (4) (2016) 1353−1360.

[93] J. Sagar, K. Sales, S. Dijk, J.W. Taanman, A. Seifalian, M. Winslet, Does doxycycline work in synergy with cisplatin and oxaliplatin in colorectal cancer? World Journal of Surgical Oncology 7 (2) (2009) 1−8.

[94] K. Ketola, et al., Salinomycin inhibits prostate cancer growth and migration via induction of oxidative stress, British Journal of Cancer 106 (1) (2012) 99−106.

[95] J.R. Jangamreddy, et al., Salinomycin induces activation of autophagy, mitophagy and affects mitochondrial polarity: differences between primary and cancer cells, Biochimica et Biophysica Acta (BBA) − Molecular Cell Research 1833 (9) (2013) 2057−2069.

[96] K. Arafat, et al., Inhibitory effects of salinomycin on cell survival, colony growth, migration, and invasion of human non-small cell lung cancer A549 and LNM35: involvement of NAG-1, PLoS One 8 (6) (2013) 1−10.

[97] T. Li, et al., Salinomycin induces cell death with autophagy through activation of endoplasmic reticulum stress in human cancer cells, Autophagy 9 (7) (2013) 1057−1068.

[98] R. Riccioni, et al., The cancer stem cell selective inhibitor salinomycin is a p-glycoprotein inhibitor, Blood Cells Molecules and Diseases 45 (1) (2010) 86−92.

[99] J.H. Kim, T.Y. Kim, H.S. Kim, S. Hong, S. Yoon, Lower salinomycin concentration increases apoptotic detachment in high-density cancer cells, International Journal of Molecular Sciences 13 (10) (2012) 13169−13182.

[100] K.H. Koo, et al., Salinomycin induces cell death via inactivation of Stat3 and downregulation of Skp2, Cell Death and Disease 4 (6) (2013) e693.

[101] D. Lu, M.Y. Choi, J. Yu, J.E. Castro, T.J. Kipps, D.A. Carson, Salinomycin inhibits Wnt signaling and selectively induces apoptosis in chronic lymphocytic leukemia cells, Proceedings of the National Academy of Sciences 108 (32) (2011) 13253−13257.

[102] S.Z. Kuo, et al., Salinomycin induces cell death and differentiation in head and neck squamous cell carcinoma stem cells despite activation of epithelial-mesenchymal transition and Akt, BMC Cancer 12 (2012) 1−14.

[103] D. Fuchs, A. Heinold, G. Opelz, V. Daniel, C. Naujokat, Salinomycin induces apoptosis and overcomes apoptosis resistance in human cancer cells, Biochemical and Biophysical Research Communications 390 (3) (2009) 743−749.

[104] F. Wang, et al., Salinomycin inhibits proliferation and induces apoptosis of human hepatocellular carcinoma cells in vitro and in vivo, PLoS One 7 (12) (2012).

[105] Q.L. Tang, et al., Salinomycin inhibits osteosarcoma by targeting its tumor stem cells, Cancer Letters 311 (1) (2011) 113−121.

[106] P.P. Phiboonchaiyanan, C. Kiratipaiboon, P. Chanvorachote, Ciprofloxacin mediates cancer stem cell phenotypes in lung cancer cells through caveolin-1-dependent mechanism, Chemico-Biological Interactions 250 (2016) 1−11.

[107] T. Miclau, M.L. Edin, G.E. Lester, R.W. Lindsey, L.E. Dahners, Effect of ciprofloxacin on the proliferation of osteoblast-like MG-63 human osteosarcoma cells in vitro, Journal of Orthopaedic Research 16 (4) (1998) 509−512.

[108] R. Koziel, J. Szczepanowska, A. Magalska, K. Piwocka, J. Duszynski, K. Zablocki, Ciprofloxacin inhibits proliferation and promotes generation of aneuploidy in Jurkat cells, Journal of Physiology and Pharmacology 61 (2) (2010) 233−239.

[109] H.S. Choi, E.H. Jeong, T.G. Lee, S.Y. Kim, H.R. Kim, C.H. Kim, Autophagy inhibition with monensin enhances cell cycle arrest and apoptosis induced by mTOR or epidermal growth factor receptor inhibitors in lung cancer cells, Tuberculosis and Respiratory Diseases 75 (1) (2013) 9−17.

[110] L. Tumova, et al., Monensin inhibits canonical Wnt signaling in human colorectal cancer cells and suppresses tumor growth in multiple intestinal neoplasia mice, Molecular Cancer Therapeutics 13 (4) (2014) 812−822.

[111] G. Pradel, M. Schlitzer, Antibiotics in malaria therapy and their effect on the parasite apicoplast, Current Molecular Medicine 10 (3) (2010) 335−349.

[112] M. Aikawa, Variations in structure and function during the life cycle of malarial parasites, Bulletin of the World Health Organization 55 (2−3) (1977) 139−156.

[113] N.F. Walker, B. Ndjm, C. JM Whitty, Malaria, Medicine (Baltimore) 46 (1) (2017) 52−58.

[114] I.B. Müller, J.E. Hyde, Folate metabolism in human malaria parasites − 75 years on, Molecular and Biochemical Parasitology 188 (1) (2013) 63−77.

[115] A. Nzila, S.A. Ward, K. Marsh, P.F.G. Sims, J.E. Hyde, Comparative folate metabolism in humans and malaria parasites (part II): activities as yet untargeted or specific to Plasmodium, Trends in Parasitology 21 (7) (2005) 334−339.

[116] J.X. Kelly, R.W. Winter, T.P. Braun, M. Osei-Agyemang, D.J. Hinrichs, M.K. Riscoe, Selective killing of the human malaria parasite *Plasmodium falciparum* by a benzylthiazolium dye, Experimental Parasitology 116 (2) (2007) 103–110.

[117] G. Lisk, M. Kang, J.V. Cohn, S.A. Desai, Specific inhibition of the plasmodial surface anion channel by dantrolene, Eukaryotic Cell 5 (11) (2006) 1882–1893.

[118] N.B. Quashie, L.C. Ranford-Cartwright, H.P. De Koning, Uptake of purines in Plasmodium falciparum-infected human erythrocytes is mostly mediated by the human equilibrative nucleoside transporter and the human facilitative nucleobase transporter, Malaria Journal 9 (1) (2010) 36.

[119] K.K. Roy, Targeting the active sites of malarial proteases for antimalarial drug discovery: approaches, progress and challenges, International Journal of Antimicrobial Agents 50 (3) (2017) 287–302.

[120] D. Burkhardt, et al., Delayed parasite elimination in human infections treated with clindamycin parallels 'delayed death' of Plasmodium falciparum in vitro, International Journal for Parasitology 37 (2007) 777–785.

[121] E.L. Dahl, J.L. Shock, B.R. Shenai, J. Gut, J.L. Derisi, P.J. Rosenthal, Tetracyclines specifically target the apicoplast of the malaria parasite plasmodium falciparum, Antimicrobial Agents and Chemotherapy 50 (9) (2006) 3124–3131.

[122] E.L. Dahl, P.J. Rosenthal, Multiple antibiotics exert delayed effects against the *Plasmodium falciparum* apicoplast, Antimicrobial Agents and Chemotherapy 51 (10) (2007) 3485–3490.

[123] C.D. Goodman, V. Su, G.I. Mcfadden, The effects of antibacterials on the malaria parasite *Plasmodium falciparum*, Molecular and Biochemical Parasitology 152 (2007) 181–191.

[124] S.A. Ralph, M.C.D. Ombrain, G.I. Mcfadden, The apicoplast as an antimalarial drug target, Drug Resistance Updates 4 (2001) 145–151.

[125] G.I. Mcfadden, D.S. Roos, Apicomplexan plastids as drug targets, Trends in Microbiology 7 (8) (1999) 6786–6791.

[126] T. Oguonu, B.O. Edelu, Challenges of Managing Childhood Malaria in a Developing Country: The Case of Nigeria, vol. 2, Intech Open, 2018, p. 64.

[127] A. Alam, et al., Novel antimalarial drug targets : hope for new antimalarial drugs, Expert Review of Clinical Pharmacology 2 (5) (2009) 469–489.

[128] S. Kumar, T.R. Bhardwaj, D.N. Prasad, R.K. Singh, Drug resistant malaria: historic to future perspectives, Biomedicine and Pharmacotherapy 104 (2018) 8–27.

[129] A.A. Amin, et al., The challenges of changing national malaria drug policy to artemisinin-based combinations in Kenya, Malaria Journal 6 (1) (2007) 72.

[130] G.E. Stein, T. Babinchak, Tigecycline: an update, Diagnostic Microbiology and Infectious Disease 75 (4) (2013) 331–336.

[131] G.E. Stein, W.A. Craig, Tigecycline : a critical analysis, Clinical Infectious Diseases 43 (2006) 518–524.

[132] J. Held, P. Zanger, S. Issifou, P.G. Kremsner, B. Mordmüller, In vitro activity of tigecycline in Plasmodium falciparum culture-adapted strains and clinical isolates from Gabon, International Journal of Antimicrobial Agents 35 (2010) 587–589.

[133] D. Ribatski-silva, et al., In vitro antimalarial activity of tigecycline against Plasmodium falciparum culture-adapted reference strains and clinical isolates from the Brazilian Amazon, Revista da Sociedade Brasileira de Medicina Tropical 47 (1) (2014) 110–112.

[134] P. Starzengruber, et al., Antimalarial activity of tigecycline , a novel glycylcycline antibiotic, Antimicrobial Agents and Chemotherapy 53 (9) (2009) 4040–4042.

[135] Q. Lin, K. Katakura, M.S. Ã, Inhibition of mitochondrial and plastid activity of Plasmodium falciparum by minocycline, FEBS Letters 515 (2002) 71–74.

[136] C.J. Beckers, Y. Cao, K.A. Joiner, Inhibition of cytoplasmic and organellar protein synthesis in *Toxoplasma gondii*. Implications for the target of macrolide antibiotics, Journal of Clinical Investigation. 95 (1) (1995) 367–376.

[137] K.N. Agwuh, A. Macgowan, Pharmacokinetics and pharmacodynamics of the tetracyclines including glycylcyclines, Journal of Antimicrobial Chemotherapy 58 (2006) 256–265.

[138] M. Grobbelaar, G.E. Louw, S.L. Sampson, P.D. Van Helden, P.R. Donald, R.M. Warren, Evolution of rifampicin treatment for tuberculosis, Infection, Genetics and Evolution 74 (April) (2019) 103937.

[139] J.A. Badejo, O.O. Abiodun, O. Akinola, C.T. Happi, A. Sowunmi, G.O. Gbotosho, Interaction between rifampicin , amodiaquine and artemether in mice infected with chloroquine resistant Plasmodium berghei, Malaria Journal 13 (1) (2014) 299.

[140] G.A. Mcconkey, M.J. Rogers, T.F. Mccutchan, Inhibition of plasmodium falciparum protein synthesis, Journal of Biological Chemistry 272 (4) (1997) 2046–2049.

[141] M. Sousa, A. Pozniak, M. Boffito, Pharmacokinetics and pharmacodynamics of drug interactions involving rifampicin , rifabutin and antimalarial drugs, Journal of Antimicrobial Chemotherapy 62 (2008) 872–878.

[142] S. Pukrittayakamee, I.C. Viravan, P. Charoenlarp, C. Yeamput, R.J.M. Wilson, Antimalarial effects of rifampin in plasmodium vivax malaria, Antimicrobial Agents and Chemotherapy 38 (3) (1994) 511–514.

[143] M.N. Bulloch, J.T. Baccas, S. Arnold, Clindamycin - induced hypersensitivity reaction, Infection 44 (3) (2016) 357–359.

[144] M. Kierzkowska, A. Majewska, K. Szymanek-majchrzak, A. Sawicka-grzelak, A. Mlynarczyk, G. Mlynarczyk, In vitro effect of clindamycin against Bacteroides and Parabacteroides isolates in Poland, Integrative Medicine Research 13 (2018) 49–52.

[145] J. Wiesner, D. Henschker, D.B. Hutchinson, E. Beck, H. Jomaa, In vitro and in vivo synergy of fosmidomycin , a novel antimalarial drug , with clindamycin, Antimicrobial Agents and Chemotherapy 46 (9) (2002) 2889–2894.

[146] M. Lanaspa, et al., Inadequate efficacy of a new formulation of fosmidomycin- clindamycin combination in Mozambican children less than three years old with uncomplicated *Plasmodium falciparum* malaria, Antimicrobial Agents and Chemotherapy 56 (6) (2012) 2923–2928.

[147] S. Borrmann, et al., Fosmidomycin-clindamycin for the treatment of *Plasmodium falciparum* malaria, The Journal of Infectious Diseases 190 (2004) 1534–1540.

[148] J.D. Walter, M. Hunter, M. Cobb, G. Traeger, P.C. Spiegel, Thiostrepton inhibits stable 70S ribosome binding and ribosome-dependent GTPase activation of elongation factor G and elongation factor 4, Nucleic Acids Research 40 (1) (2011) 360–370.

[149] S.J. Tarr, R.E.R. Nisbet, C.J. Howe, Transcript level responses of plasmodium falciparum to antimycin, Molecular and Biochemical Parasitology 179 (1) (2011) 37–41.

[150] M.N. Aminake, et al., Thiostrepton and derivatives exhibit antimalarial and gametocytocidal activity by dually targeting parasite, Antimicrobial Agents and Chemotherapy 55 (4) (2011) 1338–1348.

[151] K.C. Amorha, O.B. Ugwuowo, E.E. Ayogu, Evaluation of the hepatic effect of concomitant administration of ciprofloxacin and some antimalarial drugs in Plasmodium berghei infected mice : an in vivo study, Pakistan journal of pharmaceutical sciences 31 (5) (2018) 1805–1811.

[152] D.H. Williamson, P.R. Preiser, P.W. Moore, S. Mccready, M. Strath, R.J.M.I. Wilson, The plastid DNA of the malaria parasite *Plasmodium falciparum* is replicated by two mechanisms, Molecular Microbiology 45 (2) (2002) 533–542.

[153] P.U.E. Mayen, E.C. Udobi, M.I. Madu, Amodiaquine and ciprofloxacin combination in plasmodiasis therapy, Journal of Tropical Medicine 2015 (2015).

[154] Y.F. Falajiki, O. Akinola, O.O. Abiodun, C.T. Happi, A. Sowunmi, Amodiaquine – ciprofloxacin: a potential combination therapy against drug resistant malaria, Parasitology 152 (2015) 849–854.

[155] G.O. Gbotosho, C.T. Happi, O. Woranola, O.O. Abiodun, A. Sowunmi, A.M. Oduola, Interaction between ciprofloxacin and chloroquine in mice infected with chloroquine resistant Plasmodium berghei, Parasitology Research 110 (2012) 895–899.

[156] A.A. Andrade, et al., Enhanced activity of mefloquine and artesunic acid against Plasmodium falciparum in vitro and P . berghei in mice by combination with ciprofloxacin, European Journal of Pharmacology 558 (2007) 194–198.

[157] A.L. Baggish, D.R. Hill, Antiparasitic agent atovaquone, Antimicrobial Agents and Chemotherapy 46 (5) (2002) 1163–1173.

[158] A.B. Vaidya, I.K. Srivasta, A mechanism for the synergistic antimalarial action of atovaquone and proguanil, Antimicrobial Agents and Chemotherapy 43 (6) (1999) 1334–1339.

[159] M.R. Rameshkumar, A. Narasingam, Drug-resistant bacterial infections in HIV patients, in: Advances in HIV and AIDS Control, 2018, p. 83.

[160] C.R. Howard, N.F. Fletcher, Emerging virus diseases: can we ever expect the unexpected? Emerging Microbes and Infections 1 (1) (2012) 1–9.

[161] J.J.R. Bigna, C.S. Plottel, S. Koulla-Shiro, Challenges in initiating antiretroviral therapy for all HIV-infected people regardless of CD4 cell count, Infectious Diseases of Poverty 5 (1) (2016) 85.

[162] S.N. Holla, K. Meena Kumari, M.B. Amberkar, Pharmacotherapy of opportunistic infections in HIV patients: a review, Journal of Applied Pharmaceutical Science 6 (9) (2016) 227–236.

[163] T. Welz, C. Wyen, M. Hensel, Drug interactions in the treatment of malignancy in HIV-infected patients, Treat Oncology Research and Treatment 40 (3) (2017) 120–127.

[164] A.B. Bhatti, M. Usman, V. Kandi, Current scenario of HIV/AIDS, treatment options, and major challenges with compliance to antiretroviral therapy, Cureus 8 (3) (2016) 1–12.

[165] R. Minhajat, I. Djaharuddin, R. Halim, A.F. Benyamin, S. Bakri, Drugs hypersensitivity reaction in patient with human immunodeficiency virus (HIV) infection, Journal of Allergy and Therapy 08 (01) (2017) 2001–2004.

[166] J. Opie, Haematological complications of HIV infection, Journal of Organometallic Chemistry 102 (6) (2012) 465–468.

[167] M. Asif, Rifampin and their analogs: a development of antitubercular drugs, World Journal of Organic Chemistry 1 (2) (2013) 14–19.

[168] C. Schutz, G. Meintjes, F. Almajid, R.J. Wilkinson, A. Pozniak, Clinical management of tuberculosis and HIV-1 co-infection, European Respiratory Journal 36 (6) (2010) 1460–1481.

[169] Antibiotic resistance patterns of *Escherichia coli* isolated from HIV-sero positive adults at Mbagathi District Hospital, Nairobi, Kenya, Journal of Applied Biosciences 27 (2010) 1705–1714.

[170] Y. Khushbu, P. Satyam, Bacteriological profile of lower respiratory tract infection (LRTI) among HIV seropositive cases in central terai of Nepal, International Journal of Current Microbiology and Applied Sciences 4 (11) (2015) 431–442.

[171] M.M. Zdanowicz, The pharmacology of HIV drug resistance, American Journal of Pharmaceutical Education 70 (5) (2006) 100.

[172] S.N. Mali, P.M. Sapkal, HIV drug resistance: an overview, Human Journals 1 (1) (2015) 72–82.

[173] D.-Y.L. Ting-Ren Lu, Challenges for HIV/AIDS therapy, Advances in Pharmacoepidemiology and Drug Safety 02 (4) (2013) 2–3.

[174] A. Rothova, L.E.H. Bosch-Driessen, N.H. Van Loon, W.F. Treffers, Azithromycin for ocular toxoplasmosis, British Journal of Ophthalmology 82 (11) (1998) 1306–1308.

[175] C. Gonzalez-Martinez, et al., Azithromycin versus placebo for the treatment of HIV-associated chronic lung disease in children and adolescents (BREATHE trial): study protocol for a randomised controlled trial, Trials 18 (1) (2017) 622.

[176] D.E. Griffith, B.A. Brown, W.M. Girard, B.E. Griffith, L.A. Couch, R.J. Wallace, Azithromycin-containing regimens for treatment of *Mycobacterium avium* complex lung disease, Clinical Infectious Diseases 32 (11) (2001) 1547–1553.

[177] A. Oo, O. Oo, Effect of trimethoprim-sulfamethoxazole and azithromycin prophylaxis on antimicrobial resistance of faecal *Escherichia coli* isolated from HIV-infected and TB patients in ekiti state, American Journal of Clinical Microbiology and Antimicrobials 1 (6) (2018) 1–5.

[178] M. Kaufmann, et al., Comparing the antimicrobial in vitro efficacy of amoxicillin/metronidazole against azithromycin—a systematic review, Dental Journal 6 (4) (2018) 59.

[179] A. Alchakaki, C. Cramer, A. Patterson, A.O. Soubani, Which patients with respiratory disease need long-term azithromycin? Cleveland Clinic Journal of Medicine 84 (10) (2017) 755–758.

[180] D.J. Horne, C. Spitters, M. Narita, Experience with rifabutin replacing rifampin in the treatment of tuberculosis, International Journal of Tuberculosis and Lung Disease 15 (11) (2011) 481–482.

[181] C. Boulanger, et al., Pharmacokinetic evaluation of rifabutin in combination with lopinavir-ritonavir in patients with HIV infection and active tuberculosis, Clinical Infectious Diseases 49 (9) (2009) 1305–1311.

[182] G.J. Moyle, N.E. Buss, T. Goggin, P. Snell, C. Higgs, D.A. Hawkins, Interaction between saquinavir soft-gel and rifabutin in patients infected with HIV,, British Journal of Clinical Pharmacology 54 (2) (2002) 178–182.

[183] Y. Crabol, E. Catherinot, N. Veziris, V. Jullien, O. Lortholary, Rifabutin: where do we stand in 2016? Journal of Antimicrobial Chemotherapy 71 (7) (2016) 1759–1771.

[184] F.A. Paina, J.C. Miranda, C.F. de F. Tavares, R.H. da C. Queiroz, A.M. de Souza, WBC count and functional changes induced by co-administration of clofazimine and clarithromycin, in single and multiple doses, in Wistar rats, Brazilian Journal of Pharmaceutical Sciences 48 (3) (2012) 417–425.

[185] I.I. Salem, N. Düzgünes, Efficacies of cyclodextrin-complexed and liposome-encapsulated clarithromycin against *Mycobacterium avium* complex infection in human macrophages, International Journal of Pharmaceutics 250 (2) (2003) 403–414.

[186] A.M.T. Van Nuffel, V. Sukhatme, P. Pantziarka, L. Meheus, V.P. Sukhatme, G. Bouche, Repurposing drugs in oncology (ReDO) — clarithromycin as an anti-cancer agent, Ecancermedicalscience 9 (2015).

[187] A. Sher, IL-12 promotes drug-induced clearance of *Mycobacterium avium* infection in mice, The Journal of Immunology 160 (11) (1998) 5428–5435.

[188] M. Saidinejad, M.B. Ewald, M.W. Shannon, Transient psychosis in an immune-competent patient after oral trimethoprim-sulfamethoxazole administration, Pediatrics 115 (6) (2005) e739–e741.

[189] A.S. Hassani, B.J. Marston, J.E. Kaplan, T. Branch, R. Branch, D. Control, Assessment of the impact of cotrimoxazole prophylaxis on key outcomes among HIV-infected adults in low- and middle- income countries: a systematic review, Journal of Acquired Immune Deficiency Syndromes 68 (Suppl. 3) (2015) S257–S269.

[190] R. Senanayake, M. Mukhtar, Cotrimoxazole-induced hypoglycaemia in a patient with churg-strauss syndrome, Case Reports in Endocrinology 2013 (2013).

[191] S. Eiam-ong, N.A. Kurtzman, S.D. Sabatini, Studies on the mechanism of trimethoprim-induced hyperkalemia, Kidney International 49 (5) (1996) 1372–1378.

[192] C. Chintu, et al., Co-trimoxazole as prophylaxis against opportunistic infections in HIV-infected Zambian children (CHAP): a double-blind randomised placebo-controlled trial, Lancet 364 (9448) (2004) 1865–1871.

[193] T. Young, C.E.M. Oliphant, I. Araoyinbo, J. Volmink, Co-trimoxazole prophylaxis in HIV : the evidence, SAMJ South African Medical Journal. 98 (4) (2008).

[194] D. Kibuule, M. Mubita, E. Naikaku, F. Kalemeera, B.B. Godman, E. Sagwa, "An analysis of policies for cotrimoxazole , amoxicillin and azithromycin use in Namibia ' s public sector: findings and therapeutic implications, International Journal of Clinical Practice 71 (2) (2017) e12918.

[195] C. Manyando, E.M. Njunju, U.D. Alessandro, J. Van, Safety and efficacy of Co-trimoxazole for treatment and prevention of *Plasmodium falciparum* Malaria: a systematic review, PLoS One 8 (2) (2013) e56916.

[196] M. Bwakura-Dangarembizi, et al., A randomized trial of prolonged co-trimoxazole in HIV-infected children in Africa, New England Journal of Medicine 370 (1) (2014) 41–53.

[197] D.M. Mwenya, et al., Impact of cotrimoxazole on carriage and antibiotic resistance of Streptococcus pneumoniae and Haemophilus influenzae in HIV-infected children in Zambia, Antimicrobial Agents and Chemotherapy 54 (9) (2010) 3756–3762.

[198] K. Grimwade, G.H. Swingler, Cotrimoxazole prophylaxis for opportunistic infections in adults with HIV, Cochrane Database of Systematic Reviews 3 (2003).

[199] B. Hasse, et al., Co-trimoxazole prophylaxis is associated with reduced risk of incident tuberculosis in participants in the Swiss HIV cohort study, Antimicrobial Agents and Chemotherapy 58 (4) (2014) 2363–2368.

[200] J.A. Church, F. Fitzgerald, A.S. Walker, D.M. Gibb, A.J. Prendergast, The expanding role of co-trimoxazole in developing countries, The Lancet Infectious Diseases 15 (3) (2015) 327–339.

[201] S. Bhambri, J.Q. Del Rosso, A. Desai, Oral trimethoprim/sulfamethoxazole in the treatment of acne vulgaris, Cutis-New York 79 (6) (2007) 430–434.

[202] Y. Minato, S. Dawadi, S.L. Kordus, A. Sivanandam, C.C. Aldrich, A.D. Baughn, Mutual potentiation drives synergy between trimethoprim and sulfamethoxazole, Nature Communications 9 (1) (2018) 1003.

[203] J. Yang, et al., Multicenter study of trimethoprim/sulfamethoxazole- related hepatotoxicity: incidence and associated factors among HIV-infected patients treated for *Pneumocystis jirovecii* pneumonia, PLoS One 9 (9) (2014) e106141.

[204] Y. Yee, E.G. Rakasz, D.J. Gasper, T.C. Friedrich, L.A. Trepanier, Immunogenicity of trimethoprim/sulfamethoxazole in a macaque model of HIV infection, Toxicology 368−369 (2016) 10−18.

[205] A. Wolf, S. Schoof, S. Baumann, H.A.K.N. Kirschner, Structure − activity relationships of thiostrepton derivatives: implications for rational drug design, Journal of Computer-Aided Molecular Design 28 (2014) 1205−1215.

[206] J.C. Taganna, J.P. Quanico, R.M.G. Perono, E.C. Amor, W.L. Rivera, Tannin-rich fraction from *Terminalia catappa* inhibits quorum sensing (QS) in Chromobacterium violaceum and the QS-controlled biofilm maturation and LasA staphylolytic activity in *Pseudomonas aeruginosa*, Journal of Ethnopharmacology 134 (3) (2011) 865−871.

[207] M. Leary, S. Heerboth, K. Lapinska, S. Sarkar, Sensitization of drug resistant cancer cells: a matter of combination therapy, Cancers 10 (12) (2018) 483.

[208] V. Srivastava, H. Lee, Chloroquine-based hybrid molecules as promising novel chemotherapeutic agents, European Journal of Pharmacology 762 (2015) 472−486.

[209] F.W. Muregi, A. Ishih, Next-generation antimalarial drugs: hybrid molecules as a new strategy in drug design, Drug Development Research 71 (1) (2010) 20−32.

[210] X. Nqoro, N. Tobeka, B.A. Aderibigbe, Quinoline-based hybrid compounds with antimalarial activity, Molecules 22 (2017) 2268.

CHAPTER 3

Antibiotic Polymer for Biomedical Applications

VICTORIA OLUWASEUN FASIKU • DANIEL HASSAN • SHESAN JOHN OWONUBI •
EMMANUEL MUKWEVHO • JIMMY LOLU OLAJIDE

1 INTRODUCTION

In recent decades, the biomedical field has received great attention owing to the discovery and development of new biomaterials with several applications in biomedicine. One such biomaterial is polymers; polymers are macromolecules consisting of repeating structural subunits that are chemically covalently bonded. Polymers can either occur as natural polymers or synthetic polymers and both can be employed for different applications in biomedicine. Natural polymers can either be of plant origin or animal origin; on the other hand, synthetic polymers are human made. However, natural polymers are more attractive for use because most of them are readily available, biodegradable, biocompatible, economical, and nontoxic. Examples of natural polymers include cellulose, starch, pectin (plant origin), chitin, alginates (animal origin), while examples of synthetic polymers include poly(ethylene glycol) (PEG), poly(lactic acid), polyesters, poly(glycolic acid), *copolymer* poly(lactide-*co*-glycolide), polyurethanes, and polystyrene among others [1].

The main advantage of natural polymers over synthetic polymers is the similarity they have with the components of the extracellular matrix of the human body. In addition, natural polymers prevent the induction of toxicity and chronic immunological reactions that are associated with polymers (synthetic). Some disadvantages associated with polymers of natural (plant) origin include synthesis in small quantities and slow/expensive process of isolation. Additionally, intellectual property rights have also become an issue of concern [2]. Synthetic polymers, on the other hand, is advantageous because the chemical, structural, and mechanical properties are controllable [3]. The classification of synthetic polymers into degradable and nondegradable synthetic polymers is dependent on the method of fabrication that is based on chemical reactivity [4]. Although currently, biodegradable polymers are more explored and used in medicine, however, the initial application of polymers in treating and managing various health conditions began with the use of nondegradable polymers [5].

Generally, polymers are synthesized via polymerization techniques, and these techniques can be classified based on the types of reactions viz; addition and condensation polymerization [6]; however, polymers of animal origin (microbial origin) can be synthesized via fermentation processes [7]. Polymers have become of great interest in biomedicine when compared to other materials like metals and ceramics; this is due to the numerous advantages they offer and due to their excellent properties [3]. These include the ability to respond to external stimuli, the biodegradability of most polymers, the flexibility in the method of synthesis to suit particular applications, and the ability to fabricate into diverse forms such as fibers, films, particles, and gels [8,9]. In addition, polymers, when compared to other biomaterials, remain the most versatile as it is being used extensively in applications such as drug-delivery vehicles, implants, artificial organs, medical devices, dental materials, and tissue engineering among others [10,11]. A major area of biomedicine where polymers have been employed is in combating pathogenic microorganisms and treating infectious diseases caused by these organisms.

Microbes (bacteria, fungi, viruses, and protozoa) are the key sources of infections that untreated and managed can lead to loss of life [12,13]. Several antimicrobial agents are being used to kill or inhibit the growth of microbes; however, treatment of infectious disease is still difficult because these pathogenic microorganisms have become resistant to antibiotics and possess the ability to form biofilms [14,15]. Biofilms

Antibiotic Materials in Healthcare. https://doi.org/10.1016/B978-0-12-820054-4.00003-3

are formed by various microorganisms on various substrates, and cells mature in multicellular aggregates and are enclosed in an extracellular matrix generated by the microorganisms. Thus biofilms become difficult to eradicate and are able to resist several antimicrobial agents. Antibiotic resistance is increasingly becoming a persistent threat globally leading to high morbidity and mortality with more reports on bacterial antibiotic resistance [16]. At the moment, several limitations are faced with conventional antibiotic formulations and techniques used in fighting antibiotic resistance in bacteria; therefore, there is a need for the development of newer strategies to overcome these challenges [17].

These newer strategies should, therefore, be able to inhibit the formation of biofilms, by reducing the attachment of microbes on substrates as well as efficiently killing pathogenic microbes that are resistant to antibiotics to prevent the spread of infectious diseases. Polymers with antibiotics properties have emerged as one such new strategies that have been found relevant in this application; therefore, this chapter focuses on the application of antibiotic polymers in eradicating pathogenic microorganisms (bacteria).

2 ANTIMICROBIAL POLYMERS

Antimicrobial polymers are polymers that potentially possess inherent antimicrobial properties, that is, they do not need to be stimulated to exhibit their antimicrobial activities [18]. This class of polymers has components in the chemical structures that are involved in the antimicrobial activities of the polymer. Antimicrobial polymers were first discovered in the year 1965 and have since then become a center of attraction to both academic and industrial research [19].

3 TYPES

The classification of antimicrobial polymers into bound or leaching antimicrobial polymer is based on the component of the polymer [20]. The three basic types of antimicrobial polymers currently employed in biomedicine are biocide-releasing polymers, polymeric biocides, and biocidal polymers [19]. Polymeric biocides are antimicrobial polymers with repeating units of bioactive monomers that are covalently linked; these monomers include amino, hydroxyl, or carboxyl groups [21–23]. The antimicrobial activity of the bioactive functional monomers can either be enhanced or reduced depending on the polymerization process that was used to synthesize the polymer [20]. Various examples of polymeric biocides have been investigated for antibacterial activities and they have shown very efficient and high antibacterial efficacy against bacteria (gram negative and positive) [23–27].

Biocidal polymers, on the other hand, are antimicrobial polymers with inherent antimicrobial properties; they do not need to have repeating units of bioactive substances because the antimicrobial site is within the whole structure of the macromolecule [20]. An example of this kind of antimicrobial polymer is the antimicrobial cationic polymers that consist of two components: hydrophobic and cationic groups. They consist of cationic biocides, such as guanidinium, quaternary ammonium, tertiary sulfonium, and phosphonium. The antimicrobial activity exhibited by this set of polymer largely depends on the amount of the individual components of the polymer as well as the nature and location of the individual components [28]. In addition, the antimicrobial action of polymers that are cationic is proportional to the density of cationic groups in charge. They are able to induce bacterial cell death due to the negative charge on the surface of the bacteria that reacts with the cationic polymers [29]. These types of antimicrobial polymers are mainly of natural origin with chitosan being the most representative natural material; however, others include heparin, poly-L-lysine, and gramicidin A [19]. They are biodegradable, biocompatible, and do not exhibit toxicity, and their antimicrobial activity is highly dependent on pH value especially for chitosan. A pH value of less than pKa value will lead to an interaction between the bacterial cell wall and protonated amino groups, while a pH higher than the value of pKa leads to hydrophobic interaction and chelation effects [20].

Biocide-releasing polymers are basically polymers used as carriers for biocides. They exhibit antimicrobial activities upon the incorporation of biocidal agents such as antibiotic and/or antiseptic compounds. This can be achieved by either polymerizing the biocide-releasing molecules to the backbone of the polymeric material or via the synthesis of polymer/biocide-releasing molecules composites. These antimicrobial polymers have the ability to release biocides in a controlled manner thus, conferring advantages such as facilitating the delivery of biocides with short in vivo half-lives and high local biocide concentration close to the microorganisms [30,31]. In addition, antimicrobial polymers can also be categorized as either solution-based or surface-bound polymers. Although polymers that are surface-bound exhibit direct antimicrobial action on the polymer surface, polymers that are solution-based exhibit antimicrobial activity within solutions. Generally, polymers that are biocidal are known to be surface-bound polymers while biocide-releasing polymers are solution-based polymers that release biocides in solutions, polymeric biocides, however, can be solution-based polymers or surface-bound depending on the property of bioactive repeating units [20].

4 PROPERTIES

The properties of various natural and synthetic polymers used as antimicrobial carrier or agents possess properties that differ from one another. These properties contribute to their antimicrobial properties directly or indirectly. Table 3.1 outlines some of the common synthetic and natural antimicrobial polymers that have found relevance in the treatment and management of pathogenic microorganisms.

5 SYNTHESIS

Antimicrobial polymers are prepared via polymerization methods that are further divided into three types of polymerizations: addition polymerization, condensation polymerization, and metathesis polymerization [6]. Addition polymerization is a process of synthesizing polymers through the addition of monomers step wisely and intermediates that are reactive, which prevents the loss of small molecules. Hence, it is otherwise described as chain-growth polymerization. The reaction of monomers (acetylenes, olefins, aldehydes) leads to the formation of very high molecular weight polymers, and the pi-bond of the monomer is transformed into a sigma bond in the polymer. Addition polymerization is an exothermic reaction that greatly favors the thermodynamic chemical transformation of olefins [33].

Condensation polymerization is also called as step-growth polymerization, and it is a process that involves the combination of two different monomers by eliminating small molecules such as water, ammonia, methanol, and hydrochloric acid (HCl). The reactivity of the functional end group greatly determines the nature of the end product that results from condensation polymerization. Two main characteristics are associated with the monomers involved in condensation polymerization and this makes the monomers different from the monomers involved in addition polymerization. The monomers involved in condensation polymerization have at least two sites that are reactive and functional groups such as $-OH$, $-NH_2$, or $-COOH$ rather than possessing double bonds. This polymerization method produces polymers of high molecular weight only at high conversion with a high change in enthalpy (ΔE_a) and thus heat is required during the process. Polymers with low molecular weight are generated in cases where the monomers involved in the reaction have a single group that is reactive, and this causes the growth of the chain to terminate. On the other hand, straight polymers are produced with monomers with double end groups that are reactive while three-dimensional cross-linked polymers are the products of condensation polymerization reactions that involve monomers consisting of two or more end groups. An example technique used during condensation polymerization is the dehydration synthesis in which different monomers (two or more) are used; usually, the monomer with the hydroxyl group (-OH) is joined to a free ionizable hydrogen atom (-H). This reaction leads to the formation of the polymer and water (from the hydroxyl and hydrogen) [33].

Metathesis polymerization is another type of polymerization method that is used to synthesize polymers; the carbon—carbon double bond of the monomers remains intact within chains of the polymeric backbone and polymers formed are called as polyalkenamers [50]. This is contrary to other polymerization processes, whereby the carbon—carbon double bond cleaves away from the polymeric backbone upon the conversion of vinyl monomer into polymer. This mechanism is commonly employed for olefin metathesis, and this involves cycloaddition reaction of olefin and the transition metal alkylidene complex that leads to the formation of an intermediate material (metallocyclobutane). There are two kinds of metathesis polymerization: ring-opening metathesis polymerization and acyclic diene metathesis [51].

Furthermore, the technique of polymerization can be divided broadly into homogeneous and heterogeneous depending on the method used for its synthesis [52,53]. These are namely emulsion, suspension, bulk, solution, and precipitation polymerization techniques. In addition, the use of monomers having two double bonds cross-linking can sometimes be achieved. Table 3.2 briefly explains these other techniques involved in polymer polymerization.

6 CHARACTERIZATION

Owing to the use of polymers of both natural and synthetic origin in biomedicine, it has become necessary for various characterization to be performed to determine their applicability in diverse fields in medicine. This characterization reveals various parameters of the polymers such as strength [58], size [59], biocompatibility [60], and amorphousity [61], among others. These properties have been associated with the material performance of polymers [62]. Therefore for a better understanding and improvement of polymers, the following characterization techniques are commonly required and thus employed.

TABLE 3.1
Properties/Characteristics of Some Common Antimicrobial Polymers.

Polymer	Properties/Characteristics	References
Polycaprolactone	• It is a biodegradable polymer that degrades in physiological conditions via the hydrolysis of its ester linkages. • It has a glass transition temperature (−60°C) and low melting point (60°C). • It can be combined with a range of other materials such as starch for cost reduction and increased biodegradability. • It has been approved by the Food and Drug Administration (FDA) in particular applications, for example, as a drug delivery carrier. Ring-opening polymerization (ROP) of ε-caprolactone using dibutylzinc–triisobutylaluminum systems as a catalyst is often used to synthesize.	[32,33]
Chitosan	• It is a natural polysaccharide that is linear and consists of β-(1−4)-linked d-glucosamine (deacetylated unit) and N-acetyl-d-glucosamine (acetylated unit) distributed randomly. • It is synthesized by exposing shrimp and shells of other crustacean such as crabs and krills to alkali NaOH. • It can also be prepared by hydrolyzing the acetamide group of chitins with alkaline because of the resistance of such groups caused by the arrangement (trans) of the C2—C3 substituent in the sugar ring. • It is abundant in nature and it is a renewable polymer with outstanding properties, for instance, it is biodegradable, biocompatible, and nontoxic. • It has found relevance in biomedicine as an antibacterial agent and also as bandages to reduce bleeding. In addition, it can also be used in drug-delivery applications. Chitosan is prepared by hydrolyzing the acetamide groups of chitins. • It can be synthesized in various forms because it is very reactive in nature and it is mucoadhesive. • Its most important characteristics is that it possesses a positive charge under acidic conditions.	[34−36]
Polyethylene glycol (PEG)	• It is a polymer that is prepared via a technique known as suspension polymerization and the process is exothermic. • It is synthesized via the association of ethylene oxide and ethylene glycol oligomers, water, ethylene glycol, or an acidic or basic catalyst is used for the catalysis of the reaction. • The type of catalyst used during the process of synthesizing this polymer determines the polymerization mechanism. It can either be cationic or anionic; however, the anionic process is more preferred because it produces PEG having little polydispersity. • Its initial starting material during synthesis is ethylene glycol and its oligomers, this allows the polymer to be distributed in a narrow molecular-weight distribution thus forming a polydispersed polymer. • Its chain length is dependent on the ratio of all the reactants involved in the process of synthesis. • It has good properties such as biocompatibility and thus has found application in biomedicine as a drug-delivery biomaterial.	[37,38]
Hyaluronic acid	• It is a linear natural polymer that can as well be called as hyaluronan or hyaluronate consisting of repeating units of a disaccharide (β(1,4)-glucuronic acid (GlcUA)-β(1,3)-N-acetylglucosamine). • The structure/shape of the polymer in solution is determined by the individual disaccharide residue although they adopt the stable chair conformation. • It is a polymer with outstanding rheological properties and high molecular mass. • It is not bound in a covalent manner to the protein in the core of the proteoglycan neither is it sulfated. • It is being used as a major active component in pharmaceutical formulations for the treatment of some diseases.	[39,40]

Polymethyl methacrylate	• It is synthetic and obtained from methyl methacrylate, and it is transparent and has compactness (density of 1.17 –1.30 g/cm³), lower than half that of glass. Its maximum water absorption ratio is 0.3%–0.4% by weight and an increase in water absorption leads to a decrease in tensile strength.	[33,41,42]
	• It is generally synthesized via emulsion, solution, and bulk polymerization, and it has a relatively high thermal expansion coefficient of $5^{-10} \times 10^{-5}$ K^{-1}.	
	• It also has excellent impact strength that is greater when compared to both glass or polystyrene although this is still considerably lower than polycarbonate and few other polymers that have been engineered.	
	• Its glass transition temperature (T g) is 105°C (atactic PMMA). Although the commercial grades of PMMA have a glass transition temperature ranging from 85 to 165°C, they exist as copolymers and comonomers of other materials other than methyl methacrylate.	
	• It has the ability to swell and dissolve in various organic solvents. In addition, it is poorly resistant to many other chemicals because its ester groups can be easily hydrolyzed.	
	• It is more superior to several other polymers such as polyethylene and polystyrene, and it is environmentally stable.	
	• It is biocompatible and has been used for various medical applications such as the manufacture of intraocular lenses.	
Dextrans	• It is a polysaccharide that is branched, consisting of many glucose molecules and alpha 1,6 glycosidic linkages exists between the glucose, these glucose molecules are often of different chain lengths (from 10 to 150 kDa), thus making dextran a complex biomaterial.	[33,43]
	• Its branching often starts from alpha-1,4 linkage, alpha-1,2, and alpha-1,3 linkages.	
	• Its polyglucans are prepared from sucrose by several species of the *Lactobacillus* and *Streptococcus* genera.	
	• It (dextran B-512(F)) is soluble in solvents such as glycerol, water, ethylene glycol, 4-methylmorpholine-4-oxide, methyl sulfoxide, formamide, and hexamethylphosphoramide. Although some parts of dextran may exhibit some degree of crystallinity, this can be dissolved strong heating.	
Polylactic-co-glycolic acid	• It is a copolymer that is biocompatible/biodegradable and synthesized via polymerization, the type of PLGA formed is dependent on the lactide-glycolide ratio used for the synthesis.	[44]
	• It is synthesized via ring random ring-opening copolymerization of the cyclic 1,4-dioxane-2,5-diones (dimers) of lactic and glycolic acid.	
	• It is amorphous in nature, with a glass transition temperature of approximately 40–60°C, and degrades in the presence of water by hydrolyzing the ester linkages.	
	• Its hydrolysis leads to the production of lactic and glycolic acid (its native monomers) that are metabolic by-products. Its degradation time is associated with the ratio of the monomers employed for its preparation.	
	• Its homopolymers such as glycolic and lactic acid are poorly soluble; however, PLGA readily dissolves in a variety of solvents that include acetone, chlorinated solvents, ethyl acetate, and tetrahydrofuran.	
	• It is currently employed in various drug delivery, biomaterial, or therapeutic applications because there is minimized toxic effect associated with its usage.	

Continued

TABLE 3.1
Properties/Characteristics of Some Common Antimicrobial Polymers.—cont'd

Polymer	Properties/Characteristics	References
Alginate	• It is an anionic polysaccharide linear copolymer also known as algin or alginate that forms a viscous gum when it binds with water and upon exposure to the high acidic medium and it is widely distributed within (cell wall) brown algae. • It is formed by the reaction between the homopolymeric blocks of the residues of C-5 epimer α-L-glucuronate (G) and (1—4)-linked β-D-mannuronate (M), respectively, in a covalently linked in a sequence that is different. • It is capable of rapid water absorption, and it is a biocompatible polymer with little toxicity with moderately low cost. • It becomes mildly gel-like by the addition of Na+ and Ca^{2+} or other cations • It has found application in delivery drug and protein in a controlled manner and blood vessel, bone, cartilage, and tissue regeneration among other biomedical applications. • It is also employed in the pharmaceutical industry for diverse applications. For example, it is incorporated into tablets to increase the disintegration rate of the tablet and thus leads to a quicker release of a therapeutic substance.	[45,46]
Polyurethane	• It is a polymer consisting of chains of organic units linked by urethane. • It is not affected by heating, that is, they do melt upon exposure to heat and most polyurethane exists as thermosetting polymers • It is very flexible and have a good tear and abrasion resistance; hence, it is outstanding in performance when compared to other polymers • It has molecular structures that can be likened to that of proteins in the human body thus making it useful for a broad range of medical and biomedical applications such as applications that require biomimetic, strength biomaterial. • It is synthesized by the reaction of bischloroformates synthesized from dihydroxy compounds, abundance phosgene and diamines then by further reacting with diisocyanates with dihydroxy compounds.	[33]
Cyclodextrin	• It is a natural oligosaccharide that is cyclic in nature, which was discovered over 10 decades ago • It is made up of (α-1,4-)-linked α-d-glucopyranose units with an outer surface and central cavity that are hydrophilic and lipophilic, respectively. • It is cone-like in shape due to the chair formation of the glucopyranose units, and its secondary hydroxy groups extend from the wider edge while the primary groups from the narrow edge • It has become available as pharmaceutical excipients; it is used to enhance the solubility of drugs with poor solubility. • It has also been used to enhance the bioavailability, stability, and prolong the shelf-life of pharmaceutic compounds. • Its properties for pharmaceutical and biomedical applications can be enhanced by several chemical functionalizations.	[47—49]

TABLE 3.2
Polymerization Techniques.

Technique	Description	References
Solution polymerization	• This process involves the dissolution of monomers in a nonreactive solvent containing catalyst resulting in the formation of a soluble polymer with low molecular weights. • The reaction rate of this process is reduced by the absorption of the released heat (by the solvent) in the process. • This process is preferable for preparing polymers that are wet because of the difficulty in removing additional solvent.	[54]
Bulk (mass) polymerization	• This is a highly exothermic process that produces polymers with nonuniform molecular mass distribution catalyzed by additives and transfer agent. • This process is suitable for liquid monomers and produces higher optical clarity polymer whose properties can be controlled by varying the temperature and pressure. • This process is used to synthesize low-molecular-weight polymers.	[55]
Suspension polymerization	• This is a radical polymerization process that is heterogeneous, and it is used to produce polymers in a step-growth manner. • This involves insoluble (in liquid phase) monomers and initiators that can be dispersed in solvents such as water by stirring to form beads within the liquid matrix. These beads are pearl-like in shape; hence, this process is called *pearl polymerization.* • This reaction can be controlled because heat transfer is very efficient.	[56]
Precipitation polymerization	• This is a polymerization process that is heterogeneous and produces insoluble polymers that precipitate but can be separated by centrifugation or filtration to form gel or powder. The degree of polymerization is usually high due to lack of heat dissipation.	[33]
Emulsion polymerization	• This process involves a radical polymerization of a monomer that is liquid dispersed, in a liquid that is not soluble leading to emulsion formation. • This polymerization occurs in the latex particles consisting of several different polymer chains (100 nm in size) that form suddenly in the first few minutes of the polymerization process. The oil-in-water emulsion polymerization is the commonest of this type of polymerization.	[57]

6.1 Mechanical and Thermal Characterization

The mechanical properties of polymeric materials are largely related to their performance in any application, and they can be considered "strong," "tough," "soft," "viscous," or "ductile." For example, polymers employed as implants or coatings on formulations should possess tensile strength that is similar to the soft tissues and good adhesive properties, respectively [63,64]. The mechanical properties of polymeric materials are often determined by measuring stress-strain or viscoelasticity, and this is achieved using dynamic mechanical analysis (DMA) or oscillatory shear rheometers. Similarly, it is vital to identify the thermal properties of polymers such as crystallization, glass transition, and melting temperatures prior application, as the stability of some polymers and their dissolution profile are associated with the glass transition temperature [65]. This parameter (glass transition temperature) can be determined by using differential scanning calorimetry (DSC). In general, the following tests are used to determine polymers mechanical and thermal properties.

6.1.1 Stress-strain

This test is the most commonly employed, used to determine the mechanical properties (tensile strength) of polymers. This test is done by placing the polymer sample to one end of a loading frame and subsequently applying a force at the opposite end to achieve a displacement in a controlled manner. This leads to the formation of a stress-strain curve that is further investigated to provide information on the mechanical properties of the polymer material.

6.1.2 Viscoelasticity

when polymers are subjected to deformation, they exhibit two properties viz; viscosity (dissipated energy) and elasticity (stored energy). To describe viscoelastic properties, various models such as Kelvin-Voigt and Maxwell that incorporates different combinations of "springs" and "pistons" are used. However, owing to the complexity of the viscoelastic properties of polymers, it is required that more than one model be used for more accuracy [62].

6.1.3 Dynamic mechanical analysis

This technique is also used to determine the viscoelasticity of polymers, and it has been considered the most used technique when compared to others [66]. Usually, samples are placed in a chamber with controlled temperature and then stress is applied, the resultant strain is measured for complex modulus analysis. A completely elastic material will exhibit a stress-strain that will be perfect in phase; whereas, a completely viscous material will have a stress-strain at an angle of 90 degree. In addition, this test provides information on the storage (the elastic part, that is, stored energy), and loss (the viscous part, that is, dissipated energy) modulus, respectively. Furthermore, this technique is also more sensitive (approximately $100°$ times more than DSC) and useful in measuring transition temperatures such as the glass transition temperature (T g). A major drawback of this characterization technique is that large sample sizes are required [67].

6.1.4 Oscillatory shear rheometer

This is a very good technique for determining the dynamic mechanical properties of polymeric dispersed systems, for example, hydrogels, suspensions, and emulsions. This allows a better understanding of microstructure and formulation through the information provided from the viscosity result [68]. Rheometers are generally of two types namely, rotational and extensional rheometers but a better characterization of the complex viscoelastic properties of the polymer is provided by rotational rheometers. Compared to DMA, this technique is more superior for hydrogel characterization [69].

6.1.5 Differential scanning calorimetry

This is a great instrument for characterizing the thermal properties (temperature and heat flow) of polymers that are involved in transition temperatures. This is achieved by heating/cooling samples and inert reference in a controlled environment at a particular rate with applied heat flow to sustain the same temperature between the reference and sample. The flow of heat is measured and recorded to provide information such as crystallization, glass transition, and melting. DSC is a very efficient technique used to understand the polymer–drug interaction and the sensitivity and resolution of DSC can be enhanced by performing modulated DSC (MDSC). A major disadvantage of MDSC is a potential change in the sample as measurements often take hours [70,71].

A complementary technique to DSC is thermal gravimetrical analysis (TGA), which usually consists of a sample pan, a precision balance, and a programmable furnace used to measure the rate at which the mass of the sample changes with time and a rise or decrease in temperature. TGA supplies information about the physical and chemical changes that occur in the material. Examples of the physical and chemical information provided are vaporization, sublimation, absorption and desolvation, decomposition, oxidation, and combustion, respectively [62].

6.2 Surface and Morphological Characterization

In understanding the morphology of polymers, the macroscopic structure/shape and crystallinity/amorphousity of the material are considered. This is necessary because the structure affects the property of the system for a specific application. For instance, an increase in the porosity of a polymer used in drug delivery can result in an increase in an initial burst release of the drug [65]. In addition, the degree of crystallinity of polymers is linked to its efficiency in drug delivery applications [72]. The most appropriate techniques used to determine the shape/structure and crystallinity/amorphousity of polymeric materials are microscopy and X-ray diffraction, respectively.

6.2.1 Microscopy and X-ray diffraction

The use of microscopes such as light-, electron-, and scanning probe-based systems reveals structural images that cannot be observed with the naked eyes. Examples of optical microscopes that are widely used are fluorescent microscopy and polarized light microscopy [73] while the commonly used electron microscopes are transmission electron microscopy (TEM) and scanning electron microscopy (SEM). SEM visualizes samples down to 1 nm and smaller while TEM visualizes down to 0.1 nm and smaller; however before the discovery of these microscopy techniques, AFM and scanning tunneling microscopy are commonly used [74]. AFM provides information about the image, measurement at the nanoscale level [75]. Other techniques for crystalline and amorphous characterizations include the following.

6.2.2 XRD cryptography

This is an excellent method used to determine the crystallinity of polymers, the polymer is exposed to X-rays and this produces diffraction patterns that are subsequently used to describe crystallinity of the sample. This technique is based on Bragg's law as there is a constructive interference of the reflected X-rays from the various layers of the crystal. Another method that is used to determine the crystallinity of polymers is by measuring the density of the polymer using a density-gradient method. The region that is crystalline will be more densely packed when compared to the amorphous region [62,76].

6.2.3 Spectroscopic methods

X-ray photoelectron spectroscopy (XPS) is known to be the most popularly used spectroscopic tool for analyzing the surface of polymers as well as determining the elemental constituents of solid surfaces. The principle of operation is based on electron emission from the surface material in response to irradiation with monochromatic X-rays. Apart from XPS, some other spectroscopic methods used for surface analysis are secondary ion mass spectroscopy (SIMS), Auger electron spectroscopy, and attenuated total reflection Fourier transform infrared spectroscopy [77]. By measuring the covalent binding of molecules at the surface, SIMS provides information such as adhesiveness, wettability reactivity and detailed material surface pattern with a submicron spatial resolution [62].

6.3 Cytotoxicity Assessment

To determine the biocompatibility and safety of polymers, it is important to conduct both in vitro and in vivo cytotoxicity assessments. The in vitro cytotoxicity tests can be done by using different assays to evaluate the viability of cell lines upon exposure to polymers, examples of such assays are lactate dehydrogenase leakage assay, neutral red uptake assay, and trypan blue assay among others. Although the in vivo biocompatibility tests are conducted on animal models and it involves the evaluation of sensitivity/irritation reactivity, subchronic toxicity (subacute toxicity), systemic toxicity (acute toxicity), implantation, genotoxicity, hemocompatibility, carcinogenicity, reproductive and developmental toxicity, and immune response [78]. These assessment studies are highly required prior preclinical application of any polymeric material.

7 MECHANISMS OF ACTION

There are basically two mechanisms by which polymers exhibit their antimicrobial activities, these are passive and active actions. In passive actions, polymers prevent the adhesion of bacterial and not necessarily actively interact with the bacterial to cause death [20]. For a polymer to exhibit passive antimicrobial actions against bacteria cells, they must be water-loving, negatively charged and have low surface free energy. This will enable the repelling of microbes that are mainly hydrophobic in nature [21,79]. The most widely employed polymer with passive antimicrobial action is PEG, and this polymer has been studied extensively and has shown awesome antimicrobial by reducing protein and bacterial adhesion. This is because it possesses high chain mobility, steric hindrance effect of highly hydrated layer and large exclusion volume [79,80]. In addition, its high fouling ability hinders the growth of microorganisms.

Active action, on the other hand, involves the active/direct killing of the bacteria cell by the antimicrobial polymers. In most applications involving polymers that exhibit active antimicrobial mechanisms, the polymers are functionalized with positively charged quaternary ammonium. This allows the polymer to interact with the cell thereby destroying the cytoplasmic membrane and causes a leakage of the intracellular components that ultimately result in cell death [81]. Examples of such polymers with active antimicrobial activity include polyethylenimine, polyguanidine, and N-halamine. For instance, the electrostatic interaction between the bacterial cell membrane and polyethyleneimine leads to the rupture of the cell membrane [19].

8 APPLICATIONS

Antimicrobial polymers have found applications in diverse fields however, textiles, food and the medical industries are the three key areas it has found the greatest relevance [20]. For the purpose of this chapter, the medical application of antimicrobial polymers against pathogenic microorganisms (bacteria) are discussed therein. At the moment, the various types of antimicrobial polymers that have been developed thus far have been investigated to evaluate their antimicrobial efficacy. Majority have shown potential antimicrobial activities that can be translated into in vitro and in vivo applications against disease-causing microorganisms [28]. Some researchers have investigated the antimicrobial efficacy of polymers with inherent antibacterial properties and polymers that have antibacterial agents incorporated within the polymeric network. It is noteworthy that different factors are responsible for the antibacterial activities of polymers, these observations can be seen in various studies carried out by several research groups.

In very early studies, the effect of the molecular weight of the polymers was discovered to play a key role in their antimicrobial efficacy, this has led various research groups to conduct weight dependent investigations on antibiotic polymers. The weight-dependent antimicrobial properties of homopolymers of polyacrylates and polymethyl acrylates having biguanide groups side-chain and their copolymers with acrylamide were studied by Ikeda and cocolleagues [82,83]. It was reported that the biocidal efficacy of polymethyl acrylate with pendent biguanide groups was dependent on the molecular weight. A decrease in the molecular weight (below 5×10^4 Da) exhibited an increased antimicrobial action of the polymer when tested against *Staphylococcus aureus* while the antimicrobial activity significantly decreased with an increase in the molecular weight (above 1.2×10^5 Da) of the polymer. This may

be due to the ability of lower molecular weight polymers to penetrate the cell wall of the bacteria more easily. This same research group also evaluated the effect of molecular weight on the antimicrobial properties of (trialkylvinylbenzylammonium chloride) against *P. aeruginosa*, *S. aureus*, *E. coli*, *A. aerogenes*, and *B. subtilis* [84] and it was found that the bactericidal properties of the polymer increased with increase in the molecular weight (highest MW tested; 7.7×10^4 Da) of the polymer. However, the fractionated polymeric quaternary ammonium salts of the same polymer showed little molecular weight dependence antimicrobial activity against the same microorganisms.

In the same line, the weight dependence antimicrobial efficacy of poly(tributyl 4-vinyl benzyl phosphonium chloride) was tested against *Staphylococus aureus* by Kanazawa and coresearchers [85]. The results showed that there was an increase in the antimicrobial effect with an increase in the molecular weight (from 1.6×10^4 to 9.4×10^4 Da). This may be attributed to the individual contribution of the polymers in each bactericidal process: diffusion into the bacteria cell wall, adsorption onto the bacterial surface, disruption of the bacterial cell membrane and leakage of constituents of the cytoplasm among others. This result is similar to that obtained by Ikeda and coworkers.

It is interesting to discover that Chen and fellow workers who synthesized quaternary ammonium-functionalized poly(propylene imine) dendrimers [86] had a similar results with Tokura and coworkers [87] who investigated the effect of molecular weight on the antimicrobial properties of polymers. Both research groups discovered that the antimicrobial properties of the polymers had a parabolic dependence on the molecular weight. On the contrary, the antimicrobial efficacy of copolymers of vinylamine, methyl acrylate, and N-vinyl pyrrolidone with pendent quaternary ammonium groups synthesized by Panarin and coworkers was not dependent on the molecular weight [88]. This varying observation may be due to the bacteria gram class (gram negative; *Escherichia coli*) that was used for the study. Generally, gram-negative bacteria are known to be more resistant to antibiotics due to the nature of their cell wall. Gram-negative bacteria have an extra outer membrane in their structure that makes the penetration of foreign molecules difficult hence the observed discrepancy in these studies.

Additionally, the effect of counter-ion on the antibacterial activity of polymers have also been investigated by a couple of research groups, and both reports seem to confirm that the antibacterial properties are dependent on the presence of counter-ion. The study carried out by Chen and coresearchers showed that the bactericidal effect of polymers with bromide anions was greater than those with chloride anions [86]. Similarly, results obtained by Kanazawa and fellow workers showed that poly(tributyl(4-vinylbenzyl)phosphonium) salts had antibacterial efficacy that was affected by the structure of the counter anions. Surprisingly, again Panarin and colleague reported a contrary observation when homopolymers of vinylamine and methyl methacrylate with pendent quaternary ammonium groups were investigated for its antibacterial properties among chloride, bromide, and iodide [88]. The result showed that there was no effect of counterions on the antibacterial properties. Although, the only explanation to the reason counter-ions should affect the antibacterial properties of polymers is when they are capable of altering the solubility of their host polymers. Therefore in cases where they are unable to alter the solubility of their host polymers, little or no effect will be observed.

Another factor that has been associated with the antimicrobial efficacy of antibiotic polymers is spacer length and alkyl chain; this is because they cause a change in the conformation and charge density of the polymer; hence, the manner of polymer–bacteria interaction is affected. This has been investigated by Ikeda and coworkers [89], Sawada and coworkers [90], and Nonaka and coworkers, respectively [91]. Sawada and coworkers synthesized perfluoro-propylated and perfluoro-oxalkylated end-capped 2-(3-acrylamidopropyldimethylammonio) ethanoate (APDMAE) and investigated the effect of increasing chain length on both *S. aureus* and *P. aeruginosa*, respectively. It was noticed that prefluoro-oxalkylated APDMAE polymer with a longer chain was more effective against both microorganisms. In addition, methacryloylethyl trialkyl phosphonium chlorides/*N*-isopropyl acrylamide copolymers were prepared by Nonaka and coworkers and the antibacterial efficacy was evaluated against *Escherichia coli*. It was reported that increasing the alkyl chain length of the polymer led to an increase in the antimicrobial activity of the polymer against the tested microorganism. In summary, they all observed that there was a greater antimicrobial activity of the polymers against the microorganisms tested as the chain length increased. This may be due to the difference in the behavior of the aggregates for long and short hydrophobic materials and the availability of dual binding sites on the surface of the polymer with different binding affinities for long and short alkyl substituents [86].

Some more recent studies have also been carried on polymers with inherent antibacterial properties, one such studied by Sui et al. (2013) is a polymer made from thermoresponsive poly(*N*-isopropylacrylamide) (PNIPAM) and redox responsive poly(ferrocenylsilane) (PFS) [92]. These macromolecules were successfully synthesized via photopolymerization, and the side groups of the PFS chains and the acrylate side groups were copolymerized with NIPAM and N, N0-methylenebisacrylamide in tetrahydrofuran in the presence of a photoinitiator. These were done at a fixed ratio under ultraviolet light-emitting diode irradiation (wavelength of 365 nm). A homogeneous hydrogel hybrid (PNIPAM–PFS-based hydrogels) was smoothly formed, and it was further characterized for the equilibrium swelling ratio, rheology, and morphology properties. These hydrogels were investigated for its antibacterial efficacy by determining the zone of inhibition against *Escherichia coli*. The result showed that the hydrogel had a strong antibacterial effect on the tested organisms; at the same time, it was found to be nontoxic and thus biocompatible with human cells. This study successfully proved that polymers can serve as antimicrobial agents toward disease-causing microorganisms.

Another study that showed the potential of polymers with inherent antibacterial properties is the investigation done by Tomic' and his colleagues (2010). These groups of scientists synthesized a pH and temperature-sensitive hydrogels using hydroxyethylmethacrylate (HEMA) and itaconic acid (IA) copolymers through gamma irradiation [93]. Before carrying out the antimicrobial efficacy of the hydrogel, the in vitro biocompatibility, morphology, and mechanical and thermal properties were investigated. The biocompatibility results showed that the hydrogel is nontoxic nor hemolytic and other characterizations revealed that the hydrogel meets the requirements for hydrogels suitable for biomedical application. The synthesized hydrogel P(HEMA/IA) was used to form a dressing, and it was tested on a gram-positive and gram-negative bacteria (*Staphylococcus aureus* and *Escherichia coli* respectively). This was to determine its antimicrobial potential of the P(HEMA/IA)-based hydrogel dressing. The reported results showed that the microorganisms were unable to penetrate the hydrogel dressing. This suggested that the hydrogel has potential applications in wound dressing and skin treatment as it can be used as an excellent barrier against pathogens; hence, it can be used in the field of biomedicine.

TABLE 3.3
Summary of Other Various Antimicrobial Polymers Used for Antimicrobial Application.

Antimicrobial Polymer	Class (Type)	Microorganisms	Comments	References
Chitosan	Biocidal polymers	Bacteria, yeast, fungi	It can act independently as an antimicrobial agent or in combination with other compounds.	[96,97]
Dextrans (loaded with gentamicin)	Biocide-releasing polymers	*Staphylococcus aureus*	It prolonged the release of gentamicin as well as enhancing the drug's solubility.	[98]
Poly(*n*-vinylimidazole) modified silicone rubber	Biocidal polymers	*Pseudomonas aeruginosa, Staphylococcus aureus*	It exhibits greater antibacterial activity against gram-negative bacteria compared to gram positive (*Pseudomonas aeruginosa* > *Staphylococcus aureus*).	[99]
Cyclodextrin (loaded with silver, chitosan)	Biocide-releasing polymers	*Staphylococcus aureus, Escherichia coli*	It stabilized silver-chitosan and also improved the antimicrobial activities of silver-chitosan.	[100]
Quaternary phosphonium-modified epoxidized natural rubber	Biocidal polymers	*Staphylococcus aureus, Escherichia coli*	It showed a moderate growth inhibition in both microorganisms	[101]
Poly-L-lysine, polyethylene glycol (loaded with staphylolytic LysK enzyme)	Biocide-releasing polymers	*Staphylococcus aureus*	It enhanced the ability of LysK to lyse bacteria.	[102]
Guanylated polymethacrylate	Biocidal polymers	*Staphylococcus epidermidis, Candida albicans*	It was observed that copolymers with guanidine displayed more antimicrobial activity when compared to the amine analogues.	[103]
Poly(methyl methacrylate) complexed with silver, nanoparticles, and imidazole complex	Biocide-releasing polymers	*Staphylococcus epidermidis, Escherichia coli*	It showed a time-dependent antimicrobial activity against the test organisms.	[104]
Quaternary ammonium Polyethyleneimine	Biocidal polymers	Gram-positive and gram-negative bacteria	It showed effective antimicrobial activity; however, it was dependent on the hydrophobic and positively charged immobilized long polymeric chains.	[105]

Polycaprolactone loaded with silver	Biocide-releasing polymers	Staphylococcus aureus, Pseudomonas aeruginosa	It showed a great antimicrobial activity against planktonic bacteria and biofilms.	[106]
Ammonium ethyl Methacrylate homopolymers	Biocidal polymers	Methicillin-resistant Staphylococcus aureus, Escherichia coli	It exhibited greater antimicrobial effects against gram-positive bacteria compared to gram-negative bacteria with minimal hemolytic activity in both bacteria.	[107]
Cyclodextrin (loaded with triclosan)	Biocide-releasing polymers	Staphylococcus aureus, Escherichia coli	It significantly reduced the drug concentration required for minimum inhibitory bacterial cell growth and as well as its toxicity.	[108]
Metallo-terpyridine carboxymethyl cellulose	Biocidal polymers	Staphylococcus aureus, Streptococcus thermophilus, Escherichia coli, Saccharomyces cerevisiae	It achieved a greater than 90% microbial inhibition in all test organisms at concentration between 6 and 8 mg/L.	[109]

Furthermore, some groups of scientists have synthesized polymers with antimicrobial agents incorporated within the matrix of the polymers that have proven to be potentially capable of destroying pathogenic microorganism. An example of such a study was carried out by Kanazawa and coworkers as far back as 1993 [94]. In their study, monomers of p-vinylbenzyl tetramethylenesulfonium tetrafluoroborate and p-ethylbenzyl tetramethylenesulfonium tetrafluoroborate were synthesized, and sulfonium salt (antibacterial substance) was incorporated. This was tested against *Escherichia coli* (gram negative) and *Staphylococcus aureus* (gram positive), respectively. It was observed that the antibacterial effect was higher in gram-positive bacteria than gram-negative bacteria. However, over the years, more studies were built on the established antibacterial potential exhibited by the antimicrobial agent-releasing polymers. Another such study is the testing of polymers having quaternary ammonium and phosphonium salts within its matrix against *Escherichia coli* (gram negative), *Bacillus subtilis* PY79, and *Salmonella enterica*, respectively [95]. The polymer was formed from polyethylene (PE) and polystyrene (PS) via ozonolysis, characterized by using XPS and FTIR to confirm the surface modifications.

However, SEM and confocal microscopy were employed to monitor surface morphology and bacteria interactions. The results obtained showed that polymers with shorter quaternary ammonium and phosphonium salt polymers did not demonstrate bactericidal activity; however, antibacterial activities were noticed with increasing alkyl chain length.

Similarly, Babu and colleagues synthesized polymers capable of releasing benzimidazole (antibacterial substance). The polymer was synthesized using *N*-alkyl-2-(3-methyl-indolyl) benzimidazoles via the condensation of *N*-alkyl-2-propylbenzimidazole-phenylhydrazone [25]. Its antibacterial effect was tested toward a broad range of bacteria: *Micrococcus luteus, Staphylococcus aureus, Bacillus subtilis, Escherichia coli, Pseudomonas aeruginosa, Klebsiella planticola, and Candida albicans*. The results exhibited a good antibacterial activity on all the tested microorganisms and thus further potentiates the ability of polymers with incorporated antibacterial substance in antimicrobial applications. An excellent observation in this particular study is that a wide range of pathogens were employed and thus providing more information on the efficacy of polymers with incorporated antibacterial substances.

In addition to studies as these, Li and coresearchers synthesized a polymer from monomers of 3-(3′-acrylica-cidpropylester)-5,5-dimethylhydantoin and bonded it to a cotton fabric via a process known as electron beam irradiation [27]. To confirm the successful coating of the N-halamine precursor on the cotton, SEM, FTIR, and Energy-dispersive X-ray spectroscopy (EDX) characterization techniques were employed. Data obtained revealed the successful incorporation of the N-halamine precursor and antibacterial studies were subsequently performed on gram-positive and gram-negative bacteria (*Staphylococcus aureus* and *Escherichia coli*). A 100% bacteria inactivation was observed within contact times of 10 and 5 min, respectively, and thus showed the excellent biocidal efficacy of the polymer. These studies revealed that polymerization of monomers to form polymers actually enhances their antibacterial effect against microorganisms when compared to their monomers. In addition, a greater antibacterial activity is associated with the antimicrobial substance within the polymer matrix. Table 3.3 further summarizes other antimicrobial polymers that have been studied by various groups of researchers.

9 CONCLUSION AND FUTURE TREND

Antimicrobial polymers are more than ever before receiving great attention from researchers, and this is chiefly due to the great antimicrobial potential they possess as well as the increasing rise in the challenges associated with the treatment and management of pathogenic microorganisms. Various research groups have extensively carried out investigations to explore the antimicrobial efficacy of several polymers and results have shown the potential application of these polymers in the eradication of diverse strains of disease-causing microorganisms. Two fundamental mechanisms of action have been clearly reported; however, more precise and in-depth mechanisms of microbial killing are required to further give clarity especially for biofilm-associated mechanisms. The combination of these mechanisms will enhance the overall antimicrobial efficacy of antimicrobial polymers. In addition, the most appropriate method of synthesis that will lead to harnessing the maximum potential of these antimicrobial polymers should be employed during fabrication. Although, at the moment, the use of antimicrobial polymers appears to be very promising in the treatment of infectious pathogens/diseases. However, the fabrication of long-acting and reusable antimicrobial polymers with a broad range of antimicrobial activity should be carefully considered to overcome some of the current challenges encountered with the use of some of the existing antimicrobial polymers.

REFERENCES

[1] J. Mano, et al., Natural origin biodegradable systems in tissue engineering and regenerative medicine: present status and some moving trends, Journal of the Royal Society Interface 4 (17) (2007) 999–1030.

[2] A. Aravamudhan, et al., Natural polymers: polysaccharides and their derivatives for biomedical applications, in: Natural and Synthetic Biomedical Polymers, Elsevier, 2014, pp. 67–89.

[3] B.L. Banik, J.L. Brown, Polymeric biomaterials in nanomedicine, in: Natural and Synthetic Biomedical Polymers, Elsevier, 2014, pp. 387–395.

[4] V.P. Shastri, Non-degradable biocompatible polymers in medicine: past, present and future, Current Pharmaceutical Biotechnology 4 (5) (2003) 331–337.

[5] A. Subramaniam, S. Sethuraman, Biomedical applications of nondegradable polymers, in: Natural and Synthetic Biomedical Polymers, Elsevier, 2014, pp. 301–308.

[6] R.J. Hernandez, S.E. Selke, J.D. Culter, Plastics Packaging: Properties, Processing, Applications, and Regulations, 2000 [Doctoral dissertation, Univerza v Mariboru, Ekonomsko-poslovna fakulteta].

[7] M. Ramchandani, D. Robinson, In vitro and in vivo release of ciprofloxacin from PLGA 50: 50 implants, Journal of Controlled Release 54 (2) (1998) 167–175.

[8] D. Ramakrishna, P. Rao, Nanoparticles: is toxicity a concern? EJIFCC 22 (4) (2011) 92.

[9] C. De las Heras Alarcón, S. Pennadam, C. Alexander, Stimuli responsive polymers for biomedical applications, Chemical Society Reviews 34 (3) (2005) 276–285.

[10] J. Jagur-Grodzinski, Biomedical application of functional polymers, Reactive and Functional Polymers 39 (2) (1999) 99–138.

[11] J. Jagur-Grodzinski, Polymers for tissue engineering, medical devices, and regenerative medicine. Concise general review of recent studies, Polymers for Advanced Technologies 17 (6) (2006) 395–418.

[12] A.G. Ross, et al., Enteropathogens and chronic illness in returning travelers, New England Journal of Medicine 368 (19) (2013) 1817–1825.

[13] Y.-S. Lin, C.-H. Yang, K.-S. Huang, Biomedical devices for pathogen detection using microfluidic chips, Current Proteomics 11 (2) (2014) 116–120.

[14] C.-F. Chan, et al., Applications of nanoparticles for antimicrobial activity and drug delivery, Current Organic Chemistry 18 (2) (2014) 204–215.

[15] D. Sun, et al., Antimicrobial materials with medical applications, Materials Technology 30 (Suppl. 6) (2015) B90–B95.

[16] R. Velez, E. Sloand, Combating antibiotic resistance, mitigating future threats and ongoing initiatives, Journal of Clinical Nursing 25 (2016) 1886–1889.

[17] M. Frieri, K. Kumar, A. Boutin, Antibiotic resistance, Journal of Infection and Public Health 10 (4) (2017) 369–378.

[18] K. Yang, et al., Antimicrobial hydrogels: promising materials for medical application, International Journal of Nanomedicine 13 (2018) 2217.

[19] A. Jain, et al., Antimicrobial polymers, Advanced Healthcare Materials 3 (12) (2014) 1969−1985.

[20] K.-S. Huang, et al., Recent advances in antimicrobial polymers: a mini-review, International Journal of Molecular Sciences 17 (9) (2016) 1578.

[21] I. Francolini, et al., Antimicrobial polymers for anti-biofilm medical devices: state-of-art and perspectives, in: Biofilm-Based Healthcare-Associated Infections, Springer, 2015, pp. 93−117.

[22] V. Sedlarik, Antimicrobial modifications of polymers, Biodegradation-Life of Science (2013) 187−204.

[23] A. Martins, et al., Antimicrobial activity of chitosan derivatives containing N-quaternized moieties in its backbone: a review, International Journal of Molecular Sciences 15 (11) (2014) 20800−20832.

[24] F. Siedenbiedel, J.C. Tiller, Antimicrobial polymers in solution and on surfaces: overview and functional principles, Polymers 4 (1) (2012) 46−71.

[25] P.K. Babu, et al., Synthesis, antimicrobial, and anticancer evaluation of novel 2-(3-methylindolyl) benzimidazole derivatives, Medicinal Chemistry Research 23 (9) (2014) 3970−3978.

[26] X. Cheng, et al., Antimicrobial activity of hydrophobic cotton coated with N-halamine, Polymers for Advanced Technologies 26 (1) (2015) 99−103.

[27] X. Li, et al., Synthesis of an N-halamine monomer and its application in antimicrobial cellulose via an electron beam irradiation process, Cellulose 22 (6) (2015) 3609−3617.

[28] Y. Yang, et al., Antimicrobial cationic polymers: from structural design to functional control, Polymer Journal 50 (1) (2018) 33.

[29] D.S. Rekha, A. Kumar Sharma, P. Kumar, Cationic polymers and their self-assembly for antibacterial applications, Current Topics in Medicinal Chemistry 15 (13) (2015) 1179−1195.

[30] I. Hornyák, et al., Increased release time of antibiotics from bone allografts through a novel biodegradable coating, BioMed Research International 2014 (2014) 1−8.

[31] I. Macha, et al., Marine structure derived calcium phosphate−polymer biocomposites for local antibiotic delivery, Marine Drugs 13 (1) (2015) 666−680.

[32] M.Y. Kariduraganavar, et al., Using an additive to control the electrospinning of fibres of poly (ε-caprolactone), Polymer International 59 (6) (2010) 827−835.

[33] M.Y. Kariduraganavar, A.A. Kittur, R.R. Kamble, Polymer synthesis and processing, in: Natural and Synthetic Biomedical Polymers, Elsevier, 2014, pp. 1−31.

[34] D.K. Singh, A.R. Ray, Biomedical applications of chitin, chitosan, and their derivatives, Journal of Macromolecular Science: Part C: Polymer Reviews 40 (1) (2000) 69−83.

[35] B.G. Kozen, et al., An alternative hemostatic dressing: comparison of CELOX, HemCon, and QuikClot, Academic Emergency Medicine 15 (1) (2008) 74−81.

[36] T.A. Khan, K.K. Peh, H.S. Ch'ng, Mechanical, bioadhesive strength and biological evaluations of chitosan films for wound dressing, Journal of Pharmacy & Pharmaceutical Sciences 3 (3) (2000) 303−311.

[37] A.C. French, A.L. Thompson, B.G. Davis, High-purity discrete PEG-oligomer crystals allow structural insight, Angewandte Chemie International Edition 48 (7) (2009) 1248−1252.

[38] J.a. Kahovec, R. Fox, K. Hatada, Nomenclature of regular single-strand organic polymers (IUPAC Recommendations 2002), Pure and Applied Chemistry 74 (10) (2002) 1921−1956.

[39] C.G. Boeriu, et al., Production methods for hyaluronan, International Journal of Carbohydrate Chemistry 2013 (2013).

[40] R. Stern, Hyaluronan catabolism: a new metabolic pathway, European Journal of Cell Biology 83 (7) (2004) 317−325.

[41] M. Kutz, Handbook of Materials Selection, John Wiley & Sons, 2002, pp. 1−1499.

[42] S. Kang, et al., Optically transparent polymethyl methacrylate composites made with glass fibers of varying refractive index, Journal of Materials Research 12 (4) (1997) 1091−1101.

[43] A.N. De Belder, Dextran, in: Industrial Gums, Elsevier, 1993, pp. 399−425.

[44] M.J. Santander-Ortega, et al., Stability and physicochemical characteristics of PLGA, PLGA: poloxamer and PLGA: poloxamine blend nanoparticles: a comparative study, Colloids and Surfaces A: Physicochemical and Engineering Aspects 296 (1−3) (2007) 132−140.

[45] K.Y. Lee, D.J. Mooney, Alginate: properties and biomedical applications, Progress in Polymer Science 37 (1) (2012) 106−126.

[46] Y. Kim, G. Kim, Collagen/alginate scaffolds comprising core (PCL)−shell (collagen/alginate) struts for hard tissue regeneration: fabrication, characterisation, and cellular activities, Journal of Materials Chemistry B 1 (25) (2013) 3185−3194.

[47] V. Del, E. Martin, Cyclodextrins and their uses: a review, Process Biochemistry 39 (9) (2004) 1033−1046.

[48] A. Popielec, T. Loftsson, Effects of cyclodextrins on the chemical stability of drugs, International Journal of Pharmaceutics 531 (2) (2017) 532−542.

[49] K. Vulic, M.S. Shoichet, Affinity-based drug delivery systems for tissue repair and regeneration, Biomacromolecules 15 (11) (2014) 3867−3880.

[50] K.J. Ivin, J.C. Mol, Olefin Metathesis and Metathesis Polymerization, Elsevier, 1997, pp. 1−461.

[51] E.J. Vandenberg, J.C. Salamone, Catalysis in polymer synthesis, in: ACS Symposium Series 496, American Chemical Society, Washington, DC (United States), 1992.

[52] A.J. Peacock, A. Calhoun, Polymer Chemistry: Properties and Application, Carl Hanser Verlag GmbH Co KG, 2012, pp. 1−397.

[53] P. Bahadur, N. Sastry, Principles of Polymer Science, first ed, Alpha Science Int'l Ltd, 2002, p. 401.

[54] C.B. Friedersdorf, Processes and Apparatus for Continuous Solution Polymerization, Google Patents, 2007.

[55] T.L. Tartamella, W.M. Cole, M. Smale, Bulk Polymerization Process, Google Patents, 2008.

[56] B. Brooks, Suspension polymerization processes, Chemical Engineering & Technology 33 (11) (2010) 1737–1744.

[57] C. Chern, Emulsion polymerization mechanisms and kinetics, Progress in Polymer Science 31 (5) (2006) 443–486.

[58] I. Galeska, et al., Controlled release of dexamethasone from PLGA microspheres embedded within polyacid-containing PVA hydrogels, The AAPS Journal 7 (1) (2005) E231–E240.

[59] S. Verma, R. Gokhale, D.J. Burgess, A comparative study of top-down and bottom-up approaches for the preparation of micro/nanosuspensions, International Journal of Pharmaceutics 380 (1–2) (2009) 216–222.

[60] J.M. Morais, F. Papadimitrakopoulos, D.J. Burgess, Biomaterials/tissue interactions: possible solutions to overcome foreign body response, The AAPS Journal 12 (2) (2010) 188–196.

[61] A. Bohr, et al., Preparation of microspheres containing low solubility drug compound by electrohydrodynamic spraying, International Journal of Pharmaceutics 412 (1–2) (2011) 59–67.

[62] B. Gu, D.J. Burgess, Polymeric materials in drug delivery, in: Natural and Synthetic Biomedical Polymers, Elsevier, 2014, pp. 333–349.

[63] Y. Ikada, Challenges in tissue engineering, Journal of the Royal Society Interface 3 (10) (2006) 589–601.

[64] S.C. Porter, L.A. Felton, Techniques to assess film coatings and evaluate film-coated products, Drug Development and Industrial Pharmacy 36 (2) (2010) 128–142.

[65] A. Rawat, D.J. Burgess, Effect of ethanol as a processing co-solvent on the PLGA microsphere characteristics, International Journal of Pharmaceutics 394 (1–2) (2010) 99–105.

[66] D.S. Jones, et al., Pharmaceutical applications of dynamic mechanical thermal analysis, Advanced Drug Delivery Reviews 64 (5) (2012) 440–448.

[67] R.P. Chartoff, A.K. Sircar, Thermal analysis of polymers, Encyclopedia of Polymer Science and Technology (2002) 1–86.

[68] J.M. Morais, P.A. Rocha-Filho, D.J. Burgess, Relationship between rheological properties and one-step W/O/W multiple emulsion formation, Langmuir 26 (23) (2010) 17874–17881.

[69] T.K. Meyvis, et al., A comparison between the use of dynamic mechanical analysis and oscillatory shear rheometry for the characterisation of hydrogels, International Journal of Pharmaceutics 244 (1–2) (2002) 163–168.

[70] J.L. Ford, T.E. Mann, Fast-Scan DSC and its role in pharmaceutical physical form characterisation and selection, Advanced Drug Delivery Reviews 64 (5) (2012) 422–430.

[71] M.H. Chiu, E.J. Prenner, Differential scanning calorimetry: an invaluable tool for a detailed thermodynamic characterization of macromolecules and their interactions, Journal of Pharmacy and Bioallied Sciences 3 (1) (2011) 39.

[72] M.D. Kofron, et al., The implications of polymer selection in regenerative medicine: a comparison of amorphous and semi-crystalline polymer for tissue regeneration, Advanced Functional Materials 19 (9) (2009) 1351–1359.

[73] B. Van Eerdenbrugh, L.S. Taylor, An ab initio polymer selection methodology to prevent crystallization in amorphous solid dispersions by application of crystal engineering principles, CrystEngComm 13 (20) (2011) 6171–6178.

[74] D. Baird, A. Shew, Probing the history of scanning tunneling microscopy, Discovering the Nanoscale 2 (2004) 145–156.

[75] C. Steffens, et al., Atomic force microscopy as a tool applied to nano/biosensors, Sensors 12 (6) (2012) 8278–8300.

[76] C.H. Lindsley, Preparation of density-gradient columns, Journal of Polymer Science 46 (148) (1960) 543–545.

[77] B.D. Ratner, Advances in the analysis of surfaces of biomedical interest, Surface and Interface Analysis 23 (7-8) (1995) 521–528.

[78] V.N. Sumantran, Cellular chemosensitivity assays: an overview, in: Cancer Cell Culture, Springer, 2011, pp. 219–236.

[79] H. Zhang, M. Chiao, Anti-fouling coatings of poly (dimethylsiloxane) devices for biological and biomedical applications, Journal of Medical and Biological Engineering 35 (2) (2015) 143–155.

[80] K. Yu, et al., Engineering biomaterials surfaces to modulate the host response, Colloids and Surfaces B: Biointerfaces 124 (2014) 69–79.

[81] Y. Xue, H. Xiao, Y. Zhang, Antimicrobial polymeric materials with quaternary ammonium and phosphonium salts, International Journal of Molecular Sciences 16 (2) (2015) 3626–3655.

[82] T. Ikeda, H. Yamaguchi, S. Tazuke, New polymeric biocides: synthesis and antibacterial activities of polycations with pendant biguanide groups, Antimicrobial Agents and Chemotherapy 26 (2) (1984) 139–144.

[83] T. Ikeda, et al., Polycationic biocides with pendant active groups: molecular weight dependence of antibacterial activity, Antimicrobial Agents and Chemotherapy 30 (1) (1986) 132–136.

[84] T. Ikeda, S. Tazuke, Biocidal polycations, in: Abstracts of Papers of the American Chemical Society, 1985. American Chemical Society 1155 16th ST, NW, WASHINGTON, DC 20036.

[85] A. Kanazawa, T. Ikeda, T. Endo, A novel approach to mode of action of cationic biocides morphological effect on antibacterial activity, Journal of Applied Bacteriology 78 (1) (1995) 55–60.

[86] C.Z. Chen, et al., Quaternary ammonium functionalized poly (propylene imine) dendrimers as effective antimicrobials: structure–activity studies, Biomacromolecules 1 (3) (2000) 473–480.

[87] S. Tokura, et al., Molecular weight dependent antimicrobial activity by chitosan, in: New Macromolecular Architecture and Functions, Springer, 1996, pp. 199–207.

[88] E. Panarin, et al., Biological activity of cationic polyelectrolytes, Die Makromolekulare Chemie: Macromolecular Chemistry and Physics 9 (S19851) (1985) 25–33.

[89] T. Ikeda, S. Tazuke, Y. Suzuki, Biologically active polycations, 4. Synthesis and antimicrobial activity of poly (trialkylvinylbenzylammonium chloride) s, Die Makromolekulare Chemie: Macromolecular Chemistry and Physics 185 (5) (1984) 869–876.

[90] H. Sawada, et al., Synthesis and properties of fluoroalkylated end-capped betaine polymers, European Polymer Journal 35 (9) (1999) 1611–1617.

[91] T. Nonaka, et al., Synthesis of water-soluble thermosensitive polymers having phosphonium groups from methacryloyloxyethyl trialkyl phosphonium chlorides–N-isopropylacrylamide copolymers and their functions, Journal of Applied Polymer Science 87 (3) (2003) 386–393.

[92] X. Sui, et al., Poly (N-isopropylacrylamide)–poly (ferrocenylsilane) dual-responsive hydrogels: synthesis, characterization and antimicrobial applications, Polymer Chemistry 4 (2) (2013) 337–342.

[93] S.L. Tomić, et al., Smart poly (2-hydroxyethyl methacrylate/itaconic acid) hydrogels for biomedical application, Radiation Physics and Chemistry 79 (5) (2010) 643–649.

[94] A. Kanazawa, T. Ikeda, T. Endo, Antibacterial activity of polymeric sulfonium salts, Journal of Polymer Science Part A: Polymer Chemistry 31 (11) (1993) 2873–2876.

[95] T. Fadida, et al., Air-ozonolysis to generate contact active antimicrobial surfaces: activation of polyethylene and polystyrene followed by covalent graft of quaternary ammonium salts, Colloids and Surfaces B: Biointerfaces 122 (2014) 294–300.

[96] G.D. Mogoşanu, A.M. Grumezescu, Natural and synthetic polymers for wounds and burns dressing, International Journal of Pharmaceutics 463 (2) (2014) 127–136.

[97] K. Xing, et al., Chitosan antimicrobial and eliciting properties for pest control in agriculture: a review, Agronomy for Sustainable Development 35 (2) (2015) 569–588.

[98] R.P. Aquino, et al., Design and production of gentamicin/dextrans microparticles by supercritical assisted atomisation for the treatment of wound bacterial infections, International Journal of Pharmaceutics 440 (2) (2013) 188–194.

[99] H.I. Meléndez-Ortiz, et al., Radiation-grafting of N-vinylimidazole onto silicone rubber for antimicrobial properties, Radiation Physics and Chemistry 110 (2015) 59–66.

[100] N. Punitha, P. Ramesh, D. Geetha, Spectral, morphological and antibacterial studies of β-cyclodextrin stabilized silver–Chitosan nanocomposites, Spectrochimica Acta Part A: Molecular and Biomolecular Spectroscopy 136 (2015) 1710–1717.

[101] C. Li, et al., Preparation and antimicrobial activity of quaternary phosphonium modified epoxidized natural rubber, Materials Letters 93 (2013) 145–148.

[102] L.Y. Filatova, et al., Physicochemical characterization of the staphylolytic LysK enzyme in complexes with polycationic polymers as a potent antimicrobial, Biochimie 95 (9) (2013) 1689–1696.

[103] K.E. Locock, et al., Guanylated polymethacrylates: a class of potent antimicrobial polymers with low hemolytic activity, Biomacromolecules 14 (11) (2013) 4021–4031.

[104] O. Lyutakov, et al., Silver release and antimicrobial properties of PMMA films doped with silver ions, nano-particles and complexes, Materials Science and Engineering: C 49 (2015) 534–540.

[105] N. Beyth, et al., Antibacterial dental resin composites, Reactive and Functional Polymers 75 (2014) 81–88.

[106] P.A. Tran, D.M. Hocking, A.J. O'Connor, In situ formation of antimicrobial silver nanoparticles and the impregnation of hydrophobic polycaprolactone matrix for antimicrobial medical device applications, Materials Science and Engineering: C 47 (2015) 63–69.

[107] L.M. Thoma, B.R. Boles, K. Kuroda, Cationic methacrylate polymers as topical antimicrobial agents against *Staphylococcus aureus* nasal colonization, Biomacromolecules 15 (8) (2014) 2933–2943.

[108] A.I. Ramos, et al., Analysis of the microcrystalline inclusion compounds of triclosan with β-cyclodextrin and its tris-O-methylated derivative, Journal of Pharmaceutical and Biomedical Analysis 80 (2013) 34–43.

[109] E.A. Hassan, et al., New supramolecular metalloterpyridine carboxymethyl cellulose derivatives with antimicrobial properties, Carbohydrate Polymers 116 (2015) 2–8.

CHAPTER 4

Natural Polymeric Materials as a Vehicle for Antibiotics

WALTHER IDE • YOUSOF FARRAG

1 INTRODUCTION

Antibiotics (ABX) are a type of medication to treat bacterial infections in both humans and animals. The use of this type of medicament today has become a global concern due to the abuse committed by both the population and doctors; therefore, since 2015 the World Health Organization has been developing an action plan to improve the use of this type of medication. One of the five guidelines for mitigating the increase in antibiotic resistance is to optimize the use of antimicrobial agents [1] because pathogens adapt rapidly and evolve through various strategies to escape the action of ABX [2]. It is in this scenario where it has been sought to develop new drugs with a different action from those already known, so far without so much success [3]. Another alternative is to develop new materials capable of being more selective and efficient when delivering pharmacological therapy with current ABX. For more than a decade this has been considered a global priority [4].

Biomaterials are classified as a wide family of materials from a diverse origin but born with the same purpose: to solve health problems and also to be friendly with the environment. This has generated a sustainable and safe alternative for transporting substances, especially for ABX. Broadly, there are two types of biomaterials, those of natural and synthetic origin. Although this classification generates controversies [5−7], it is the most accepted. This chapter will be focused on the first type.

Biomaterials of natural origin have been widely studied for medical applications because they have been shown to present versatility in their source of origin, low immunogenicity, and toxicity [8]. For medical purposes, they can be classified as metallic, ceramic, and polymeric [9]. Some examples of metal biomaterials are the ionic metal particles of gold, silver, iron, and copper. In the case of ceramics, hydroxyapatite and zinc or aluminum oxides can be mentioned.

Finally, some examples of polymers are alginate, hyaluronic acid, polylactic acid, chitosan, and starch; the latter has a variety of forms in terms of obtaining them, as well as in terms of their modifications and functionality.

Polymeric biomaterials are produced from polymers obtained from living organisms, which can be classified according to their origin as animal, plant, bacteria, fungi, or synthesized from basic biological systems [10] This can be seen summarized in Table 4.1, where natural polymers with broad biomedical applications are mentioned, for example, starch and chitosan, which have also been shown to have high compatibility with ABX [11−13].

2 POLYMERIC BIOMATERIALS

The study of natural polymeric materials was born by its direct application in medicine. These materials have found their development and boom mainly in tissue engineering and bone regeneration, where hydrogels, composites, and nanoparticles are the main actors. This is due to the biomimetic ability to resemble the physical and mechanical characteristics of biological tissues, allowing the formation of a suitable substrate for the gradual recovery of cellular components [17]. Hydrogels are a good example of how research and technology have improved a natural polymer, mainly through cross-linking, forming materials with varied uses, such as tissue engineering, administration of active compounds (ABX), and biosensors [18]. Another advantage of natural polymeric biomaterials is that they have great similarity with the biochemical characteristics of the molecules in the human body, are abundant in nature and relatively easy to isolate, and have the ability to undergo chemical modifications, which can satisfy the technological needs to be produced on a larger scale [19].

Antibiotic Materials in Healthcare. https://doi.org/10.1016/B978-0-12-820054-4.00004-5

TABLE 4.1

Classification of Biomaterials According to Their Origin [14–16].

Natural Polymers	Synthetic Polymers
Proteins and polypeptides • Albumin • Casein • Zein • Gelatin • Collagen • Fibrinogen • Silk • Elastin • Resilin • Polylysine	Aliphatic polyesters • Polylactic acid (PLA) • Poly lactic-co-glycoside (PLGA) • Poly(β,γ,δ-hydroxyalkanoates) (PHAs) • Poly ε-caprolactone (PCL)
Polysaccharides • Starch • Alginic acid • Cellulose • Pectin • Chitin and chitosan • Carrageenan • Hyaluronic acid • Konjac • Gums • Pullulan • Elsinan • Scleroglucan • Levan • Xanthan • Gelan • Dextran • Lignin	Poly(ether-ester) • Polydioxanone (PDO or PDS) Polyamides • Polyamide 1010 (PA 1010) Poly(ester amide)s • Poly(butylene adipate-co-caproamide) Polycarbonates • Poly(ethylene carbonate) (PEC) Poly alkylcyanoacrylate (PACA) Vinyl polymers • Poly(vinyl alcohol) (PVA)
Cells and viruses	

Hydrogels are two- or three-dimensional network structures [20], characterized by having adjustable properties according to their possible use. It can be said that their main characteristic is to absorb important amounts of water, or also pore size, molecular weight, and stiffness [21]. They can be synthesized starting with various biopolymers, such as chitosan, alginate, and starch. Once the polymer matrix is selected, the type of cross-linking can be chosen: by radical polymerization, by complementary groups chemical reaction, by ionic interactions, or by crystallization [18,21]. There are two types of cross-linking, each will allow forming characteristic interactions, either chemical cross-linking by covalent bonds, or physical cross-linking by weaker bonds as hydrogen bonds, van der Waals interactions,

or physical entanglements [20]. Depending on the degree of cross-linking, the delivery of a certain type of ABX will be favored, according to its size and its interaction with the polymer matrix. As described, antibacterial hydrogels can be classified according to their matrix and are divided into three types: (1) inorganic hydrogels containing nanoparticles, (2) antibacterial hydrogels, and (3) hydrogels with an inherent antibacterial capacity [22]. According to this classification in the first type, there are hydrogels that include mainly metal ions (Ag^+, Au^+, Cu^+) or metal oxide nanoparticles (ZnO, TiO_2) [20]. A recent example demonstrates the versatility of these biomaterials where a self-adhesive hydrogel has an antibacterial effect thanks to silver-lignin nanoparticles [23]. The second type is antibacterial hydrogels that take advantage of the local administration matrix, a high surface/volume ratio and control of the structural design. Apart from the previous benefits, some proven substances could be added, such as ABX, managing to decrease the dose by exerting a more localized effect [22]. A proven case is the binding of a polyvinyl alcohol (PVA) hydrogel with ciprofloxacin [24]. Another example is the hydrogel from purified mucin, a biopolymer of bacterial origin, which was loaded with fluoroquinolone against the pathogen *Pseudomonas aeruginosa* [25]. The third type is hydrogels with inherent antibacterial capacity. Within this category, there are those that do not need a stimulus to act and potential antibacterial polymers, which need an external stimulus to exert their action against pathogens [22]. Polymers such as PNIPAM with poly(ferrocenyl silane) molecules were found to show strong antibacterial activity with high biocompatibility [26] Another example is the formation of a hydrogel of poly-(3,4-ethylenedioxythiophene): poly(styrenesulfonate) (PEDOT: PSS) and agarose that when activated by photothermal conversion, using infrared light of intensity 2 Wcm^2 for 2 min, destroyed almost all pathogen bacteria of *Escherichia coli* and *Staphylococcus aureus* [27].

Polymeric composites can be formed by the interaction between two or more types of materials, which can be polymers of natural or synthetic origin; these, in turn, are in different phases and due to their union, they have significantly altered their modulus of elasticity compared to the starting materials [28,29]. The properties of composite depend on the form in the heterogenicity, the fraction of volume occupied by them, and the interface of their constituents [29]. Then, composite will improve their properties synergistically, both in terms of their design requirements for a specific function and in their biocompatibility and in the loading and delivery of active substances, such as some ABX.

Particularly, chitosan scaffolds, which have macropores to carry gentamicin and release this drug in a controlled way in bone recovery treatments, have been designed with β-tricalcium phosphate [30]. Similarly, there are multiple composites of high potential for the release of drugs in the regeneration of bone tissue, but one that generates high expectations is hydroxyapatite. Thus a recent study where hydroxyapatite reinforced with aluminum oxide was used to load ampicillin showed that there are different release profiles in different formulations of the same material. This could be used in medical devices that require different release times for treatment in bone regeneration [31]. Another example is chitosan nanocomposite polymeric films. These contain silver nanoparticles, which have been shown to be effective against microorganisms such as *E. coli, Pseudomonas, Staphylococcus, Micrococcus, Candida albicans*, and *P. aeruginosa*. Therefore they could be useful in the prevention and/or treatment of infectious diseases [32].

Polymeric nanoparticles (PNPs) are defined as solids or colloids of 10–1000 nm in diameter [33]. They can be synthesized from materials of natural or synthetic origin and by various techniques, such as nanoprecipitation, emulsification-solvent evaporation, dialysis, salting out, cross-linked polymerization, ionic gelation, and supercritical fluids [33–35]. The choice in the design of the nanoparticles will depend on the therapeutic application, the target site (tissues, organs, cells), and routes of administration [36]. The term "polymeric particles" is usually associated with nanospheres and nanocapsules [33]. In addition, the chemistry of PNPs and their loads affects their stability, biodegradability, biocompatibility, biodistribution, and cellular and subcellular destination [36]. The technique for the elaboration of the particles must be chosen by consensus, especially when these particles need to be loaded, for example, with an ABX to achieve better results in biomedical applications. In Table 4.2, the techniques and some recent applications in the area of PNPs are summarized.

It is due to these considerations that PNPs must be studied further, to gain a greater understanding of the mechanisms of synthesis, size distribution, and morphology of nanoparticles. The potential offered by natural sources PNPs, from the point of view of their biocompatibility, is more than optimistic for biomedical applications. Therefore researchers declare that naturally occurring nanoparticles can be more efficient and effective in the administration of drugs via oral and intravenous [8,37,38].

3 POLYMERIC BIOMATERIALS IN DRUG DELIVERY

A controlled drug-delivery system allows to provide proteins, peptides, nucleic acids, or active molecules, such as medicines (ABX among others) in low concentrations over a specific period of time, which can vary between hours, days, weeks, or months. The main mechanisms by which biomaterials act are controlled release systems, such as controlled diffusion matrixes and reservoirs, chemically controlled biodegradable and bioerodible materials, hydrogels, and solvent-activated osmotic pumps [58,59]. In addition, liposomes, dendrimers, transfersomes, niosomes, microspheres, micelles, and polymeric nanospheres were all developed as controlled release systems [58,60].

As it was already mentioned, polymeric biomaterials have great versatility in medical applications, mainly due to the low immunogenicity and toxicity for in vivo models, that is, these materials have high biocompatibility [61]. However, when designing a material that can be a suitable vehicle for a specific ABX, which in turn is the most effective for a given infection, the characteristics of the microbial agent, the anatomical location, and the dynamics of the therapeutic release of the drug must also be considered [21,62]. Binding between polymeric materials and antibiotic compounds can occur in different ways, which may lead to an increase or decrease in the release rate of active molecules, that is, by modifying the binding bonds, by changing the formulation and the chemical composition of the polymer [63]. Thus the design of an optimal pharmacological vehicle for the release of ABX should always take into account the following considerations for an adequate treatment: (1) increasing the local concentration of the drug at the site of infection; (2) minimizing the accumulation of ABX in other organs; (3) improving the efficacy of antimicrobial treatment; (4) reducing the risks of toxicity and exposure of commensal microflora to sublethal doses of ABX, which may promote the development of antimicrobial resistance [64].

Hydrogels, considered as "intelligent" drug-delivery systems [20,65], have a high capacity to absorb water, and this is due to the presence of numerous hydrophilic groups such as $-OH$, $-COOH$, $-NH_2$, $-CONH$, and $-CONH_2$. These high absorption materials have allowed the development of intelligent drug-delivery systems, such as the applications of peptide delivery and genes in biological models [21,66]. Hydrogels from biomaterials have increased their development by the ease of loading active substances, such as ABX

TABLE 4.2
Methods of Synthesis of Polymeric Nanoparticles (PNPs) and Biomedical Applications.

Synthesis Method	Biomedical Application	Ref.
Emulsification-solvent evaporation	The emulsification solvent evaporation technique is the most common method used for the preparation of polymeric nanoparticles. However, new synthesis methods based on magnetic nanoparticles for drug delivery and diagnosis of cancer therapy are used.	[39–42]
Nanoprecipitation	This technique involves the precipitation of the PNPs by slow addition of an antisolvent to the polymer solution in an adequate solvent. The antisolvent has to be miscible with the polymer solvent but not a solvent for the polymer. Starch and chitosan nanoparticles obtained with this technique were loaded with model drugs, showing good levels of encapsulation and drug release. One of the advantages of this technique is avoiding the usage of surfactants in most of the cases.	[35,43–45]
Emulsification polymerization	Pickering emulsions is an old and recent method for the synthesis of nanoparticles. It is currently used for its stability and versatility in medical, pharmaceutical, and cosmetic applications. Some of them include nanomaterials encapsulation hydrophilic substances and drug delivery.	[46,47]
Dialysis	The dialysis method is a simple way of charged nanoparticles synthesized with a narrow range of size, and modified chitosan nanoparticles loaded with different active molecules including model anticancer drugs have been showed effective release and high cytotoxicity on cell culture of cancer.	[48–50]
Salting out	It is considered a modification of the emulsion technique, which is based on the separation of a water-miscible solvent from an aqueous solution by the effect of salting out. In this case, alginate and chitosan nanoparticles were formed by the displacement of an electrolyte solution with PVA over an acetone solution with stirring polymers, where large amounts of water were dispersed allowing the diffusion of acetone.	[35,51]
Cross-linking polymerization	This method allows the formation of nanoparticles during polymerization by adding a cross-linking agent. Sodium alginate nanoparticles have been synthesized with a 2-acrylamido-2-methylpropane-1-sulfonic acid (AMPS) with the high absorption capacity of methylene blue and heavy metals such as lead, cadmium, copper, and nickel.	[35,52]
Ionotropic gelation	Ionic gelation occurs in a polysaccharide and ion of opposite charge, an example may be chitosan that interacts with sodium triphosphate (TPP) to form nanoparticles that have been loaded with growth factors (b-FGF) to be included on scaffolds of chitosan and gelatin for biomedical applications.	[34,53,54]
Supercritical fluids	This synthesis avoids the use of organic solvents through the use of CO_2 and H_2O in the gaseous state. Some recent applications are natural polymeric bioaerogels—such as pectin and alginate—which were loaded with ketoprofen proving to have release sensitivity to the pH of the medium.	[34,55–57]

[22]. The ability of these biomaterials to sequester, stabilize, and then release bioactive ABX represents a great advantage over traditional implants and controlled release pumps [67]. Hydrogels of animal origin have been developed, as is the case with hydrogels of a tripeptide (Leu- Phe-Phe), through its interaction with ciprofloxacin, which is part of its structural assembly, where the formation of noncovalent bonds allows the integration of the molecule and subsequent release [68]. In the same context, a hydrogel has been developed from gelatin, which has incorporated thermosensitive silver nanoparticles with antibacterial properties [69]. Another option is the use of hydrogels of plant origin, using polysaccharides such as chitosan with inherent antibacterial activity, by developing a hydrogel together with polyvinyl pyrrolidone that is pH sensitive for the release of amoxicillin, which was released at 73% after 3 h at pH 1 [70]. Hydrogels have also been developed from starch loaded with silica-coated copper nanoparticles, and these presented antimicrobial activity against pathogens such as *E. coli* and *S. aureus* [71].

During the last 50 years, fundamental concepts have been found in the field of drug-delivery systems. It is known that the efficacy of these is directly related to the chemical structure of the material, the shape and size, which can affect the properties and interactions with the immune system because their participation with the whole body is high, even when they are not designed for that [58]. During this time the field of nanotechnology has presented great advances in controlled drug-release systems, so there are several types of materials that can be used for the delivery of drugs in a controlled manner. They have been described with many terminologies, but one way of classifying them is according to their composition: (1) polymer systems, (2) lipid systems, (3) peptides or proteins, and (4) metal structures [72]. Within this classification, there are nanoparticles of natural origin that can be encapsulated in various ways. These in turn can be classified into three groups according to their synthesis mechanism: (1) those that use physiochemical methods (coacervation) [73,74], emulsification methods [75,76], methods with supercritical fluids [77,78], and thermal gelation [79], (2) chemical methods (interfacial and in situ polycondensation) [80], gelation [81,82], and polymerization [83]), and (3) physical methods (solvent evaporation) [84] and spray drying [85]). A final subclassification is the PNP-based systems used par excellence as transporters of active substances because these have favorable characteristics for the drug, such as improving its

stability, increasing the duration of the therapeutic effect and various routes of administration, allowing to minimize the degradation of the active substance [8,10,17,29,33,34,37]. Polymeric controlled release systems have been described as a broad spectrum of possibilities determined by their polymer–polymer, drug–polymer, polymer–solvent, or physiological medium polymer interactions [86].

As mentioned, some materials may have intrinsic antibacterial properties, but others need some substance with this capacity. It is at this point that the concepts of drug loading and drug encapsulation efficiency become important because these parameters act as true quality control for materials with aspirations in biomedicine, especially when we refer only to PNPs. The success of these materials is defined as the high loading capacity to reduce the amount of vehicle required for administration [87]. The loading of drugs can be carried out during or after the synthesis of the materials by incubating them in a solution with the drug to be incorporated. Once the active substance has been acquired within the material, the encapsulation efficiency can be evaluated by the following formula [86,88,89]:

Encapsulation efficiency (%) = ((Concentration of drug input − Concentration of free drug)/Concentration of drug input) × 100

In this same way, you can calculate drug loading:

Drug loading (%) = ((Concentration of drug input − Concentration of free drug)/Concentration of material) × 100

In the case of PNPs, it has been stated that the release depends on (1) desorption of the drug bound/adsorbed to the surface; (2) diffusion from the nanoparticle; (3) erosion of the PNPs; and (4) a combination of the diffusion/erosion process [86].

4 CHITOSAN IN DRUG DELIVERY

In recent decades, chitosan as a material has gained great relevance due to the physical, chemical, and biological properties it possesses. Its precursor, the chitin molecule, is the second most abundant polysaccharide, after cellulose. Once deacetylated, chitosan can be obtained: a more soluble form in water and other organic solvents. This presents the opportunity to be widely chemically modified [19,90]. Chitosan is a cationic polysaccharide under conditions of neutral or basic pH, leaving amino groups free, being insoluble in water. However, in acidic conditions, it increases its solubility by protonating its amino groups [91,92].

Chitosan is biocompatible with human tissues because when it decomposes it forms amino sugars that are completely harmless to the body [93].

Chitosan has taken particular importance as a starting material for tissue engineering and regenerative medicine [94], especially because of its capacity as a drug-delivery system [95]. Its versatility in mucoadhesive systems has stood out due to the ease with which chemical modifications can be made, obtaining derivatives such as carboxymethyl chitosan, glycol chitosan, trimethyl chitosan, chitosan-EDTA, chitosan acrylated, chitosan-catechol, methyl-pyrrolidone chitosan, cyclodextrin chitosan, and quaternized chitosan with oleoyl, thereby improving physiological solubility, stability, and controlled release [96]. Among the materials developed for the delivery of drugs in a controlled manner, always evaluating the conditions in which the drug will be delivered, the methods that use chitosan stand out for their variety in options. These can be [97] microspheres [98], microparticles [99], nanoparticles [100], membranes [101], sponges [102], scaffolding [103], and hydrogels [70,86,104]. The most common methodologies for the synthesis of chitosan nanoparticles are emulsion-based methods, coacervation/precipitation-based methods, ionic gelation, and self-assembly [94]. However, it is the chitosan-DNA nanoparticles that focus attention the most on controlled delivery systems with this material, developed for various research and clinical applications, such as transfections, vaccines, and medical applications [105–108]. Nanometric systems, as it has already been mentioned, are those that generate the greatest expectation when developing controlled drug-delivery systems because they take advantage of the safety properties of biomaterials such as chitosan as well as the cellular interactions that occur at this level, favoring synergism in the delivery of drugs such as ABX.

A recent retrospective article refers to the extensive use of chitosan PNPs as an ABX vehicle for the controlled delivery of cefazolin, rifampicin, daptomycin, ofloxacin, amphotericin B, vancomycin, amoxicillin, isoniazid, and tetracycline [109]. Sponges have been developed from chitosan for the local treatment of wound infections; these were loaded with amikacin and vancomycin, showing a significant decrease of *P. aeruginosa* and *S. aureus* bacteria in the wound [102]. Using the ionic gelation technique with tripolyphosphate combined with a spray-drying method, Ngyuen et al. obtained nanoparticles from chitosan of different molecular weights, which were also loaded with amoxicillin to evaluate their antibacterial activity. The results showed that nanoparticles of lower

molecular weight presented greater antibacterial activity and if they were also loaded with an ABX, they had a synergistic effect against *Streptococcus pneumoniae* [110]. Similarly, Rossi et al. obtained chitosan ascorbate nanoparticles loaded with amoxicillin for the treatment of atrophic vaginitis, which presented antibacterial and mucoadhesive properties in vitro, where they propose that polycationic chitosan nanoparticles with high surface charge density can interact with bacterial cells to a greater extent than free polymer, and with this exert significantly greater antibacterial activity [111]. Recently, Patel et al. have found that when functionalizing with DNase-I and alginate lyase the chitosan nanoparticles loaded with ciprofloxacin exert an effect on the development of the biofilm of *P. aeruginosa* of great importance in cystic fibrosis [112,113]. Marei et al. evaluated other sources of chitosan by obtaining chitin from desert insects. Then, by means of the ionic cross-linking method, they obtained chitosan nanoparticles, which loaded with ciprofloxacin HCl, showed improved antibacterial activity against *E. coli* and *S. aureus* methicillin-resistant compared to uncharged nanoparticles, uncharged marine nanoparticles, and ABX [114]. The decrease in ABX necessary to exert an antibacterial effect is one of the advantages of PNPs use, maintaining safety conditions. This was demonstrated by Changxuan et al., who developed vancomycin-loaded superparamagnetic chitosan nanoparticles for the treatment of chronic pyogenic osteoarthritis, where they managed to maintain a controlled release inadequate dose for 10 days [115].

5 STARCH IN DRUG DELIVERY

Starch is one of the most abundant polysaccharides in nature; it is of plant origin, being in the form of a granule of different sizes according to its origin in seeds (wheat, corn, rice, legumes), tubers (potatoes), and roots (tapioca) [116]. Starch is white and is insoluble in cold water, due to polymerization in its structure and hydrogen bonds between adjacent chains [117]. This polysaccharide is composed of two main structures: the amylose, which is mainly linear, represents most of the time between 10% and 20% of starch, and the amylopectin that is branched and is the main component of starch constituting 80%–90% [116,117]. In biomedical applications and controlled delivery systems, starch is usually presented in two ways: native and modified, that is, from direct extraction of the natural source and modified once extracted.

Starch is one of the polymers of greatest interest to researchers in the area of materials for biomedicine due to its biocompatibility and its ability to biodegrade [118]. There are several results that propose it as a substrate for cell culture, scaffolds in tissue engineering, drug-delivery systems, and bone replacement implants [119]. However, due to the high hydrophilic nature of starch, its low solubility, and—above all—its low mechanical resistance, it has been necessary to modify starch for the development of applications where it is used as a biomaterial substrate. Several techniques have been used for this purpose: physical, chemical, enzymatic and genetic, and addition of additives or a combination of these modification techniques, all of them aim at three available hydroxyl groups of starch copolymer [120,121]. Chemical modifications are preferably used to improve the properties of starches; they consist of adding functional groups to native starch, improving the performance and increasing efficiency in drug-delivery applications. Recent studies propose oxidation of the surface of starch granules as an ideal technique for clinical materials [120]. Other examples, where starch has been modified, allowed generating new applications in the delivery of drugs from matrices, reservoirs, hydrogels, film coating, nanocomposite, and conjugation to polymers that make up starch [121]. Recent cases are starch nanohydrogels with broad projections in drug delivery where new preparation techniques have been developed [122]. In addition, in drug delivery, as is the case of starch-based magnetite hydrogels in stimulus-sensitive administration systems [123]. Some vehicles of recent interest are starch-based aerogels, which due to their porosity have the ability to load medications such as ketoprofen, using impregnation methods [124]. Magnetic nanoparticles coated with a modified starch-based film have been developed as vehicles for immobilization of bioligands, such as albumin in catalysis and pharmaceutical analysis, through starch amination, which is the highest interaction between the nanoparticles and the ligand [125]. In this context, Farrag et al. have developed a simple nanoprecipitation method to obtain starch nanoparticles loaded with quercetin as a model of transport and release of an antioxidant molecule, observing significant differences in release depending on the starch plant origin [43]. Another option that has been measured was to load starch nanoparticles with essential oils with antibacterial properties. Consequently, Qui et al. demonstrated extended antioxidant activity against E. coli and S. aureus, once loaded the particles with oil menthol extracts [126].

Among the various applications of starch, its use in the delivery of ABX can be found, and this polymer has been used as a matrix to attract ABX in capsules and tablets [72]. However, recently researchers have taken advantage of the versatility of this material to develop different kinds of vehicles, such as a hydrogel from modified starch with palmitoyl chains to load levofloxacin. The method of starch modification was by esterification using palmitoyl chloride with dimethylformamide in the presence of pyridine. Once this was done, the hydrogel was cross-linked with citric acid and loaded with levofloxacin, showing suitable ABX-release profiles in vitro environments [127], while Avval et al. chemically modified starch by carboxymethylation and then added grafts of synthetic polymers such as acrylamide and hydroxyethyl acrylate monomers, which were loaded with cephalexin and tested for release at different pH [128]. Furthermore, Thiele et al. oxidized starch obtaining anionic copolymers of glucuronic acid and glucose, and these were attached to a cationic thioether of β-cyclodextrin obtaining 130 nm nanoparticles that loaded with drug dihydroxyanthraquinone, thereby proposing a versatile synthesis for cyclodextrins loaded with poorly soluble drugs such as some ABX [129]. In this same area, Ismali et al. were able to synthesize starch nanoparticles in an easy way by means of emulsion nanoprecipitation, which were loaded with penicillin and streptomycin, showing an improved antimicrobial effect against S. pyogenes and E. coli using the disk diffusion method [130].

6 CHITOSAN-STARCH IN ANTIBIOTIC DELIVERY

It is important to mention that the combination of these polymeric materials of natural origin (chitosan and starch) was also investigated in recent years, obtaining promising results in the field of drug-delivery systems. This was especially true for ABX, such as a hydrogel of starch-chitosan polymer mixture obtained by cross-linking, which was loaded with ciprofloxacin and tetracycline, showing high sorption efficiencies compared to substrates separately [131]. In contrast, Shehabeldine et al. recently developed a nanocomposite material from hydrolyzed starch and chitosan, to which ciprofloxacin hydrochloride was incorporated to evaluate its behavior against pathogens such as Enterobacter aerogenes and P. aeruginosa. The results showed that both the antimicrobial behavior and the profile of drug release have been improved [132].

Balmayor et al. developed spherical microparticles of starch-conjugated chitosan for the controlled release of gentamicin in the treatment of bone infections, obtaining a drug load up to 27% and a sustained release profile up to 30 days in vitro with 82% release. Moreover, its antimicrobial effect against *S. aureus* was evaluated, achieving 70%–100% relative activity [133]. Similarly, Yasar et al. developed a chitosan-starch complex polymeric system in nanoparticles form, which showed high encapsulation efficiency for cationic antibiotics. Through the use of oxidation-modified starch in the form of nanoparticles, it was made to interact with antibiotics, which in turn were coated with PVA-stabilized chitosan. This led to the formation of the polymeric nanocomplex that showed the same efficacy as antibiotics such as tobramycin and colistin, in their free form, against *E. coli* and *P. aeruginosa* [134].

7 CONCLUSIONS AND FUTURE PERSPECTIVES

As has been reviewed in this chapter the biomedical applications based on natural polymers have become more important due to their versatility, immunogenicity, biocompatibility, availability, and relatively easy obtaining. However, it remains a challenge to find materials that can come from renewable resources that maintain the balance of the biomedical and engineering aspects. Finding a unique solution for the delivery of any drugs, especially antibiotics, is effectively not possible through the use of polymeric materials. Instead, it has been proposed that an optimal vehicle should be developed for each drug and specific for a type of pathology, respecting the nature of the human body and the interactions that these vehicles generate with it. To achieve this, multidisciplinary teams must work to be able to cover the complexity of the human body and to consider the technical-technological needs from the clinical to the commercial point of view [135]. These considerations become more relevant when referring to antibiotics, as the resistance of microorganisms to these kinds of drugs can only be fought by promoting new organized strategies, by mutual agreement, and the development of new materials that can provide an alternative to existing antibiotics. This work aims to update the latest advances in biomaterials of natural origin designed for the transport and delivery of antibiotics, with special attention on hydrogels, composite materials, and nanoparticles, which are useful as a tool to make better decisions when looking for an antimicrobial strategy.

ACKNOWLEDGMENTS

Centro de Investigación de Polímeros Avanzados (CIPA), CONICYT Regional and GORE BIO-BIO R17A10003, FONDEF-IDeA ID16i10425 and CONICYT Proyects I+D Ciencia-Territorio R18F10016. To Ph.D.(c) Francisca Espinoza for her constant support and advice.

REFERENCES

[1] World Health Organization, Antibiotic resistance, 2018. https://www.who.int/news-room/fact-sheets/detail/antibiotic-resistance.

[2] T. Pham, P. Loupias, A. Dassonville-Klimpt, P. Sonnet, Drug delivery systems designed to overcome antimicrobial resistance, Medicinal Research Reviews (2019), https://doi.org/10.1002/med.21588.

[3] S. Torrades, Uso y abuso de los antibióticos, OffFarm 20 (2001) 82–93.

[4] R.S. Kalhapure, N. Suleman, C. Mocktar, N. Seedat, T. Govender, Nanoengineered drug delivery systems for enhancing antibiotic therapy, Journal of Pharmaceutical Sciences 104 (2015) 872–905, https://doi.org/10.1002/jps.24298.

[5] S. Ramakrishna, M. Ramalingam, T.S. Kumar, W.O. Soboyejo, Biomaterials: A Nano Approach, CRC press, 2016, https://doi.org/10.1080/10426914.2014.950068.

[6] F.-M. Chen, X. Liu, P. Polym, S. Author, Advancing biomaterials of human origin for tissue engineering HHS Public Access Author manuscript, Progress in Polymer Science 53 (2016) 86–168, https://doi.org/10.1016/j.progpolymsci.2015.02.004.

[7] J.F. Mano, G.A. Silva, H.S. Azevedo, P.B. Malafaya, R.A. Sousa, S.S. Silva, L.F. Boesel, J.M. Oliveira, T.C. Santos, A.P. Marques, N.M. Neves, R.L. Reis, Natural origin biodegradable systems in tissue engineering and regenerative medicine: present status and some moving trends, Journal of The Royal Society Interface 4 (2007) 999–1030, https://doi.org/10.1098/rsif.2007.0220.

[8] S. Bhatia, Natural polymers vs synthetic polymer, in: Natural Polymer Drug Delivery Systems, Springer, 2016, pp. 95–118, https://doi.org/10.1007/978-3-319-41129-3_3.

[9] W.P. Limited, Elena P. Ivanova, Kateryna Bazaka, Russell J. Crawford, New functional biomaterials for medicine and healthcare, MRS Bulletin 39 (2014) 824, https://doi.org/10.1557/mrs.2014.209.

[10] R. Rebelo, M. Fernandes, R. Fangueiro, Biopolymers in medical implants: a brief review, Procedia Engineering 200 (2017) 236–243, https://doi.org/10.1016/j.proeng.2017.07.034.

[11] S.P. Noel, H. Courtney, J.D. Bumgardner, W.O. Haggard, Chitosan films: a potential local drug delivery system for antibiotics, Clinical Orthopaedics and Related Research 466 (2008) 1377–1382, https://doi.org/10.1007/s11999-008-0228-1.

[12] J. Marto, A. Duarte, S. Simões, L.M. Gonçalves, L.F. Gouveia, A.J. Almeida, H.M. Ribeiro, Starch-based pickering emulsions as platforms for topical antibiotic delivery: in vitro and in vivo studies, Polymers (Basel) 11 (2019), https://doi.org/10.3390/polym11010108.

[13] I. Björck, M. Nyman, B. Pedersen, M. Siljeström, N.G. Asp, B.O. Eggum, On the digestibility of starch in wheat bread — studies in vitro and in vivo, Journal of Cereal Science 4 (1986) 1—11, https://doi.org/10.1016/S0733-5210(86)80002-9.

[14] A. Agrawal, S. Rellegadla, S. Jain, Biomedical applications of PLGA particles, Biomaterials for Biomedical Engineering 4 (2019) 87—129, https://doi.org/10.1016/B978-0-12-816913-1.00004-0.

[15] D.L. Kaplan, Introduction to biopolymers from renewable resources, Biopolymers from Renewable Resources (1998) 1—29, https://doi.org/10.1007/978-3-662-03680-8_1.

[16] M. Niaounakis, 1 - Introduction to biopolymers, in: M. Niaounakis (Ed.), Biopolymers Reuse, Recycling, and Disposal, William Andrew Publishing, Oxford, 2013, pp. 1—75, https://doi.org/10.1016/B978-1-4557-3145-9.00001-4.

[17] W. He, R. Benson, 8 Polymeric Biomaterials, Elsevier Inc., 2017.

[18] F. Khan, M. Tanaka, S.R. Ahmad, Fabrication of polymeric biomaterials: a strategy for tissue engineering and medical devices, Journal of Materials Chemistry B 3 (2015) 8224—8249, https://doi.org/10.1039/c5tb01370d.

[19] N.B. Shelke, R. James, C.T. Laurencin, S.G. Kumbar, Polysaccharide biomaterials for drug delivery and regenerative engineering, Polymers for Advanced Technologies 25 (2014) 448—460, https://doi.org/10.1002/pat.3266.

[20] N.A.N. Hanafy, S. Leporatti, M.A. El-Kemary, Mucoadhesive hydrogel nanoparticles as smart biomedical drug delivery system, Applied Sciences 9 (2019), https://doi.org/10.3390/app9050825.

[21] J.D. Caplin, A.J. García, Implantable antimicrobial biomaterials for local drug delivery in bone infection models, Acta Biomaterialia 93 (2019) 2—11, https://doi.org/10.1016/j.actbio.2019.01.015.

[22] S. Li, S. Dong, W. Xu, S. Tu, L. Yan, C. Zhao, J. Ding, X. Chen, Antibacterial hydrogels, Advancement of Science 5 (2018), https://doi.org/10.1002/advs.201700527.

[23] D. Gan, W. Xing, L. Jiang, J. Fang, C. Zhao, F. Ren, L. Fang, K. Wang, X. Lu, Plant-inspired adhesive and tough hydrogel based on Ag-Lignin nanoparticles-triggered dynamic redox catechol chemistry, Nature Communications 10 (2019) 1—10, https://doi.org/10.1038/s41467-019-09351-2.

[24] S. Manju, M. Antony, K. Sreenivasan, Synthesis and evaluation of a hydrogel that binds glucose and releases ciprofloxacin, Journal of Materials Science 45 (2010) 4006—4012, https://doi.org/10.1007/s10853-010-4474-8.

[25] T. Samad, J.Y. Co, J. Witten, K. Ribbeck, Mucus and mucin environments reduce the efficacy of polymyxin and fluoroquinolone antibiotics against *Pseudomonas aeruginosa*, ACS Biomaterials Science & Engineering 5 (2019) 1189—1194, https://doi.org/10.1021/acsbiomaterials.8b01054.

[26] X. Sui, X. Feng, D. Luca, C.A. Van Blitterswijk, L. Moroni, Poly (N-Isopropylacrylamide)—Poly (Ferrocenylsilane) Dual Responsive Hydrogels: Synthesis, Characterization and Antimicrobial Applications, 2012, pp. 3—6, https://doi.org/10.1039/c2py20431b.

[27] Y. Ko, J. Kim, H.Y. Jeong, G. Kwon, D. Kim, M. Ku, J. Yang, Y. Yamauchi, H.Y. Kim, C. Lee, J. You, Antibacterial poly (3,4-ethylenedioxythiophene):poly(styrenesulfonate)/agarose nanocomposite hydrogels with thermo-processability and self-healing, Carbohydrate Polymers 203 (2019) 26—34, https://doi.org/10.1016/j.carbpol.2018.09.026.

[28] N. Pettinelli, S. Rodríguez-Llamazares, V. Abella, L. Barral, R. Bouza, Y. Farrag, F. Lago, Entrapment of chitosan, pectin or κ-carrageenan within methacrylate based hydrogels: effect on swelling and mechanical properties, Materials Science and Engineering: C 96 (2019) 583—590, https://doi.org/10.1016/j.msec.2018.11.071.

[29] T. Lechelmayr, T. Preuss, I. Jumpertz, F. Novotný, M. Eblenkamp, E. Wintermantel, Nanocomposites as biomaterials, Kunststoffe International 101 (2011) 50—52.

[30] Y. Zhang, M. Zhang, Calcium phosphate/chitosan composite scaffolds for controlled in vitro antibiotic drug release, Journal of Biomedical Materials Research 62 (2002) 378—386, https://doi.org/10.1002/jbm.10312.

[31] J.M.C. Teixeira, J.S.V. Alburquerque, E.B. Duarte, S.A. Silva, R.E.F.Q. Nogueira, In vitro drug release study from hydroxyapatite-alumina composites, Journal of Sol-Gel Science and Technology 89 (2019) 521—530, https://doi.org/10.1007/s10971-018-4888-3.

[32] K. Vimala, M.M. Yallapu, K. Varaprasad, N.N. Reddy, S. Ravindra, N.S. Naidu, K.M. Raju, Fabrication of curcumin encapsulated chitosan-PVA silver nanocomposite films for improved antimicrobial activity, Journal of Biomaterials and Nanobiotechnology 02 (2011) 55—64, https://doi.org/10.4236/jbnb.2011.21008.

[33] J.P. Rao, K.E. Geckeler, Polymer nanoparticles: preparation techniques and size-control parameters, Progress in Polymer Science 36 (2011) 887—913, https://doi.org/10.1016/j.progpolymsci.2011.01.001.

[34] V.J. Mohanraj, Y. Chen, Nanoparticles — a review, Tropical Journal of Pharmaceutical Research 5 (2007) 561—573, https://doi.org/10.4314/tjpr.v5i1.14634.

[35] B.V.N. Nagavarma, H.K.S. Yadav, A. Ayaz, L.S. Vasudha, H.G. Shivakumar, Different techniques for preparation of polymeric nanoparticles- A review, Asian Journal of Pharmaceutical and Clinical Research 5 (2012) 16—23.

[36] M. Elsabahy, K. Wooley, Design of polymeric nanoparticles for biomedical delivery applications, Chemical Society Reviews 41 (2012) 1−7, https://doi.org/10.1039/c2cs15327k.

[37] S. Sundar, J. Kundu, S.C. Kundu, Biopolymeric nanoparticles, Science and Technology of Advanced Materials 11 (2010) 014104, https://doi.org/10.1088/1468-6996/11/1/014104.

[38] J. Yang, S. Han, H. Zheng, H. Dong, J. Liu, Preparation and application of micro/nanoparticles based on natural polysaccharides, Carbohydrate Polymers 123 (2015) 53−66, https://doi.org/10.1016/j.carbpol.2015.01.029.

[39] Y. Farrag, B. Montero, M. Rico, L. Barral, R. Bouza, Preparation and characterization of nano and micro particles of poly(3-hydroxybutyrate-co-3-hydroxyvalerate) (PHBV) via emulsification/solvent evaporation and nanoprecipitation techniques, Journal of Nanoparticle Research 20 (2018), https://doi.org/10.1007/s11051-018-4177-7.

[40] R. Jenjob, T. Phakkeeree, F. Seidi, M. Theerasilp, D. Crespy, Emulsion techniques for the production of pharmacological nanoparticles, Macromolecular Bioscience 19 (2019) 1−13, https://doi.org/10.1002/mabi.201900063.

[41] S. Khoee, M. Soleymani, Janus arrangement of smart polymer on magnetite nanoparticles through solvent evaporation from emulsion droplets, Applied Surface Science 494 (2019) 805−816, https://doi.org/10.1016/j.apsusc.2019.07.215.

[42] H.Y. Yang, Y. Li, D.S. Lee, Multifunctional and stimuli-responsive magnetic nanoparticle-based delivery systems for biomedical applications, Advances in Therapy 1 (2018) 1800011, https://doi.org/10.1002/adtp.201800011.

[43] Y. Farrag, W. Ide, B. Montero, M. Rico, S. Rodríguez-Llamazares, L. Barral, R. Bouza, Preparation of starch nanoparticles loaded with quercetin using nanoprecipitation technique, International Journal of Biological Macromolecules 114 (2018) 426−433, https://doi.org/10.1016/j.ijbiomac.2018.03.134.

[44] A.G. Luque-Alcaraz, J. Lizardi-Mendoza, F.M. Goycoolea, I. Higuera-Ciapara, W. Argüelles-Monal, Preparation of chitosan nanoparticles by nanoprecipitation and their ability as a drug nanocarrier, RSC Advances 6 (2016) 59250−59256, https://doi.org/10.1039/c6ra06563e.

[45] T. Tóth, É. Kiss, A method for the prediction of drug content of poly(lactic-co-glycolic)acid drug carrier nanoparticles obtained by nanoprecipitation, Journal of Drug Delivery Science and Technology 50 (2019) 42−47, https://doi.org/10.1016/j.jddst.2019.01.010.

[46] C.L.G. Harman, M.A. Patel, S. Guldin, G.L. Davies, Recent developments in Pickering emulsions for biomedical applications, Current Opinion in Colloid and Interface Science 39 (2019) 173−189, https://doi.org/10.1016/j.cocis.2019.01.017.

[47] Y. Chevalier, M.A. Bolzinger, Emulsions stabilized with solid nanoparticles: pickering emulsions, Colloids and Surfaces A: Physicochemical and Engineering Aspects 439 (2013) 23−34, https://doi.org/10.1016/j.colsurfa.2013.02.054.

[48] T. Woraphatphadung, W. Sajomsang, T. Rojanarata, T. Ngawhirunpat, P. Tonglairoum, P. Opanasopit, Development of chitosan-based pH-sensitive polymeric micelles containing curcumin for colon-targeted drug delivery, AAPS PharmSciTech 19 (2018) 991−1000, https://doi.org/10.1208/s12249-017-0906-y.

[49] J.H. Kim, Y.S. Kim, S. Kim, J.H. Park, K. Kim, K. Choi, H. Chung, S.Y. Jeong, R.W. Park, I.S. Kim, I.C. Kwon, Hydrophobically modified glycol chitosan nanoparticles as carriers for paclitaxel, Journal of Controlled Release 111 (2006) 228−234, https://doi.org/10.1016/j.jconrel.2005.12.013.

[50] J.W. Nah, Y.W. Paek, Y.I. Jeong, D.W. Kim, C.S. Cho, S.H. Kim, M.Y. Kim, Clonazepam release from poly(DL-lactide-co-glycolide) nanoparticles prepared by dialysis method, Archives of Pharmacal Research 21 (1998) 418−422, https://doi.org/10.1007/BF02974636.

[51] E. Allémann, R. Gurny, E. Doelker, Preparation of aqueous polymeric nanodispersions by a reversible salting-out process: influence of process parameters on particle size, International Journal of Pharmaceutics 87 (1992) 247−253, https://doi.org/10.1016/0378-5173(92)90249-2.

[52] Z. jian Shao, X. lian Huang, F. Yang, W. feng Zhao, X. zhi Zhou, C. sheng Zhao, Engineering sodium alginate-based cross-linked beads with high removal ability of toxic metal ions and cationic dyes, Carbohydrate Polymers 187 (2018) 85−93, https://doi.org/10.1016/j.carbpol.2018.01.092.

[53] S. Azizian, A. Hadjizadeh, H. Niknejad, Chitosan-gelatin porous scaffold incorporated with Chitosan nanoparticles for growth factor delivery in tissue engineering, Carbohydrate Polymers 202 (2018) 315−322, https://doi.org/10.1016/j.carbpol.2018.07.023.

[54] B. Siddhardha, J. Rajkumari, Preface. Fungal Nanobionics Principles Applications, 2018, https://doi.org/10.1007/978-981-10-8666-3.

[55] F.P. Soorbaghi, M. Isanejad, S. Salatin, M. Ghorbani, S. Jafari, H. Derakhshankhah, Bioaerogels: synthesis approaches, cellular uptake, and the biomedical applications, Biomedicine and Pharmacotherapy 111 (2019) 964−975, https://doi.org/10.1016/j.biopha.2019.01.014.

[56] C.A. García-González, M. Jin, J. Gerth, C. Alvarez-Lorenzo, I. Smirnova, Polysaccharide-based aerogel microspheres for oral drug delivery, Carbohydrate Polymers 117 (2015) 797−806, https://doi.org/10.1016/j.carbpol.2014.10.045.

[57] K. Byrappa, S. Ohara, T. Adschiri, Nanoparticles synthesis using supercritical fluid technology − towards biomedical applications, Advanced Drug Delivery

Reviews 60 (2008) 299–327, https://doi.org/10.1016/j.addr.2007.09.001.

[58] M.W. Tibbitt, J.E. Dahlman, R. Langer, Emerging frontiers in drug delivery, Journal of the American Chemical Society 138 (2016) 704–717, https://doi.org/10.1021/jacs.5b09974.

[59] D.S. Kohane, R. Langer, Polymeric biomaterials in tissue engineering, Pediatric Research 63 (2008) 487–491, https://doi.org/10.1203/01.pdr.0000305937.26105.e7.

[60] R. Costa, L. Santos, Delivery systems for cosmetics - from manufacturing to the skin of natural antioxidants, Powder Technology 322 (2017) 402–416, https://doi.org/10.1016/j.powtec.2017.07.086.

[61] J.M. Morais, F. Papadimitrakopoulos, D.J. Burgess, Biomaterials/tissue interactions: possible solutions to overcome foreign body response, The AAPS Journal 12 (2010) 188–196, https://doi.org/10.1208/s12248-010-9175-3.

[62] D. Campoccia, L. Montanaro, P. Speziale, C.R. Arciola, Antibiotic-loaded biomaterials and the risks for the spread of antibiotic resistance following their prophylactic and therapeutic clinical use, Biomaterials 31 (2010) 6363–6377, https://doi.org/10.1016/j.biomaterials.2010.05.005.

[63] N.D. Stebbins, M.A. Ouimet, K.E. Uhrich, Antibiotic-containing polymers for localized, sustained drug delivery, Advanced Drug Delivery Reviews 78 (2014) 77–87, https://doi.org/10.1016/j.addr.2014.04.006.

[64] R. Canaparo, F. Foglietta, F. Giuntini, C.D. Pepa, F. Dosio, L. Serpe, Recent developments in antibacterial therapy: focus on stimuli-responsive drug-delivery systems and therapeutic nanoparticles, Molecules 24 (2019), https://doi.org/10.3390/molecules24101991.

[65] C. Chang, B. Duan, J. Cai, L. Zhang, Superabsorbent hydrogels based on cellulose for smart swelling and controllable delivery, European Polymer Journal 46 (2010) 92–100, https://doi.org/10.1016/j.eurpolymj.2009.04.033.

[66] J.K. Kim, H.J. Kim, J.Y. Chung, J.H. Lee, S.B. Young, Y.H. Kim, Natural and synthetic biomaterials for controlled drug delivery, Archives of Pharmacal Research 37 (2014) 60–68, https://doi.org/10.1007/s12272-013-0280-6.

[67] E.M. Pritchard, T. Valentin, B. Panilaitis, F. Omenetto, D.L. Kaplan, Antibiotic-releasing silk biomaterials for infection prevention and treatment, Advanced Functional Materials 23 (2013) 854–861, https://doi.org/10.1002/adfm.201201636.

[68] S. Marchesan, Y. Qu, L.J. Waddington, C.D. Easton, V. Glattauer, T.J. Lithgow, K.M. McLean, J.S. Forsythe, P.G. Hartley, Self-assembly of ciprofloxacin and a tripeptide into an antimicrobial nanostructured hydrogel, Biomaterials 34 (2013) 3678–3687, https://doi.org/10.1016/j.biomaterials.2013.01.096.

[69] B. Manjula, K. Varaprasad, R. Sadiku, K. Ramam, G.V.S. Reddy, K.M. Raju, Development of microbial resistant thermosensitive Ag nanocomposite (gelatin) hydrogels via green process, Journal of Biomedical Materials Research Part A 102 (2014) 928–934, https://doi.org/10.1002/jbm.a.34780.

[70] M.V. Risbud, A.A. Hardikar, S.V. Bhat, R.R. Bhonde, pH-sensitive freeze-dried chitosan–polyvinyl pyrrolidone hydrogels as controlled release system for antibiotic delivery, Journal of Controlled Release 68 (2000) 23–30, https://doi.org/10.1016/S0168-3659(00)00208-X.

[71] M.E. Villanueva, A.M.D.R. Diez, J.A. González, C.J. Pérez, M. Orrego, L. Piehl, S. Teves, G.J. Copello, Antimicrobial activity of starch hydrogel incorporated with copper nanoparticles, ACS Applied Materials and Interfaces 8 (2016) 16280–16288, https://doi.org/10.1021/acsami.6b02955.

[72] M.L. Bruschi, Strategies to Modify the Drug Release from Pharmaceutical Systems, Woodhead Publishing, 2015, pp. 87–194, https://doi.org/10.1016/B978-0-08-100092-2.00006-0.

[73] G. Wang, K. Siggers, S. Zhang, H. Jiang, Z. Xu, R.F. Zernicke, J. Matyas, H. Uludağ, Preparation of BMP-2 containing bovine serum albumin (BSA) nanoparticles stabilized by polymer coating, Pharmaceutical Research 25 (2008) 2896–2909, https://doi.org/10.1007/s11095-008-9692-2.

[74] Q. Gan, T. Wang, Chitosan nanoparticle as protein delivery carrier-Systematic examination of fabrication conditions for efficient loading and release, Colloids and Surfaces B: Biointerfaces 59 (2007) 24–34, https://doi.org/10.1016/j.colsurfb.2007.04.009.

[75] M.J. Santander-Ortega, T. Stauner, B. Loretz, J.L. Ortega-Vinuesa, D. Bastos-González, G. Wenz, U.F. Schaefer, C.M. Lehr, Nanoparticles made from novel starch derivatives for transdermal drug delivery, Journal of Controlled Release 141 (2010) 85–92, https://doi.org/10.1016/j.jconrel.2009.08.012.

[76] C. Pan, J. Qian, J. Fan, H. Guo, L. Gou, H. Yang, C. Liang, Preparation nanoparticle by ionic cross-linked emulsified chitosan and its antibacterial activity, Colloids and Surfaces A: Physicochemical and Engineering Aspects 568 (2019) 362–370, https://doi.org/10.1016/j.colsurfa.2019.02.039.

[77] P. Trucillo, R. Campardelli, Production of solid lipid nanoparticles with a supercritical fluid assisted process, The Journal of Supercritical Fluids 143 (2019) 16–23, https://doi.org/10.1016/j.supflu.2018.08.001.

[78] F.J. Caro-León, W. Argüelles-Monal, E. Carvajal-Millán, Y.L. López-Franco, F.M. Goycoolea-Valencia, J. San Román del Barrio, J. Lizardi-Mendoza, Production and characterization of supercritical CO_2 dried chitosan nanoparticles as novel carrier device, Carbohydrate Polymers 198 (2018) 556–562, https://doi.org/10.1016/j.carbpol.2018.06.102.

[79] A. Masotti, G. Ortaggi, Chitosan micro-and nano-spheres: fabrication and applications for drug and DNA delivery, Nanotechnology Mini Reviews in Medicinal Chemistry 4 (2009) 463–469, https://doi.org/10.2174/138955709787847976.

[80] C. Vauthier, K. Bouchemal, Methods for the preparation and manufacture of polymeric nanoparticles,

Pharmaceutical Research 26 (2009) 1025−1058, https://doi.org/10.1007/s11095-008-9800-3.

[81] A. Grenha, B. Seijo, C. Remuñán-López, Microencapsulated chitosan nanoparticles for lung protein delivery, European Journal of Pharmaceutical Sciences 25 (2005) 427−437, https://doi.org/10.1016/j.ejps.2005.04.009.

[82] J. Yang, F. Li, M. Li, S. Zhang, J. Liu, C. Liang, Q. Sun, L. Xiong, Fabrication and characterization of hollow starch nanoparticles by gelation process for drug delivery application, Carbohydrate Polymers 173 (2017) 223−232, https://doi.org/10.1016/j.carbpol.2017.06.006.

[83] K. Landfester, Miniemulsion polymerization and the structure of polymer and hybrid nanoparticles, Angewandte Chemie International Edition 48 (2009) 4488−4508, https://doi.org/10.1002/anie.200900723.

[84] S.D. DeVeaux, C.T. Gomillion, Assessing the potential of chitosan/polylactide nanoparticles for delivery of therapeutics for triple-negative breast cancer treatment, Regenerative Engineering and Translational Medicine 5 (2019) 61−73, https://doi.org/10.1007/s40883-018-0089-4.

[85] C. Arpagaus, A. Collenberg, D. Rütti, E. Assadpour, S.M. Jafari, Nano spray drying for encapsulation of pharmaceuticals, International Journal of Pharmaceutics 546 (2018) 194−214, https://doi.org/10.1016/j.ijpharm.2018.05.037.

[86] N. Bhattarai, J. Gunn, M. Zhang, Chitosan-based hydrogels for controlled, localized drug delivery, Advanced Drug Delivery Reviews 62 (2010) 83−99, https://doi.org/10.1016/j.addr.2009.07.019.

[87] X.Y. Lu, D.C. Wu, Z.J. Li, G.Q. Chen, Polymer Nanoparticles, vol. 104, Elsevier Inc., 2011, pp. 299−323, https://doi.org/10.1016/B978-0-12-416020-0.00007-3.

[88] R. Justin, B. Chen, Characterisation and drug release performance of biodegradable chitosan-graphene oxide nanocomposites, Carbohydrate Polymers 103 (2014) 70−80, https://doi.org/10.1016/j.carbpol.2013.12.012.

[89] S.A. Agnihotri, S.S. Jawalkar, T.M. Aminabhavi, Controlled release of cephalexin through gellan gum beads: effect of formulation parameters on entrapment efficiency, size, and drug release, European Journal of Pharmaceutics and Biopharmaceutics 63 (2006) 249−261, https://doi.org/10.1016/j.ejpb.2005.12.008.

[90] A.K. Sailaja, P. Amareshwar, P. Chakravarty, Different techniques used for the preparation of nanoparticles using, International Journal of Pharmacy and Pharmaceutical Sciences 3 (2011) 45−50.

[91] S.A. Agnihotri, N.N. Mallikarjuna, T.M. Aminabhavi, Recent advances on chitosan-based micro- and nanoparticles in drug delivery, Journal of Controlled Release 100 (2004) 5−28, https://doi.org/10.1016/j.jconrel.2004.08.010.

[92] M. Rinaudo, Chitin and chitosan: properties and applications, Progress in Polymer Science 31 (2006) 603−632, https://doi.org/10.1016/j.progpolymsci.2006.06.001.

[93] C. Peniche, W. Argüelles-Monal, H. Peniche, N. Acosta, Chitosan: an attractive biocompatible polymer for microencapsulation, Macromolecular Bioscience 3 (2003) 511−520, https://doi.org/10.1002/mabi.200300019.

[94] S.M. Ahsan, M. Thomas, K.K. Reddy, S.G. Sooraparaju, A. Asthana, I. Bhatnagar, Chitosan as biomaterial in drug delivery and tissue engineering, International Journal of Biological Macromolecules 110 (2018) 97−109, https://doi.org/10.1016/j.ijbiomac.2017.08.140.

[95] J. Li, C. Cai, J. Li, J. Li, J. Li, T. Sun, L. Wang, H. Wu, G. Yu, Chitosan-based nanomaterials for drug delivery, Molecules 23 (2018) 1−26, https://doi.org/10.3390/molecules23102661.

[96] T.M.M. Ways, W.M. Lau, V.V. Khutoryanskiy, Chitosan and its derivatives for application in mucoadhesive drug delivery systems, Polymers (Basel) 10 (2018), https://doi.org/10.3390/polym10030267.

[97] E.B. Denkbaş, R.M. Ottenbrite, Perspectives on: chitosan drug delivery systems based on their geometries, Journal of Bioactive and Compatible Polymers 21 (2006) 351−368, https://doi.org/10.1177/0883911506066930.

[98] V.R. Sinha, A.K. Singla, S. Wadhawan, R. Kaushik, R. Kumria, K. Bansal, S. Dhawan, Chitosan microspheres as a potential carrier for drugs, International Journal of Pharmaceutics 274 (2004) 1−33, https://doi.org/10.1016/j.ijpharm.2003.12.026.

[99] J.A. Ko, H.J. Park, S.J. Hwang, J.B. Park, J.S. Lee, Preparation and characterization of chitosan microparticles intended for controlled drug delivery 249 (2002) 165−174, https://doi.org/10.1016/S0378-5173(02)00487-8.

[100] K.A. Janes, M.P. Fresneau, A. Marazuela, A. Fabra, J. Alonso, Chitosan nanoparticles as delivery systems for doxorubicin, Journal of Controlled Release 73 (2−3) (2001) 255−267.

[101] S. Şenel, G. Ikinci, S. Kaş, A. Yousefi-Rad, M.F. Sargon, A.A. Hincal, Chitosan films and hydrogels of chlorhexidine gluconate for oral mucosal delivery, International Journal of Pharmaceutics 193 (2000) 197−203, https://doi.org/10.1016/S0378-5173(99)00334-8.

[102] D.J. Stinner, S.P. Noel, W.O. Haggard, J.T. Watson, J.C. Wenke, Local antibiotic delivery using tailorable chitosan sponges: the future of infection control? Journal of Orthopaedics and Trauma 24 (2010) 592−597, https://doi.org/10.1097/BOT.0b013e3181ed296c.

[103] S.E. Kim, J.H. Park, Y.W. Cho, H. Chung, S.Y. Jeong, E.B. Lee, I.C. Kwon, Porous chitosan scaffold containing microspheres loaded with transforming growth factor-β1: implications for cartilage tissue engineering, Journal of Controlled Release 91 (2003) 365−374, https://doi.org/10.1016/S0168-3659(03)00274-8.

[104] H. Gholizadeh, S. Cheng, M. Pozzoli, E. Messerotti, D. Traini, P. Young, A. Kourmatzis, H.X. Ong, Smart thermosensitive chitosan hydrogel for nasal delivery of ibuprofen to treat neurological disorders, Expert Opinion on Drug Delivery 16 (2019) 453−466, https://doi.org/10.1080/17425247.2019.1597051.

[105] M. Iqbal, W. Lin, I. Jabbal-Gill, S.S. Davis, M.W. Steward, L. Illum, Nasal delivery of chitosan-DNA plasmid expressing epitopes of respiratory syncytial virus (RSV) induces protective CTL responses in BALB/c mice, Vaccine 21 (2003) 1478–1485, https://doi.org/10.1016/S0264-410X(02)00662-X.

[106] L.M. Bravo-Anaya, K.G. Fernández-Solís, J. Rosselgong, J.L.E. Nano-Rodríguez, F. Carvajal, M. Rinaudo, Chitosan-DNA polyelectrolyte complex: influence of chitosan characteristics and mechanism of complex formation, International Journal of Biological Macromolecules 126 (2019) 1037–1049, https://doi.org/10.1016/j.ijbiomac.2019.01.008.

[107] S. Mao, W. Sun, T. Kissel, Chitosan-based formulations for delivery of DNA and siRNA, Advanced Drug Delivery Reviews 62 (2010) 12–27, https://doi.org/10.1016/j.addr.2009.08.004.

[108] K. Roy, H.Q. Mao, S.K. Huang, K.W. Leong, Oral gene delivery with chitosan-DNA nanoparticles generates immunologic protection in a murine model of peanut allergy, Nature Medicine 5 (1999) 387–391, https://doi.org/10.1038/7385.

[109] S. Naskar, K. Koutsu, S. Sharma, Chitosan-based nanoparticles as drug delivery systems: a review on two decades of research, Journal of Drug Targeting 27 (2019) 379–393, https://doi.org/10.1080/1061186X.2018.1512112.

[110] T.V. Nguyen, T.T.H. Nguyen, S.L. Wang, T.P.K. Vo, A.D. Nguyen, Preparation of chitosan nanoparticles by TPP ionic gelation combined with spray drying, and the antibacterial activity of chitosan nanoparticles and a chitosan nanoparticle–amoxicillin complex, Research on Chemical Intermediates 43 (2017) 3527–3537, https://doi.org/10.1007/s11164-016-2428-8.

[111] S. Rossi, B. Vigani, A. Puccio, M.C. Bonferoni, G. Sandri, F. Ferrari, Chitosan ascorbate nanoparticles for the vaginal delivery of antibiotic drugs in atrophic vaginitis, Marine Drugs 15 (2017), https://doi.org/10.3390/md15100319.

[112] K.K. Patel, M. Tripathi, N. Pandey, A.K. Agrawal, S. Gade, M.M. Anjum, R. Tilak, S. Singh, Alginate lyase immobilized chitosan nanoparticles of ciprofloxacin for the improved antimicrobial activity against the biofilm associated mucoid *P. aeruginosa* infection in cystic fibrosis, International Journal of Pharmaceutics 563 (2019) 30–42, https://doi.org/10.1016/j.ijpharm.2019.03.051.

[113] K.K. Patel, A.K. Ashish, M. Anjum, M. Tripathi, N. Pandey, DNase - I functionalization of ciprofloxacin – loaded chitosan nanoparticles overcomes the biofilm – mediated resistance of *Pseudomonas aeruginosa*, Applied Nanoscience (2019), https://doi.org/10.1007/s13204-019-01129-8.

[114] N. Marei, A.H.M. Elwahy, T.A. Salah, Y. El Sherif, E.A. El-Samie, Enhanced antibacterial activity of Egyptian local insects' chitosan-based nanoparticles loaded with ciprofloxacin-HCl, International Journal of Biological Macromolecules 126 (2019) 262–272, https://doi.org/10.1016/j.ijbiomac.2018.12.204.

[115] W. Changxuan, C. Chaode, Q. Xiangyang, W. Jian, L. Cong, L. Ming, L. Xing, J. Dianming, J. Linjun, W. Jun, Superparamagnetic chitosan nanoparticles for a vancomycin delivery system: optimized fabrication and in vitro characterization, Journal of Biomedical Nanotechnology 15 (2019) 2121–2129, https://doi.org/10.1166/jbn.2019.2831.

[116] L. Avérous, Biodegradable multiphase systems based on plasticized starch: a review, Journal of Macromolecular Science: Part C: Polymer Reviews 44 (2004) 231–274, https://doi.org/10.1081/MC-200029326.

[117] D.R. Lu, C.M. Xiao, S.J. Xu, Starch-based completely biodegradable polymer materials, Express Polymer Letters 3 (2009) 366–375, https://doi.org/10.3144/expresspolymlett.2009.46.

[118] A. Reis, M. Guilherme, T. Moaia, L. Mattoso, E. Muniz, E. Tambourgi, Synthesis and characterization of a starch-modified hydrogel as potential carrier for drug delivery system, Journal of Polymer Science 46 (2008) 2567–2574, https://doi.org/10.1002/pola.22588.

[119] F.G. Torres, S. Commeaux, O.P. Troncoso, Starch-based biomaterials for wound-dressing applications, Starch Staerke 65 (2013) 543–551, https://doi.org/10.1002/star.201200259.

[120] N. Masina, Y.E. Choonara, P. Kumar, L.C. du Toit, M. Govender, S. Indermun, V. Pillay, A review of the chemical modification techniques of starch, Carbohydrate Polymers 157 (2017) 1226–1236, https://doi.org/10.1016/j.carbpol.2016.09.094.

[121] U. Shah, F. Naqash, A. Gani, F.A. Masoodi, Art and science behind modified starch edible films and coatings: a review, Comprehensive Reviews in Food Science and Food Safety 15 (2016) 568–580, https://doi.org/10.1111/1541-4337.12197.

[122] J. Chen, L. Chen, F. Xie, X. Li, Drug Delivery Applications of Starch Biopolymer Derivatives, Springer, Singapore, 2019, pp. 41–99, https://doi.org/10.1007/978-981-13-3657-7_4.

[123] N. Poorgholy, B. Massoumi, M. Jaymand, A novel starch-based stimuli-responsive nanosystem for theranostic applications, International Journal of Biological Macromolecules 97 (2017) 654–661, https://doi.org/10.1016/j.ijbiomac.2017.01.063.

[124] F. Zhu, Starch based aerogels: production, properties and applications, Trends in Food Science and Technology 89 (2019) 1–10, https://doi.org/10.1016/j.tifs.2019.05.001.

[125] M. Ziegler-Borowska, Magnetic nanoparticles coated with aminated starch for HSA immobilization- simple and fast polymer surface functionalization, International Journal of Biological Macromolecules 136 (2019) 106–114, https://doi.org/10.1016/j.ijbiomac.2019.06.044.

[126] C. Qiu, R. Chang, J. Yang, S. Ge, L. Xiong, M. Zhao, M. Li, Q. Sun, Preparation and characterization of essential oil-loaded starch nanoparticles formed by short glucan chains, Food Chemistry 221 (2017) 1426–1433, https://doi.org/10.1016/j.foodchem.2016.11.009.

[127] A. Uliniuc, T. Hamaide, M. Popa, S. Băcăiță, Modified starch-based hydrogels cross-linked with citric acid and their use as drug delivery systems for levofloxacin, Soft Materials 11 (2013) 483−493, https://doi.org/10.1080/1539445X.2012.710698.

[128] M.E. Avval, P.N. Moghaddam, A.R. Fareghi, Modification of starch by graft copolymerization: a drug delivery system tested for cephalexin antibiotic, Starch Staerke 65 (2013) 572−583, https://doi.org/10.1002/star.201200189.

[129] C. Thiele, D. Auerbach, G. Jung, L. Qiong, M. Schneider, G. Wenz, Nanoparticles of anionic starch and cationic cyclodextrin derivatives for the targeted delivery of drugs, Polymer Chemistry 2 (2011) 209−215, https://doi.org/10.1039/c0py00241k.

[130] N.S. Ismail, S.C.B. Gopinath, Enhanced antibacterial effect by antibiotic loaded starch nanoparticle, Journal of the Association of Arab Universities for Basic and Applied Sciences 24 (2017) 136−140, https://doi.org/10.1016/j.jaubas.2016.10.005.

[131] A.U. Itodo, I.S. Eneji, T.T. Weor, Chitosan-starch polymeric blend hydrogels as scavengers of antibiotics from simulated effluent: sorbent characterization and sorption kinetic studies, Journal of Chemical Society of Nigeria 43 (2018) 667.

[132] A. Shehabeldine, M. Hasanin, Green synthesis of hydrolyzed starch−chitosan nano-composite as drug delivery system to gram negative bacteria, Environmental Nanotechnology, Monitoring and Management 12 (2019) 100252, https://doi.org/10.1016/j.enmm.2019.100252.

[133] E.R. Balmayor, E.T. Baran, H.S. Azevedo, R.L. Reis, Injectable biodegradable starch/chitosan delivery system for the sustained release of gentamicin to treat bone infections, Carbohydrate Polymers 87 (2012) 32−39, https://doi.org/10.1016/j.carbpol.2011.06.078.

[134] H. Yasar, D.K. Ho, C. De Rossi, J. Herrmann, S. Gordon, B. Loretz, C.M. Lehr, Starch-chitosan polyplexes: a versatile carrier system for anti-infectives and gene delivery, Polymers (Basel) 10 (2018) 1−21, https://doi.org/10.3390/polym10030252.

[135] M. Vert, Polymeric biomaterials: strategies of the past vs. strategies of the future, Progress in Polymer Science 32 (2007) 755−761, https://doi.org/10.1016/j.progpolymsci.2007.05.006.

CHAPTER 5

Biodegradable Antibiotic Importers in Medicine

GBOLAHAN JOSEPH ADEKOYA • EMMANUEL ROTIMI SADIKU •
YSKANDER HAMAM • SUPRAKAS SINHA RAY • EHIGIE DAVID ESEZOBOR •
BABATUNDE BOLASODUN • WAKUFWA BONEX MWAKIKUNGA •
OLUWASEGUN CHIJIOKE ADEKOYA • JIMMY LOLU OLAJIDE •
OLADIPO FOLORUNSO • OLUSESAN FRANK BIOTIDARA • ABAYOMI AWOSANYA •
AHAMDU GEORGE APEH • EBIOWEI MOSES YIBOWEI • UGONNA KINGSLEY UGO •
OMONEFE JOY ODUBUNMI • OMONDI VINCENT OJIJO •
KEHINDE WILLIAMS KUPOLATI • OLUYEMI OJO DARAMOLA •
IDOWU DAVID IBRAHIM

1 INTRODUCTION

A significant threat that invariably endangers human existence is bacterial infections that often lead to severe diseases that include tuberculosis pneumonia, meningitis, sepsis, and cholera. In addition, the source of major clinical failures of medical implants and devices, such as artificial hip, dental, and knee implants, caters; food and water contamination is attributed to bacteriological infections. Bacterial infection on either animate or inanimate surfaces poses a significant risk on the quality of life, specifically on human health with substantial economic consequences. Their ability to form biofilms on biotic and abiotic surfaces further worsens their looming threat against human health [1]. Over 700,000 people were reported dead from bacterial infections [2]. Moreover, the economic impact of antimicrobial resistance AMR (from any microbe) on public health is estimated to range from US$21,832 per case to over $3 trillion in GDP loss between 2013 and 2015 [3]. A cumulative cost of US$100 trillion will be incurred globally from bacterial infection by 2050 if the current trend of the problem persists [2].

However, to address these prevailing threats and prevent massive loss in human lives and global economic capital, antibiotics are continuously researched, developed, and administered to combat the growing threat of unfriendly bacteria. They are employed in treating subjects suspected, infected, or at the risk of infection. These antibiotics exist as either natural or synthetic organic compounds with a propensity to inhibit or annihilate target bacterial at low concentrations [4]. There are different classes of antibiotics, including β-lactam antibiotics (e.g., penicillins and cephalosporins), tetracyclines, macrolides, sulfonamides, fluoroquinolones, and aminoglycosides. Although these drugs are useful for mitigating the growth and spread of diseases caused by microbes, excessive and ineffective use of antibiotics have even led to a more challenging situation with the evolution of antibiotic-resistant bacteria. Their resistance to conventional antibiotic therapies emerged from their ability to adapt to a variety of different ecological niches, their intrinsic immune responses to antimicrobial drugs, and their ability to rapidly acquire resistance mechanisms such as generation of biofilms, development of persister cells, sequestration of antibiotics in the bacterial periplasm, and reduction of bacterial metabolic activity [5–7]. Antibiotic-resistant bacteria cause chronic infections that threaten human survival; however, some critical solutions to addressing this inherent problem involves reducing the dosage of antibiotics while increasing their bioavailability concurrently [2] and combining different antibiotics for practical disruptive impact against the formation of biofilm resistance. Enzymatic disruption, sugar metabolites, D-amino acids, and activation of the quorum-sensing system are among the few ways to prevent the formation of stable biofilm and disruption of established ones [8].

Antibiotic Materials in Healthcare. https://doi.org/10.1016/B978-0-12-820054-4.00005-7

In the case of wound healing, treatment of bacterial infection involves complex coordination among cells of diverse types in the different microenvironments (e.g., pH or bacterial enzymes). High protease activity, for instance, including persistent infection, hypoxia, and excess inflammation often hinders the rapid healing of chronic wounds. Immense research has been conducted to find new and better ways to tackle and manage severe skin wound infections, especially chronic wounds. Four separate phases are identified in the wound healing process, which includes hemostasis, inflammation, proliferation, and remodeling. Nevertheless, with special consideration given to friendly bacterial in vivo and the intricacy of wound healing, it is imperative to develop antibiotic carriers with a unique mechanism to target pathogenic infections, disrupt the formation of resistant biofilm, control release of drugs, and accelerate wound cure while minimizing side effects. To meet these needs, antibiotic biopolymer importers loaded with drugs or antibiotics and having high efficacy, tunable properties, and better release kinetics are of primary interest.

Antibiotic polymers are an essential class of materials that play a vital role in the delivery of antibiotics to inhibit or annihilate target bacterial and similar microbes. These polymers are employed as a potential matrix/carrier to control the release of antibiotic and for antibacterial treatment, particularly in drug delivery and in the case of wound care where absorbent functionality, moisture permeability, and nonallergenic property of the wound dressing materials for the treatments of dry and moist wounds are essential. They either have inherent antibacterial properties (with moieties such as quaternary ammonium groups, polycations, phosphonium groups, antimicrobial peptides (AMPs), and their synthetic mimics) or have their structures changed (conjugated/embedded with antibiotic) to show antibacterial abilities. However, these carrier systems are composed of petroleum-based polymers. The problem with fossil-based polymers is that it must be retrieved from the body after serving its intended purpose. A promising sustainable alternative to petroleum-based antibacterial polymers is the biodegradable polymers because of the additional benefits they provide.

Biodegradable polymers have an essential advantage of being bioassimilable and expunged after serving their intended function. They are mimic after antimicrobial peptide (AMP). Polysaccharides, proteins/polypeptides, polyurethanes, and polyesters make up significant classes of the biodegradable-antibacterial polymer. Understanding the properties of these polymers are highly essential in the development of carrier systems with desirable biocompatibility, stability, and degradable properties, as well as chemical and physical functionalities. This chapter supplies information on the properties of two major classes of biodegradable antibiotic polymers, namely biobased/natural antibiotic polymeric importers and synthetic (nonbiobased) biodegradable antibiotic polymeric importers. It explains the recent development around the use of these materials, including being loaded with antibacterial agents for multifunctional applications.

2 BIODEGRADABLE ANTIBIOTIC IMPORTERS

Bacterial infections pose severe challenges to human health. Antibiotic resistance bacteria, such as *Carbapenem-resistant Pseudomonas aeruginosa*, *Enterobacteriaceae*, and *Carbapenem-resistant Acinetobacter baumannii*, are among the deadly enlisted pathogens by WHO [8a]. However, from the different drug-release systems developed for combating these infections, the polymer-based system has garnered immersed medical research, specifically, antimicrobial biopolymers carrier system. This is because biodegradable polymers offer favorable benefits it provides over other carrier platforms that include biocompatibility, biodegradable, sustainable, cheap, nontoxic, and safe approach to drug delivery and wound dressing. Biodegradable antibiotic polymeric materials have proven effective in treating critical infections of the nervous system, wound, burn, bloodstream urinary, and respiratory tracts [9]. Other benefits include high encapsulation efficiency of antibiotic agents, prevention of drug against physiochemical degradation, tunability of their physiochemical properties, control over polymer matrix degradation, versatility, and potency to control bacterial infections from food packaging, water treatment systems, biomedical devices, and cosmetic products with the ability to improve wound healing caused by pathogens [10]. Most of these degradable and biodegradable biopolymers contain hydrolyzable linkages, viz: ester, urethane, anhydride, orthoester, amide, urea, and carbonate, in their backbone chains [11].

2.1 Classification of Biodegradable Antibiotic Importers

Biodegradable antibiotic importers are biodegradable and biocompatible carrier systems employed in delivering antibiotics in vivo and in vitro to target pathogens. They can be classified based on origin, chemical composition, and structure. Based on the origin of the material, they are classified into biobased or biopolymers or natural polymeric importers (biologically derived polymers) and nonbiobased or synthetic biodegradable polymeric importers (Fig. 5.1). However, based on chemical composition, they are grouped into intrinsic antimicrobial biomaterials and antibiotic-containing biomaterials. Antibiotic-containing importers are further classified into conjugated and encapsulated antibiotic biodegradable Polymers. Based on physical structure, there are polymeric micelles, vesicles, and hydrogels (macrogels, microgels, and nanogels).

3 BIOBASED ANTIBIOTIC POLYMERIC IMPORTERS

These are bioassimilable polymeric materials obtained from renewable sources such as biomass of plant and animals and are used to transport antibiotics to target bacteria. Polysaccharides, protein, biobased polyesters, and biobased polyurethanes are the major classes of biobased polymers for the delivery of antibiotics.

3.1 Polysaccharides-Based Antibiotic Importers

Polysaccharides are biopolymers composed of long chains of monosaccharide units linked through glycosidic bonds. Sugars are monosaccharides or disaccharides while polysaccharides are macromolecules with a larger number of repeating units [12]. Based on functionality, they are grouped as structural (e.g., cellulose

and chitin), storage (e.g., the starch in plant and glycogen in animal) [13], and gel-forming polysaccharides [14]. Moreover, they have a broad spectrum of structures ranging from linear to highly branched. Polysaccharides are insoluble in water and heterogenous in structure due to the high number of hydrogen bonds and a slight modification in the repeat unit. However, when the monosaccharide units of the polysaccharide are the same in structure, homopolysaccharide or homoglycan is formed. Polysaccharides are an abundant source of biopolymer for medical applications. Owing to their biodegradability, availability, and property tuneability, they are suitable materials for antimicrobial biobased materials. The essential classes of polysaccharides for drug delivery and wound care include starch, cellulose, chitin, chitosan, alginate, carrageenan, pectin, cyclodextrin, pullulan, heparin, and agarose.

3.1.1 Starch

Starch is an interesting polysaccharide due to its cost competitiveness, abundance, biodegradability, biocompatibility, and its renewable source. It is tasteless, white, and odorless powder, sourced mainly from photosynthetic tissues and various plant storage organs such as seeds (maize, wheat, oat, millet, barley, sorghum, and rice), tubers, roots (e.g., potato, yam, cassava, canna, and cocoyam), and fruits. Naturally, starch is hydrophilic granules composed of two main polysaccharides, namely linear water-soluble amylose (which forms about 20% of the starch mass and may contain some

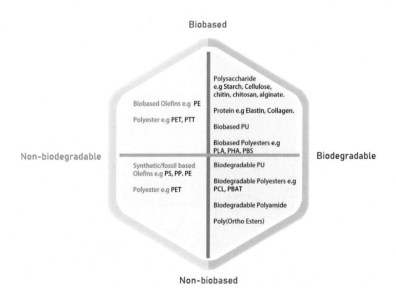

FIG. 5.1 Classification of biodegradable polymeric antibiotic importers.

α(-1,6-) branching point) and branched water-insoluble amylopectin (about 80% of the mass of starch) [15] (Fig. 5.2) with minute compounds such as protein and lipids. The properties and characteristics of starch, specifically the molecular structure of its granule including the shape, size composition, constituent, and biosynthetic mechanism varies among starches of different botanic sources. Native starch granules lack inherent antimicrobial properties and the propensity to be processed directly into polymer carriers for antibacterial delivery. This weak mechanical property is due to the inherent low share resistance, retrogradation, thermal resistance, and degradation [16]. To overcome these challenges, starch granules are processed into useful polymer matrix with the addition of fillers such as metallic nanofillers, for example, copper nanoparticle (CuNP) [17] or by blending with other biopolymers like poly(lactic acid) (PLA) for enhanced mechanical strength [14].

Moreover, for antibacterial treatment and wound dressing, starch is chemically modified by conjugating antibacterial moieties into the chemical backbone chain of the polymer. For instance, cationic groups can be introduced by graft copolymerization, etherification, or a combination of both [14]. Alternatively, they can be embedded with antimicrobial agents for delivery onto target infections [18,19]. As a result, owing to the inherent limitation of starch for antimicrobial purposes, Zhijun and colleagues prepared different ratios of a ternary blend of starch/polyvinyl alcohol (PVA)/citric acid film. $S/P/C^{3:1:0.08}$ and $S/P/C^{3:3:0.08}$ blends displayed effective antimicrobial activity against the food-borne pathogenic bacteria—*Listeria monocytogenes* and *Escherichia coli* [20]. Shaghayegh and colleagues developed hydrogel based on PVA, starch (St), and chitosan (Cs) with the incorporation of nano zinc oxide (nZnO). The resulting hydrogel membrane displayed in vivo wound healing effect, which makes it a suitable bandage

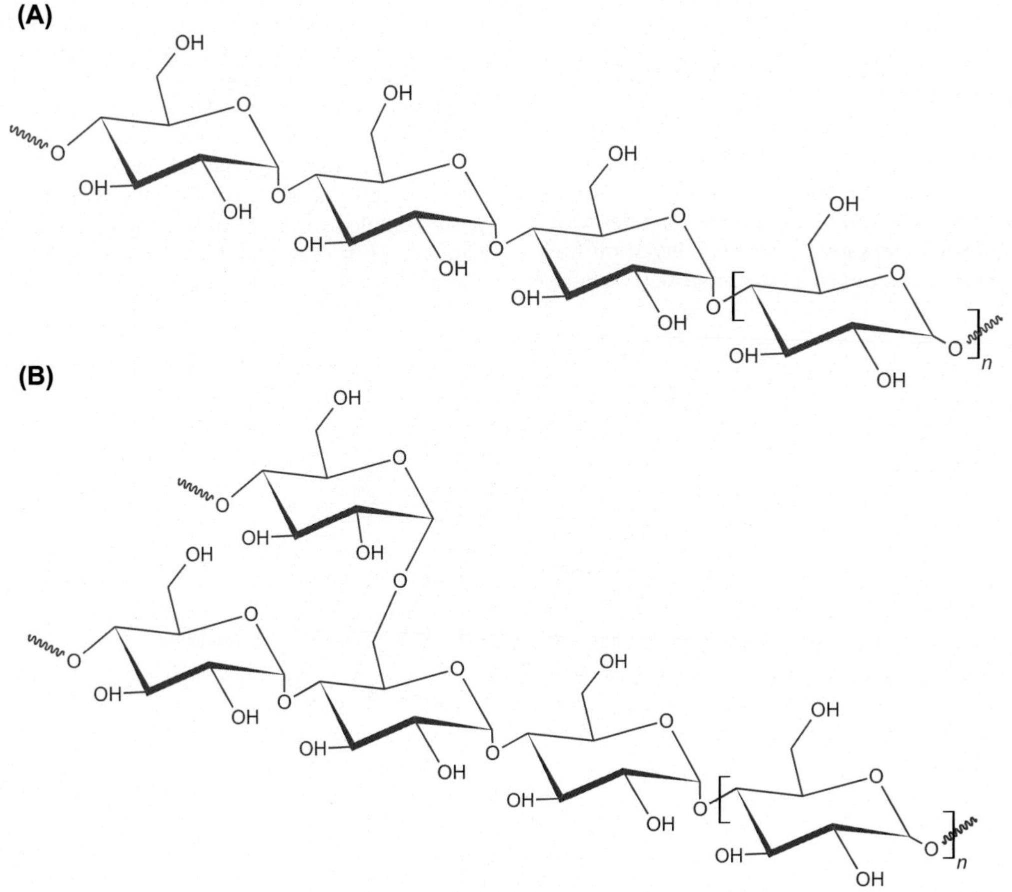

FIG. 5.2 Chemical structures of **(A)** Amylose and **(B)** Amylopectin.

material for wound dressing [19]. In contrast, for a much higher potential, antibiotics can be nanostructurally hybridized and encapsulated to the polymer matrix. Consequently, hydrogels based on cassava starch was developed and loaded with a hybrid of silver nanoparticles (AgNPs) and cysteamine (cysa) for controlled release systems against *Escherichia coli* and *Staphylococcus aureus*. The inhibiting activity of the nanostructured hybrid systems was found to be most effective against *E. coli* than a single antibiotic system [21].

The incorporation of the antibiotic nanoparticle is proven as an effective strategy to improve the mechanical properties and the antibacterial efficacy of starch against resistant pathogens. Nor [22], on the one hand, encapsulated different amounts of antibiotics into a soluble form of starch nanoparticle (SNP) to target *Streptococcus pyogenes*. The starch was prepared by dissolving it in urea alkaline solution and precipitated in the ethanolic emulsion to produce a nonclumping nanoparticle with a larger surface area. The antibacterial property displayed by the SNP was highly effective for drug delivery. On another hand, the concentration of essential oil was found to influence the antimicrobial efficacy of starch. Majid and colleagues [23] investigated the antimicrobial effect of corn starch bioactive film against *Escherichia coli*, *Staphylococcus aureus*, *Listeria monocytogenes*, and *Pseudomonas aeruginosa*. The bioactive film was prepared with *Zataria multiflora* and *Bunium persicum* essential oils having a concentration ranging from 1 to 20 mg/mL. The concentration at 20 mg/mL for *Zataria multiflora* and *Bunium persicum* essential oils show the most effective antibacterial activity against *S. aureus* with an inhibition zone of 31.3 and 22.36 mm, respectively. However, the effect on *E. coli* was less pronounced due to the high resistance of the bacteria against the oils. The inhibition zones recorded for *E. coli* for *Zataria multiflora* and *Bunium persicum* were 26.13 and 19 mm, respectively. In a recent study, the orange essential oil was used to prepare potato starch bioactive having high antimicrobial activity against *S. aureus and L. monocytogenes* [24]. However, while essential oil increases the antimicrobial efficacy of starch film, it also has a detrimental effect on the mechanical properties. Potato starch–based bioactive composite films prepared using three antimicrobial agents, clove oil, cinnamon oil, and potassium sorbate with a constant concentration of zinc oxide nanoparticles were reported to exhibit antibacterial activities that depend on the concentration of the antimicrobial agents. However, increasing the concentration of the antimicrobial agents led to a decrease in the tensile strength of the films [25].

3.1.2 Cellulose

Cellulose is a linear homopolysaccharide of β-(1,4)-D-glucose units, as shown in Fig. 5.3. It is one of the most abundant biomaterials on the earth [26] and makes up the main component of the plant cell wall. It forms as microfibrils that encapsulate the cell with a mesh-like structure, and controls, along with hemicellulose, pectin, and lignin, the mechanical properties of the plants [12]. These sturdy structural properties of cellulose are as a result of the β-(1,4) that forms a rigid and stable bond. Each chain bonds to the neighboring chain through intrachain (intramolecular) and interchain (intermolecular) hydrogen bonding and firmly holds the chain side by side forming crystalline microfibrils with high mechanical strength. Cellulose is mainly synthesized by the plant. However, some bacteria such as *Acetobacter xylinum* also synthesize cellulose. It is synthesized by cellulose synthase, a complex enzyme that forms the cell membrane in plant and membrane protein in the cell wall of bacteria. The enzyme catalyzes the direct polymerization of glucose from the intermediate substrate, Uridine Diphosphate-glucose into cellulose.

Cellulose obtained from bacteria is purer than the plant-based cellulose. It also exhibits better wáter retention properties of about 1000% of the cellulose weight. Plant cellulose, on the other hand, exhibits only about 60%. Owing to the superior water retention property of bacterial-based cellulose, it is used for medical applications [26]. Other such applications include biobased edible antimicrobial packaging systems [27] for food, filtration, pregnancy tests, cosmetics, and healthcare due to its nontoxicity, low-cost, hydrophilicity, biocompatibility, and biodegradation [28]. Cellulose can be broken down by a β-glucosidase enzyme, also known as cellulase, which attacks the β-linkages. Humans lack this enzyme, therefore, making digestion of cellulose impossible. Some ruminant animals, however, possess symbiotic bacteria in the flora of their rumen, which produces cellulase that degrades cellulose [29].

Cellulose, different from starch, is rigid and exhibits high mechanical strength that makes it processible into films for drug delivery and wound care. However, just like starch, it lacks antimicrobial properties. To enhance

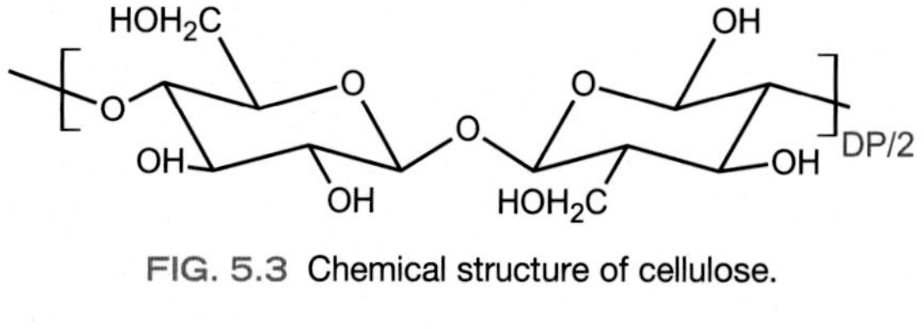

FIG. 5.3 Chemical structure of cellulose.

this property, cellulose is encapsulated or conjugated with antibacterial agents such as metal salts, metal oxides, and essential oils. Cyril and fellow workers [30] explored photosensitizer, a protoporphyrin IX (PpIX) amino derivative to prepare cellulose material with an antibacterial activity that eradicates *Staphylococcus aureus*. It was successfully done by grafting protoporphyrin IX units covalently onto the cellulose backbone chain. This approach renders cellulose as a promising exemplar of biomaterial for application in bacteria decontamination. A similar work functionalized hairy nanocrystalline cellulose with aldehyde groups to promote the adhesion of lysozyme and nisin against *Bacillus subtilis* and *Staphylococcus aureus*. The conjugated sterically stabilized nanocrystalline cellulose effectively retains the activity of the antibacterial agents for a long period [31]. Using this method of immobilizing antibacterial agent unto the backbone chain of the polymer, Zhai and coworkers in (2018) extended the experiment to *Escherichia coli, Listeria monocytogenes, Yersinia enterocolitica, Aspergillus niger, and Saccharomyces cerevisiae* [32].

Although by chemically modifying the chemical structure of cellulose, one can induce biocidal property on the biopolymer against a wide range of bacteria species [33]. Copolymerization and etherification of cationized nanofibrillated cellulose with quaternary ammonium compounds were reported to inhibit Gram-positive and Gram-negative bacteria. This is attributed to the polymer's ability to penetrate the thick cell wall of Gram-positive bacteria after been functionalized, followed by the destabilization of the cellular membrane with no evidence of cytotoxicity against human cells [34].

Another study adopted a different approach to enhance antibacterial activity against multidrug-resistant infections. Hui and fellow workers prepared cellulose/TiO_2/β-CD hydrogel to evaluate its high photocatalytic antibacterial activity and sustained release of the drug against *Escherichia coli* and *Staphylococcus aureus*. The composite hydrogel displayed high antibacterial efficacy with a complete release of curcumin after 120 h. The only limitation associated with this approach is that the antibiotic activity of the composite is negligible in the absence of light [35].

The incorporation of antibacterial agents such as metallic nanoparticles (for example, CuO, TiO_2, ZnO, Al_2O_3, SiO_2, Fe_2O_3, and CeO_2) is no longer an uncommon approach to induce antibacterial properties in noninherent antimicrobial biomaterial like cellulose. For instance, antibacterial hybrid nanomaterials were prepared by in situ syntheses of silver (Ag), copper oxide (CuO), or zinc oxide (ZnO) nanoparticles (NPs) during regeneration of cellulose from microcrystalline

cellulose and cotton linter. The nanoparticles imparted thermal stability with potent antibacterial activity against *E. coli* and *L. monocytogenes* [36]. Silver nanoparticles (AgNPs) [37−39] are known to exhibit good biocompatibility and antimicrobial properties. However, the extensive use of Ag NPs possess some severe threats to the environment and specifically human health. Aqueous Ag NPs often degenerate into ions when left for a more extended period. These ions increase cell toxicity with an increasing quantity of silver ions in the dispersion. Studies show that Ag NPs display toxic side effects, In vitro and in vivo, in mammalian cells found in the brain, skin, lung, liver, vascular system, and other tissues. Therefore, in place of Ag NPs, Au NPs have been shown to exhibit much higher biocompatibility without increasing cell toxicity. Consequently, this promising property has lent Au NPs its extensive use in the development of new drug delivery systems [28]. A recent study used silver-coated gold nanoparticles (Au−Ag NPs) [28] composited with cellulose for the development of antibacterial drug-delivery systems.

Bacterial cellulose is highly hydrophilic, which makes it a desirable biomaterial for wound care by providing the constant moisturizing effect to the wound. It is biocompatible with living tissues and possesses excellent mechanical properties with a porous structure that mimics the extracellular matrix of the skin. In addition to its ability to hold incorporated drugs, other modifiers such as silk sericin that has cytoprotective and mitogenic effects [40], collagen [41], and gelatin [42] can modulate its properties for an enhanced antimicrobial activity for rapid wound healing [43].

3.1.3 Chitin and chitosan

Chitin, a linear biopolymer of β1,4-linked N-acetylglucosamine [44] (Fig. 5.4A), is made up of repeat unit of aminomonosaccharide. It is a polysaccharide found in the exoskeleton of insects and crustaceans, such as crab, lobster, krill, prawn (*Macrobrachium nipponense*) [45], and shrimp (*Parapenaeus longirostris*) [46] shells, as well as the cell wall of yeasts and fungi. N-acetylglucosamine structure of chitin is similar to the β-(D)-glucose structure of cellulose with acetylated amino group replacing the hydroxyl group on the C_2 of glucose [47]. Chitosan, however, a deacetylated derivative of chitin [48] (Fig. 5.4B), is the second most abundant polysaccharide after cellulose. It is a linear biopolymer with its structure made up of the random distribution of β-(1 → 4)-glycosidic linkage of D-glucosamine (deacetylated unit) and N-acetyl-D-glucosamine (acetylated unit). Moreover, it can be synthesized from

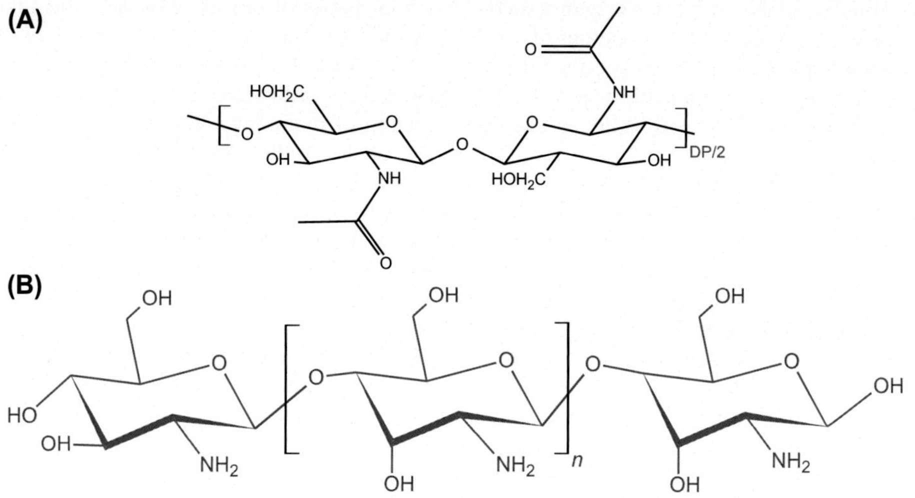

(A)

(B)

FIG. 5.4 Chemical structures of **(A)** chitin and **(B)** chitosan.

chitin through partial or complete deacetylation using alkaline compounds such as sodium hydroxide, NaOH. Different from starch, chitosan has inherent antimicrobial properties that qualify it as a suitable biopolymer for medical applications for drug delivery. This is due to the presence of an amino-functional group present in the structure of chitin and chitosan that grants them several interesting properties such as antibacterial activity and antitumor activity [49]. Consequently, it is mostly applied for wound dressing as bandages to reduce bleeding and inhibit the growth of bacteria and fungus. Moreover, it has an antimicrobial effect against a broad range of microbes. For instance, it is reported to reduce the rate of infection of *S. aureus* [18]. Chitosan can realize this function due to the protonation resulting from the interaction of the amino group in its chemical structure with an organic acid such as lactic acid. Therefore, in a neutral solution, chitosan becomes positively charged hemostatic agent, which then binds electrostatically with the cell membrane of erythrocytes (negatively charged microbes). First, it separates the cell wall from the cell membrane and reacts with the anionic phosphate group of phospholipids present on the bacterial cell membrane. This alters the permeability of the cell membrane, causing the leakage of it and eventual annihilation of the bacteria [50]. In addition, it is able to interact with DNA to inhibit the synthesis of RNA [14,18].

Furthermore, the introduction of functional groups such as alkyl or carboxymethyl groups can enhance the solubility of chitosan at neutral and alkaline pH microenvironment without altering its cationic behavior. Functionalization of chitosan is imperative for the

synthesis of a new bifunctional biopolymer. This is essential for retaining the inherent physicochemical and biochemical properties while introducing new, improved properties without changing the backbone structure of chitosan. Depending on the functional group introduced, a broad spectrum of chemical and biological properties can be obtained, ranging from osteoinductive, bactericidal, bioabsorbable, biocompatible, and chelating metal properties. Phosphorylated functionalized chitin and chitosan have been prepared by reaction of chitin or chitosan with orthophosphoric acid and urea in N,N-dimethylformamide (*DMF*), with phosphorous pentoxide in methane sulfonic acid, with phosphorous acid and formaldehyde, with diethyl chlorophosphate, NaOH, *n*-hexane and with H3PO4/Et3PO4/P2O5/hexanol. Besides, chitosan conjugated with α-aminophosphonates was reported to be used as inhibitors of serine hydrolase, enzyme inhibitors, herbicidal, antiviral, antibacterial, peptide mimics, antifungal, anti-HIV, anticancer, antibiotics, and in several other applications [51].

Chitosan can also be loaded with antibacterial agents such as silver nanoparticles [52], cinnamaldehyde, and sulfur nanoparticles (SNP) [53] for target delivery to infectious pathogens. Chitosan nanoparticle was conjugated with antitransferrin and antibradykinin B2 antibody for effective delivery of siRNA through the blood−brain barrier to target human immunodeficiency virus (HIV)-infected brain astrocytes and to inhibit the replication of HIV in the neural system [54]. A different author blended chitosan with PLA to serve as a suitable carrier for lamivudine−an inhibitor for type 1 and type 2 HIV [55]. Although chitosan is susceptible and biodegradable by human enzymes such as lysozyme, however, to enhance the properties of chitosan for better mechanical strength, the specific physicochemical, biological, and degradation properties present an efficient way to blend with other synthetic biodegradable polymers.

In this regard, the cross-linking of chitosan plays a vital role in the control of drug release to target microbes. Other factors that affect the elution of antibiotics from chitosan include the composition and concentration of drugs carried and the size of the implant. Cross-linked chitosan proves to take extended time in the delivery of antibiotics when compared to noncross-linked counterparts. Cross-linked chitosan loaded with gentamicin took 8 weeks to be delivered with high efficacy in vivo. Cross-linking helps to prolong the degradation of this biodegradable polysaccharide. In general, owing to the multifunctional ability of chitosan-based biodegradable materials, they have shown high antibacterial effectiveness against gram-negative and gram-positive bacteria [16].

Furthermore, for wound dressing, chitin scaffold based on chitin/silk fibroin/TiO2 nanocomposite with interconnected microporous biodegradable and biocompatible property can be explored. It was reported to be suitable for fast wound healing application, owing to the well-defined interconnected porous construction with high porosity of above 90%, swelling and water uptake (of about 93%), and hemostatic potential for higher blood clotting ability. The nanocomposite also exhibited inhibitory activity against the growth of S. aureus, E. coli, and C. albicans [56].

3.1.4 Alginate

Alginate is a linear polysaccharide of (1–4) linked β-D-mannuronic acid (M), and its C-5 epimer α-L-guluronic acid (G) (Fig. 5.5) extracted from Brown algae (such as certain species of *Lessonia*, *Durvillea*, and *Laminaria*) and bacteria (for instance *Azotobacter vinelandii* and *Pseudomonas aeruginosa*). It is a nontoxic anionic biopolymer that can be synthesized through the postpolymerization reaction of mannurate. The reaction involves the inversion of the configuration of uronic acid at epimerases C_5 without disrupting the glycosidic bond. Alginate is a favorable choice of biodegradable polymer for drug delivery and wound care due to its biocompatibility, low cost, biodegradability, and nontoxicity to human cells [16].

Kumar and colleagues [9] recently reported an inexpensive one-step route to synthesize antimicrobial alginate from sodium alginate using a different amount of aminoglycoside. Among the various aminoglycoside, kanamycin disulfate was observed to produce the highest amount and the most malleable biopolymer. Neomycin sulfate equally gave a high yield of biopolymer. Other aminoglycosides employed include gentamicin sulfate, tobramycin sulfate, kanamycin sulfate, and streptomycin sulfate. The researchers also studied the controlled release of antibiotics of the synthesized biopolymer against E. coli. It was observed to be highly effective in inhibiting E. coli through the slow release of antibiotics.

Alginate is a widely used biopolymer in medicine. It was found to be effective against multidrug-resistant infections. Infections associated with gram-negative opportunistic *Pseudomonas* sp. have a severe effect on human health, causing disease ranging from nonhealing skin wounds to chronic respiratory disease. For instance, infection of the lung with *Pseudomonas aeruginosa* in cystic fibrosis often causes inflammation and deterioration of lung function, increasing morbidity, and mortality [7]. The infection starts with the formation of a nonmucoid *Pseudomonas aeruginosa* that later plummet into a mucoid phenotype with excessive

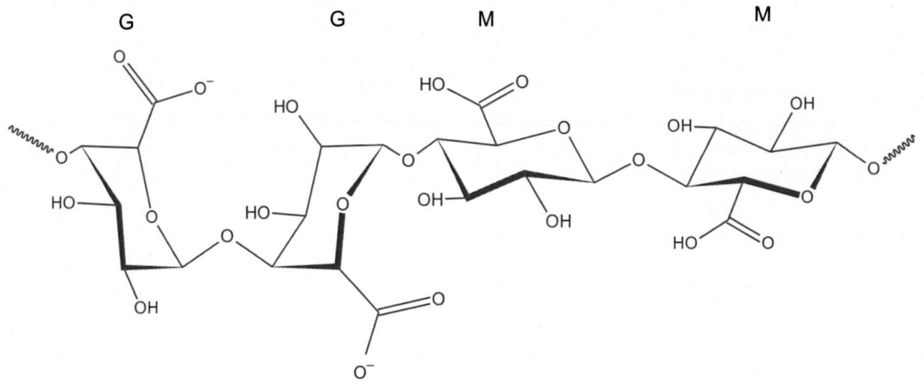

FIG. 5.5 Chemical structure of alginate.

production of exopolysaccharide alginate through the mutation in *mucA*. This bacterial extracellular polymeric substance (EPS) is composed of polysaccharides, lipid proteins, and nucleic acids, which is responsible for biofilm formation and maturation [57,58]. The bacterial achieve this biofilm formation through the production of two polysaccharides, namely Pel and Psi. Pel is a positive exopolysaccharide that contains partially acetylated 1−4 glycosidic linkage of N-acetylglucosamine and N-acetylgalactosamine sugars [59] while Psl, a neutral pentasaccharide is made up of D-mannose, L-rhamnose, and D-glucose [59,60]. Within this well-formed biofilm, the bacteria show much higher resistance to drugs. However, with the administration of a low molecular weight alginate oligomer (Mn = 3200 $gmol^{-1}$) to the biofilm structure of mucoid *Pseudomonas aeruginosa*, a significant reduction in the structural quantities of EPS polysaccharides, and extracellular (e)DNA was observed, thereby establishing antibiotic efficacy against mucoid *Pseudomonas aeruginosa* [7]. This is because alginate is a negatively charged biopolymer composed of guluronic and mannuronic acid [59,61]. However, in terms of stability, viscosity, and drug release, 0.625% was found as the best alginate concentration for the formation of hydrogel carrier [62].

Alginate has not only proven effective against multiresistant drugs but has also gained attention in preparing scaffold for wound dressing and tissue regeneration. The inclusion of silver nanoparticles (AgNPs)-doped collagen in alginate further enhanced the antibacterial activity of the biocomposite against *Staphylococcus aureus* and *Escherichia coli* and improved its propensity for wound dressing [63].

3.1.5 Carrageenan

Another important biopolymer of the family of polysaccharides is the carrageenan. They are gel-forming and viscose polysaccharides, which are obtained by extraction from certain species of red seaweeds (Rhodophyta). It is a linear hydrophilic sulfate polysaccharide composed of alternating units of D-galactose and 3,6-anhydrogalactose (3,6-AG) joined by α-1,3 and β-1,4-glycosidic bond. There are six basic types of Carrageenan namely kappa (κ)-, iota (ι)-, lambda (λ)-, nu (υ)-, mu (μ)-, and theta (θ)-carrageenan (Fig. 5.6) [64]. κ-carrageenan is often employed as an antibacterial material with additional properties such as to improve the moisture barrier and general tensile properties of a bio blend, although except elongation property. Although ι-carrageenan often impaired most of these properties [65], λ-carrageenan is used as a potent inhibitor of many

viruses. However, owing to the general multifunctional properties, namely antiviral, antitumor, immunomodulatory, antihyperlipidemic, and anticoagulant, carrageenan is employed in a medical application specifically for wound care and virus inhibition.

Luo and colleagues investigated the potential inhibition role of λ-carrageenan (P32) against Rabies virus (RABV) infection. It was found to exhibit antiviral behavior by inhibiting viral internalization and glycoprotein-mediated cell fusión. This makes λ-carrageenan a promising biomaterial for the development of novel anti-RABV drugs [66]. Recently, Diogo and coworkers extended the application of λ-carrageenan to the inhibition of animal viruses belonging to the *Alphaherpesvirinae* subfamily, SuHV-1 (suid herpesvirus type 1) strain Bartha and BoHV-1 (bovine herpesvirus type 1) strain Cooper [67].

The most essential attribute of κ- and ι-type carrageenans are gel formation [68]. Through chemical modification, the properties can be enhanced for antibacterial application. For instance, κ-carrageenans can be oxidized through hydrogen peroxide and copper sulfate redox system to exhibit a detrimental effect on the bacterial cell wall and cytoplasmic membrane. This enhanced the antibacterial activity of oxidized κ-carrageenans can effectively suppress the growth of both gram-negative and gram-positive bacteria [69]. Blending with PVA through ionic cross-linking in an aqueous medium with Borax (Bx) and K+ ions at room temperature has the potential to display nonthrombogenic and noncytotoxic behaviors. With minocycline drug loaded into carrageenans film, the film showed excellent antimicrobial behavior [70].

Alternatively, antibacterial agents such as metallic nanoparticles (such as ZnO and CuO nanoparticles [71]) and essential oils can be encapsulated into κ-Carrageenan-based films. A study reveals that a concentration as low as 2% Cinnamon essential oil will produce an inhibitory zone of about 1.59 and 12.64 mm against *Escherichia coli* and *Staphylococcus aureus*, respectively [72]. Nouri and colleagues in 2018 used rosemary extract in place of cinnamon oil with the inclusion of 3% nanoclay. The resulting carrageenan film with 3% extract exhibit >99% inhibition against *E. coli*, *B. cereus*, *S. aureus*, and *P. aeruginosa* [73].

Meanwhile, Roy and colleagues prepared a composite of Carrageenan and zinc oxide nanoparticle (Carr/mZnONP) by reacting zinc acetate with KOH the presence of melanin, a capping agent. The biocomposite film developed shows higher thermal stability with better mechanical and water vapor barrier properties. Although the transparency of the film was slightly

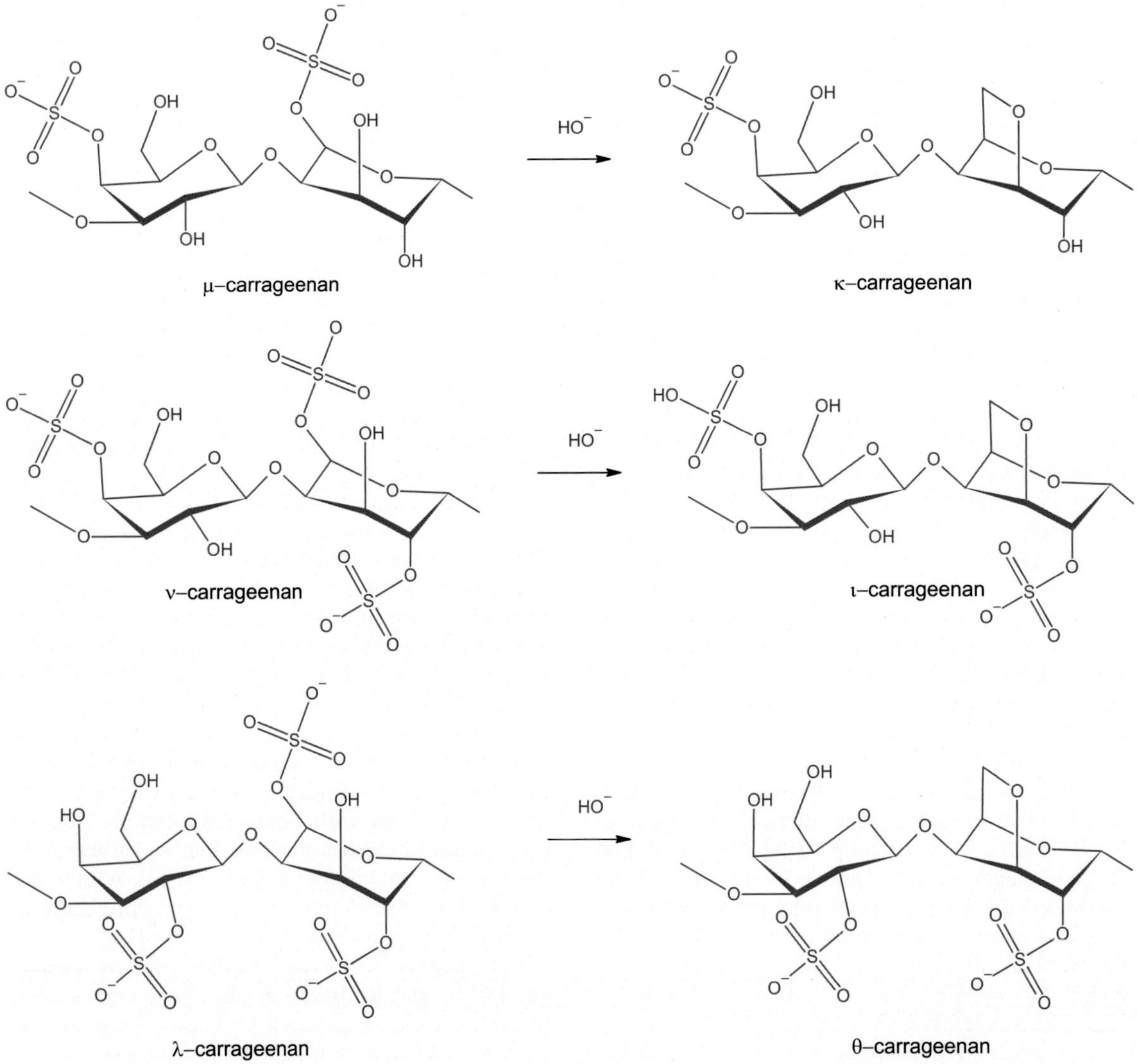

FIG. 5.6 Chemical structures of kappa (κ)-, iota (ι)-, lambda (λ)-, nu (υ)-, mu (μ)-, and theta (θ)-carrageenans.

impaired, the ultraviolet (UV) blocking property was improved with the exhibition of intense antibacterial activity against gram-negative bacteria, *E. coli*; however, the antibacterial activity against gram-positive bacteria, *L. monocytogenes* was less pronounced [74,75].

3.1.6 Pectin

Pectin is a heterogeneous anionic polysaccharide extracted mainly from citrus and apple pomace. The citrus pectin such as orange peel and grapefruit contain about 30% of wet pectin while the later-pomace has about 15% of wet pectin [76]. Pectin is made up of poly α-(1−4)-galacturonic acids (Fig. 5.7) found in the cell wall of most higher plants and possesses good gas permeability properties and gel-forming capability. The structure of pectin varies in composition based on the source and environmental factors such as conditions of extraction and location. Generally, it is composed of homogalacturonan with two types of alternating highly branched rhamnogalacturonans. Xylogalacturonans, arabinogalactans, and arabinans formed other classes of structural classes of pectin [77]. Owing to its biodegradability, biocompatibility, edibility, and bioavailability, pectin makes a suitable

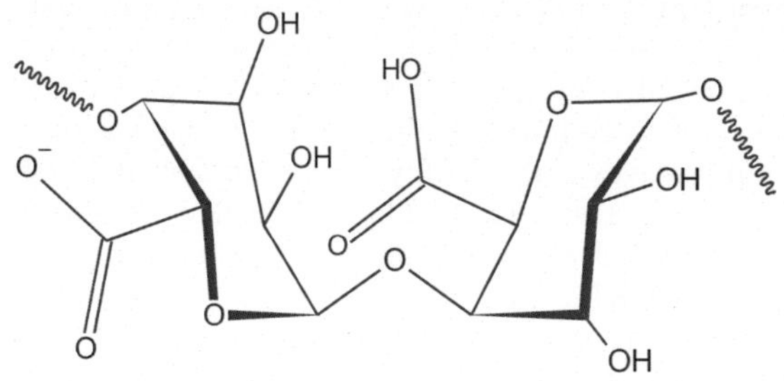

FIG. 5.7 Chemical structure of pectin.

matrix for the delivery of antibiotics and accelerated wound healing. In addition, pectin is usually cross-linked with ions such as divalent ions (Ca^{2+}, Mg^{2+}, Zn^{2+}), glutaraldehyde, and diamines to impact antimicrobial activity in the polymer and improve its stability in water and the physical properties [78]. It has vast applications, including pharmaceutics, food packaging medical products, bioimplants, cosmetics, tissue engineering, herbicides, textile, and coatings [14].

The bioactivity of pectin and pectin hydrolysates extract from local fruit, *Spondias dulcis*, were investigated against four strains of *Salmonellae*. Three of the extracts namely crude pectin pH 2.5, pH 4, 12-h hydrolysate, and pH 4, 1-h hydrolysate had minimum bactericidal concentration (MBC) and minimum inhibitory concentration (MIC) against all four Salmonellae strains with MBC ranging from 11.36 to 44.45 μg/mL and MIC from 5.68 to 44.45 μg/mL. Three treatments, *viz*: the pH 4−12 h, hydrolysate at 10 mg/kg and 5 mg/kg, and the pH 4−1 h, hydrolysate at 10 mg/kg, had therapeutic influence against *Salmonella* infection in vivo. In contrast, just one of the extracts, namely crude pectin pH 2.5, was active against all the *Salmonellae* through well diffusion with growth inhibition varying from 12 to 15 mm at 100 μg/m [76]. Pectin coating is also reported to demonstrate the antimicrobial effect on *Campylobacter jejuni* infection [79].

Pectin can be grouped as high methoxyl or low methoxyl pectin based on the degree of esterification with methanol [80]. The degree of esterification of pectin is an important factor to be considered during the loading of antibiotics on the pectin capsule. Hui Wang and fellow workers in 2017 studied the effects of the degree of esterification of pectin on Nisin-loaded encapsulation antibiotic properties. It was observed that low methoxyl pectin nanocapsules prepared had higher loading capacity with smaller particle size and encapsulation efficiency, while high methoxyl pectin nanocapsules display opposite effect [81] *Listeria monocytogenes*. A novel biodegradable film based on

bacterial cellulose (BC) modified with high methoxylated pectin (HMP) (from 0.1% to 2.0%) and human serum albumin (HSA) BC−HMP−HSA was developed for drug delivery. The films were encapsulated with an antibiotic, levofloxacin (Levo) having a maximum drug payload of 6.23 mg g^{-1} against *Staphylococcus aureus*. The release kinetics HMP and HSA displayed a hyperbolic profile with sustained drug release [82]. Aside from transporting levofloxacin and nisin antibiotics having broad antibacterial spectrum against a wide range of infections, a cinnamon essential oil also demonstrated significant antimicrobial activity against typical foodborne pathogens namely *Staphylococcus aureus*, *Escherichia coli*, and *Listeria monocytogenes* when processed with pectin biopolymer [83]. After that, Nisar and colleagues in 2018 incorporate clove bud essential oil (0.5%, 1.0%, and 1.5%) into the citrus pectin to develop edible antimicrobial film for food packaging [84]. Recently, pectin biofilm was prepared with β-carotene (0.015%, 0.03%) and *Carum copticum* essence (0.25%, 0.5%) as additives. The resulting biofilm showed effective antibacterial properties against *B. cereus* more than *E. coli*. The effects of orange oil [85] and moringa extracts [86] have also been investigated on the antimicrobial activity of pectin film.

Pectin has also been blended with other polymers to expedite wound healing. This healing effect is due to the physicochemical properties of pectins such as its hydrophilicity that facilitate the expungement of the extrudate; it has acidic pH nature that prevents bacteria and fungi and the ability to bind with bioactive agents and prevent their degradation. Pectin has been used in hydrogels fabricated with gum arabic and Ca^{2+}. The bioinspired hydrogels with essential fibroblast growth factor (bFGF) showed, in vivo and in vitro, significant enhancement of cell proliferation, collagen deposition, wound reepithelialization, and contraction in the absence of any visible toxicity and inflammation as compared to the hydrogels not having bFGF and other commercial wound curing products [87]. Kim and colleagues, on the other hand, blended pectin with polyvinyl alcohol PVA to develop a hydrogel that promotes wound healing [88]. However, recently, Yu and fellow workers reported the synthesis of antibacterial pectin from *Akebia trifoliata* var. australis waste to promote faster healing [89].

3.1.7 Cyclodextrin

Cyclodextrins are cyclic oligosaccharides (Fig. 5.8) with a toroid cone structures having a lipophilic core and hydrophilic outer surface. It is composed of glucopyranose units formed by α (1−4) linkage. It occurs naturally as α (having 6 glucose units), β (7 glucose units), ϒ (8 glucose units) [90], methylated β-cyclodextrin, and sulfobutyl ether β-cyclodextrin. Moreover, they are synthesized from the cyclic starch reaction of linear chains of glucopyranose by cyclodextrin glucosyltransferase. This enzyme is known to produce a mixture of γ-, β-, and γ-cyclodextrins. Cyclodextrins are nontoxic, biodegradable compounds with the potency to complex and stabilize a wide range of substances. Owing to its low toxicity and immunogenicity, cyclodextrin is employed in drug delivery and wound care [91]. The unique cavity structure of cyclodextrin can be imbedded with antimicrobial agents for enhancing solubility, improved stability of the antimicrobial agent and drug-delivery efficiency, minimized local irritation and toxicity and prolonged degradation. For instance, β-cyclodextrin was employed to encapsulate and enhance the aqueous solubility of hydrophobic molecules such as Carvacrol and Thyme essential oil. These are suitable natural antimicrobial and antioxidant agent. By forming host–guest complexes with cyclodextrin through photon interaction, their reactivity and volatility were remarkably reduced. Moreover, they were able to display improved controlled release of antimicrobial agents against the growth of *Escherichia coli* and *Salmonella enterica serovar Typhimurium* at minimal concentration [92,93]. Rakmai and colleagues reported the efficacy of guava leaf oil incorporated in β-cyclodextrin against the growth of *Staphylococcus aureus* and *Escherichia coli* employing a similar mechanism [94]. Blending with another biopolymer can significantly increase the hydrophilicity and mechanical properties of cyclodextrin composites. Liu Y and coworkers encapsulated cinnamaldehyde into a blend of β-cyclodextrin (β-CD), and PLA to synthesize nanofibers composite

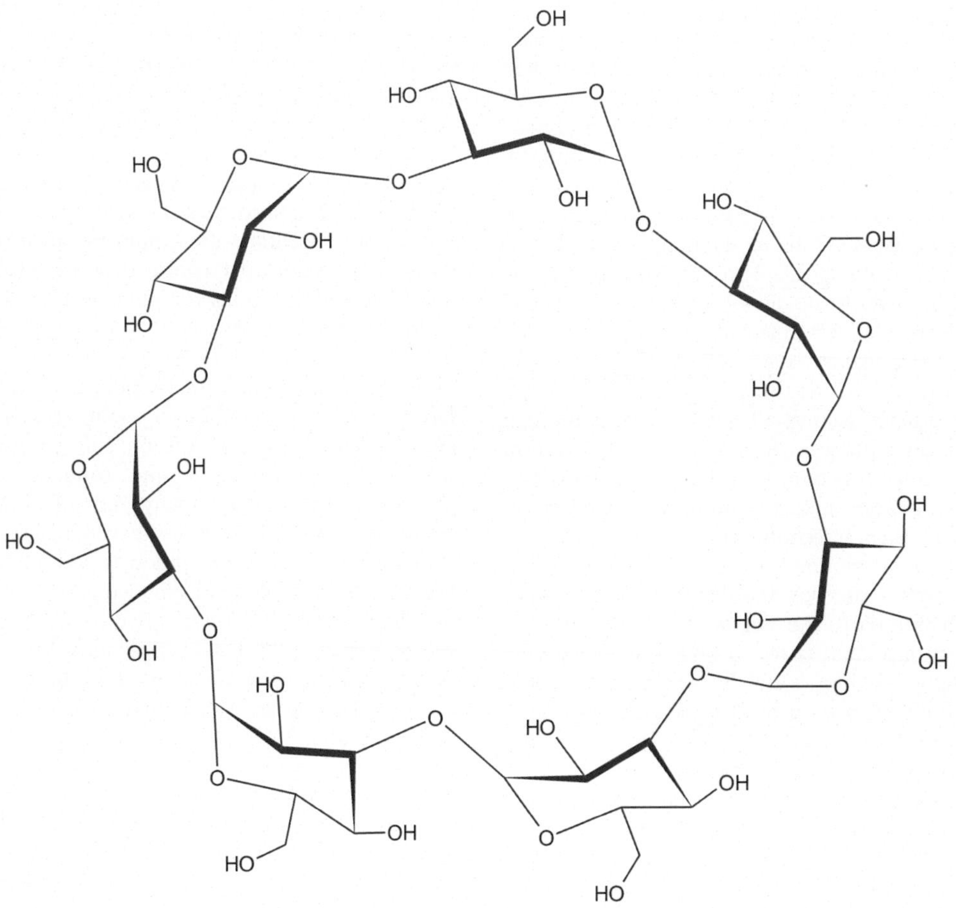

FIG. 5.8 Chemical structure of cyclodextrin.

via electrospinning. The nanofiber displayed low cyto-toxicity with excellent antibacterial effects against *Escherichia coli* and *Staphylococcus aureus* [95].

Additionally, chemical tuning of cyclodextrin structure can also be tailored for various applications such as food preservation, wound healing, and drug delivery. For example, the β-cyclodextrin-carvacrol complexes mentioned earlier could have significant applications in food systems due to their storage stability and enhanced antimicrobial activity [93].

3.1.8 Agarose

Another interesting biopolymer is agarose. It is a linear polysaccharide obtained from seaweed and consisting of 1,3-linked β-D-galactopyranose and 1,4-linked 3,6-anhydro-α-L-galactopyranose repeating units (Fig. 5.9) with an average molecular mass of 120,000 Da. Agarose is gel-forming macromolecules involving a porous rigid network that has high gel strength at low concentration and employed for drug and cell encapsulation [12]. Agarose can be chemically modified to enhance its antibacterial efficacy against gram-positive and gram-negative bacteria. A study by Satar and colleagues conjugated agarose nanoparticles (ANPs) with poly(-quaternary ammonium) in an aqueous system and on agar plates. Nanoprecipitation technology was adopted in synthesizing ANPs, and an investigation was carried out on the antibacterial potency of the synthesized ANPs. Results revealed that in comparison with poly(-quaternary ammonium) modified agarose nanoparticles (mANPs), unmodified ANPs displayed a lower inhibitory effect against *E. coli* and *S. aureus*. With ANPs at 0.2 and 0.4 mM, the zone of inhibition (ZOI) observed was 2.9 and 3.8 cm for *E. coli* and at 3.1 and 4.0 cm for *S. aureus*. In contrast, for modified ANPs at similar conditions, the ZOI increased to 3.2 and 3.8 cm for *E. coli* and 3.5 and 4.1 cm for *S. aureus* [96]. On another hand, with near-infrared (NIR), one

can induce photothermal annihilation of pathogenic bacterial in agarose, specifically to target antibiotic-resistant bacteria. Ko and colleagues investigated the effect of NIR light absorption on a blend of PEDOT:PSS/agarose. When the blend was exposed to the light for 100 s, a sharp rise in temperature of about 24.5°C was observed in the blend. The approach proved useful in annihilating about 100% of the deadly bacteria within 2 min of exposure to NIR irradiation [97].

Moreover, combining agarose hydrogel with Bioglass was used as copper delivery systems. The inclusion of the bioglass positively influences the hydrophilicity and the swelling profile of agarose hydrogel and improves its antibacterial effectiveness against *Bacillus* sp. and *Pseudomonas aeruginosa* sp [98].

3.1.9 Hyaluronic acid

Hyaluronic acid is a linear anionic glycosaminoglycan also known as hyaluronan. It is an essential heteropolysaccharide with linear chains of repeating units of disaccharides of β-(D)-glucuronic acid and *N*-acetyl-β-(D)-glucosamine [12] as shown in Fig. 5.10. It is commercially obtained from animal tissue. A high concentration of HA resides in the synovial fluid and the ECM of cartilage. Owing to its noninflammatory, biodegradability, biocompatibility, nontoxic, nonimmunogenic, homeostatic, hygroscopic, and the stimulation of cytokine activity, HA is employed in wound healing and drug delivery [99–101]. Other application of HA includes osteoarthritis, anesthetic medicine, and tissue regeneration ophthalmology. There are two types of hyaluronic acid biopolymers used in burn management—Hyalomatrix and Hyalosafe. Hyalomatrix is a hyaluronic acid-based dermal substitute administered for immediate wound closure. Hyalosafe, on another hand, is hyaluronic acid-based film employed the treatment of superficial wounds [102]. Hyaluronic Acid can be modified and functionalized through a variety of

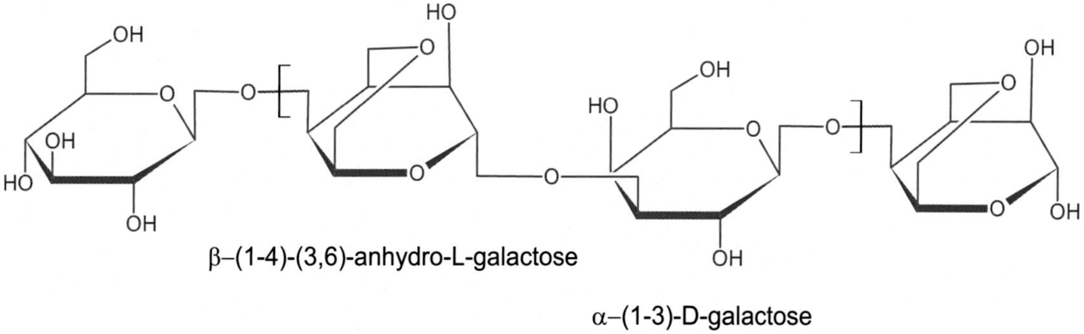

β–(1-4)-(3,6)-anhydro-L-galactose

α–(1-3)-D-galactose

FIG. 5.9 Chemical structure of agarose.

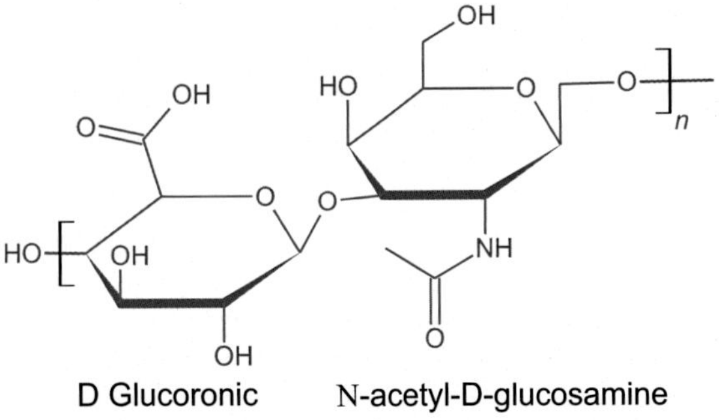

D Glucoronic   N-acetyl-D-glucosamine

FIG. 5.10 Chemical structure of hyaluronan.

chemical techniques to enhance its physical and mechanical properties. The hydroxyl functional group can be subjected to esterification or etherification processes. Although the carboxyl group can be modified through esterification, amidation, or oxidation. The acetyl group may be functionalized through amidation or deacetylation. It can equally be modified with peptides to improve cell attachment, spreading, and proliferation. Owing to its rapid degradation, HA is usually covalently cross-linked to enhance its physical properties and extend its biodegradation. Carbodiimide (water-soluble agents) or photo cross-linking agents such as glycidylmethacrylate or methacrylic anhydride may be used to cross-link HA [12].

3.2 Protein/Polypeptide-Based Antibiotic Importers

Protein is another class of sustainable biobased biopolymer for delivery of antibiotics and wound care. They are simple, linear, high-molecular-weight biopolymer made up of amino acid monomers linked covalently by amide(peptide) bonds. The peptide is a linear chain of amino acid residues. When the peptide contains about 4—10 amino acid residues, it is termed *oligopeptide*. A polypeptide is a peptide composed of several amino acid residues having molecular weight usually <5000. Protein, however, is composed of at least a single polypeptide with a molecular weight ranging from ≈5000 to about 40,000,000 [47]. Significant groups of antimicrobial protein-based biopolymers include structural or fibrous proteins such as keratin, collagen, gelatin and elastin, and plasma or globular or functional proteins such as fibrin, albumin, caseinates, and fibrinogen [12] When protein is entirely made up of only amino acids, it is called homoprotein or simple proteins or unconjugated proteins. However, when it has a nonprotein portion, called a prosthetic group, it is known as heteroprotein or conjugated or complex protein. Protein-based biopolymers with drug-delivery functionality have attracted immense attention for biomedical and packaging applications such as bone healing, neuroinflammation, antibiotic release, corneal regeneration, diabetes, wound healing, and cancer treatments [103].

3.2.1 Collagen

Collagen is structural proteins extracted only from animals such as rat tails and bovine tendons, calfskin, human placenta, and bovine cartilage. It forms 20%—30% of body protein that makes it the most abundant protein in the human body. There are at least 28 different types with types I—V being the major types in existence. Types I—IV are the fibrous protein that supplies structural support for the tissues, while type V nonfibrillar forms a two-dimensional (2D) network of all basal laminae. Collagen obtained from human cadavers, or cultured human dermal fibroblasts is allogeneic. However, those extracted from the bovine dermis, for instance, are xenogeneic [12]. Fig. 5.11 shows the secondary structure of collagen.

Collagen is an excellent biomaterial for biomedical applications due to their abundance, low antigenicity, availability, formability, biodegradability, and biocompatibility. Consequently, it can quickly be processed into solid forms such as sheets, hydrogels, and sponges. They are excellent biopolymers for drug delivery and

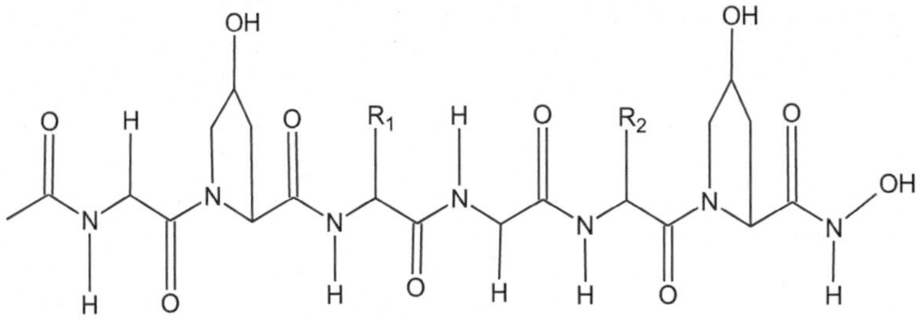

FIG. 5.11 Secondary structure of collagen.

wound treatment. Collagen has been employed as sponges for dressing wounds and burns, a matrix for protein, gene, and transdermal delivery [47]. Collagen was loaded with naringin, a Chinese herbal medicine that could prevent bacterial infection and promote osseointegration and coated onto titanium for an orthopedic implant. The attachment, osteogenic differentiation, proliferation, and mineralization of mesenchymal stem cells on the coating were significantly enhanced. Additionally, the antibacterial efficacy against *Staphylococcus aureus* was equally promoted. The release kinetics of collagen displayed a synergistic effect of an initial burst release and subsequent slow-release stage, which is subject to the concentration of naringin [104].

Recently, Liming Ge and colleagues in a simple one-step solution-immersion approach synthesized collagen/DXG-AgNP composite dressings. The AgNPs were homogeneously dispersed into the matrix of collagen and simultaneously cross-linked with dialdehyde xanthan gum (DXG). The antibacterial activity and physicochemical properties of collagen were effectively improved with this simple approach. Moreover, the composite further exhibited excellent blood compatibility, shape memory, cytocompatibility properties, and effectively promote full-thickness burn healing without scar formation [105].

3.2.2 Keratin

Keratin is a fibrous cysteine-rich structural protein with high mechanical strength. They are found in wool, human and animal hair, skin, feather, horn, hooves, and fingernails. The structure of keratin is depicted in Fig. 5.12.

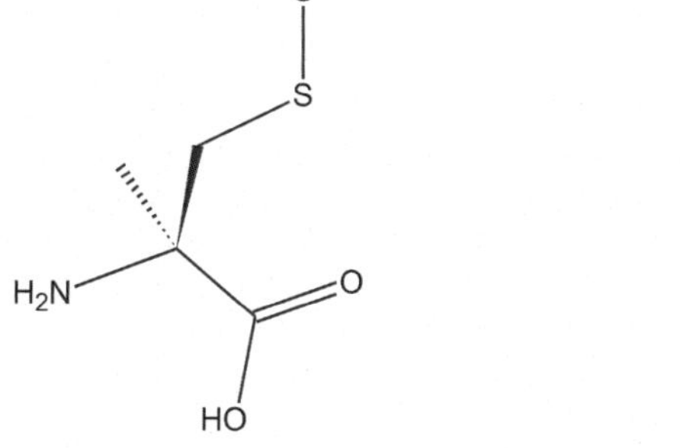

FIG. 5.12 Chemical structure of keratin.

The presence of disulfide bonds renders keratins insoluble in aqueous acidic, neutral, or alkaline solutions. This inherent property makes the extraction and dissolution of keratins extremely difficult. Keratins are, however, combined with other materials to improve its properties. They are often tethered with hydrophilic and hydrophobic moieties to modify its degradation rate. For instance, Yi Hu and coworkers subjected keratin fibers to surface modification through thiol-ene click reaction and studied the antibacterial, antistatic, and hydrophilicity properties. First, they applied tris(2-carboxyethyl) phosphine hydrochloride wool samples to generate thiols through a controlled reduction of cystine disulfide bonds present in keratin. Then thiol-acrylate quaternary ammonium salt with divinyl terminated was conjugated unto the reduced keratin through thiol-ene click chemistry and simultaneously accompanied by a step-growth dithiol-diacrylate reaction. The resulting wool fabric exhibited good antibacterial, hydrophilicity, and antistatic properties [106]. Plasticizers, cross-linking agents, and polymers are usually added to keratin to enhance its mechanical properties. Keratins are biodegradable, biocompatible and can be formed into hydrogels, film, scaffolds, and fibers. These properties make them suitable biopolymer for biomedical applications such as for control of drug delivery, nerve regeneration, burns, and wound care. For instance, a commercial keratin product-Keramatrix is employed in the treatment of burns, acute, and chronic wounds. Other biomedical applications include ocular regeneration, bone, and cartilage tissue engineering, and skin regeneration [47].

To promoting wound healing and skin regeneration, keratin was blended with chitosan. The developed porous scaffold was evaluated for cell proliferation, antibacterial activity, cytotoxicity, and mechanical strength. Although the presence of chitosan enhanced the mechanical and biocidal properties of Keratin, the proliferation of the fibroblast cells were found to be independent of the concentration of chitosan [107]. This experiment was later repeated with the inclusion of nano zinc oxide, *n*-ZnO. The resulting biocomposite hydrogel displayed rapid skin cell reconstruction with collagen [108]. The encapsulation of metallic nanoparticles such as silver [109], gold [110], and ZnO nanoparticles have been reported to promote bioactivity, biocompatibility, and biocidal behavior in Keratin for wound dressing. When keratin hydrogel comes in contact with a contaminated bacterial media, it swells release the nanoparticles at a controlled rate. A recent report shows that 5% of zinc oxide nanoparticles concentration is the optimum for wound dressing applications [111].

3.2.3 Gelatin

Gelatin, a derivative of collagen, is a linear polypeptide obtained by partial hydrolysis of porcine or bovine collagen. The hydrolysis processes break down the molecular bond holding individual strands of collagen. Gelatin consists of 18 amino acids [112] (Fig. 5.13), having biologically active sites that promote cell viability, adhesion, migration, and proliferation [113]. When hydrolyzed in an acid type A gelatin is formed. However, when it is subjected to basic hydrolysis, type B gelatin is produced. Type A is preferably used for developing scaffold because it possesses more carboxylic groups than type B [114]. Gelatin is extracted from the bones, skin, and connective tissues of animals such as domesticated cattle, chicken, pigs, and fish. It is insoluble in cold water, but at elevated temperature it becomes soluble, and upon cooling, it forms a gel. Gelatin, like its parent macromolecule, is sterile, pyrogen-free, biodegradable and biocompatible and has found use as a matrix for biocomposites, and scaffolds [47]. Moreover, it has high tensile strength when electrospun into a fiber. However, owing to its solubility, it is mechanically weak and requires chemical modification or the addition of other materials to form hydrogels of high performance [114].

To induce bactericidal functionality in gelatin against gram-positive and gram-negative bacteria, Junying and colleagues, under alkaline conditions (pH 10–11),

modified gelatin with epoxy silicone quaternary ammonium salt. With this approach, bioactive sites, including silyl and quaternary ammonium groups, were grafted unto gelatin backbone chain, making the biopolymer to possess inhibitory property against gram-positive, and gram-negative bacteria [115].

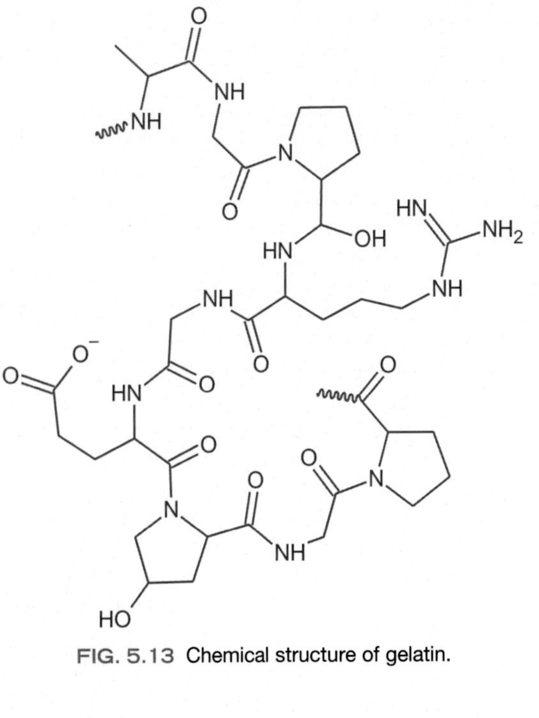

Furthermore, reinforcing biopolymers with metallic nanoparticles is a common practice to increase their biofunctionality. These include metallic nanoparticles such as metal hybrid, salts, and oxides. For example, AuNPs, AgNPs, and AuAgNPs have been reported to exhibit potent antioxidant activity with gelatin. Meanwhile, gelatin-based composite films with CuNPs, AgNPs, with their respective alloy, TiO_2NPs, and Fe_3O_4NPs displayed strong antibacterial efficacy against both gram-positive and gram-negative pathogenic [116–118]. Gelatin-ZnO NPs have been demonstrated to show antibacterial and antiangiogenesis effects with active inhibitory properties against the growth of fungus *Candida albicans* biofilm at 50 µg/mL. Moreover, it equally reduced the viability of hepatocarcinoma cancer cell lines at 100 µg/mL [119–121].

Gelatin is a suitable biopolymer matrix for drug and gene delivery. To promote the proliferation of fibroblasts cells for accelerating wound healing, enhance antibacterial properties to target multiresistant bacteria, improved cytocompatibility, and hemostatic activity, gelatins are usually loaded with antibacterial agents such antibiotics and essential oil extracts (e.g., oregano and lavender essential oils). Studies show that gelatin demonstrates sustained release profile for vancomycin hydrochloride, ibuprofen, ciprofloxacin, vitamin A palmitate, D-α-Tocopheryl polyethylene glycol 1000 succinate (vitamin E TPGS), tannins, and platelet-rich plasma [122–126].

3.2.4 Elastin

Elastin is a fibrous bioelastic protein synthesized and secreted by vascular smooth muscle cell (SMC) and fibroblasts [127]. It has a structure (Fig. 5.14) characterized by pentapeptide sequence (Val-Pro-Gly-Val-Gly)$_n$ with n ranging from 10 to 13. It is responsible for tissue and organ to regain their initial shape after being subjected to strain. Elastin is generated as ECM protein and makes up 70% of the elastic ligaments, 50% of large arteries, 30% of lungs, and 2%–4% of the skin. It is made insoluble in water by cross-linking chains of water-soluble tropoelastin molecules composed of alternating hydrophobic and hydrophilic stretches of 50–70 kDa [47]. Elastin has attractive features such as elasticity, mechanical strength, stability, and self-assembly properties that make it a suitable matrix for biomedical application. It can be processed into a

FIG. 5.13 Chemical structure of gelatin.

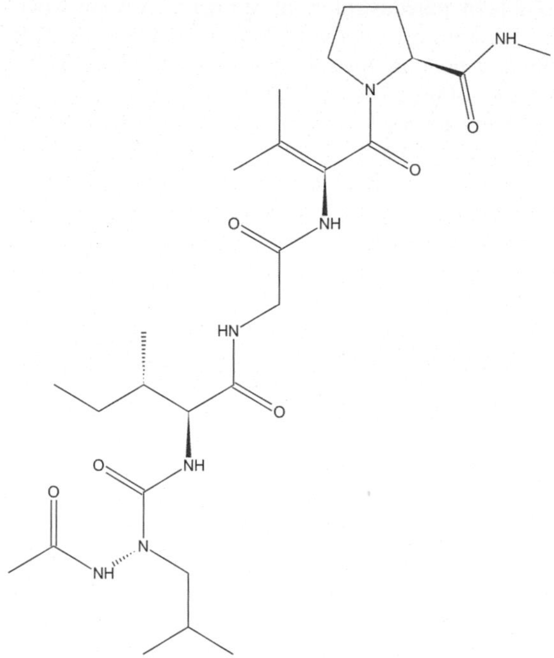

FIG. 5.14 Chemical structure of elastin.

different solid structure such as fibers, hydrogels, sponges, and sheets for wound care and drug importation applications. Elastin fibers are composed of two regions—amorphous and crystalline. The amorphous domain is characterized by highly cross-linked elastin protein that is about 100 nm thick and makes up 90% of the core of the fiber, while the remaining 10% is composed of the fibrillar mantle of about 5 nm thick microfibrils [127]. Elastin biopolymers are initially employed for medical applications in similar ways with conventional polymeric carriers. However, due to their unique properties, specifically, their ability to self-assemble, make them an excellent choice of material for advanced systems.

Recombinant technology allows complex smart macromolecular engineering and biological synthesis of recombinant protein-based polymers (rPBPs) having a precise size and composition control. Moreover, it is possible with recombinant technology to incorporate functional bioactive domains present in natural protein such as AMPs. The two most essential rPBPs are elastin-like recombinases (ELRs) and silk-elastin like proteins. There are two soluble types of elastin, namely α-elastin(a-cid-based) and Ҟ-elastin(alkaline-based). They are used as smart biopolymers in tissue regeneration and drug-delivery systems due to their mechanical properties, biodegradability, biocompatibility, and thermostability [128]. Functionalized ELRs displayed antibacterial potential

against several gram-positive and gram-negative bacterial strains [129]. Using decellularized elastin-containing tissue, synthetic tropoelastin or elastin-containing materials and stimulation of de novo elastin deposition, it is possible to use elastin for tissue recreation [128].

3.2.5 Fabrin

Fibrin is a fibrous protein-based biopolymer that is a part of the clotting cascade and formed by polymerization of the soluble plasma protein fibrinogen. Fibrinogen is a dimeric, elongated, 340 kDa glycoprotein macromolecule consisting of two sets of three peptide chains (Aα, Bβ, and γ) linked by 29 disulfide bridges. The three chains are covalently bonded by disulfide bridging at their N-termini in the central domain of the molecule, which is contained in the E-domain. The inner E-domain is connected on either side to two outer D-domains by a coiled-coil segment. Following the stimulation of the coagulation cascade, thrombin activates fibrinogen, which binds to adjacent fibrin(ogen) molecules to produce an insoluble fibrin matrix. This fibrin network is the fundamental protein component in clots and provides a scaffold for infiltrating cells during tissue repair. Owing to the role of fibrin in hemostasis and tissue repair, it is often used as a tissue sealant [47], adipose tissue, cardiac tissue, muscle tissue, cartilage, nervous tissue, respiratory tissue, ocular tissue, skin, ligaments, and tendons. Moreover, it is widely employed as a carrier for antibacterial agents.

Platelet concentrates are widely employed for medical use to stimulate soft- and hard-tissue recreation. It achieves this by producing a large number of growth factors that help the wound to heal. It also acts as an antiinfective host defense, while PRP, PRF, and I-PRF concentrates have been reported to exhibit antibacterial activity [130]. Only PRP displayed inhibitory activity against periodontal pathogens such as *P. gingivalis, P. intermedia, F. nucleatum,* and *A. actinomycetemcomitans* [131]. Although, a study showed that when PRF is conjugated with Leukocytes, a potent inhibition against *P. gingivalis* is displayed. However, it lacks an inhibitory effect against *P. intermedia, F. nucleatum,* and *A. actinomycetemcomitans* [132].

Khorshidi and coworkers try to enhance the antibacterial activity of L-PRF by incorporating SNP into it. The modified L-PRF matrix with SNP did exhibit antibacterial efficacy against pneumonia and *Streptococcus viridans* for a period of 24 h [133]. In addition, PEGylated fibrin containing silver sulfadiazine loaded into the chitosan microsphere significantly reduced infections caused by bacteria and inflammatory cytokine (TNF-α) expression. Besides, it was found to enhance neovascularization, granulation tissue formation, wound closure, and collagen maturation in burn wound [134,135].

Adjusting the concentration of the antibiotic as well as the duration exposure proves as a better approach to combating the deadly pathogenic infections. Consequently, studies show that fibrin clots, sealants, glue, nanoparticles, and hydrogels loaded with teicoplanin, vancomycin, fluconazole, ceftazidime, ciprofloxacin [136], moxifloxacin, PA5 bacteriophages, and lomefloxacin antibiotics [137] can effectively deliver antibiotics to infected sites [137] and inhibit the growth of methicillin-resistant *Staphylococcus aureus, Escherichia coli (E.coli), Staphylococcus epidermidis, Streptococcus pneumoniae, Pseudomonas aeruginosa,* and *Candida albicans (C. albicans)* [138,139].

3.2.6 Silk

Silks are a fibrous protein with the core of these proteins consisting of sequences of repeating units of alanine—glycine with serine or tyrosine (Fig. 5.15) and coated with a glue-like nonfibroin protein called sericin. They are primarily obtained from Lepidoptera larvae such as silkworms and spiders. The biosynthesis of silk often takes place in epithelial cells and stored in the lumen of specialized glands, from where they spun into fibers. The silk of different biological origin varies in structure, amino acid composition, and properties. The most widely researched and cited silks in the literature are those synthesized by spiders namely *Araneus diadematus* and *Nephila clavipes* and by silkworm namely *Bombyx mori. Bombyx mori* silks are polypeptides composed of light (25-kDa) and heavy kDa chains of fibroin proteins; whereas, spider protein is characterized by polyalanine and glycine domains ranging from 70 to 700 kDa. They possess excellent environmental stability with better mechanical strength than synthetic material such as Kevlar. The ultimate tensile strength of *B. mori* and spider silks are 740 and 950 MPa, respectively [12].

The silk fiber is composed of a highly crystalline hydrophobic β-sheets region linked through hydrogen bonding. They are insoluble in common solvents such as water. However, the fibroin core of silk is made soluble in aqueous solutions or organic solvents when the nonfibroin component (sericin) is removed. This makes it possible for silk to be processed into fibers, sponges, films, tubes, and gels [12].

Owing to the biodegradable, biocompatible, and high mechanical strength silks, they are used for biomedical applications. Although they lack intrinsic antibacterial properties and are hygroscopic [140,141], they are, however, chemically modified to enhance their functionality. Chenglong and colleagues reported the development of antibacterial and absorbable silk-based materials loaded with gentamicin for fixation screw used in orthopedic internal fixation surgery. The material in vivo displayed excellent biocompatibility, degradation rate, and antimicrobial activity. The water-absorbing capacity reduced drastically, making the modified silk a promising biomaterial for orthopedic implants and eliminating the need for the second surgeries [142].

For enhanced UV protection, bioactivity, cytocompatibility, stem cell differentiation potential, mineralization, and antibacterial activities against antibiotic-resistant bacteria, in wound care, bone tissue engineering, and functional textile applications, silks and their blends are often functionalized with metallic nanoparticles such as AgNPs [141,143—145], Cu NPs [143,146], ZnO NPs [147], and natural antibacterial and fungal extracts such as *Thermomyces* [148] and *Cinnamomum camphora* [149].

3.2.7 Caseinates

Caseinate is an attractive animal protein for food packaging and culinary applications. It is an edible biopolymer that is soluble in water and function as an emulsifier. Caseinates can form a transparent film with good barrier performance against carbon dioxide, oxygen, and aroma compounds. Owing to its plasticity, elasticity, flexibility, and film-forming ability, caseinates are employed as a suitable matrix for antimicrobial delivery in food applications. Caseinate has interesting antimicrobial properties that help to increase the bioavailability of antibacterial agents at the target site at the minimal payload. This is achievable with caseinates because of their tendency to slow down the diffusion of

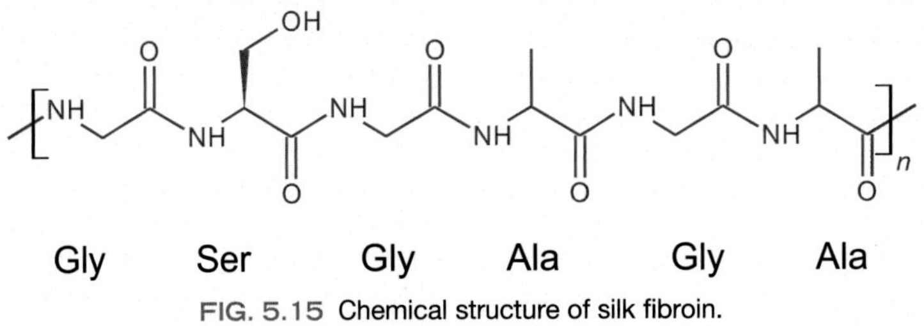

Gly   Ser   Gly   Ala   Gly   Ala

FIG. 5.15 Chemical structure of silk fibroin.

antibacterial agents through the matrix, thereby enriching the delivery of drugs at the target site [14].

Zhang S and colleague evaluated rutin-loaded zein-sodium caseinate nanoparticles for antibacterial and antioxidant applications [150]. Zhang and coworkers, on another hand, encapsulated thymol in sodium caseinate. The interaction of thymol with the biopolymer promoted strong bonding with tyrosine and with other amino acid residues away from tryptophan of caseins. The resulting material exhibited similar antibacterial activity as that of free eugenol against *Escherichia coli, Listeria monocytogenes Scott A, and Salmonella enteritidis* [151].

3.3 Polyurethanes and Polyurethane-Based Antibiotic Importers

Another unique family of biopolymers for biomedical application are polyurethanes. Owing to their excellent physicochemical, mechanical, and biocompatibility, polyurethane is used as carriers for antimicrobial delivery. Their relative stability within the human body makes them be grouped into either biostable (bioinert) and biodegradables. Biodegradable Polyurethane (PU) is employed as drug-delivery carriers and for tissue engineering; whereas, biostable PUs have found use as a potential material for artificial organs and medical devices [152]. The method and monomers employed in the production of PU largely determine its properties. PU is prepared in a two-step process. The first step involves the reaction between polyisocyanate (such as diisocyanate) and polyol (macrodiols) to form a prepolymer. Here, the reaction between the isocyanate group and the hydroxyl group in polyol leads to the formation of urethane linkage. At this stage, the addition of the diol chain extender

further results in additional urethane functional groups. However, when diamine is used, the urea functional group is formed (Fig. 5.16). The chain extenders are low molecular weight hydroxyl- and amine-terminated compounds and provide segregation between the hard and soft segments of PU. The hard segment that is made up of —NH—COO— and the chain extender impart mechanical strength; whereas, flexibility is imparted by the soft segment that is mainly made of polyol. Examples of chain extender are ethylenediamine, putrescine, 1,3-propanediol, and 1,4-butanediol (BDO) [152].

Monomers used in the synthesis of PU play an essential role in determining the resulting properties of PU. Polyol for bio-based PU is dihydroxyl-terminated polyester, polyether, or polycarbonates obtained from natural resources [153]. Others include pentaerythritol or N-BOC-serinol. Primary sources of polyester polyol include diacids (such as succinic acid, adipic acid, glutaric acid, and azelaic acid) and di- or trifunctional polyols (such as ethylene glycol, 1,6-hexanediol, 1,4-butanediol, glycerol, 1,2-propanediol, and 1,10-dodecanediol). PU-ester is used as biodegradable carriers for drug delivery. The polyester polyol used in biodegradable PU includes polylactide, polycaprolactone, and polyglycolide. However, in comparison to PU-ether, PU-ester is more responsive to hydrolytic cleavage. However, with the inclusion of bulky alkyl to the hydroxyl-terminal of the polyester, PU-ester becomes more hydrolytically stable. Polyether polyol on another hand is obtained from glucose, sucrose, fructose, and glycerol. Vegetable oils such as soy, rapeseed, sunflower, castor, and linseed oils are the sources of most biobased polyols used in the production of biobased PUs [154].

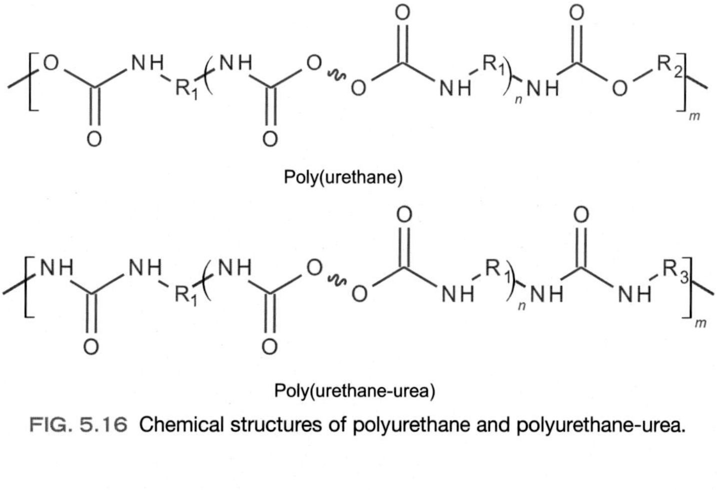

Poly(urethane)

Poly(urethane-urea)

FIG. 5.16 Chemical structures of polyurethane and polyurethane-urea.

Polyisocyanate is either aliphatic or aromatic but is nonbiobased. PU synthesized using aliphatic isocyanates tend to be biocompatible; hence, it is used for drug delivery in vivo. Example of such aliphatic isocyanates includes hexamethylene diisocyanate, butane diisocyanate, lysine diisocyanate ethyl ester, and lysine diisocyanate methyl ester [154]; whereas, aromatic isocyanate-based PU often breaks down to form aromatic amines that are toxic; hence, carcinogenic. PU is a suitable matrix for antimicrobial application. This is achieved by (1) modifying the surface of the PU with targeting ligands via covalent bonding or electrostatic charge. For instance, PU is coated with polyclonal IgG and β-lactam antibiotics to reduce bacterial adhesion. In addition, (2) through antibiotic encapsulation [155]. The rate at which the polymer releases drug can be controlled by tuning the chemistry of the polymer. Different from nonbiodegradable PU that releases the drug by diffusion through the pores of the polymer matrix, biodegradable PU depends on the composition, initial drug loading, swelling, and degradation rate. PU-ester degrades with the ester bond undergoing rapid but controlled hydrolysis, followed by the degradation of urethane bonds into smaller fractions [152,156].

3.4 Biobased Polyesters Antibiotic Importers

Polyester derived from a renewable source (such as carbohydrate) is a particular class of biopolymers. They can be synthesized through a polycondensation reaction of hydroxy acid/diacid/anhydride with secondary alcohol or be synthesized through ring-opening polymerization [157]. Based on chemical structure, biobased polyesters are classed as aromatic or aliphatic polyesters. Aliphatic biobased polyesters including polyhydroxyalkanoate (PHA), PLA, poly(caprolactone), and poly(glycolic acid) (PGA) are extensively exploited for medical and pharmaceutical applications, for example, in surgical sutures, drug-delivery systems, implants, wound closure, and tissue regeneration [14].

3.4.1 Polylactic acid

PLA is a thermoplastic aliphatic polyester that is chemically synthesized from simple sugars that are obtained from renewable sources, such as corn, tapioca, potatoes, sugarcane, beet, and other plant root starches. The synthesis starts with bacterial fermentation of starch or sugar to the lactic acid monomer. Polymerization of lactic acid directly into PLA is impossible because polycondensation of two lactic acids monomer generates one molecule of water. The water generated degrades the oligomer chain and leads to the formation of low-molecular-weight lactide. Two lactic acid molecules then undergo single esterification and get catalytically cyclized to form a cyclic dilactate ester. There are, however, several routes to obtain PLA from lactic acid. The most common method employed at the industry to produce PLA is through ring-opening polycondensation (ROP) of cyclic lactide dimer (Fig. 5.17) with metal catalysts (such as tin octoate) with the elimination of water at a relatively high temperature of about 200°C. PLA, at the industrial level, can be processed with the same conventional processing technology used for traditional petroleum-based thermoplastics such as melt blending, extrusion, injection molding, thermoforming, film forming, and electrospinning to obtain fibers, films, sheets, and molded articles [11]. PLA is mostly used for biomedical heterogeneous implants for the treatments of the orthopedic sufferers. However, it possesses a weak antibacterial property. To enhance this property, it is usually encapsulated with antibacterial agents.

Chen and colleagues studied the controlled release of ampicillin sodium salt and the vancomycin hydrochloride from the surfaces of PLA disks against *E. coli* and *S. aureus*. The bioactivity test on those microorganisms reveals that the in vitro drug-release profile of the

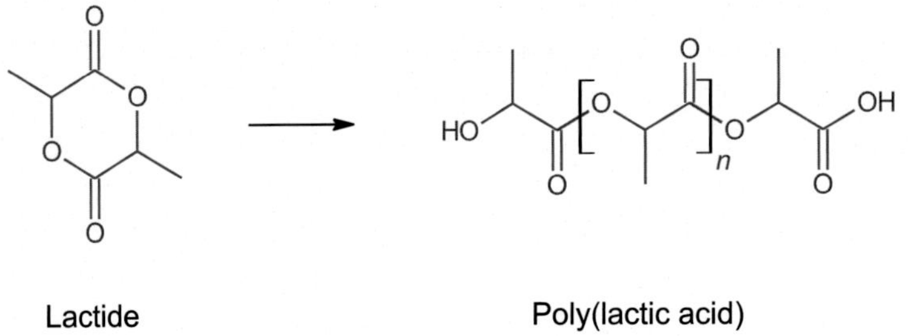

Lactide   Poly(lactic acid)

FIG. 5.17 Chemical structure of poly(lactic acid) synthesized by ring opening of lactide.

antibiotic agent can inhibit *E. coli* and *S. aureus* for over 80% up to 28 days [158].

In recent studies, nano-silver (nano-Ag), nano-zinc oxide (nano-ZnO) [159,160], nano-TiO₂ [161], and capsicum oleoresin-impregnated nanoporous silica (SiCO) [162] have been used to enhance the properties of PLA, specifically the antibacterial property against gram-positive pathogenic bacteria such as *Staphylococcus aureus* and *Bacillus* and the gram-negative pathogenic bacteria including *Escherichia coli* and *Salmonella enterica*.

3.4.2 Poly(hydroxyalkanoate)

PHA is a biobased polyester synthesized as intracellular carbon- and energy storage compounds in both gram-negative (such as those belonging to the genera *Azobacter*, *Pseudomonas Bacillus*, and *Alcaligenes*) and gram-positive bacteria (those belonging to the genera *Nocardia*, *Streptomyces* and *Rhodococcus*). PHA is a high molecular weight, semicrystalline, biocompatible, and biodegradable thermoplastic polyester. It exhibits thermal and mechanical strength that is comparable to most common olefins such as polystyrene and polypropylene. Poly(3-hydroxybutyrate) (PHB) (Fig. 5.18) is the most basic and widely investigated PHAs for medical applications such as drug delivery, tissue engineering, and implants, including sutures, repair patches slings, orthopedic pins. Another common PHA is poly(3-hydroxybutyrate-co-3-hydroxy valerate). PHAs, however, have high hydrophobicity, poor mechanical, and thermal stability that limit their use in many biomedical applications. For instance, PHB has a melting temperature of around 170−180°C that is close to its degradation temperature at 270°C. This makes PHB less resistant to thermal degradation [14,16].

However, PHB has been reported to demonstrate a lack of physiological toxicity to the skin, good sterilizing effect, and promotes wound healing. Ma L and colleagues investigated the antibacterial and antifungal properties of poly(3-hydroxybutyric acid) (PHB)

oligomer. The findings show that PHB possesses inherent antimicrobial that includes mechanisms such as disruption of biofilm, bacterial wall/membrane, leakage of the intracellular content, inhibition of protein activity, and change in the transmembrane potential [163]. It is often, however, chemically modified to improve its properties for biomedical use.

A recent study investigated the properties of a functionalized PHB. PHB was subjected to transesterification reaction under reflux conditions in the presence of 1,4-butanediol to produce telechelic PHB-diol. It was further modified by the reaction of PHB-diol with acryloyl chloride to form PHB-diacrylate and then conjugated with amino compounds such as piperazine, 1,4-butanediamine, 1,3-propanediamine, cyclohexylamine, 1,2-ethylenediamine, 2,2′-(ethane-1,2-diylbis(oxy)) diethanami morpholine through Michael-type addition reaction. The resulting modified PHB-ethylenediamine displayed enhanced antibacterial activity against *Staphlococcus aureus*, *Escherichia coli*, *Klebsiella pneumoniae*, and *Pseudomonas aeruginosa* [164].

3.4.3 Poly(butylene succinate)

Another interesting natural biodegradable polyester is poly(butylene succinate) (Fig. 5.19). PBS is one of the most popular poly(alkylene dicarboxylate) polymers due to its attractive features, including biodegradability. It is synthesized through polycondensation of succinic acid and BDO monomers. Using biotechnological routes, PBS is crystalline polyester that can be entirely bio-based. Succinic acid based on renewable feedstocks, for instance, is obtained from refined sugar-based biomass feedstock such as glucose, sucrose, and bio-based glycerol. An alternative route involves extracting succinic from microorganisms, including fungi or bacteria such as *Anaerobiospirillum succiniciproducens* and *Actinobacillus succinogenes* due to their ability to produce a relatively large amount of succinic acid. Renewable 1,4-butanediol (BDO from renewable resources) is a liquid, colorless, di-alcohol. It is synthesized by fermentation based on renewable feedstocks (such as dextrose). BDO can also be produced directly from biomass or indirectly from bio-based succinic acid through its catalytic hydrogenation [14]. Owing to its

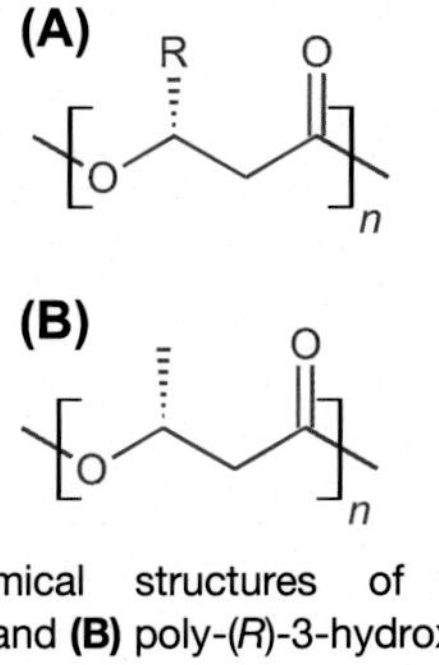

(A)

(B)

FIG. 5.18 Chemical structures of **(A)** poly-(*R*)-3-hydroxyalkanoate and **(B)** poly-(*R*)-3-hydroxybutyrate.

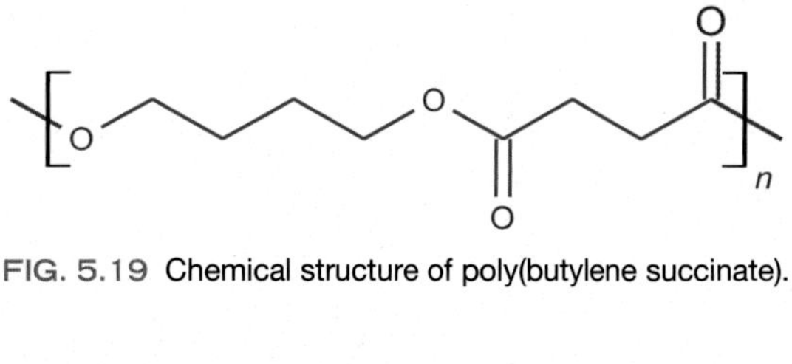

FIG. 5.19 Chemical structure of poly(butylene succinate).

biodegradability and biocompatibility, it is used for delivering antibacterial agents to target deadly bacterial.

PBS-based composites developed with the inclusion of nanosilver-coated carbon black (AgCB) exhibited enhanced antimicrobial activity against *E. coli* and *C. albicans* [165]. Moreover, when carvacrol essential oil was introduced into PBS. Effective inhibitory response against *S. aureus* was observed as 4 wt% of carvacrol; whereas, 10 wt% of carvacrol was needed to effectively inhibit *E. coli* growth [166].

4 BIODEGRADABLE ANTIBIOTIC POLYMERIC IMPORTERS

These are fossil-based antibiotic polymeric carriers that are biodegradable and biocompatible. They are often employed in treating infected tissues. This class of material includes polyamides, biodegradable polyester, phosphorus-containing polymers, polyphosphoesters poly(rotho esters), biodegradable PU, and polyanhydrides. A few of these biopolymers are explained in the following.

4.1 Polyamides

Polyamides are macromolecules with an amide group as part of the polymer chain. Biopolyamides are both biobased and fossil based biodegradable polyamides synthesized from polycondensation of (1) diamines and dicarboxylic acids (e.g., from castor oil), (2) ω-amino carboxylic acids as bifunctional monomers and (3) α-amino carboxylic acids as bifunctional monomers [154]. Examples of biodegradable polyamides include polyamide 4,10 (PA 4,10), polyamide 6,10 (PA 6,10), polyamide 10,10 (PA 10,10), and polyphthalamides (PPAs); whereas, that produced by the ring-opening polymerization of ε-caprolactam is polyamide 11 (PA 11) [154].

4.2 Poly(Ortho Esters)

Poly(ortho esters) are biodegradable polymers used in drug-delivery systems. They are biocompatible, bioerodible, and hydrophobic. There are four types of POEs, namely POE I, POE II, POE III, and POE IV. POE I is synthesized by trans-esterification of diethoxytetrahydrofuran with diols. Although POE II is produced from polycondensation of 3,9-bis(ethylidene-2,4,8,10-tetraoxaspiro [5] undecane) with di- and tripolyols. A linear POE II is obtained when diols are used, while with the use of triol a cross-linked POE II is synthesized. POE III is prepared by polycondensation of alkyl orthoacetate with flexible triol. POE IV, which possesses all required properties for commercial medical application, is polymerized through polycondensation of alkylorthoacetate with rigid triol [154].

5 CONCLUSION

Modern healthcare is faced with a daunting threat from antibiotic-resistant bacteria, which in turn incite the development of biodegradable antibiotic polymeric carrier systems. This chapter summarizes the biodegradable polymers used in the delivery of antibiotics to target antibiotic-resistance bacterial infections. These materials were classified based on their origin, such as biobased and nonbiobased. Most of these polymers are either conjugated or loaded with antibiotics and other antibacterial agents such as metallic nanoparticles and plant extract. Moreover, they have been proven to demonstrate enhanced bacterial recognition toward pathogenic bacteria, controlled release of antibiotic drugs in the microenvironment (e.g., acidity or bacterial enzymes) of bacterial infection sites, and eradication of bacterial biofilms. Other biodegradable polymeric importers show inherent bactericidal behavior due to the presence of antibacterial moieties (such as the quaternary ammonium groups, phosphonium groups, antimicrobial peptides) in their backbone chains. Moreover, their application has assisted in slowing down the growth of bacteria on implants and infected tissues, including the disruption of bacterial biofilms. Although biodegradable polymers and biotechnology have paved better pathways in managing extreme conditions posed by multidrug-resistant bacteria, they however still have limitations. With research moving at a rapid pace to engineer high performing biodegradable nanocarriers, the future heralds the promise that biodegradable polymers are potentially the suitable matrix for delivery of antibiotics.

REFERENCES

[1] M. Konai, et al., Recent progress in polymer research to tackle infections and antimicrobial resistance, Biomacromolecules 19 (2018).

[2] X. Ding, et al., Biodegradable antibacterial polymeric nanosystems: a new hope to cope with multidrug-resistant bacteria, Small 15 (2019) 1900999.

[3] N.R. Naylor, et al., Estimating the burden of antimicrobial resistance: a systematic literature review, Antimicrobial Resistance and Infection Control 7 (2018) 58.

[4] J. Chen, et al., Antibacterial polymeric nanostructures for biomedical applications, Chemical Communications 50 (93) (2014) 14482–14493.

[5] M. Arzanlou, W.C. Chai, H. Venter, Intrinsic, adaptive and acquired antimicrobial resistance in Gram-negative bacteria, Essays in Biochemistry 61 (1) (2017) 49–59.

[6] O. Ciofu, et al., Antimicrobial resistance, respiratory tract infections and role of biofilms in lung infections in cystic fibrosis patients, Advanced Drug Delivery Reviews 85 (2014).

[7] L.C. Powell, et al., Targeted disruption of the extracellular polymeric network of *Pseudomonas aeruginosa* biofilms by alginate oligosaccharides, Npj Biofilms and Microbiomes 4 (1) (2018) 13.

[8] B. Remy, et al., Interference in bacterial quorum sensing: a biopharmaceutical perspective, Frontiers in Pharmacology 9 (2018) 203.

[8a] S. Tomczyk, et al., Control of Carbapenem-resistant Enterobacteriaceae, Acinetobacter baumannii, and Pseudomonas aeruginosa in healthcare facilities: a systematic review and reanalysis of quasi-experimental studies, Clinical Infectious Diseases: An Official Publication of the Infectious Diseases Society of America 68 (5) (2019) 873–884, https://doi.org/10.1093/cid/ciy752.

[9] L. Kumar, et al., Antimicrobial biopolymer formation from sodium alginate and algae extract using aminoglycosides, PloS One 14 (3) (2019) e0214411.

[10] C.d.C. Spadari, L.B. Lopes, K. Ishida, Potential use of alginate-based carriers as antifungal delivery system, Frontiers in Microbiology 8 (97) (2017).

[11] C. Zhang, Biodegradable Polyesters: Synthesis, Properties, Applications, 2015, pp. 1–24.

[12] D.F. Williams, Chapter 36 – Hydrogels in regenerative medicine, in: A. Atala, et al. (Eds.), Principles of Regenerative Medicine, third ed., Academic Press, Boston, 2019, pp. 627–650.

[13] Y. Xu, Chapter 19 – Hierarchical materials, in: R. Xu, Y. Xu (Eds.), Modern Inorganic Synthetic Chemistry, second ed., Elsevier, Amsterdam, 2017, pp. 545–574.

[14] A. Muñoz-Bonilla, et al., Bio-based polymers with antimicrobial properties towards sustainable development, Materials 12 (2019) 641.

[15] M. Nasrollahzadeh, et al., Chapter 5 – Green nanotechnology, in: M. Nasrollahzadeh, et al. (Eds.), Interface Science and Technology, Elsevier, 2019, pp. 145–198.

[16] R. Song, et al., Current development of biodegradable polymeric materials for biomedical applications, Drug Design, Development and Therapy 12 (2018) 3117–3145.

[17] G. Dinda, et al., Green synthesis of copper nanoparticles and their antibacterial property, Journal of Surface Science and Technology 31 (2015) 117–122.

[18] M. El-Husseiny, et al., Biodegradable antibiotic delivery systems, Journal of Bone and Joint Surgery British Volume 93 (2011) 151–157.

[19] S. Baghaie, et al., Wound healing properties of PVA/starch/chitosan hydrogel membranes with nano zinc oxide as antibacterial wound dressing material, Journal of Biomaterials Science, Polymer Edition 28 (18) (2017) 2220–2241.

[20] Z. Wu, et al., Preparation and application of starch/polyvinyl alcohol/citric acid ternary blend antimicrobial functional food packaging films, Polymers 9 (2017) 102.

[21] M. Palencia, T. Lerma, E. Combatt, Hydrogels based in cassava starch with antibacterial activity for controlled release of cysteamine-silver nanostructured agents, Current Chemical Biology 11 (2017) 28–35.

[22] N.S. Ismail, S.C.B. Gopinath, Enhanced antibacterial effect by antibiotic loaded starch nanoparticle, Journal of the Association of Arab Universities for Basic and Applied Sciences 24 (2017) 136–140.

[23] M. Aminzare, et al., Antibacterial activity of corn starch films incorporated with Zataria multiflora and Bonium persicum essential oils, Annual Research and Review in Biology 19 (2017) 1–9.

[24] J.A. do Evangelho, et al., Antibacterial activity, optical, mechanical, and barrier properties of corn starch films containing orange essential oil, Carbohydrate Polymers 222 (2019) 114981.

[25] P. Raigond, et al., Antimicrobial activity of potato starch-based active, Biodegradable Nanocomposite Films 62 (1) (2019) 69–83.

[26] C. Brigham, Chapter 3.22 – Biopolymers: biodegradable alternatives to traditional plastics, in: B. Török, T. Dransfield (Eds.), Green Chemistry, Elsevier, 2018, pp. 753–770.

[27] J. Padrão, et al., Bacterial cellulose-lactoferrin as an antimicrobial edible packaging, Food Hydrocolloids 58 (2016) 126–140.

[28] T.-T. Tsai, et al., Antibacterial cellulose paper made with silver-coated gold nanoparticles, Scientific Reports 7 (1) (2017) 3155.

[29] M.C. Tanzi, S. Farè, G. Candiani, Chapter 4 – Biomaterials and applications, in: M.C. Tanzi, S. Farè, G. Candiani (Eds.), Foundations of Biomaterials Engineering, Academic Press, 2019, pp. 199–287.

[30] C. Ringot, et al., Antibacterial activity of a photosensitive hybrid cellulose fabric, Photochemical and Photobiological Sciences 17 (11) (2018) 1780–1786.

[31] M. Tavakolian, et al., Developing antibacterial nanocrystalline cellulose using natural antibacterial agents, ACS Applied Materials and Interfaces 10 (40) (2018) 33827–33838.

[32] P. Bayazidi, H. Almasi, A.K. Asl, Immobilization of lysozyme on bacterial cellulose nanofibers: characteristics, antimicrobial activity and morphological properties, International Journal of Biological Macromolecules 107 (2018) 2544–2551.

[33] Y. Pan, et al., Cellulose fibers modified with nano-sized antimicrobial polymer latex for pathogen deactivation, Carbohydrate Polymers 135 (2016) 94–100.

[34] K. Littunen, et al., Synthesis of cationized nanofibrillated cellulose and its antimicrobial properties, European Polymer Journal 75 (2016) 116–124.

[35] H. Zhang, et al., Study on photocatalytic antibacterial and sustained-release properties of cellulose/TiO$_2$/β -CD composite hydrogel, Journal of Nanomaterials 2019 (2019) 1–12.

[36] S. Saini, M.N. Belgacem, J. Bras, Effect of variable aminoalkyl chains on chemical grafting of cellulose nanofiber and their antimicrobial activity, Materials Science and Engineering: C 75 (2017) 760–768.

[37] W. Shao, et al., Synthesis and antimicrobial activity of copper nanoparticle loaded regenerated bacterial cellulose membranes, RSC Advances 6 (70) (2016) 65879–65884.

[38] Y. Xu, et al., Review of silver nanoparticles (AgNPs)-Cellulose antibacterial composites, BioResources 13 (2018) 2150–2170.

[39] L. Zhai, et al., Synthesis, characterization, and antibacterial property of eco-friendly Ag/cellulose nanocomposite film, International Journal of Polymeric Materials and Polymeric Biomaterials 67 (7) (2018) 420–426.

[40] L. Lamboni, et al., Silk sericin-functionalized bacterial cellulose as a potential wound-healing biomaterial, Biomacromolecules 17 (9) (2016) 3076–3084.

[41] P. Moraes, et al., Bacterial cellulose/collagen hydrogel for wound healing, Materials Research 19 (2016).

[42] Y. Pei, et al., Effectively promoting wound healing with cellulose/gelatin sponges constructed directly from a cellulose solution, Journal of Materials Chemistry B 3 (38) (2015) 7518–7528.

[43] J. Kucińska-Lipka, I. Gubanska, H.J.P.B. Janik, Bacterial cellulose in the field of wound healing and regenerative medicine of skin: recent trends and future prospectives, Polymer Bulletin 72 (9) (2015) 2399–2419.

[44] P. Orlean, D. Funai, Priming and elongation of chitin chains: implications for chitin synthase mechanism, The Cell Surface 5 (2019) 100017.

[45] N. Tanha 1, K. Karimzadeh, A. Zahmatkesh, A Study on the Antimicrobial Activities of Chitin and Chitosan Extracted from Freshwater Prawn Shells (*Macrobrachium nipponense*), vol. 2017, 2017.

[46] J. Hafsa, et al., Antioxidant and antimicrobial proprieties of chitin and chitosan extracted from Parapenaeus Longirostris shrimp shell waste, Annales Pharmaceutiques Françaises 74 (1) (2016) 27–33.

[47] E. Sproul, S. Nandi, A. Brown, 6 – Fibrin biomaterials for tissue regeneration and repair, in: M.A. Barbosa, M.C.L. Martins (Eds.), Peptides and Proteins as Biomaterials for Tissue Regeneration and Repair, Woodhead Publishing, 2018, pp. 151–173.

[48] A. Bose, T.W. Wong, Oral colon cancer targeting by chitosan nanocomposites, Applications of Nanocomposite Materials in Drug Delivery (2018) 409–429.

[49] K. Divya, et al., Antimicrobial properties of chitosan nanoparticles: mode of action and factors affecting activity, Fibers and Polymers 18 (2) (2017) 221–230.

[50] Y.-C. Chung, C.-Y. Chen, Antibacterial characteristics and activity of acid-soluble chitosan, Bioresource Technology 99 (2008) 2806–2814.

[51] E.-R.S. Kenawy, M.M. Azaam, K.M. Saad-Allah, Synthesis and antimicrobial activity of α-aminophosphonates containing chitosan moiety, Arabian Journal of Chemistry 8 (3) (2015) 427–432.

[52] R. Kalaivani, et al., Synthesis of chitosan mediated silver nanoparticles (Ag NPs) for potential antimicrobial applications, Frontiers in Laboratory Medicine 2 (1) (2018) 30–35.

[53] S. Shankar, J.-W. Rhim, Preparation of sulfur nanoparticle-incorporated antimicrobial chitosan films, Food Hydrocolloids 82 (2018) 116–123.

[54] J. Gu, K. Al-Bayati, E.A. Ho, Development of antibody-modified chitosan nanoparticles for the targeted delivery of siRNA across the blood-brain barrier as a strategy for inhibiting HIV replication in astrocytes, Drug Delivery and Translational Research 7 (4) (2017) 497–506.

[55] A. Dev, et al., Preparation of poly(lactic acid)/chitosan nanoparticles for anti-HIV drug delivery applications, Carbohydrate Polymers 80 (2010) 833–838.

[56] M.G. Mehrabani, et al., Chitin/silk fibroin/TiO$_2$ bio-nanocomposite as a biocompatible wound dressing bandage with strong antimicrobial activity, International Journal of Biological Macromolecules 116 (2018) 966–976.

[57] H.C. Flemming, J. Wingender, The biofilm matrix, Nature Reviews Microbiology 8 (9) (2010) 623–633.

[58] P.M. Bales, et al., Purification and characterization of biofilm-associated EPS exopolysaccharides from ESKAPE organisms and other pathogens, PloS One 8 (6) (2013) e67950.

[59] L.K. Jennings, et al., Pel is a cationic exopolysaccharide that cross-links extracellular DNA in the *Pseudomonas aeruginosa* biofilm matrix, Proceedings of the National Academy of Sciences of the United States of America 112 (36) (2015) 11353–11358.

[60] M.J. Franklin, et al., Biosynthesis of the *Pseudomonas aeruginosa* extracellular polysaccharides, alginate, Pel, and Psl, Frontiers in Microbiology 2 (2011) 167.

[61] M.S. Byrd, et al., Genetic and biochemical analyses of the *Pseudomonas aeruginosa* Psl exopolysaccharide reveal overlapping roles for polysaccharide synthesis enzymes in Psl and LPS production, Molecular Microbiology 73 (4) (2009) 622–638.

[62] N. Kristo, et al., Immediate treatment of burn wounds with high concentrations of topical antibiotics in an alginate hydrogel using a platform wound device, Advanced Wound Care 9 (2) (2020) 48–60.

[63] H. Zhang, et al., Silver nanoparticles-doped collagen–alginate antimicrobial biocomposite as potential wound dressing, Journal of Materials Science 53 (21) (2018) 14944–14952.

[64] P.S. Shukla, et al., Carrageenans from red seaweeds as promoters of growth and elicitors of defense response in plants, Frontiers in Marine Science 3 (81) (2016).

[65] G.A. Paula, et al., Development and characterization of edible films from mixtures of κ-carrageenan, ι-carrageenan, and alginate, Food Hydrocolloids 47 (2015) 140–145.

[66] Z. Luo, et al., λ-Carrageenan P32 is a potent inhibitor of Rabies virus infection, PloS One 10 (2015).

[67] J.V. Diogo, et al., Antiviral activity of lambda-carrageenan prepared from red seaweed (Gigartina

skottsbergii) against BoHV-1 and SuHV-1, Research in Veterinary Science 98 (2015) 142−144.

[68] J.N. BeMiller, 13 − Carrageenans, in: J.N. BeMiller (Ed.), Carbohydrate Chemistry for Food Scientists, third ed., AACC International Press, 2019, pp. 279−291.

[69] M. Zhu, et al., Preparation, characterization and antibacterial activity of oxidized κ-carrageenan, Carbohydrate Polymers 174 (2017) 1051−1058.

[70] S.K. Bajpai, et al., Water absorption and antimicrobial behavior of physically cross linked poly (vinyl alcohol)/carrageenan films loaded with minocycline, Designed Monomers and Polymers 19 (7) (2016) 630−642.

[71] A.A. Oun, J.-W. Rhim, Carrageenan-based hydrogels and films: effect of ZnO and CuO nanoparticles on the physical, mechanical, and antimicrobial properties, Food Hydrocolloids 67 (2017) 45−53.

[72] S. He, et al., Antimicrobial activity and preliminary characterization of k-carrageenan films containing cinnamon essential oil, Advance Journal of Food Science and Technology 9 (2015) 523−528.

[73] A. Nouri, et al., Biodegradable k-carrageenan/nanoclay nanocomposite films containing Rosmarinus officinalis L. extract for improved strength and antibacterial performance, International Journal of Biological Macromolecules 115 (2018) 227−235.

[74] S. Roy, J.-W. Rhim, Carrageenan-based antimicrobial bionanocomposite films incorporated with ZnO nanoparticles stabilized by melanin, Food Hydrocolloids 90 (2019) 500−507.

[75] S. Roy, S. Shankar, J.-W. Rhim, Melanin-mediated synthesis of silver nanoparticle and its use for the preparation of carrageenan-based antibacterial films, Food Hydrocolloids 88 (2019) 237−246.

[76] D. Zofou, et al., In vitro and in vivo anti-Salmonella evaluation of pectin extracts and hydrolysates from "Cas Mango" (Spondias dulcis), Evidence-Based Complementary and Alternative Medicine 2019 (2019) 1−10.

[77] C. Lara-Espinoza, et al., Pectin and pectin-based composite materials: beyond food texture, Molecules 23 (4) (2018).

[78] S. Ye, et al., Facile and green preparation of pectin/cellulose composite films with enhanced antibacterial and antioxidant behaviors, Polymers 11 (2019) 57.

[79] B.R. Wagle, et al., Pectin or chitosan coating fortified with eugenol reduces Campylobacter jejuni on chicken wingettes and modulates expression of critical survival genes, Poultry Science 98 (3) (2018) 1461−1471.

[80] P.J.P. Espitia, et al., Edible films from pectin: physical-mechanical and antimicrobial properties − a review, Food Hydrocolloids (2014) 287−296.

[81] H. Wang, Pectin-chitosan polyelectrolyte complex nanoparticles for encapsulation and controlled release of nisin, American Journal of Polymer Science and Technology 3 (2017) 82.

[82] M.L. Cacicedo, et al., Hybrid bacterial cellulose−pectin films for delivery of bioactive molecules, New Journal of Chemistry 42 (9) (2018) 7457−7467.

[83] D. Sharma, D. Dhanjal, B. Mittal, Development of edible biofilm containing cinnamon to control food-borne pathogen, Journal of Applied Pharmaceutical Science 7 (2017) 160−164.

[84] T. Nisar, et al., Characterization of citrus pectin films integrated with clove bud essential oil: physical, thermal, barrier, antioxidant and antibacterial properties, International Journal of Biological Macromolecules 106 (2018) 670−680.

[85] P. Jantrawut, et al., Enhancement of antibacterial activity of orange oil in pectin thin film by microemulsion, Nanomaterials (Basel, Switzerland) 8 (2018), https://doi.org/10.3390/nano8070545.

[86] R. Sanaa, A. Rohaim, A.M.F.E.a.Y.M.A., Incorporation of Moringa leaves extract in Pectin-based edible coating as antimicrobial agent, Current Science International 07 (04) (2018) 602−615.

[87] X. Zhang, et al., Stimulation of wound healing using bioinspired hydrogels with basic fibroblast growth factor (bFGF), International Journal of Nanomedicine 13 (2018) 3897−3906.

[88] J. Kim, C.-M. Lee, Wound healing potential of a polyvinyl alcohol-blended pectin hydrogel containing Hippophae rahmnoides L. extract in a rat model, International Journal of Biological Macromolecules 99 (2017).

[89] N. Yu, et al., Development of antibacterial pectin from Akebia trifoliata var. Australis waste for accelerated wound healing, Carbohydrate Polymers (2019) 217.

[90] S. Pund, A. Joshi, Chapter 23 − Nanoarchitectures for neglected tropical Protozoal diseases: challenges and state of the art, in: A.M. Grumezescu (Ed.), Nano- and Microscale Drug Delivery Systems, Elsevier, 2017, pp. 439−480.

[91] 6 - Drug delivery systems, in: M.L. Bruschi (Ed.), Strategies to Modify the Drug Release from Pharmaceutical Systems, Woodhead Publishing, 2015, pp. 87−194.

[92] C.L. Del Toro-Sánchez, et al., Controlled release of antifungal volatiles of thyme essential oil from β-cyclodextrin capsules, Journal of Inclusion Phenomena and Macrocyclic Chemistry 67 (3) (2010) 431−441.

[93] E.H. Santos, et al., Characterization of carvacrol beta-cyclodextrin inclusion complexes as delivery systems for antibacterial and antioxidant applications, Lebensmittel-Wissenschaft und -Technologie- Food Science and Technology 60 (1) (2015) 583−592.

[94] J. Rakmai, et al., Antioxidant and antimicrobial properties of encapsulated guava leaf oil in hydroxypropyl-beta-cyclodextrin, Industrial Crops and Products 111 (2018) 219−225.

[95] Y. Liu, et al., Fabrication of electrospun polylactic acid/cinnamaldehyde/β-cyclodextrin fibers as an antimicrobial wound dressing, Polymers 9 (2017) 464.

[96] R. Satar, et al., Investigating the antibacterial potential of agarose nanoparticles synthesized by nanoprecipitation technology, Polish Journal of Chemical Technology 18 (2016).

[97] Y. Ko, et al., Antibacterial poly (3,4-ethylenedioxythiophene):poly(styrene-sulfonate)/agarose

nanocomposite hydrogels with thermo-processability and self-healing, Carbohydrate Polymers 203 (2019) 26–34.

[98] E. Wers, B. Lefeuvre, New hybrid agarose/Cu-Bioglass® biomaterials for antibacterial coatings, Korean Journal of Chemical Engineering 34 (8) (2017) 2241–2247.

[99] D. Ficai, et al., Chapter 13 – Advances in the field of soft tissue engineering: from pure regenerative to integrative solutions, in: A.M. Grumezescu (Ed.), Nanobiomaterials in Soft Tissue Engineering, William Andrew Publishing, 2016, pp. 355–386.

[100] T.K. Giri, 5 - Nanoarchitectured polysaccharide-based drug carrier for ocular therapeutics, in: A.M. Holban, A.M. Grumezescu (Eds.), Nanoarchitectonics for Smart Delivery and Drug Targeting, William Andrew Publishing, 2016, pp. 119–141.

[101] M. Nurunnabi, et al., Chapter 14 - Polysaccharide based nano/microformulation: an effective and versatile oral drug delivery system, in: E. Andronescu, A.M. Grumezescu (Eds.), Nanostructures for Oral Medicine, Elsevier, 2017, pp. 409–433.

[102] C. Weller, C. Weller, V. Team, 4 - Interactive dressings and their role in moist wound management, in: S. Rajendran (Ed.), Advanced Textiles for Wound Care, second ed., Woodhead Publishing, 2019, pp. 105–134.

[103] D. Jao, et al., Protein-based drug-delivery materials, Materials (Basel) 10 (5) (2017) 517.

[104] M. Yu, et al., Controlled release of naringin in metal-organic framework-loaded mineralized collagen coating to simultaneously enhance osseointegration and antibacterial activity, ACS Applied Materials and Interfaces 9 (23) (2017) 19698–19705.

[105] L. Ge, et al., Fabrication of antibacterial collagen-based composite wound dressing, ACS Sustainable Chemistry and Engineering 6 (7) (2018) 9153–9166.

[106] Y. Hu, et al., Surface modification of keratin fibers through step-growth dithiol-diacrylate thiol-ene click reactions, Materials Letters 178 (2016) 159–162.

[107] H.B. Tan, et al., Fabrication and evaluation of porous keratin/chitosan (KCS) scaffolds for effectively accelerating wound healing, Biomedical and Environmental Sciences 28 (3) (2015) 178–189.

[108] M. Zhai, et al., Keratin-chitosan/n-ZnO nanocomposite hydrogel for antimicrobial treatment of burn wound healing: characterization and biomedical application, Journal of Photochemistry and Photobiology B: Biology 180 (2018) 253–258.

[109] O. Shanmugasundaram, M. Ramkumar, Characterization and study of physical properties and antibacterial activities of human hair keratin–silver nanoparticles and keratin–gold nanoparticles coated cotton gauze fabric 47 (5) (2018) 798–814.

[110] C.D. Tran, F. Prosenc, M. Franko, Facile synthesis, structure, biocompatibility and antimicrobial property of gold nanoparticle composites from cellulose and keratin, Journal of Colloid and Interface Science 510 (2018) 237–245.

[111] M.E. Villanueva, et al., Smart release of antimicrobial ZnO nanoplates from a pH-responsive keratin hydrogel, Journal of Colloid and Interface Science 536 (2019) 372–380.

[112] S. Mohamadi, M. Hamidi, Chapter 3 – The new nanocarriers based on graphene and graphene oxide for drug delivery applications, in: E. Andronescu, A.M. Grumezescu (Eds.), Nanostructures for Drug Delivery, Elsevier, 2017, pp. 107–147.

[113] J.K. Williams, J.J. Yoo, A. Atala, Chapter 59 – Regenerative medicine approaches for tissue engineered heart valves, in: A. Atala, et al. (Eds.), Principles of Regenerative Medicine, third ed., Academic Press, Boston, 2019, pp. 1041–1058.

[114] F. Boccafoschi, et al., Biological grafts: surgical use and vascular tissue engineering options for peripheral vascular implants, in: R. Narayan (Ed.), Encyclopedia of Biomedical Engineering, Elsevier, Oxford, 2019, pp. 310–321.

[115] J. Li, et al., Preparation and antibacterial properties of gelatin grafted with an epoxy silicone quaternary ammonium salt, Journal of Biomaterials Science, Polymer Edition 27 (10) (2016) 1017–1028.

[116] S. Shankar, et al., Gelatin-based dissolvable antibacterial films reinforced with metallic nanoparticles, RSC Advances 6 (71) (2016) 67340–67352.

[117] Q. He, et al., Fabrication of gelatin–TiO_2 nanocomposite film and its structural, antibacterial and physical properties, International Journal of Biological Macromolecules 84 (2016) 153–160.

[118] N. Cai, et al., Tailoring mechanical and antibacterial properties of chitosan/gelatin nanofiber membranes with Fe_3O_4 nanoparticles for potential wound dressing application, Applied Surface Science 369 (2016) 492–500.

[119] M. Divya, et al., Biopolymer gelatin-coated zinc oxide nanoparticles showed high antibacterial, antibiofilm and anti-angiogenic activity, Journal of Photochemistry and Photobiology B: Biology 178 (2018) 211–218.

[120] J. Lin, et al., Antibacterial zinc oxide hybrid with gelatin coating, Materials Science and Engineering: C 81 (2017) 321–326.

[121] S. Shankar, et al., Preparation, characterization, and antimicrobial activity of gelatin/ZnO nanocomposite films, Food Hydrocolloids 45 (2015) 264–271.

[122] H. Li, et al., Electrospun gelatin nanofibers loaded with vitamins A and E as antibacterial wound dressing materials, RSC Advances 6 (55) (2016) 50267–50277.

[123] B. Lu, et al., Healing of skin wounds with a chitosan–gelatin sponge loaded with tannins and platelet-rich plasma, International Journal of Biological Macromolecules 82 (2016) 884–891.

[124] A.G. Krishnan, et al., Evaluation of antibacterial activity and cytocompatibility of ciprofloxacin loaded gelatin–hydroxyapatite scaffolds as a local drug delivery system

for osteomyelitis treatment, Tissue Engineering Part A 21 (7−8) (2015) 1422−1431.

[125] J. Rivadeneira, et al., Evaluation of the antibacterial effects of vancomycin hydrochloride released from agar−gelatin−bioactive glass composites, Biomedical Materials 10 (1) (2015) 015011.

[126] H. Li, et al., Preparation, characterization, antibacterial properties, and hemostatic evaluation of ibuprofen-loaded chitosan/gelatin composite films, Applied Polymer 134 (42) (2017) 45441.

[127] T.C. Gasser, Chapter 8 − Aorta, in: Y. Payan, J. Ohayon (Eds.), Biomechanics of Living Organs, Academic Press, Oxford, 2017, pp. 169−191.

[128] J.C. Rodríguez-Cabello, et al., Elastin-like polypeptides in drug delivery, Advanced Drug Delivery Reviews 97 (2016) 85−100.

[129] A.M. Pereira, et al., P-265 − Development of antimicrobial protein-based polymers for biomedical applications, Free Radical Biology and Medicine 120 (2018) S125.

[130] P. Kour, et al., Comparative evaluation of antimicrobial efficacy of platelet-rich plasma, platelet-rich fibrin, and injectable platelet-rich fibrin on the standard strains of porphyromonas gingivalis and aggregatibacter actinomycetemcomitans, Contemporary Clinical Dentistry 9 (Suppl. 2) (2018) S325−s330.

[131] P.S. Badade, et al., Antimicrobial effect of platelet-rich plasma and platelet-rich fibrin, Indian Journal of Dental Research 27 (3) (2016) 300−304.

[132] A.B. Castro, et al., Antimicrobial capacity of leucocyte- and platelet rich fibrin against periodontal pathogens, Scientific Reports 9 (1) (2019) 8188.

[133] H. Khorshidi, et al., Does adding silver nanoparticles to leukocyte- and platelet-rich fibrin improve its properties? BioMed Research International 2018 (2018) 1−5.

[134] J. Banerjee, et al., Delivery of silver sulfadiazine and adipose derived stem cells using fibrin hydrogel improves infected burn wound regeneration, PloS One 14 (6) (2019) e0217965.

[135] J. Gil, et al., A PEGylated fibrin hydrogel-based antimicrobial wound dressing controls infection without impeding wound healing, International Wound Journal 14 (6) (2017) 1248−1257.

[136] J. Chotitumnavee, et al., In vitro evaluation of local antibiotic delivery via fibrin hydrogel, Journal of Dental Sciences 14 (1) (2019) 7−14.

[137] R. Tan, et al., Antibacterial effect of antibiotic-saturated fibrin sealant; in vitro study, Journal of Wound Management and Research 14 (2018) 12−17.

[138] B.M. Alphonsa, et al., Antimicrobial drugs encapsulated in fibrin nanoparticles for treating microbial infested wounds, Pharmaceutical Research 31 (5) (2014) 1338−1351.

[139] s. Kara, et al., Fibrin sealant as a carrier for sustained delivery of antibiotics, Journal of Clinical and Experimental Investigations 5 (2014) 194−199.

[140] J. Kaur, et al., Facts and myths of antibacterial properties of silk, Biopolymers 101 (3) (2014) 237−245.

[141] V.G. Nadiger, S.R. Shukla, Antibacterial properties of silk fabric treated with silver nanoparticles, Journal of the Textile Institute 107 (12) (2016) 1543−1553.

[142] C. Shi, et al., An antibacterial and absorbable silk-based fixation material with impressive mechanical properties and biocompatibility, Scientific Reports 6 (2016) 37418.

[143] P.K. Baruah, et al., Antibacterial effect of silk treated with silver and copper nanoparticles synthesized by pulsed laser ablation in distilled water, AIP Conference Proceedings 1953 (1) (2018) 030064.

[144] G. Li, et al., Preparation of antibacterial degummed silk fiber/nano-hydroxyapatite/polylactic acid composite scaffold by degummed silk fiber loaded silver nanoparticles, Nanotechnology 30 (29) (2019) 295101.

[145] S. Patil, N. Singh, Antibacterial silk fibroin scaffolds with green synthesized silver nanoparticles for osteoblast proliferation and human mesenchymal stem cell differentiation, Colloids and Surfaces B: Biointerfaces 176 (2019) 150−155.

[146] P. Chitichotpanya, P. Pisitsak, C. Chitichotpanya, Sericin−copper-functionalized silk fabrics for enhanced ultraviolet protection and antibacterial properties using response surface methodology 89 (7) (2019) 1166−1179.

[147] K.V. Rani, B. Sarma, A. Sarma, Plasma pretreatment on tasar silk fabrics coated with zno nanoparticles against antibacterial activity, Surface Review and Letters 26 (5) (2019) 1850193.

[148] P. Manickam, G. Thilagavathi, Development of antibacterial silk sutures using natural fungal extract for healthcare applications, Journal of Textile Science and Engineering 06 (2016).

[149] A. Khan, Ultraviolet Protection and Antibacterial Properties of Silk Fabric Dyed with Cinnamomum Camphora Plant Leaf Extract, 2018.

[150] S. Zhang, Y. Han, Preparation, characterisation and antioxidant activities of rutin-loaded zein-sodium caseinate nanoparticles, PloS One 13 (3) (2018) e0194951.

[151] Y. Zhang, K. Pan, Q.J.F.B. Zhong, Eugenol nanoencapsulated by sodium caseinate: Physical, Antimicrobial, and Biophysical Properties 13 (1) (2018) 37−48.

[152] A. Basu, et al., 8 − Polyurethanes for controlled drug delivery, in: S.L. Cooper, J. Guan (Eds.), Advances in Polyurethane Biomaterials, Woodhead Publishing, 2016, pp. 217−246.

[153] A.S. Dutta, 2 − Polyurethane foam chemistry, in: S. Thomas, et al. (Eds.), Recycling of Polyurethane Foams, William Andrew Publishing, 2018, pp. 17−27.

[154] M. Niaounakis, 1 − Introduction to biopolymers, in: M. Niaounakis (Ed.), Biopolymers Reuse, Recycling, and Disposal, William Andrew Publishing, Oxford, 2013, pp. 1−75.

[155] I. Francolini, A. Piozzi, 12 − Antimicrobial polyurethanes for intravascular medical devices, in: S.L. Cooper, J. Guan (Eds.), Advances in Polyurethane Biomaterials, Woodhead Publishing, 2016, pp. 349−385.

[156] L.C. Xu, C.A. Siedlecki, 9 — Antibacterial polyurethanes, in: S.L. Cooper, J. Guan (Eds.), Advances in Polyurethane Biomaterials, Woodhead Publishing, 2016, pp. 247–284.

[157] K.M. Zia, et al., Recent developments and future prospects on bio-based polyesters derived from renewable resources: a review, International Journal of Biological Macromolecules 82 (2016) 1028–1040.

[158] C.-H. Chen, et al., Long-term antibacterial performances of biodegradable polylactic acid materials with direct absorption of antibiotic agents, RSC Advances 8 (29) (2018) 16223–16231.

[159] Z. Chu, et al., Characterization of antimicrobial poly (lactic acid)/nano-composite films with silver and zinc oxide nanoparticles, Materials (Basel, Switzerland) 10 (6) (2017) 659.

[160] I. Restrepo, P. Flores, S. Rodríguez-Llamazares, Antibacterial nanocomposite of poly(lactic acid) and ZnO nanoparticles stabilized with poly(vinyl alcohol): thermal and morphological characterization, Polymer-Plastics Technology and Materials 58 (1) (2019) 105–112.

[161] W. Li, et al., Development of antimicrobial packaging film made from poly(lactic acid) incorporating titanium dioxide and silver nanoparticles, Molecules 22 (7) (2017).

[162] L. Techawinyutham, et al., Antibacterial and thermomechanical properties of composites of polylactic acid modified with capsicum oleoresin-impregnated nanoporous silica, Journal of Applied Polymer Science 136 (31) (2019) 47825.

[163] L. Ma, et al., A new antimicrobial agent: poly (3-hydroxybutyric acid) oligomer, Macromolecular Bioscience 19 (2019) 1800432.

[164] M.A. Abdelwahab, et al., Evaluation of antibacterial and anticancer properties of poly(3-hydroxybutyrate) functionalized with different amino compounds, International Journal of Biological Macromolecules 122 (2019) 793–805.

[165] C. Veranitisagul, et al., Antimicrobial, conductive, and mechanical properties of AgCB/PBS composite system, Journal of Chemistry 2019 (2019) 1–14.

[166] S. Wiburanawong, N. Petchwattana, S. Covavisaruch, Carvacrol as an antimicrobial agent for poly(butylene succinate): tensile properties and antimicrobial activity observations, Advanced Materials Research 931–932 (2014) 111–115.

CHAPTER 6

Biodegradable Antibiotics in Wound Healing

DANIEL HASSAN • VICTORIA OLUWASEUN FASIKU • SHADRACK JOEL MADU • JAMILU MUAZU

1 INTRODUCTION

Skin remains the largest organ in the human body and is damaged easily by injuries or chronic diseases. According to the World Health Organization, annually, about 265,000 deaths occur owing to burns and lack of patients' access to appropriate treatments including proper skin substitutes [1]. Wound healing remains a challenge due to healing complexity [2,3] that starts at the onset of injury [4] and affects millions of people causing billions of economic loss [5]. Wound healing still remains a serious complication that negatively affects the patient's socioeconomic life [6], quality of life [7], and involved a complicated biological process including numerous physiological factors [8]. Although wound healing has been in practice since ancient times, modern wound healing provides basic protection and accelerated healing process. These wound healing processes involved hemostatis, inflammation, proliferation/tissue formation (angiogenesis and tissue remodelling), and remodelling stages. The remodelling stages; a series of four overlapping, complex, dynamic, and sequential process is associated [9,10] viz hemostasis, inflammation, proliferation [11], and remodeling causing damage to the cells [12], promoting aesthetic repair due to nontoxic, transmit water vapor, sustained enhancement, and prolonged shelf life [13–15]. Several researchers have suggested different steps and processes involved in wound healing [16]. Currently, wound healing limitations have shown to promote the moist environment, rapid wound healing, mechanical protection, noncytotoxic to healthy tissue, antimicrobial/antifungal effect, easy use, and acceptance to patients with the needs to develop innovative wound healing [17]. Although wound healing occurs within a short time, wound especially from diabetic patients could interpose from weeks to even months [14]. The tissue healing process in this type of wound involves direct cell-to-cell and cell–matrix interaction, and the prevention of microbial infection and transepidermal water loss can be accelerated by appropriate internal and external indications depending on the extracellular properties leading to skin barrier restoration [18].

Several years ago, the fabrication of bioactive materials based on natural polymers has been used extensively for wound healing [19]. Effective wound healing materials such as plant fibers, animal fats, and honey pastes are known to be suitable for complete healing to prevent infections [20]. Several studies have suggested that good wound healing materials must present significant properties such as biocompatibility, nontoxicity [21], injectability, excessive swelling, and difficulty to conform to deep, narrow wounds or wounds of irregular shapes [2]. According to Buzarovska et al. (2019), they suggested that wound healing materials should have a high level of water-uptake ability to help accumulate excess fluid from the wound governed by the porosity and thickness of the biodegradable materials [22].

Several disadvantages such as duration of suturing, inability to bind adequately in wet areas, low tensile strength, high cost, and low mechanical strength are also found to be associated with the wound healing process [16]. To overcome these problems, numerous wound healing biodegradable materials with the use of nanotechnology to synthesize films, hydrogels, foams, and hydrocolloids have been developed to accelerate wound healing [23]. Nanotechnology is an emerging technology extensively used in wound healing. This technology process offers several advantages such as an organized network in coagulation, inflammation at wound sites, the proliferation of aggregating factors, and restructuring of tissues [4]. Different methods of wound healing are available such as suture,

Antibiotic Materials in Healthcare. https://doi.org/10.1016/B978-0-12-820054-4.00006-9

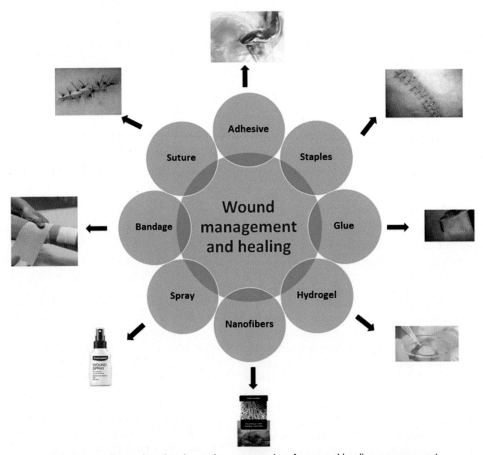

FIG. 6.1 An illustration showing various approaches for wound healing management.

adhesive, bandage, spray, hydrogel, glue, staples, and nanofibers as seen in Fig. 6.1. Furthermore, the use of biodegradable materials in wound healing is becoming at the age front of nanotechnology.

Scaffolding carrier is one of the drug-delivery technologies for the sustained release of molecules at the local targeted site. Several types of scaffolds such as celluloses, gellan, curdlan, hyaluronic acid [24], pullulan, and gelatin scaffolds have been widely used for wound dressings due to high hydrophilicity, biocompatibility and less toxicity compared to synthetic polymers. During wound healing, they attach tightly to a receiver on the fibroblast's surface and lead to the generation of an extracellular matrix [25]. However, they show disadvantageous for the wound healing in terms of wound contraction, scar formation, and poor integration with host tissue, resulting in imperfect healing or even the cosmetic concern [26]. Nowadays, the incorporation of various antibiotics and nanoparticles into electrospun nanofibers is one of the practical choices for wound healing [9].

Biodegradable materials have been widely studied for their applications in biomedical fields including wound healing. They are specifically preferred in the development of therapeutic devices, including temporary implants and three-dimensional scaffolds for tissue engineering [27]. This is evident as the characters displayed by the biodegradable materials make them exceptional, thereby replacing other material classes such as metals, alloys, and ceramics in biomedical applications [27,28]. The increasing relevance of biodegradable materials in wound healing is associated but not limited to their long-term biocompatibility and bioactivity particularly common with natural biodegradable materials, permeability, and processability for its intended application, and mechanical properties that promote regeneration and degradation products are nontoxic and easily metabolized and cleared from the body [29]. The origin of biodegradable materials can be natural or synthetic depending on their source and the presence of naturally occurring extracellular matrix. Medicines topically administered via the skin fall into two general groups: those applied for local action and those meant to exert systemic effects [30]. The foremost biodegradable materials dominating the wound care industry include cellulose, alginate, hyaluronate, collagen, and chitosan [28]. Several biodegradable materials have been reported for effective wound healing applications. The following section focuses on the use of biodegradable materials such as polyvinyl alcohol (PVA), for preparation/formulation and application in wound healings. Various studies have reported different materials, foams, films, hydrocolloids, hydrogels, and hydrofibers to achieve wound healing [8]. It is well established that an optimal environment over the wound to accelerate wound healing acts as a barrier against elements, against the invasion of the pathogenic organism, and be able to absorb wound exudates to prevent tissue maceration [31].

2 BIODEGRADABLE POLYMERS IN WOUND HEALING

Wound healing is a complicated process that regularly requires treatment with antibiotics to bypass infection, which affects the tissue regeneration process. It has been demonstrated that excellent scaffold materials for wound healing should combine extracellular matrix and antibacterial properties for tissue growth to avoid bacterial colonization [32]. Despite the progress made in developing wound healing materials such as films [33], hydrogels [23,33], hydrofibers [34], and foams [35]. Several disadvantages are associated with the materials and their composition; for example, cytotoxicity, nonbiodegradability, and perfect healing ability by the materials have demonstrated limitations. These limitations have called for the development of novel materials that are due to their biodegradable, nontoxic, and flexibility for enhanced wound healing. These scaffolds include but are not limited to a range of dermal matrices as well as materials and composites such as chitosan/PLA, PVA/alginate, and cellulose/collagen. These biodegradable materials have good biocompatibility [36]. Infections in wound healing from infections still remain a major challenge complication for wound management in public health management [37].

The function of biodegradable wound healing materials is to stop bleeding, absorb exudates, and augment repair due to their inherent wound healing properties [38]. However, several natural biodegradable polymers such as fibronectin [39], hydroxyapatite [40,41], polypeptides [42,43], hyaluronic acid [44], chitosan [45], alginates [46], gellan gum [47], and polypeptides [42,43] have been used in wound healing application as shown in Table 6.1. These biomaterials

can enhance wound healing and give huge supports for skin development [3], and they have also demonstrated advantages such as low toxicity, low chronic inflammatory response [38], biocompatibility, and biodegradability. However, these properties do not make them perfect candidates for wound healing. These natural biodegradable polymers have suffered from a few disadvantages such as low toxicity, low inflammatory response [38], slow degradation velocity [48], and adhesive strength [36].

Interestingly, the combination of natural and synthesized biodegradable polymers includes polyurethane (PLE) [49], poly(γ-glutamic acid) (γ-PGA) [50], polylactic acid (PLA) [51], poly(L-lactic acid)-co-poly-(epsilon-caprolactone) [5], parylene (poly(p-xylylene) [52], poly lactic-co-glycolic acid (PLGA) [53], and PVA [25]. The combination of both synthesized and natural biodegradable polymers have demonstrated several advantages such as rapid adhesion [54], accelerated wound healing [55], protection against infections, improve mechanical properties [3], good adhesive strength [56], and good biocompatibility [36] for potential wound healing.

2.1 Structure, Synthesis, and Properties of Biodegradable

The synthesis, structures, and properties of biodegradable polymers are very important characteristics to consider in wound healing application. Table 6.1 identifies various biodegradable polymers for wound healing.

3 APPLICATION OF BIODEGRADABLE POLYMERS IN WOUND HEALING

Novel antibiotic eluting biodegradable for wound healing based on a polyglyconate mesh and a porous poly-(DL-lactic-co-glycolic acid) (PDLGA) binding matrix was developed and studied. Their study investigated the release profiles of gentamicin and ceftazidime from wound healing and on their effect on bacterial inhibition. Release profiles from dressings derived from the emulsions stabilized with BSA demonstrated a higher burst release of the antibiotics, followed by a gradual release at a decreasing rate over time. However, release profiles from dressings derived from emulsions stabilized with Span demonstrated a low burst release followed by a lower rate of release. Their studies showed that wound healing is highly effective [60]. In another study by Dzikowski et al. (2017) blended with poly-(ε-caprolactone) and poly-D,L-(lactic acid) (PCL/PDLLA) polymers containing ciprofloxacin (CIF) by the jet-spraying method. The incorporation CIF dispersed the antibiotic in the polymer solutions. The result demonstrated that crystal CIF was distributed in

TABLE 6.1
Structure of Biodegradable Polymers Used in Wound Healing Application.

S/N	Structures	Name	Properties	Ref
1		Polyurethane	• Biocompatible • Good barrier properties • Better oxidation stability	[49]
2		Poly (γ-glutamic acid)	• Biocompatible • Nontoxic	[57]
3		Polylactic acid (PLA)	• Biocompatible • Good barrier capacity to oxygen	[51]
4		Poly lactic-co-glycolic acid (PLGA)	• Biocompatible	[53]
5		Hyaluronic acid	• Biocompatible • Nontoxicity • Nonantigenicity • Cell surface interactions • Cell migration • Proliferation • Angiogenesis	[44]

	Polymer	Structure	Properties	Reference
6	Polyvinyl alcohol (PVA)		• Nontoxic • Biocompatible • Better fiber-forming	[25]
7	Chitosan		• Nontoxic • Biocompatible • Antibacterial • Antifungal • Antiviral activity	[58]
8	Alginates		• Biocompatible • High adsorption • Gas permeability • High flame retardancy • High permeability for gases	[46]
9	Gellan gum		• Biocompatible • Low cytotoxicity	[3]
10	Polyethylene glycol		• Nontoxic • Nonimmunogenic • Nonantigenic	[59]

the fibers and embedded to control the release of the ciprofloxacin (CIP) through the degradation of the PCL/PDLLA rate. The highly porous structures of the biodegradable fibers swift cellular invasion and colonization causing no detrimental effect from the antibiotic [15]. Interestingly, Nafee et al. (2013) synthesized a novel amphiphilic diblock copolymers (PCL-PEG) to develop nanocarrier for the encapsulation of hypericin a natural antimicrobial potent photosensitizer and assessed against biofilm and planktonic cells of methicillin-resistant *Staphylococcus aureus* on infected wounds in rats. The nanoparticles were around 45 nm in diameter and demonstrated higher inhibition of biofilm over planktonic cells with faster wound healing, better epithelialization, keratinization, and development of collagen fibers [61]. Another novel nanosystem synthesized prepared by the lyophilization process is Ag/ZnO sponge. The nanosystems were used to enhance antibacterial activity due to low cytotoxicity. The sponge-like nanosystem (Ag/ZnO)-loaded chitosan were evaluated against drug-resistant pathogen bacteria. The in vitro and in vivo evaluation in mice revealed that the chitosan-Ag/ZnO nanoparticles enhanced wound healing and promoted reepithelialization and collagen deposition [62]. To combat such limitations, Li et al. 2019 synthesized chitosan-glyoxal/Polyvinylprrolidone/MoS$_2$ (CSG/PVP/MoS$_2$) via the hydrothermal-ultrasonic method. The results revealed that the antibiotics release kinetics from the nanosystems (CSG/PVP/MoS$_2$) followed the pseudo-first-order model, thereby demonstrating the photodegradation of the diclofenac sodium mechanism [63]. This section discusses various polymers that have been studied and employed in wound healing.

3.1 Polyurethane

PLE is a synthetic material consisting of repeated block segments that provide unique properties and a wide range of applications [49]. Biodegradable polyurethanes are a class of biomaterials that have received much attention due to potential applications in tissue engineering, regenerative medicine, controlled drug delivery, wound healing dressings, and bone fixation [64]. Owing to their tunable physicochemical properties, they are processed into several forms, including sheets, sponges, and foam. The absorbency, thickness, good barrier properties, elasticity, oxygen permeability, thermal stimulation [65], and high-temperature resistance [66] provide adequate mechanisms for wound healing area [67] and pore size for desired application [14] due to excellent physical and mechanical performance [68,69]. The bilayered structure of polyurethane is composed of the microporous top layer and highly

porous sponge-like sublayer that prevents dehydration, bacterial penetration, and bullae formation [70].

In a study by Khodabakhshi et al. (2019), polyurethane foams were fabricated by solvent casting/salt leaching method and investigated for wound healing. The prepared wound dressings foams were later coated with water extract of propolis (an antibacterial agent). The in vitro study revealed that wound healing coated with water extract of propolis enhanced cytocompatibility (cell viability), which is biocompatible and applicable for wound dressing [71].

Despite advantages such as excellent water absorption capability, optimal mechanical properties, and unequaled economic loss, limitations such as low bioactivity and poor healing capability surround polyurethane. To solve these limitations associated with polyurethane in wound healing, Song et al. (2017) fabricated hybridized bioactive silica nanoparticles with polyurethane. The hybridization with silica did not affect the intrinsically porous microstructure of polyurethane foams with silica contents and was effective for wound healing [72]. This is due to the polyurethane antibacterial properties, natural composition with excellent fluid handling, and healing properties [22].

3.2 Poly(γ-Glutamic Acid)

Poly-γ-glutamic acid (γ-PGA) is a natural material [50] produced by several *Bacillus* species derived from soya beans fermentation, and it acts as a protective barrier against bacteriophages and host immune systems and also as a nutrient reservoir [73,74]. γ-PGA is rich in glutamate composed of D- and L-glutamate monomers [75] with a molecular mass of 9−25 kD that are linked by amide bonds between the α-amino and γ-carboxylic acid groups [76] and have a strong affinity for hydroxyapatite (HAp) and enamel [50]. Owing to its nontoxic nature, biodegradability [75], large amount of free carboxyl groups in its main chain [5], and water-solubility of γ-PGA, it can be used in a variety of fields, including the food industries [77], sensing [78], medicine, cosmetic industries [76,79], drug delivery, wound dressing [25], and tissue engineering [78].

Wound healing using biodegradable material such as poly(γ-glutamic acid) has attracted researchers' attention globally especially novel multifunctionalγ-PGA/gelatin hydrogel. Furthermore, water-soluble γ-PGA salts of sodium and calcium forms are chosen for their good biocompatibility, biodegradability, and water absorption capacity [80]. γ-PGA-chitosan-based was prepared by Kim et al. (2018) with porous matrices through dehydration and freeze gelation. These matrices enhanced serum protein adsorption and cytocompatibility. The nanosystems revealed good stability, minimal macrophage uptake in vitro, and enhanced permeation and retention effect in vivo for wound healing [81].

In a study by Wang et al. (2012), nanosystems for wound healing were fabricated using the injectable γ-PGA-based hydrogels with low molecular weight chitosan and high molecular weight chitosan through electrostatic interaction and hydrophobic nanofibers for biomedical applications. The study revealed that γ-PGA prepared with either low molecular weight chitosan or high molecular weight chitosan did not form a hydrogel. However, the combination of low molecular weight chitosan and high molecular weight chitosan produced physically thermosensitive hydrogel, and it was suitable for wound healing due to its unnoticeable acute biocompatibility issue [78].

3.3 Polylactic Acid

PLA is the most extensively studied and industrialized biopolymer [51], among the most widespread biodegradable plastics, due to high mechanical performance, biodegradability. It has found relevance biocompatibility in several fields such as food packaging, tissue engineering [82], smart delivery systems [83], drug delivery [84] surgical healing products, orthopedic devices, and implant cell carriers [51]. However, PLA suffers from low ductility and toughness, low crystallization rate [51], limited tensile deformation, uncontrollable degradation rate, poor hydrophilic properties, and poor swelling properties, and this makes it irrelevant in wound dressings [85].

Thus, ideal wound dressing material is characterized by high porosity and high surface-to-volume ratio required for oxygen permeability, absorption of exudates [18], biodegradable to fabricate a wound healing with high elastic, controllable degradation rate, high hydrophilicity, and swelling properties by electrospinning [85,86]. For this purpose, devices developed that could be used for wound dressing at wound sites where a water-resistant dressing is required to aid in the prevention of bacterial contamination as a primary dressing [86].

3.4 Poly Lactic-Co-Glycolic Acid

PLGA is a biodegradable copolymer generated from poly(lactic acid) and poly(glycolic acid) that offers a potential substrate for growing cells for use in regenerative therapies [87] and has recently attracted more attention in wound healing [88]. They are widely reported as a scaffold biomaterial due to its good biosafety, biocompatibility [89], biodegradability [90], wound healing ability [91], and controlled drug delivery [92]. However, unmodified PLGA substrates are poor substrates for cellular attachment and growth due to low surface energy and high hydrophobicity [87].

The application of exogenous lactate released from PLGA polymer accelerated angiogenesis and wound healing processes [93]. Wound dressings should ideally be able to maintain high humidity, remove excess wound exudate, permit thermal insulation, provide certain mechanical strength, and in some cases deliver antibiotics to prevent infections. Until now, none of the existing wound dressing products can meet all these requirements [94]. Wound treatment and its medical complications remain one of the most prevalent and economically challenging healthcare issues in the world. Acute wounds heal in a very orderly, timely and efficient manner characterized by four distinct, strictly connected but overlapping phases: hemostasis, inflammation, proliferation, and remodeling [53].

Hlaing et al. (2018) developed S-nitrosoglutathione nitric oxide donor-loaded PLGA microparticles (GSNO-MPs) that releases NO over a prolonged period, to accelerate wound healing with less frequent dosing. The GSNO was successfully encapsulated into PLGA microparticles by a solid-in-oil-in-water emulsion solvent evaporation method. Furthermore, GSNO-MPs had good antibacterial efficacy and were found to accelerate wound healing in a mouse model. In this study, GSNO-MPs effectively enhanced wound healing in a mouse model via the delivery system with a sustained NO release [95].

To design a wound dressing, a study by Lui et al. (2018) formulated CIP an antibiotic, loaded electrospun hydrophobic PLGA fibrous mats modified with hydrophilic sodium alginate (ALG) microparticles. Their results showed that ALG could improve the wettability, water absorption capacity, and enhance the burst release rate of ciprofloxacin from the PLGA fibrous mats. This study demonstrated the potential of combining hydrophilic that improved the wettability and hydrophobic polymers to design the desired wound dressings via the electrospinning process [94]. In another study, the biodegradable was used to prepare bead-free curcumin-loaded PLGA nanofibers (CPNF). The CPNF shows smooth nanofiber with an average diameter of 100−300 nm with high yield and exhibits high drug encapsulation efficiency. The in vitro curcumin release studies from the CPNF indicate that curcumin was released in a sustained manner over a prolonged period with no burst release. The cell viability assay demonstrated the efficacy of the curcumin nanofibers on skin cancer (squamous carcinoma) cell lines. The CPNF is capable of delivering curcumin over a long period in a controlled manner [92].

Curcumin is reported as a well-known topical wound healing agent. A study by Chereddy et al. (2013) encapsulated curcumin PLGA nanoparticles to potentially accelerate wound healing with a twofold higher wound healing activity compared to that of PLGA or curcumin. The encapsulated curcumin-PLGA-nanoparticles offered numerous benefits from light degradation, enhanced water solubility, and showed a sustained release of curcumin over a period of 8 days. They demonstrated and suggested the additive effect of lactic acid from PLGA and encapsulated curcumin for the active healing of wounds because of its innate lactic acid activity and sustained drug release [93].

The successful development of optimal PLGA/ginsenoside coating on stainless steel substrate brings up a potential alternative to the current drug-eluting stent development. This composition demonstrates a drug-release profile governed by the combination of diffusion and swelling-controlled of PLGA with drug release in a month that is desirable to provide optimum ginsenoside dose over a period. This study suggests the potential application of PLGA/ginsenoside coating for developing novel herbal-based drug-eluting stents [96].

3.5 Polyvinyl Alcohol

PVA is a semicrystalline polymer produced by the hydrolysis of polyvinyl acetate and has attracted the researcher's attention [97]. PVAs are reported to be water-soluble, biodegradable polymer [98], highly biocompatible, excellent hydrophilic [99], nontoxic, good hydrophilicity [100], low price [98], biologically stable polymer that possesses excellent film-forming, emulsifying, adhesive, and flexible mechanical properties [13]. The addition of a synthetic polymer like PVA biodegradable and biocompatible polymer improves the mechanical properties of the material and aids in regulating the release of the encapsulated biomolecules and for various wound healing and tissue engineering applications [9,12]. For instance, PVA/CS/nZnO nanocomposite hydrogel wound dressings were successfully designed and produced via the response surface methodology (RSM) method. The results of the toxicity and antibacterial activity study revealed that the hydrogels were biocompatible and nontoxic and were significantly able to protect the wounds against microorganisms [101]. In another study by Ahmed et al. (2018), chitosan/PVA/ZnO nanofiber membranes were synthesized by the electrospinning technique. The samples of chitosan/PVA and chitosan/PVA/ZnO nanofiber membranes also manifested higher antioxidant properties compared to the control (ascorbic acid) that makes them promising candidates for applications in the wounds healing [102].

In addition, nanocomposite sponges of sodium alginate/graphene oxide/polyvinyl alcohol were prepared by a freeze-thawing cyclic process and freeze-dried molding for effective wound healing process [103]. Furthermore, PVA hydrogel blended with thermosensitive sodium alginate-g-N-isopropyl acrylamide (SA-g-NIPAM) to produce thermosensitive hydrogel (PVA/SA-g-NIPAM) has been reported to control response to reinfected wounds healing [104]. In addition, poly(vinyl alcohol)/dextran-aldehyde (PVA/DA) hydrogel dressing formulated by Zheng et al. (2019) exhibited good fluid absorbability and moisture vapor transmission ability and highly porous microstructure. All the hydrogels enhanced the wound healing process, and they also showed good biocompatibility proved by very low hemolytic potential and cytotoxicity [105]. The PVA/DA hydrogels had a three-dimensional, highly porous structure with uniformly distributed pores of $5-10$ μm, strong tensile strength of 5.6 MPa, ability to absorb fluid of 6 times its weight and suitable water vapor transmission rate of 2100 g m^{-2} day^{-1} to keep a moist environment and good biocompatibility [105]. In another study by Alipour et al. (2019), AgNPs/PVA/PVP/PEC/MF nanofibers were prepared and used as dressing for the wounds. It was shown that the healing rate depended on the Ag content of the nanofibers. The higher the Ag content, the more the inhibitory effect on the gram-negative and gram-positive bacteria. Ag$^+$ release studies, on the other hand, showed that the rate of release mainly followed a diffusion method. This is one of the few reports on the application of AgNPs/PVA/PVP/PEC/MF nanofibers as a reliable wound dressing [9].

In another study by Gao et al. (2018), AgSD/PVA hydrogels were prepared via e-beam irradiation and their potential use in antimicrobial wound dressing was evaluated. The cytotoxicity of AgSD/PVA-WS hydrogels was lower compared to AgSD/PVA-AS hydrogels. However, they observed no obvious difference between the AgSD/PVA-WS and AgSD/PVA-AS samples either AgSD contents of 0.75% or 1% (w/w). Hydrogels loaded with 0.75% (w/w) AgSD showed the highest tensile strength and elongation at break of 0.13 MPa, 203.82% for AgSD/PVA-WS and 0.21 MPa, 195.83% for AgSD/PVA-AS, respectively [97]. Furthermore, Abdel-Mohsen et al. (2019) synthesized a novel wound dressing nanofibrous mats from nanocomposite polyvinyl alcohol/hyaluronan-silver nanoparticles (PVA/HA-AgNPs) with controlled size and shape of AgNPs by safe and green processing. Furthermore, a study by Abdel-Mohsen demonstrated that the novel nanofibrous mats (PVA/HA-AgNPs) nanocarriers for in vitro and in vivo applications remonstrated a massive reduction of coagulation and from the immune system in the wound healing process. The new nanocomposite nanofibrous mats might be useful in a wide variety of medical and biological purposes that take advantage of the biological activates of HA and antimicrobial properties of AgNPs [106].

3.6 Chitosan/Poly(Lactic Acid)

PLA is a thermoplastic biopolymer obtained from renewable natural sources. It has demonstrated several properties such as biodegradability ideal for multiple applications [107]. PLA has been extensively studied to fabricate scaffolds because it is a renewable resource that is biocompatible, biodegradable, highly porous, and controllable, exhibiting excellent thermal and mechanical properties. In addition, it has been approved by the FDA for various applications [8,108]. It is cost-effective [109] and nontoxic [110]. On the other hand, chitosan a biopolymer is biocompatible and biodegradable and possesses antimicrobial activity ideal for wound healing application [107]. To harness these properties from each polymer, nanofibers of poly(lactic acid)/chitosan were prepared with different weight ratios via electrospinning [111] and investigated for potential wound healing. The PLA-CS nanofibers showed a two-stage release behavior of curcumin drug with an initial burst release followed by a sustained release of curcumin for potential wound healing. Hence, it has potential applications in some biomedical areas such as wound dressing and drug delivery [112].

In another study by Siqueira et al. (2015), fiber mats of blends of PLA/CS were prepared with a submicron range of diameters, from different concentrations of two polymers. The materials were investigated concerning their chemical composition, morphology, thermal, swelling, and mechanical and surface properties for wound healing application. The unusual morphology of PLA fiber mats was probably influenced by the rate of solvent evaporation, as well as polymer concentration. It was supported by an electrostatic interaction, which was proved through the swelling behavior of the blend fiber mats. The results observed a strong influence PLA/CS on the diameter distribution of the nanofibers, surface hydrophobicity, and mechanical properties on wound healing ability [113].

4 FORMS OF APPLYING BIODEGRADABLE POLYMERS IN WOUND HEALING

Although these are reports in the literature by which therapeutic materials are used in wound healing management [114]. However, these methods suffer from several disadvantages such as inability to bind adequately in wet areas, low tensile strength, high cost, low mechanical, scar formation, and poor integration with host tissue, resulting in imperfect healing or even the cosmetic concern [26] that need to be tackled with either modified biodegradable or natural polymers. This section introduces novel forms of explicating biodegradable polymers that are efficient, peculiar, and nonexpensive use in wound healing.

4.1 Biodegradable Ointments for Wound Healing

According to the International Pharmacopoeia (2019), ointments are homogeneous, semisolid preparations intended for external application to the skin or mucous membranes. They are used as emollients for the application of active ingredients to the skin for protective, therapeutic, or prophylactic purposes with a degree of occlusion [115,116]. Biodegradable ointments for wound healing are commonly applied to wounds before dressing with first aid dressing to maintain a moist environment as one of the prerequisites for cosmetic wound healing. This accelerating inflammatory and proliferative phases in wound healing, more appealing scar is formed [117]. Similar to normal topical ointments, biodegradable ointments are semisolid formulations that usually act as viscoelastic materials when shear stress is applied. They can be medicated or nonmedicated for topical application to intact or broken skin or to mucous membranes [118]. Biodegradable materials have been used in topical preparations and are recently being considered in the formulation of ointments employed for wound healing due to their unique characteristics. These include the protection of unstable drugs from degradation, decreased side effects leading to controlled release, and enhanced cutaneous penetration of drugs across the skin barrier by increasing concentration gradient. Although several organic tissue adhesives designed to facilitate wound healing are gaining great attention in diverse clinical applications, they present significant inherent limitations, such as rejection, infection toxicity, or excessive swelling. Furthermore, Lu et al. (2018) fabricated mesoporous silica nanoparticles (Ag-MSNs) with antibacterial and degradable properties for wound healing through the nanobridging effect and assessed in mouse skin wound model. Their results demonstrated that Ag-MSNs had a higher failure force between two gelatin films compared to the control. Furthermore, the Ag-MSNs nanosystems exhibited higher antibacterial activity against gram-negative (*E. coli*) and gram-positive (*S. aureus*) and achieved esthetic wound healing in the infected rat model without inducing side effects [2]. Another alternative in wound healing is the design of nanosystems that combined the properties of biodegradable (PLGA) and topical intrinsic antibiotics (gentamicin) to provide controlled release of the antibiotics [119].

These characteristics help in reducing the most common side effects of topical treatments that include irritation, thinning of the skin, and dilated blood vessels [120–122]. The advantage of using biodegradable materials over nonbiodegradable materials is that they are generally nonreactive when they come in contact with the human body and can be broken down or metabolized and removed from the body via normal metabolic pathways. In contrast, nonbiodegradable materials can accumulate in various body tissues and cause irritation [122] either by mechanical incorporation or fusion methods. The adjuncts (ointment base, preservatives, chelating agents, antioxidants, and perfumes) used in the preparation of ointments [118,123] are similar to these employed preparing biodegradable ointments only with a variation in the addition of a biodegradable material and an active pharmaceutical ingredient if it is designed to be medicated.

Moist wound healing conditions are beneficial in a professional wound care setting as well as for self-treatment of acute, superficial wounds. To create a moist healing environment in superficial wounds, dressings and ointments are utilized. Ointments form a semiocclusive, breathable protective film that moisturizes and prevents the wound from drying out [63]. External influences are kept to a minimum, and the healing process is promoted [117]. Superficial cutaneous wounds treated with the novel wound healing ointment displayed a significant improvement of wound healing with an earlier onset of reepithelization, faster wound closure, and a better cosmetic outcome [117]. In addition, Kuhlmann et al. (2019) carried out a randomized, controlled, and investigator-blinded investigation to determine the local tolerability, wound healing efficacy, and cosmetic outcome of a novel wound healing ointment in an intra-individual comparison of four treatment regimens in an abrasive wound model [117]. The wound healing ointment exhibited excellent local tolerability with superior assessments in comparison to treatment utilizing only dressings without ointment. Significant differences between area under the curve (AUC) values for reepithelization and overall wound healing efficacy were demonstrated in favor of treatment with the wound healing ointment in comparison to dry wound healing conditions. Wounds treated with the wound healing ointment showed a faster onset of healing and the cosmetic outcome was rated as being superior for the wound healing ointment both by the investigator and the subject [117].

4.2 Biodegradable Spray for Wound Healing

Biodegradable sprays are novel dosage forms employed in wound healing. Spray formulation can be in the form of aerosol formulation or nonaerosol spray. Aerosols require special containers and propellants while nonaerosolized sprays are difficult to spray [124]. Pharmaceutical aerosol/spray is a suspension of fine solid drug particles or liquid drug droplets in a carrier gas/propellant. Biodegradable materials can be added to serve as drug carriers for wound healing sprays in which case they usually form a film around the wound [125,126]. The film applied directly to the wound with a known drug dose (typically metered sprays) provides physical protection and prevents the entry of microorganisms from the external environment. The flexibility of the film permits adaptation to body contours, and the film mechanical properties may be increased by repeated spraying to build up several overlapping layers of the film. Importantly, the control over the permeation of air and moisture is also feasible to maintain a favorable environment for wound healing and to avoid maceration due to excess moisture accumulation [124,126].

Biodegradable sprays have several benefits when compared with other delivery systems such as plasters and bandages for wound dressing and cream or gel for wound treatment. The benefits and the innovativeness of the spray compared to the technology currently available are as follows: (a) the spray is easily applied onto any size of wound area without touching the wound physically compared to cream or gel, (b) compared to the plasters, sprays do not have limited size are not singly packaged and do not cause sufferings and pains to the patient during the dressing and removal processes, (c) the sprays are easily contaminated when removed from the pack, (d) in the aerosol system, the product remains sterile because no foreign materials can go into the container [127]. Some examples of biodegradable materials used in the formulation of sprays include hydroxypropyl methylcellulose, carboxymethylcellulose sodium (CMC Sodium), acacia, tragacanth, chitosan, and gelatine, all at concentrations of 2% [127].

Interestingly, the release of the antibiotic was active in inhibiting bacterial growth on wound healing [15]. Although poor mechanical properties, low adhesion to native tissue and lack of antimicrobial properties are associated with hydrogel-based bioadhesives. However, Annabi et al. (2018) introduced a novel sprayable, elastic, and biocompatible composite hydrogel with broad-spectrum antimicrobial activity for chronic wound treatment. Their results also revealed that the hydrogels implanted via in vivo subcutaneous implantations in rats are biocompatible and biodegradable [128].

4.3 Biodegradable Lotions for Wound Healing

Lotions are dilute aqueous suspensions, emulsions, or solutions meant for topical application to inflamed ulcerated skin [129]. Lotion cools the skin by evaporation of solvents and shares many characteristics with creams [130]. According to the Report of Pharmacopoeia forum (2009), topical suspensions are liquid preparations that contain solid particles dispersed in a liquid vehicle intended for application to the skin. Some suspensions labeled as lotions fall into this category; conversely, topical solutions are liquid preparations that are usually aqueous but often contain other solvents such as alcohol and polyols that contain one or more dissolved chemical substances intended for topical application to the skin [131]. Depending on the solubility of biodegradable material employed in the formulation of a lotion, it could exist as a suspension or solution.

Lotions presented in the form of emulsions are recently presented as microemulsions that are transparent solutions obtained by titrating a normal coarse emulsion with medium-chain alcohols. Microemulsions for topical use have proven to offer some advantages such as spontaneous formation, ease of manufacturing and scale-up, thermodynamic stability, improved drug solubilization of hydrophobic drugs, and bioavailability. In addition, microemulsions that have an inverse micellar structure may be less comedogenic than either creams or solutions [132]. Some additives commonly employed in the preparation of lotions include bentonite (used as suspending agent in the preparation of lotion), methylcellulose, or sodium carboxymethyl cellulose (a biodegradable material that holds the active ingredient in contact with the affected site and at the same time easily rinsed off with water), alcohol (the drying and cooling effect of a lotion may be accentuated by adding alcohol to the formula), and humectant [glycerin keeps the skin moist for considerable period of time] [130].

4.4 Biodegradable Patches for Wound Healing

Transdermal drug delivery systems, also recognized universally as "patches," are fundamentally a collection of technologies designed to deliver therapeutically effective amounts of the drug to the body across patient's skin [133]. Some patches may be utilized topically to promote wound healing [134]. Transdermal patches were developed in the 1970s, and the first was approved by the FDA in 1979 for the treatment of motion sickness (a 3-day patch that delivered scopolamine). In 1981, patches for nitroglycerin were approved, and today there exist a number of patches for drugs such as clonidine, fentanyl, lidocaine, nicotine, nitroglycerin, oestradiol, oxybutynin, scopolamine, and testosterone [134].

There are four major types of transdermal patches documented. These are single-layer drug-in-adhesive, multilayer drug-in-adhesive, drug matrix-in-adhesive, and drug reservoir-in-adhesive [134]. The components of patches as the delivery system include biodegradable materials, active pharmaceutical ingredients, permeation enhancers, and other adjuncts [135,136]. Biodegradable materials have proven to be successful in developing wound healing patches with respect to patient compliance physiologically, economically, and psychologically [137,138]. They also serve as a reservoir in which active pharmaceutical ingredients are enclosed, thereby regulating their release thus making them good candidates in the formulation of patches meant for wound healing [139].

Some biodegradable materials from various sources commonly employed in the formulation of biodegradable patches include cellulose derivatives, chitosan, carrageenan, polyacrylates, polyvinyl alcohol, polyvinylpyrrolidone, and silicones [140]. Transdermal dosage forms, though a costly alternative to conventional formulations, are becoming popular because of their unique advantages. Controlled absorption, more uniform plasma levels, improved bioavailability, reduced side effects, painless and simple application, and flexibility of terminating drug administration by simply removing the patch from the skin are some of the potential advantages of transdermal drug delivery [141,142].

Factors limiting the success of transdermal technology include local skin irritation and other adverse reactions associated with certain drugs and formulation, limitation on the dose of a drug that can be delivered transdermally, a lag time associated with the delivery of the drug across the skin, resulting in a delay in onset of action. Other limitations include a variation of absorption rate based on site of application, skin disease (absorption may be delayed, especially in the case of water-soluble compounds), and variation in adhesive effectiveness in different individuals, the inability of the adhesives to adhere well to all types of skin and discomfort. In addition, the high cost of the product is also a major drawback for the wide acceptance of this product [143]. Biodegradable patches employed for wound healing can be evaluated by the following test after formulation: compatibility between the drug and biodegradable material using FTIR, some physical evaluation parameters include film thickness, percentage flatness, folding endurance, tensile strength, patch thickness, elongation break test, weight uniformity, drug content, percentage moisture content, percentage of moisture uptake, water vapor transmission rate, content uniformity test, uniformity of dosage unit test, and polariscope examination; some adhesive studies that can be carried out are shear adhesion test, peel adhesion test and tack properties, and lastly in vitro skin permeation and release kinetic studies [144].

Wound healing patch proposed might include an ultraconcentrated formula infused into a small dermal patch that adheres to the skin surface. This advanced technology introduces the formulation as a controlled delivery during wound healing based on their physiological properties. Nowadays, the use of biodegradable polymers of natural origin is quite fruitful in developing wound healing patches not only with respect to the environment but also to patient compliance physiologically, economically, and psychologically. The study by Asthana et al. (2015) proposed to develop a biodegradable patch of curcuminoid for wound healing that can counter issues pertaining to the variability of natural action and the frequency of physician's visit ensuring complete wound care during skin's healing process [145]. Niranjan et al. (2019) synthesized a novel nanocurcumin by slow vaporization technique incorporated into PLA and chitosan and obtained as PVA/Chi/Cur patch by the gel casting method. The nanosystem was evaluated for swelling rate, evaporation rate, effect with blood, cell biocompatibility, and activity against bacteria. The nanosystems (PVA/Chi/Cur) patch demonstrated exceptional wound healing ability and future application in treating other epidermal damage [100]. Similarly, Biswas et al. (2018) developed a novel scaffold (patch) using an old herbal drug for wound healing with better patient compliance. The diameter of the novel nanosystem was 40 nm with positively charged zeta potential. The in vivo study in albino rats demonstrated enhanced wound healing by the thymoquinone loaded PLA/cellulose acetate nanofiber with higher wound area, contraction, minimum inflammation, faster epithelialization, and vacuolization compared to the thymoquinone [146].

5 CONCLUSION

In this chapter, wound healing processes, biodegradable polymers, and their applications in wound healing have been discussed. Biodegradable antibiotics loaded polymers for wound healing applications are gaining exponential potential and popularity globally. This is due to their biodegradability, biocompatibility, low cost, low chronic inflammatory response, low toxicity, rapid adhesion, protection against infections, improved mechanical properties, good adhesive strength, accelerated wound healing, and convenience. Important biodegradable polymers agents ranging from natural, synthetic, and a combination of both have been discussed. In addition, antimicrobial biodegradable polymers in the form of ointments, spray, lotions, and patches for wound healing have been covered. Furthermore, this chapter has highlighted several ranges of biodegradable-drug-loaded polymeric for wound healing both in vitro and in vivo have been highlighted. Multidisciplinary fields such as formulation and drug delivery, microbiology, polymer and materials sciences, histochemistry, physiology, and biopharmaceutics are urgently needed to effectively develop biodegradable antibiotics to enhanced wound healing.

REFERENCES

[1] M.T. Khorasani, A. Joorabloo, A. Moghaddam, H. Shamsi, Z. MansooriMoghadam, Incorporation of ZnO nanoparticles into heparinised polyvinyl alcohol/chitosan hydrogels for wound dressing application, International Journal of Biological Macromolecules 114 (2018) 1203–1215.

[2] M.-m. Lu, J. Bai, D. Shao, J. Qiu, M. Li, X. Zheng, Y. Xiao, Z. Wang, Z.-m. Chang, L. Chen, Antibacterial and biodegradable tissue nano-adhesives for rapid wound closure, International Journal of Nanomedicine 13 (2018) 5849.

[3] N.A. Ismail, K.A.M. Amin, F.A.A. Majid, M.H. Razali, Gellan gum incorporating titanium dioxide nanoparticles biofilm as wound dressing: physicochemical, mechanical, antibacterial properties and wound healing studies, Materials Science and Engineering: C 103 (2019) 109770.

[4] A. Parveen, N. Kulkarni, M. Yalagatti, V. Abbaraju, R. Deshpande, In vivo efficacy of biocompatible silver nanoparticles cream for empirical wound healing, Journal of Tissue Viability 27 (2018) 257–261.

[5] Y. Fang, X. Zhu, N. Wang, X. Zhang, D. Yang, J. Nie, G. Ma, Biodegradable core-shell electrospun nanofibers based on PLA and γ-PGA for wound healing, European Polymer Journal 116 (2019) 30–37.

[6] W.N. Hozzein, G. Badr, B.M. Badr, A. Allam, A. Al Ghamdi, M.A. Al-Wadaan, N.S. Al-Waili, Bee venom improves diabetic wound healing by protecting functional macrophages from apoptosis and enhancing Nrf2, Ang-1 and Tie-2 signaling, Molecular Immunology 103 (2018) 322–335.

[7] Y. Shao, M. Dang, Y. Lin, F. Xue, Evaluation of wound healing activity of plumbagin in diabetic rats, Life Sciences 231 (2019) 116422.

[8] H. Li, G.R. Williams, J. Wu, H. Wang, X. Sun, L.-M. Zhu, Poly (N-isopropylacrylamide)/poly (l-lactic acid-co-

ε-caprolactone) fibers loaded with ciprofloxacin as wound dressing materials, Materials Science and Engineering: C 79 (2017) 245–254.

[9] R. Alipour, A. Khorshidi, A.F. Shojaei, F. Mashayekhi, M.J.M. Moghaddam, Skin wound healing acceleration by Ag nanoparticles embedded in PVA/PVP/Pectin/Mafenide acetate composite nanofibers, Polymer Testing (2019) 106022.

[10] W. Wentao, Z. Tao, S. Bulei, Z. Tongchang, Z. Qicheng, W. Fan, Z. Ninglin, S. Jian, Z. Ming, S. Yi, Functionalization of polyvinyl alcohol composite film wrapped in am-ZnO@ CuO@ Au nanoparticles for antibacterial application and wound healing, Applied Materials Today 17 (2019) 36–44.

[11] R. Rajakumari, T. Volova, O.S. Oluwafemi, R. Kumar, S. Thomas, N. Kalarikkal, Nano formulated proanthocyanidins as an effective wound healing component, Materials Science and Engineering: C (2019) 110056.

[12] A. Satish, R. Aswathi, J.C. Maria, P.S. Korrapati, Triiodothyronine impregnated alginate/gelatin/polyvinyl alcohol composite scaffold designed for exudate-intensive wound therapy, European Polymer Journal 110 (2019) 252–264.

[13] T.-Y. Huang, G.-S. Wang, C.-C. Tseng, W.-T. Su, Epidermal cells differentiated from stem cells from human exfoliated deciduous teeth and seeded onto polyvinyl alcohol/silk fibroin nanofiber dressings accelerate wound repair, Materials Science and Engineering: C (2019) 109986.

[14] H.J. Choi, T. Thambi, Y.H. Yang, S.I. Bang, B.S. Kim, D.G. Pyun, D.S. Lee, AgNP and rhEGF-incorporating synergistic polyurethane foam as a dressing material for scar-free healing of diabetic wounds, RSC Advances 7 (2017) 13714–13725.

[15] M. Dzikowski, N. Castanié, A. Guedon, B. Verrier, C. Primard, J. Sohier, Antibiotic incorporation in jet-sprayed nanofibrillar biodegradable scaffolds for wound healing, International Journal of Pharmaceutics 532 (2017) 802–812.

[16] P. Lalhmangaihzuali, B.D. Latha, N. More, M. Choppadandi, G. Kapusetti, Natural fiber reinforced biodegradable staples: novel approach for efficient wound closure, Medical Hypotheses 126 (2019) 60–65.

[17] C.K. Byrnes, F.H. Khan, P.H. Nass, C. Hatoum, M.D. Duncan, J.W. Harmon, Success and limitations of a naked plasmid transfection protocol for keratinocyte growth factor-1 to enhance cutaneous wound healing, Wound repair and regeneration 9 (5) (2001) 341–346.

[18] A. Kumar, X. Wang, K.C. Nune, R. Misra, Biodegradable hydrogel-based biomaterials with high absorbent properties for non-adherent wound dressing, International Wound Journal 14 (2017) 1076–1087.

[19] G. Suarato, R. Bertorelli, A. Athanassiou, Borrowing from Nature: biopolymers and biocomposites as smart wound care materials, Frontiers in Bioengineering and Biotechnology 6 (2018).

[20] G. Majno, The Healing Hand: Man and Wound in the Ancient World, Harvard University Press, 1991.

[21] R. Lalani, L. Liu, Electrospun zwitterionic poly (sulfobetaine methacrylate) for nonadherent, superabsorbent,

and antimicrobial wound dressing applications, Biomacromolecules 13 (2012) 1853–1863.

[22] A. Bužarovska, S. Dinescu, A.D. Lazar, M. Serban, G.G. Pircalabioru, M. Costache, C. Gualandi, L. Avérous, Nanocomposite foams based on flexible bio-based thermoplastic polyurethane and ZnO nanoparticles as potential wound dressing materials, Materials Science and Engineering: C (2019) 109893.

[23] M. Chen, J. Tian, Y. Liu, H. Cao, R. Li, J. Wang, J. Wu, Q. Zhang, Dynamic covalent constructed self-healing hydrogel for sequential delivery of antibacterial agent and growth factor in wound healing, Chemical Engineering Journal 373 (2019) 413–424.

[24] P. Picone, M.A. Sabatino, A. Ajovalasit, D. Giacomazza, C. Dispenza, M. Di Carlo, Biocompatibility, hemocompatibility and antimicrobial properties of xyloglucan-based hydrogel film for wound healing application, International Journal of Biological Macromolecules 121 (2019) 784–795.

[25] M.R. Safaee-Ardakani, A. Hatamian-Zarmi, S.M. Sadat, Z.B. Mokhtari-Hosseini, B. Ebrahimi-Hosseinzadeh, J. Rashidiani, H. Kooshki, Electrospun Schizophyllan/polyvinyl alcohol blend nanofibrous scaffold as potential wound healing, International Journal of Biological Macromolecules 127 (2019) 27–38.

[26] T. Siritienthong, J. Ratanavaraporn, P. Aramwit, Development of ethyl alcohol-precipitated silk sericin/polyvinyl alcohol scaffolds for accelerated healing of full-thickness wounds, International Journal of Pharmaceutics 439 (2012) 175–186.

[27] R. Song, M. Murphy, C. Li, K. Ting, C. Soo, Z. Zheng, Current development of biodegradable polymeric materials for biomedical applications, Drug Design, Development and Therapy 12 (2018) 3117.

[28] T. Sahana, P. Rekha, Biopolymers: applications in wound healing and skin tissue engineering, Molecular Biology Reports 45 (2018) 2857–2867.

[29] R. Song, M. Murphy, C. Li, K. Ting, C. Soo, Z. Zheng, Current development of biodegradable polymeric materials for biomedical applications, Drug design, development and therapy 12 (2018) 3117.

[30] S.A. Eming, T. Krieg, J.M. Davidson, Inflammation in wound repair: molecular and cellular mechanisms, Journal of Investigative Dermatology 127 (2007) 514–525.

[31] J. Aragón, C. Costa, I. Coelhoso, G. Mendoza, A. Aguiar-Ricardo, S. Irusta, Electrospun asymmetric membranes for wound dressing applications, Materials Science and Engineering: C (2019) 109822.

[32] C. Tonda-Turo, F. Ruini, C. Ceresa, P. Gentile, P. Varela, A.M. Ferreira, L. Fracchia, G. Ciardelli, Nanostructured scaffold with biomimetic and antibacterial properties for wound healing produced by 'green electrospinning', Colloids and Surfaces B: Biointerfaces 172 (2018) 233–243.

[33] L. Jaiswal, S. Shankar, J.-W. Rhim, Carrageenan-based functional hydrogel film reinforced with sulfur nanoparticles and grapefruit seed extract for wound healing application, Carbohydrate Polymers 224 (2019) 115191.

[34] P.G. Bowler, D. Parsons, Combatting wound biofilm and recalcitrance with a novel anti-biofilm Hydrofiber® wound dressing, Wound Medicine 14 (2016) 6–11.

[35] Z. Li, H. Wang, B. Yang, Y. Sun, R. Huo, Three-dimensional graphene foams loaded with bone marrow derived mesenchymal stem cells promote skin wound healing with reduced scarring, Materials Science and Engineering: C 57 (2015) 181–188.

[36] M. Rottmar, M. Richter, X. Mäder, K. Grieder, K. Nuss, A. Karol, B. von Rechenberg, E. Zimmermann, S. Buser, A. Dobmann, In vitro investigations of a novel wound dressing concept based on biodegradable polyurethane, Science and Technology of Advanced Materials 16 (2015) 034606.

[37] R.J. Hay, N.E. Johns, H.C. Williams, I.W. Bolliger, R.P. Dellavalle, D.J. Margolis, R. Marks, L. Naldi, M.A. Weinstock, S.K. Wulf, The global burden of skin disease in 2010: an analysis of the prevalence and impact of skin conditions, Journal of Investigative Dermatology 134 (2014) 1527–1534.

[38] T. Anilkumar, J. Muhamed, A. Jose, A. Jyothi, P. Mohanan, L.K. Krishnan, Advantages of hyaluronic acid as a component of fibrin sheet for care of acute wound, Biologicals 39 (2011) 81–88.

[39] A.P. Serezani, G. Bozdogan, S. Sehra, D. Walsh, P. Krishnamurthy, E.A.S. Potchanant, G. Nalepa, S. Goenka, M.J. Turner, D.F. Spandau, IL-4 impairs wound healing potential in the skin by repressing fibronectin expression, Journal of Allergy and Clinical Immunology 139 (2017), 142–151. e145.

[40] X. Xu, X. Liu, L. Tan, Z. Cui, X. Yang, S. Zhu, Z. Li, X. Yuan, Y. Zheng, K.W.K. Yeung, Controlled-temperature photothermal and oxidative bacteria killing and acceleration of wound healing by polydopamine-assisted Au-hydroxyapatite nanorods, Acta Biomaterialia 77 (2018) 352–364.

[41] R. Shukla, S.K. Kashaw, A.P. Jain, S. Lodhi, Fabrication of Apigenin loaded gellan gum–chitosan hydrogels (GGCH-HGs) for effective diabetic wound healing, International Journal of Biological Macromolecules 91 (2016) 1110–1119.

[42] S.-K. Choi, J.-K. Park, J.-H. Kim, K.-M. Lee, E. Kim, K.-S. Jeong, W.B. Jeon, Integrin-binding elastin-like polypeptide as an in situ gelling delivery matrix enhances the therapeutic efficacy of adipose stem cells in healing full-thickness cutaneous wounds, Journal of Controlled Release 237 (2016) 89–100.

[43] A. Song, A.A. Rane, K.L. Christman, Antibacterial and cell-adhesive polypeptide and poly (ethylene glycol) hydrogel as a potential scaffold for wound healing, Acta Biomaterialia 8 (2012) 41–50.

[44] H. Ying, J. Zhou, M. Wang, D. Su, Q. Ma, G. Lv, J. Chen, In situ formed collagen-hyaluronic acid hydrogel as biomimetic dressing for promoting spontaneous wound healing, Materials Science and Engineering: C 101 (2019) 487–498.

[45] N. Islam, I. Dmour, M.O. Taha, Degradability of chitosan micro/nanoparticles for pulmonary drug delivery, Heliyon 5 (5) (2019) e01684.

[46] M. Abbas, T. Hussain, M. Arshad, A.R. Ansari, A. Irshad, J. Nisar, F. Hussain, N. Masood, A. Nazir, M. Iqbal, Wound healing potential of curcumin cross-linked chitosan/polyvinyl alcohol, International Journal of Biological Macromolecules 140 (2019) 871–876.

[47] M.T. Cerqueira, L.P. da Silva, T.C. Santos, R.P. Pirraco, V.M. Correlo, A.P. Marques, R.L. Reis, Human skin cell fractions fail to self-organize within a gellan gum/hyaluronic acid matrix but positively influence early wound healing, Tissue Engineering Part A 20 (9-10) (2014) 1369–1378.

[48] J. Santerre, K. Woodhouse, G. Laroche, R. Labow, Understanding the biodegradation of polyurethanes: from classical implants to tissue engineering materials, Biomaterials 26 (2005) 7457–7470.

[49] S. Arévalo-Alquichire, C. Ramírez, L. Andrade, Y. Uscategui, L.E. Diaz, J.A. Gómez-Tejedor, A. Vallés-Lluch, M.F. Valero, Polyurethanes from modified castor oil and chitosan: synthesis, characterization, in vitro degradation, and cytotoxicity, Journal of Elastomers and Plastics 50 (2018) 419–434.

[50] Z. Qamar, Z.B.H.A. Rahim, G.S. Neon, H.P. Chew, T. Zeeshan, Effectiveness of poly-γ-glutamic acid in maintaining enamel integrity, Archives of Oral Biology (2019) 104482.

[51] T. Sango, G. Stoclet, N. Joly, A. Marin, A.M.C. Yona, L. Duchatel, M.K. Ndikontar, J.M. Lefebvre, Water–soluble extracts from banana pseudo–stem as functional additives for polylactic acid: thermal and mechanical investigations, European Polymer Journal 112 (2019) 466–476.

[52] J.B. Fortin, T.-M. Lu, A model for the chemical vapor deposition of poly (para-xylylene)(parylene) thin films, Chemistry of Materials 14 (2002) 1945–1949.

[53] K.K. Chereddy, C.-H. Her, M. Comune, C. Moia, A. Lopes, P.E. Porporato, J. Vanacker, M.C. Lam, L. Steinstraesser, P. Sonveaux, PLGA nanoparticles loaded with host defense peptide LL37 promote wound healing, Journal of Controlled Release 194 (2014) 138–147.

[54] N. Raghuwanshi, P. Kumari, A.K. Srivastava, P. Vashisth, T.C. Yadav, R. Prasad, V. Pruthi, Synergistic effects of Woodfordia fruticosa gold nanoparticles in preventing microbial adhesion and accelerating wound healing in Wistar albino rats in vivo, Materials Science and Engineering: C 80 (2017) 252–262.

[55] Y.-m. Cao, M.-y. Liu, Z.-w. Xue, Y. Qiu, J. Li, Y. Wang, Q.-k. Wu, Surface-structured bacterial cellulose loaded with hUSCs accelerate skin wound healing by promoting angiogenesis in rats, Biochemical and Biophysical Research Communications 516 (2019) 1167–1174.

[56] H. Wu, F. Li, S. Wang, J. Lu, J. Li, Y. Du, X. Sun, X. Chen, J. Gao, D. Ling, Ceria nanocrystals decorated mesoporous silica nanoparticle based ROS-scavenging tissue adhesive for highly efficient regenerative wound healing, Biomaterials 151 (2018) 66–77.

[57] D. Kim, Q.-V. Le, Y.B. Kim, Y.-K. Oh, Safety and photochemotherapeutic application of poly (γ-glutamic acid)-based biopolymeric nanoparticle, Acta Pharmaceutica Sinica B 9 (2019) 565–574.

[58] M. Alavi, A. Nokhodchi, An overview on antimicrobial and wound healing properties of ZnO nanobiofilms, hydrogels, and bionanocomposites based on cellulose, chitosan, and alginate polymers, Carbohydrate Polymers (2019) 115349.

[59] A.P.S. de Melo Fiori, P.H. Camani, D. dos Santos Rosa, D.J. Carastan, Combined effects of clay minerals and polyethylene glycol in the mechanical and water barrier properties of carboxymethylcellulose films, Industrial Crops and Products 140 (2019) 111644.

[60] J.J. Elsner, I. Berdicevsky, M. Zilberman, In vitro microbial inhibition and cellular response to novel biodegradable composite wound dressings with controlled release of antibiotics, Acta Biomaterialia 7 (2011) 325–336.

[61] N. Nafee, A. Youssef, H. El-Gowelli, H. Asem, S. Kandil, Antibiotic-free nanotherapeutics: hypericin nanoparticles thereof for improved in vitro and in vivo antimicrobial photodynamic therapy and wound healing, International Journal of Pharmaceutics 454 (2013) 249–258.

[62] Z. Lu, J. Gao, Q. He, J. Wu, D. Liang, H. Yang, R. Chen, Enhanced antibacterial and wound healing activities of microporous chitosan-Ag/ZnO composite dressing, Carbohydrate Polymers 156 (2017) 460–469.

[63] X. Li, Z. Zhang, A. Fakhri, V.K. Gupta, S. Agarwal, Adsorption and photocatalysis assisted optimization for drug removal by chitosan-glyoxal/Polyvinylpyrrolidone/MoS$_2$ nanocomposites, International Journal of Biological Macromolecules 136 (2019) 469–475.

[64] L. Li, X. Liu, Y. Niu, J. Ye, S. Huang, C. Liu, K. Xu, Synthesis and wound healing of alternating block polyurethanes based on poly (lactic acid)(PLA) and poly (ethylene glycol)(PEG), Journal of Biomedical Materials Research Part B: Applied Biomaterials 105 (2017) 1200–1209.

[65] M. Li, J. Chen, M. Shi, H. Zhang, P.X. Ma, B. Guo, Electroactive anti-oxidant polyurethane elastomers with shape memory property as non-adherent wound dressing to enhance wound healing, Chemical Engineering Journal (2019) 121999.

[66] L. Gao, C. Li, C. Wang, J. Cui, L. Zhou, S. Fang, Structure and luminescent property of polyurethane bonded with Eu^{3+}-complex, Journal of Luminescence 212 (2019) 328–333.

[67] L. Mishnaevsky Jr., J. Sütterlin, Micromechanical model of surface erosion of polyurethane coatings on wind turbine blades, Polymer Degradation and Stability 166 (2019) 283–289.

[68] N. Namviriyachote, V. Lipipun, Y. Akkhawattanangkul, P. Charoonrut, G.C. Ritthidej, Development of polyurethane foam dressing containing silver and asiaticoside for healing of dermal wound, Asian Journal of Pharmaceutical Sciences 14 (2019) 63–77.

[69] X. Zhou, H. Wang, J. Zhang, Z. Zheng, G. Du, Lightweight biobased polyurethane nanocomposite foams reinforced with pineapple leaf nanofibers (PLNFs), Journal of Renewable Materials 6 (2018) 68–74.

[70] S.M. Lee, I.K. Park, Y.S. Kim, H.J. Kim, H. Moon, S. Mueller, Y.-I. Jeong, Physical, morphological, and wound healing properties of a polyurethane foam-film dressing, Biomaterials Research 20 (2016) 15.

[71] D. Khodabakhshi, A. Eskandarinia, A. Kefayat, M. Rafienia, S. Navid, S. Karbasi, J. Moshtaghian, In vitro and in vivo performance of a propolis-coated polyurethane wound dressing with high porosity and antibacterial efficacy, Colloids and Surfaces B: Biointerfaces 178 (2019) 177–184.

[72] E.-H. Song, S.-H. Jeong, J.-U. Park, S. Kim, H.-E. Kim, J. Song, Polyurethane-silica hybrid foams from a one-step foaming reaction, coupled with a sol-gel process, for enhanced wound healing, Materials Science and Engineering: C 79 (2017) 866–874.

[73] T. Hachiya, S. Hase, Y. Shiwa, H. Yoshikawa, Y. Sakakibara, S.L.T. Nguyen, K. Kimura, Poly-γ-glutamic acid production of Bacillus subtilis (natto) in the absence of DegQ: a gain-of-function mutation in yabJ gene, Journal of Bioscience and Bioengineering 128 (2019) 690–696.

[74] Z. Csikós, E. Fazekas, D. Rózsa, J. Borbély, K. Kerekes, Crosslinked poly-γ-glutamic acid based nanosystem for drug delivery, Journal of Drug Delivery Science and Technology 48 (2018) 478–489.

[75] B. Halmschlag, X. Steurer, S.P. Putri, E. Fukusaki, L.M. Blank, Tailor-made poly-γ-glutamic acid production, Metabolic Engineering 55 (2019) 239–248.

[76] W.J. Jang, S.-Y. Choi, J.M. Lee, G.H. Lee, M.T. Hasan, I.-S. Kong, Viability of Lactobacillus plantarum encapsulated with poly-γ-glutamic acid produced by Bacillus sp. SJ-10 during freeze-drying and in an in vitro gastrointestinal model, Lebensmittel-Wissenschaft und -Technologie 112 (2019) 108222.

[77] M.I. Bajestani, S. Mousavi, S. Mousavi, A. Jafari, S. Shojaosadati, Purification of extra cellular poly-γ-glutamic acid as an antibacterial agent using anion exchange chromatography, International Journal of Biological Macromolecules 113 (2018) 142–149.

[78] S. Wang, X. Cao, M. Shen, R. Guo, I. Bányai, X. Shi, Fabrication and morphology control of electrospun poly (γ-glutamic acid) nanofibers for biomedical applications, Colloids and Surfaces B: Biointerfaces 89 (2012) 254–264.

[79] T. Thaweechai, A. Kaewvilai, Benzoxazine grafted poly (γ-Glutamic acid) functional material: synthesis, characterization and photophysical properties, Materials Chemistry and Physics 227 (2019) 117–122.

[80] W.-C. Liu, H.-Y. Wang, T.-H. Lee, R.-J. Chung, Gamma-poly glutamate/gelatin composite hydrogels crosslinked by proanthocyanidins for wound healing, Materials Science and Engineering: C 101 (2019) 630–639.

[81] W. Kim, M. Kim, G. Tae, Injectable system and its potential application for the delivery of biomolecules by using thermosensitive poly (γ-glutamic acid)-based physical hydrogel, International Journal of Biological Macromolecules 110 (2018) 457–464.

[82] M. Wang, P. Favi, X. Cheng, N.H. Golshan, K.S. Ziemer, M. Keidar, T.J. Webster, Cold atmospheric plasma (CAP) surface nanomodified 3D printed polylactic acid (PLA) scaffolds for bone regeneration, Acta Biomaterialia 46 (2016) 256–265.

[83] R. Scaffaro, A. Maio, F. Lopresti, Effect of graphene and fabrication technique on the release kinetics of carvacrol from polylactic acid, Composites Science and Technology 169 (2019) 60−69.

[84] P.E. Alves, B.G. Soares, L.C. Lins, S. Livi, E.P. Santos, Controlled delivery of dexamethasone and betamethasone from PLA electrospun fibers: a comparative study, European Polymer Journal 117 (2019) 1−9.

[85] F. Zou, X. Sun, X. Wang, Elastic, hydrophilic and biodegradable poly (1, 8-octanediol-co-citric acid)/polylactic acid nanofibrous membranes for potential wound dressing applications, Polymer Degradation and Stability 166 (2019) 163−173.

[86] I.J. Macha, M.M. Muna, J.L. Magere, In vitro study and characterization of cotton fabric PLA composite as a slow antibiotic delivery device for biomedical applications, Journal of Drug Delivery Science and Technology 43 (2018) 172−177.

[87] A.R. Adhikari, T. Geranpayeh, W.K. Chu, D.C. Otteson, Improved cellular response of ion modified poly (lactic acid-co-glycolic acid) substrates for mouse fibroblast cells, Materials Science and Engineering: C 60 (2016) 151−155.

[88] X. Liu, L.H. Nielsen, H. Qu, L.P. Christensen, J. Rantanen, M. Yang, Stability of lysozyme incorporated into electrospun fibrous mats for wound healing, European Journal of Pharmaceutics and Biopharmaceutics 136 (2019) 240−249.

[89] Z. Gu, H. Yin, J. Wang, L. Ma, Y. Morsi, X. Mo, Fabrication and characterization of TGF-β1-loaded electrospun poly (lactic-co-glycolic acid) core-sheath sutures, Colloids and Surfaces B: Biointerfaces 161 (2018) 331−338.

[90] H. Shi, S. Yang, S. Zeng, X. Liu, J. Zhang, T. Wu, X. Ye, T. Yu, C. Zhou, J. Ye, Enhanced angiogenesis of biodegradable iron-doped octacalcium phosphate/poly (lactic-co-glycolic acid) scaffold for potential cancerous bone regeneration, Applied Materials Today 15 (2019) 100−114.

[91] L.C. Da, Y.Z. Huang, H.Q. Xie, Progress in development of bioderived materials for dermal wound healing, Regenerative biomaterials 4 (5) (2017) 325−334.

[92] M. Sampath, R. Lakra, P. Korrapati, B. Sengottuvelan, Curcumin loaded poly (lactic-co-glycolic) acid nanofiber for the treatment of carcinoma, Colloids and Surfaces B: Biointerfaces 117 (2014) 128−134.

[93] K.K. Chereddy, R. Coco, P.B. Memvanga, B. Ucakar, A. des Rieux, G. Vandermeulen, V. Préat, Combined effect of PLGA and curcumin on wound healing activity, Journal of Controlled Release 171 (2013) 208−215.

[94] X. Liu, L.H. Nielsen, S.N. Kłodzińska, H.M. Nielsen, H. Qu, L.P. Christensen, J. Rantanen, M. Yang, Ciprofloxacin-loaded sodium alginate/poly (lactic-co-glycolic acid) electrospun fibrous mats for wound healing, European Journal of Pharmaceutics and Biopharmaceutics 123 (2018) 42−49.

[95] S.P. Hlaing, J. Kim, J. Lee, N. Hasan, J. Cao, M. Naeem, E.H. Lee, J.H. Shin, Y. Jung, B.-L. Lee, S-Nitrosoglutathione loaded poly (lactic-co-glycolic acid) microparticles for prolonged nitric oxide release and enhanced healing of methicillin-resistant Staphylococcus aureus-infected wounds, European Journal of Pharmaceutics and Biopharmaceutics 132 (2018) 94−102.

[96] Z. Miswan, S.K. Lukman, F.A.A. Majid, M.F. Loke, S. Saidin, H. Hermawan, Drug-eluting coating of ginsenoside Rg1 and Re incorporated poly (lactic-co-glycolic acid) on stainless steel 316L: physicochemical and drug release analyses, International Journal of Pharmaceutics 515 (2016) 460−466.

[97] D. Gao, X. Zhou, Z. Gao, X. Shi, Z. Wang, Y. Wang, P. Zhang, Preparation and characterization of silver sulfadiazine−loaded polyvinyl alcohol hydrogels as an antibacterial wound dressing, Journal of Pharmaceutical Sciences 107 (2018) 2377−2384.

[98] N. Golafshan, R. Rezahasani, M.T. Esfahani, M. Kharaziha, S. Khorasani, Nanohybrid hydrogels of laponite: PVA-alginate as a potential wound healing material, Carbohydrate Polymers 176 (2017) 392−401.

[99] Q. Chen, J. Wu, Y. Liu, Y. Li, C. Zhang, W. Qi, K.W. Yeung, T.M. Wong, X. Zhao, H. Pan, Electrospun chitosan/PVA/bioglass nanofibrous membrane with spatially designed structure for accelerating chronic wound healing, Materials Science and Engineering: C (2019) 110083.

[100] R. Niranjan, M. Kaushik, J. Prakash, K. Venkataprasanna, C. Arpana, P. Balashanmugam, G.D. Venkatasubbu, Enhanced wound healing by PVA/Chitosan/Curcumin patches: in vitro and in vivo study, Colloids and Surfaces B: Biointerfaces 182 (2019) 110339.

[101] M.T. Khorasani, A. Joorabloo, H. Adeli, Z. Mansoori-Moghadam, A. Moghaddam, Design and optimization of process parameters of polyvinyl (alcohol)/chitosan/nano zinc oxide hydrogels as wound healing materials, Carbohydrate Polymers 207 (2019) 542−554.

[102] R. Ahmed, M. Tariq, I. Ali, R. Asghar, P.N. Khanam, R. Augustine, A. Hasan, Novel electrospun chitosan/polyvinyl alcohol/zinc oxide nanofibrous mats with antibacterial and antioxidant properties for diabetic wound healing, International Journal of Biological Macromolecules 120 (2018) 385−393.

[103] R. Ma, Y. Wang, H. Qi, C. Shi, G. Wei, L. Xiao, Z. Huang, S. Liu, H. Yu, C. Teng, Nanocomposite sponges of sodium alginate/graphene oxide/polyvinyl alcohol as potential wound dressing: in vitro and in vivo evaluation, Composites Part B: Engineering 167 (2019) 396−405.

[104] A. Montaser, M. Rehan, M.E. El-Naggar, pH-Thermosensitive hydrogel based on polyvinyl alcohol/sodium alginate/N-isopropyl acrylamide composite for treating re-infected wounds, International Journal of Biological Macromolecules 124 (2019) 1016−1024.

[105] C. Zheng, C. Liu, H. Chen, N. Wang, X. Liu, G. Sun, W. Qiao, Effective wound dressing based on Poly (vinyl alcohol)/Dextran-aldehyde composite hydrogel, International Journal of Biological Macromolecules 132 (2019) 1098−1105.

[106] A. Abdel-Mohsen, D. Pavliňák, M. Čileková, P. Lepcio, R. Abdel-Rahman, J. Jančář, Electrospinning of hyaluronan/polyvinyl alcohol in presence of in-situ silver nanoparticles: preparation and characterization,

International Journal of Biological Macromolecules 139 (2019) 730−739.

[107] L. Lizárraga-Laborín, J. Quiroz-Castillo, J. Encinas-Encinas, M. Castillo-Ortega, S. Burruel-Ibarra, J. Romero-García, J. Torres-Ochoa, D. Cabrera-Germán, D. Rodríguez-Félix, Accelerated weathering study of extruded polyethylene/poly (lactic acid)/chitosan films, Polymer Degradation and Stability 155 (2018) 43−51.

[108] S. Chen, X. Zhao, C. Du, Macroporous poly (l-lactic acid)/chitosan nanofibrous scaffolds through cloud point thermally induced phase separation for enhanced bone regeneration, European Polymer Journal 109 (2018) 303−316.

[109] S. Kasirajan, D. Umapathy, C. Chandrasekar, V. Aafrin, M. Jenitapeter, L. Udhyasooriyan, A.S.B. Packirisamy, S. Muthusamy, Preparation of poly (lactic acid) from Prosopis juliflora and incorporation of chitosan for packaging applications, Journal of Bioscience and Bioengineering 128 (2019) 323−331.

[110] Z. Guo, D. Bo, Y. He, X. Luo, H. Li, Degradation properties of chitosan microspheres/poly (L-lactic acid) composite in vitro and in vivo, Carbohydrate Polymers 193 (2018) 1−8.

[111] Y. Li, F. Chen, J. Nie, D. Yang, Electrospun poly (lactic acid)/chitosan core−shell structure nanofibers from homogeneous solution, Carbohydrate Polymers 90 (2012) 1445−1451.

[112] S. Afshar, S. Rashedi, H. Nazockdast, M. Ghazalian, Preparation and characterization of electrospun poly (lactic acid)-chitosan core-shell nanofibers with a new solvent system, International Journal of Biological Macromolecules 138 (2019) 1130−1137.

[113] N.M. Siqueira, K.C. Garcia, R. Bussamara, F.S. Both, M.H. Vainstein, R.M. Soares, Poly (lactic acid)/chitosan fiber mats: investigation of effects of the support on lipase immobilization, International Journal of Biological Macromolecules 72 (2015) 998−1004.

[114] R.Z. Murray, Z.E. West, A.J. Cowin, B.L. Farrugia, Development and use of biomaterials as wound healing therapies, Burns and Trauma 7 (2019) 2.

[115] G. Navas, R. Buitrago, H. Schmidt, S. Kopp, Development of ivermectin and ivermectin tablets monograph for the International Pharmacopoeia, Farma Journal 4 (2019) 194.

[116] W.H. Organization, The International Pharmacopoeia, World Health Organization, 2006.

[117] M. Kuhlmann, W. Wigger-Alberti, Y. Mackensen, M. Ebbinghaus, R. Williams, F. Krause-Kyora, R. Wolber, Wound healing characteristics of a novel wound healing ointment in an abrasive wound model: a randomised, intra-individual clinical investigation, Wound Medicine 24 (2019) 24−32.

[118] A.E. Al-Snafi, Chemical constituents and pharmacological importance of agropyron repens−a review, Research Journal of Pharmacy and Technology 1 (2015) 37−41.

[119] J.J. Elsner, D. Egozi, Y. Ullmann, I. Berdicevsky, A. Shefy-Peleg, M. Zilberman, Novel biodegradable composite wound dressings with controlled release of antibiotics: results in a Guinea pig burn model, Burns 37 (2011) 896−904.

[120] Z. Zhang, P.C. Tsai, T. Ramezanli, B.B. Michniak-Kohn, Polymeric nanoparticles-based topical delivery systems for the treatment of dermatological diseases, Wiley Interdisciplinary Reviews: Nanomedicine and Nanobiotechnology 5 (2013) 205−218.

[121] L. Naldi, C. Griffiths, Traditional therapies in the management of moderate to severe chronic plaque psoriasis: an assessment of the benefits and risks, British Journal of Dermatology 152 (2005) 597−615.

[122] A. Ammala, Biodegradable polymers as encapsulation materials for cosmetics and personal care markets, International Journal of Cosmetic Science 35 (2013) 113−124.

[123] R. Asija, S. Asija, D. Sharma, P.C. Dhaker, N. Nama, Topical ointment: an updated review, Journal of Drug Discovery and Therapeutics 3 (2015) 47−51.

[124] B. Sulekha, G. Avin, Topical spray of silver sulfadiazine for wound healing, Journal of Chemical and Pharmaceutical Research 8 (2016) 492−498.

[125] E. Algin-Yapar, Ö. İnal, Transdermal spray in hormone delivery, Tropical Journal of Pharmaceutical Research 13 (2014) 469−474.

[126] R. Sritharadol, T. Nakpheng, P. Wan Sia Heng, T. Srichana, Development of a topical mupirocin spray for antibacterial and wound-healing applications, Drug Development and Industrial Pharmacy 43 (2017) 1715−1728.

[127] N.A. Febriyenti, S. Baie, Formulation of aerosol concentrates containing haruan (Channa striatus) for wound dressing, Malaysian Journal of Pharmaceutical Sciences 6 (2008) 43−58.

[128] N. Annabi, D. Rana, E.S. Sani, R. Portillo-Lara, J.L. Gifford, M.M. Fares, S.M. Mithieux, A.S. Weiss, Engineering a sprayable and elastic hydrogel adhesive with antimicrobial properties for wound healing, Biomaterials 139 (2017) 229−243.

[129] P. Bakker, N. Wieringa, V. Gooskens, H. Van Doorne, Dermatological Preparations for the Tropics, University of Groningen, 1990.

[130] R.-K. Chang, A. Raw, R. Lionberger, L. Yu, Generic development of topical dermatologic products: formulation development, process development, and testing of topical dermatologic products, The AAPS Journal 15 (2013) 41−52.

[131] E. Marguí, C. Fontas, A. Buendía, M. Hidalgo, I. Queralt, Determination of metal residues in active pharmaceutical ingredients according to European current legislation by using X-ray fluorescence spectrometry, Journal of Analytical Atomic Spectrometry 24 (2009) 1253−1257.

[132] N. Grampurohit, P. Ravikumar, R. Mallya, Microemulsions for topical use−a review, Indian Journal of Pharmaceutical Education 45 (2011) 100−107.

[133] T. Tanner, R. Marks, Delivering drugs by the transdermal route: review and comment, Skin Research and Technology 14 (2008) 249−260.

[134] D. Patel, S.A. Chaudhary, B. Parmar, N. Bhura, Transdermal drug delivery system: a review, The Pharma Innovation 1 (2012).

[135] S. Parivesh, D. Sumeet, D. Abhishek, Design, evaluation, parameters and marketed products of transdermal

patches: a review, Journal of Pharmacy Research 3 (2010) 235–240.

[136] P. Arunachalam, Inclusive Growth in India, Serials Publications, 2010.

[137] S. Toshkhani, G. Shilakari, A. Asthana, Advancements in wound healing biodegradable dermal patch formulation designing, Inventi Rapid: Pharm Tech (2013) 1–11.

[138] A. Sood, M.S. Granick, N.L. Tomaselli, Wound dressings and comparative effectiveness data, Advances in Wound Care 3 (2014) 511–529.

[139] S. Saghazadeh, C. Rinoldi, M. Schot, S.S. Kashaf, F. Sharifi, E. Jalilian, K. Nuutila, G. Giatsidis, P. Mostafalu, H. Derakhshandeh, Drug delivery systems and materials for wound healing applications, Advanced Drug Delivery Reviews 127 (2018) 138–166.

[140] C. Valenta, B.G. Auner, The use of polymers for dermal and transdermal delivery, European Journal of Pharmaceutics and Biopharmaceutics 58 (2004) 279–289.

[141] K. Ezhumalai, P. Ilavarsan, R.M. Mugundhan, U. Sathiyaraj, A. Rajalakshmi, Transdermal patches in novel drug delivery system, International Journal of Pharmacy and Technology 3 (2011) 2402–2419.

[142] G. Aggarwal, S. Dhawan, Development, fabrication and evaluation of transdermal drug delivery system-a review, Pharmaceutical Reviews 7 (2009) 1–28.

[143] P. Minghetti, F. Cilurzo, A. Casiraghi, Measuring adhesive performance in transdermal delivery systems, American Journal of Drug Delivery 2 (2004) 193–206.

[144] F. Cilurzo, C.G. Gennari, P. Minghetti, Adhesive properties: a critical issue in transdermal patch development, Expert Opinion on Drug Delivery 9 (2012) 33–45.

[145] A. Asthana, G.S. Asthana, Polymeric sustained biodegradable patch formulation for wound healing, world academy of science, engineering and technology, International Journal of Medical, Health, Biomedical, Bioengineering and Pharmaceutical Engineering 9 (2015) 577–580.

[146] A. Biswas, M. Amarajeewa, S. Senapati, M. Sahu, P. Maiti, Sustained release of herbal drugs using biodegradable scaffold for faster wound healing and better patient compliance, Nanomedicine: Nanotechnology, Biology and Medicine 14 (2018) 2131–2141.

CHAPTER 7

Antibiotics Encapsulated Scaffolds as Potential Wound Dressings

B. BUYANA • S. ALVEN • X. NQORO • BLESSING A. ADERIBIGBE

1 INTRODUCTION

A wound is described as a disruption of the mucosa or epithelial lining of the skin resulting from thermal or physical damage [1]. The skin is the most exposed human organ to injuries, impairment, burns, and scratches. If the connective and epithelium structures are damaged, the protection that is provided by the human body from the outer environment becomes weakened. Wounds are categorized as chronic or acute wound depending on the nature and duration of the healing process [2]. A chronic wound is a wound that fails to heal through the usual stages of healing and cannot be restored in a timely and orderly manner [1]. Chronic wounds include leg ulcers and burn wounds [3]. Factors that contribute to these chronic wounds include diabetes, prolonged bed rest, obesity, smoking, and age. An acute wound results from the damage to the skin from a surgical procedure or accidental injury. It is repaired at an expected period of 2−3 months depending on the depth, size, and extent of injury in the dermis and epidermis layer of the skin [4,5]. It is reported that this burden of treating chronic wounds contribute an increase in healthcare cost worldwide and an excess of US $25 billion is spent annually for the management of chronic wounds [6]. Skin injuries represent a severe health risk to the human body because it exposes the body to bacterial invasion that can be fatal [7].

Wound healing is a complex and dynamic process of tissue formation, and it occurs in four various stages: hemostasis stage is the first phase of wound healing; followed by inflammatory phase; the third phase is the proliferation phase in which new blood vessels and tissue are formed; lastly, the maturation phase, which is the final phase of wound healing (transformation of new tissues takes place) [8]. There are several wound dressings that can be used during the process of wound healing. The available wound dressings are hydrogel, foam, films, fibers, hydrocolloids, etc. [1].

The aforementioned types of wound dressings act as a barrier against the infiltration of bacteria to the wound's biological environment [9,10]. Wound dressings also allow coagulation and terminate the loss of tissue fluid, protecting the wound from damage during healing. There are several limitations of wound dressing depending on their types: some of them require frequent change of dressing, and they are not good for dry wounds and low exudate wounds because they depend on exudates for its treatment [11]; some of them are not suitable for neuropathic ulcers and they are used as secondary dressings [12]. Some wound dressings are not biocompatible with the skin, have poor absorption of wound exudate, are unable to maintain a moist environment, delay the wound healing process, and exhibit poor gas exchange between the wound and the environment; some are toxic and lack effectiveness against bacterial infection [13]. Selected wound dressings promote exudate accumulation at the wound surface leading to bacterial proliferation that forms a bad smell on the wound [14]. Therefore, the encapsulation of antibiotics into wound dressings is a good approach that can be employed to improve and accelerate wound healing [15].

There are several studies that revealed the use of different antibiotics in wound healing resulting in wound closing although their positive effects are not yet noticed. Most of the numerous antibiotics are known to be effective against infection-producing microorganisms, such as aminoglycosides, tetracyclines, quinolones, and cephalosporins. Antibiotics inhibit some metabolic path or functions of bacteria by blocking their main metabolic pathways, interfering on protein synthesis, inhibition of bacterial cell wall synthesis and nucleic acid synthesis. Furthermore, an improper or high dosage use of antibiotics can result in the bacteria developing resistance [16]. Long-term and high dosage of antibacterial agent does not only

Antibiotic Materials in Healthcare. https://doi.org/10.1016/B978-0-12-820054-4.00007-0

increase the risk of drug resistance but also cause an adverse impact on wound repair [17] such as their ability to kill the good bacteria that keeps the body healthy and cause stomach upset that can lead to fatal diarrhea and its use is not cost effective. This chapter reports the comprehensive in vivo and in vitro overview of wound dressing scaffolds incorporated with antibiotics for accelerated wound healing.

2 MECHANISMS OF WOUND HEALING
2.1 Wound Healing Process
The wound healing process is a complex process that involves an interaction of different types of growth factors, cells, extracellular matrix components, and proteinases [18]. When the skin is injured, the body undergoes certain stages that return the skin to its normal form. The wound healing process involves four stages, namely hemostasis, inflammation, proliferation, and maturation phase (Fig. 7.1). These phases occur at a constant sequence and sometimes in an overlapping manner [19]. The wound healing process is promoted by wound dressings and some antibiotics to prevent infection at the injured site.

Hemostasis is the first stage of wound healing and acts as the first response to injuries. The process of vasoconstriction (constriction to slow blood loss) occurs immediately after an injury and platelets are released at the wound surface to promote blood clotting by the fibrous protein, the fibrin. Blood clotting is formed by a cluster of platelets and fibrin, which closes the broken blood vessel and impedes blood loss. The hemostasis process lasts up to 2 or more days, depending on the gravity of the wound and using a bandage or gauze also promotes blood clotting [20].

The second phase of wound healing is the inflammatory phase, where there is localized swelling with continued controlled bleeding to prevent infection. The white blood cells (exudate) are responsible for the swelling, warmth, and redness of the injured site. The phagocytic cells release proteases and reactive oxygen species to eliminate debris at the wounded site and to shield it from bacterial infection [21]. The damaged blood vessels are rebuilt when the white blood cells are transformed into tissue macrophages

liberating cytokines and growth factors, engaging with endothelial cells, fibroblast, and keratinocytes [21]. The epithelial cells move toward the wound site to replace the dead cells.

Proliferation is the third phase of wound healing in which there is a formation of new tissue. The newly formed tissue is usually red or pink in color when it covers the wound site. The epithelium covers the wound with the formation of granulation tissue. The final phase of wound healing is the maturation phase where there is tissue remodeling and the wound fully closes. The surface of the wound is covered with fibroblasts as a new layer of the skin [21,22].

2.2 Classification of Wound Dressings
Wound, be it a major or a minor skin breakdown, proper medical care should be exercised with the aim of infection prevention. The process of medical care implemented on cuts or skin breakdown is characterized as a wound dressing that can be classified as (1) traditional or passive, (2) skin substitutes, (3) interactive materials, and (4) bioactive dressings (Fig. 7.2) [1,23].
1. Traditional/passive wound dressings are designed to protect the wound from contamination or foreign attack, cover the wound and stop bleeding, provide a dry environment and cushion the wound, and absorb exudates [23,24]. However, during the healing process, repeated changing of wound dressing is required because of the absorbed exudate that causes pain and reskins damage [1,25,26]. Examples of traditional/passive wound dressings include gauze, plaster, bandages, and wool dressings [1,23,24].
2. Skin substitute dressings are designed to replace the damaged skin and are made up of two tissues dermal and epidermal layers made from fibroblasts and keratinocytes on collagen matrix [1,27]. Skin substitutes application is limited resulting from host rejection, limited survival time on wound site, and possibilities of disease infection and transmission; examples include allograft, autografts and acellular xenografts [23,27].
3. Interactive dressings including hydrogels, sponges, foams, films, and spray act as a barrier against bacterial infection, modify the physiology of the wound environment, provide a wet environment for the wound, enhance granulation and reepithelialization, and improve water vapor transmission rate with good tensile strength [1,23,24,26,28,29]. They are prepared from either synthetic or biopolymers such as gelatin, chitosan, and alginate [23].
4. Bioactive dressings are wound dressings responsible for the delivery of active materials such as antibiotics.

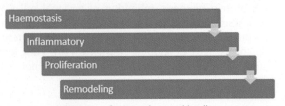

FIG. 7.1 Stages of wound healing.

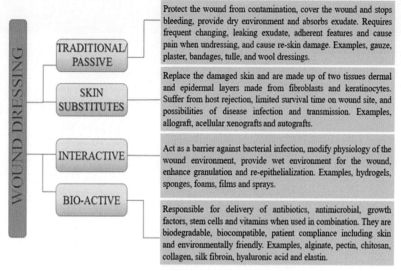

FIG. 7.2 Classification of wound dressings.

They are sometimes incorporated with antimicrobials, vitamins, growth factors, stem cells, etc., such as nanofibers, wafers, foams, sponges, hydrogels, membranes, and films to enhance their biological activity [23,30]. These dressings are derived from a variety of naturally occurring biopolymers or artificial sources such as alginate, pectin, chitosan, collagen, silk fibroin, hyaluronic acid, and elastin [1,24,27,30]. Bioactive dressings reported in current studies are biodegradable, biocompatible, and patient compliance including skin and environmentally friendly characteristics [1,31]. Biopolymers can be used solely or in combination with other biologically active materials like antibiotics for direct delivery to the site of action, and this chapter is focused on such biological studies.

3 SCAFFOLD WOUND DRESSINGS

3.1 Hydrogels

Hydrogels are greatly hydrated three-dimensional porous networks of cross-linked polymers, capable of absorbing wound exudates without dissolving [32–36]. Their advantages compared to other absorbents are that they provide easy application, accelerated healing, biodegradability, high storage capacity, high adsorption capacity of biological fluids, prevention of fluid loss from wound, biocompatibility, and fast adsorption kinetics, and their capability as a barrier against microorganisms and specific environmental stimuli-responsiveness, for example, temperature, pH, and ionic strength promote drug release into the infected wound area continuously, thereby reducing the dosing frequency [32,34,36,37]. The incorporation of bioactive substances like natural biodegradable polymers and polysaccharides such as chitosan, alginate, and nano-silver into hydrogel network enhances its activity making them effective at controlling the release of the incorporated agents [38]. Chitosan is the most commonly reported biopolymer used in the design of hydrogels employed in wound dressing with outstanding immune-stimulatory activity, good biodegradability, scar prevention, low toxicity, and excellent biocompatibility, favorable contribution to hemostatic and bacteriostatic wounds [39–43].

The introduction of antibiotics to wound treatment has been reported to be effective in protecting wounds from bacterial infections [44]. Fluoroquinolone antibiotics offer several advantages such as readily availability, good therapeutic efficacy, and a broad spectrum of activity [45]. Ciprofloxacin, a fluoroquinolone antibiotic, is widely used in wound treatment of infections and it is active against a wide range of both gram-negative and gram-positive bacterial spectrums. It acts by binding itself to DNA gyrase, blocking bacterial DNA duplication, leading to double-stranded splits in the bacterial chromosomes reducing the bacterial ability to develop resistance [35,44]. Levofloxacin also belongs to the fluoroquinolone antibiotic family, and it is active against a wide range of gram-positive bacterial spectrums [46]. It acts by inhibiting DNA gyrase and topoisomerase IV leading to bacterial death. There is a wide range of antibiotics used in wound treatment (Table 7.1).

TABLE 7.1
Classification and Mechanism of Action of Antibiotics.

Antibiotic	Classification	Mechanism of Action	References
Ciprofloxacin	Fluoroquinolone	Binds to bacteria's DNA gyrase, blocking its DNA duplication. Active against gram-positive and gram-negative strains.	[35,44]
Levofloxacin		Inhibiting DNA gyrase and topoisomerase IV causing bacterial death. Active against gram-positive bacterial strains.	[46]
Norfloxacin		Inhibit synthesis of DNA gyrase blocking DNA replication of the bacteria. Active against gram-positive and gram-negative bacterial strains.	[95]
Moxifloxacin		Inhibit synthesis of topoisomerase IV and DNA gyrase enzymes essential in bacterial DNA repair and replication. Have both gram-positive and gram-negative bacterial activity.	[96]
Vancomycin	Glycopeptide	Prevent peptidoglycan cross-linking, inhibit cell wall phospholipid synthesis of the bacteria. Active against gram-positive microorganisms.	[57,97]
Cefuroxime	β-Lactam	Inhibit synthesis of bacterial cell wall resulting to cell death. Selectively active against gram-positive bacterial strains.	[40]
Ampicillin		Inhibit bacterial cell-wall formation by irreversible inhibiting transpeptidase enzyme. Effective on both gram-positive and gram-negative selected bacterial strains.	[56]
Tetracycline hydrochloride	Tetracycline	Inhibiting synthesis of protein and ribosome, enzymatic reactions, and alter synthesis of cytoplasm membrane. Have gram-positive and gram-negative activity.	[55]
Gentamicin	Aminoglycoside	Inhibits proteins and ribosome affecting bacterial growth. Selectively active against gram-positive and gram-negative bacterial strains.	[31,97—99]
Amikacin		Diffuse through the outer membrane of the bacterial cell wall preventing formation of ribosomes. Selectively active against gram-positive and gram-negative bacteria.	[97]

Hanna et al. prepared ciprofloxacin (CIP)-encapsulated chitosan-based hydrogels for drug delivery. Their findings revealed that the hydrogels highly loaded with CIP exhibited faster and higher drug-release rates when compared to hydrogels with less amount of ciprofloxacin. At a maximum concentration of 250 µg/mL, a high percentage of 96.1% drug release was observed with a drug encapsulation efficiency percentage of 93.8% at pH 7.4. The rate of drug released was dependent on the concentration of the drug loaded. The hydrogels not loaded with ciprofloxacin did not display significant inhibition against the tested bacterial strains. However, the hydrogels loaded with CIP exhibited a significant inhibition with high antibacterial activity against both gram-negative and gram-positive tested bacterial strains similar to the reference drugs. CIP exhibited an inhibition zone of $(67.0 \pm 1.0 \text{ mm})$ and $(64.3 \pm 0.6 \text{ mm})$ when compared to the reference drug (gentamicin) with an inhibition zone of $(28 \pm 0.5 \text{ mm})$ and $(25 \pm 0.5 \text{ mm})$ against E. coli and K. pneumonia, respectively. Nonetheless, a greater inhibition zone was observed against S. aureus $(65.3 \pm 0.6 \text{ mm})$ and Streptococcus mutans $(41.0 \pm 1.0 \text{ mm})$ as compared to the reference drug (ampicillin) with $(22 \pm 0.1 \text{ mm})$ and $(28 \pm 0.5 \text{ mm}$, respectively) [47]. These results confirm that CIP has a wide range of bacterial activity against both gram-positive and gram-negative strains. However, despite the increase in drug concentration causing an increase in drug release and drug activity, negative observation was witnessed in cytotoxicity evaluation. The percentage of cell viability decreased with increased drug concentration, where at 50 µg/mL $(97 \pm 0.5\%)$ and at 300 µg/mL $(80.3 \pm 0.9\%)$ was observed on the prepared hydrogels. The decrease in cell viability with an increase in drug concentration can be linked to the toxicity of the drug when used in high dosage due to high drug release. Regardless of the decreased percentage of the maximum concentration, it is still considered biocompatible and nontoxic. The antibacterial activity of ciprofloxacin is dependent on the release kinetics of the prepared hydrogels, and the solvent PBS (pH 7.4) also played a significant role.

Choipang et al. reported PLGA hydrogels loaded with (CIP) hydrochloride nanoparticles. Their release characteristics were examined using a dialysis bag in PBS solution within 4 days. After the fourth day, an increased amount of ciprofloxacin released from the hydrogel was 94.63%, 79.75%, and 43.38% for 244 mg, 39 mg, and 6.25 mg concentrations of CIP, respectively. The results suggested that the release kinetics of the drug was dependent on the concentration of the drug loaded in the hydrogel, with the higher concentration of 244 mg exhibiting a higher drug release. They measured the antibacterial activity of the gels to determine the minimum bactericidal concentration (MBC) value and minimum inhibition concentration (MIC) value against E. coli and S. aureus. The MIC value and MBC values were 0.325 µg/mL, 0.625 µg/mL, and 0.5 µg/mL, for the concentrations 224 mg, 39 mg, and 6.25 mg, respectively against E. coli showing a greater inhibition when compared to S. aureus with 0.3 µg/mL, 0.5 µg/mL, and 1.5 µg/mL, respectively. The findings indicate that at greater CIP concentrations, the decrease in bacterial growth was significant in vitro against E. coli and S. aureus. The hydrogels were more active against E. coli strain when compared to S. aureus bacterial strain. Incubation for 24 h further resulted in the reduction of E. coli bacterial growth by 96.07% and 99.69% in 50 µg/mL polyvinyl alcohol (PVA) hydrogels loaded with 39 and 244 mg CIP-PLGA nanoparticles, whereas 500 µg/mL PVA hydrogels achieved reductions of 92.64% and 99.98%. A comparison of CIP-PLGA-loaded PVA hydrogels against S. aureus revealed that 50 µg/mL of PVA hydrogels loaded with the three concentrations of CIP-PLGA nanoparticles achieved reduction of 99.99%, 99.93%, and 99.93%, for the concentrations 224 mg, 39 mg, and 6.25 mg, respectively and 500 µg/mL of PVA hydrogels loaded with the three concentrations of CIP-PLGA nanoparticles achieved reductions of 100%, 94.64%, 99.92%, respectively [48]. These results are in agreement with the previously reported findings by Ref. [47], where a higher concentration of CIP exhibited enough antibacterial activity to totally clear the bacterial colony-forming units (cfu). The cytotoxicity results against human dermal fibroblast cells were in excess of 75% cell viability for the three concentrations that are considered nontoxic and can be regarded for further clinical tests.

Pawar et al. prepared cefuroxime-conjugated chitosan hydrogels (CS-CEF-5, CS-CEF-10, and CS-CEF-20) for the treatment of infected wound, and their drug-release studies observed that CS-CEF-5, CS-CEF-10, and CS-CEF-20 exhibited 58%, 51%, and 43% of cefuroxime release in 25 days in phosphate buffer (pH 7.4), respectively. In comparison with phosphate buffer containing esterase enzyme and sodium carbonate buffer (pH 10), CEF was released more rapidly through hydrolysis of the ester. These results suggest that the release of CEF is not dependent on the concentration of the drug, but it is, however, influenced by the enzyme and the pH of the incubation medium. The CS-CEF-10 exhibited almost a 100% release over a period of 23 days via chemical and enzymatic hydrolysis [40]. CS-CEF-20 and CS-CEF-5 hydrogels exhibited a slow drug release in all the tested incubation medium. The CS-CEF-20 hydrogel exhibited 99% and 89% drug release in phosphate buffer with an enzyme and sodium carbonate buffer, respectively, over a period of 25 days. The slow release of CH-CEF-20 is attributed to a high drug loading that resulted in a denser polymer matrix. The antibacterial activity of the unloaded chitosan hydrogel showed a small inhibition zone that can be linked to the antimicrobial activity of chitosan. The antibacterial inhibition zones for CS-CEF-5, CS-CEF-10, and CS-CEF-20 were reported to increase (6–15 mm), (8–18 mm), and (9–20 mm), respectively. Expectedly, the inhibition zone against the tested bacterial strains showed an increased inhibition zone with an increase in the drug concentration of the antibiotic. In a period of 24 h, a reduction of 5-log was revealed for the hydrogels CS-CEF-10 and CS-CEF-5 but CS-CEF-20 hydrogel presented a reduction of 6-log suggesting an increase in concentration results in an increase in the gelation rate. A strong antibacterial effect of the prepared hydrogels was observed in all three hydrogels that exhibited a 6-log reduction even after reinoculation using unchanged concentrations of bacteria (1×106 cfu/mL). A higher drug loaded in the hydrogel (CS-CEF-20) is responsible for the first burst release compared to the other two hydrogels with minimal drug loading concentration. Furthermore, based on the results, the hydrolyzed CEF released from the CS-CEF-5, CS-CEF-10, and CS-CEF-20 hydrogels displayed an effective antibacterial activity over a longer period of time against *S. aureus*. Thus the conjugation of CEF to the chitosan polymer did not change the antibacterial efficacy of the CEF. It is notable that the concentration of CEF in the hydrogels played a significant role against the rate of drug release. The prepared hydrogels exhibited good hemocompatibility and cell compatibility [40]. The antibiotic concentration and other biochemical properties such as porosity, pH, and enzyme influenced the activity and release of the antibiotic from the hydrogel formulations.

Mohammed et al. prepared hydrogels encapsulated with gentamicin and moxifloxacin antibiotics, and tested their drug release kinetics over a period of 4 h. In the first 30 min, the antibiotics already displayed a greater absorbance unit (a.u) values than the control (0.09 ± 0.01 a.u) where gentamicin was (0.52 ± 0.04 a.u) and moxifloxacin (1.53 ± 0.01 a.u) enough to impact bacterial survival against *S. aureus*. A continual increase in the release of the antibiotics was observed with gentamicin (1.22 ± 0.01 a.u) and moxifloxacin (6.18 ± 0.36 a.u) both displaying a greater inhibition than the control levels ($P < 0,00$). Gentamicin-loaded gels displayed 2.44 ± 2.1 cfu and moxifloxacin 0.66 ± 0.70 cfu in 30 min, which was not a complete inhibition; however, it was considerably lower when compared to the unloaded gels 1121 ± 575 cfu. After culturing bacteria with both antibiotics, gentamicin and moxifloxacin, no bacterial colony forming units were visible because both drugs significantly inhibited bacterial growth. Similar results were observed when they loaded the antibiotics into the gels [49].

Esposito and coworkers prepared polymeric hydrogels encapsulated with neomycin and observed that a fast initial burst of the drug was released within the first 8 h of the 48 h period and a constant drug release was maintained until the 48th hour. They drug-release profile from the hydrogel is attributed to its fluid withholding capability [36]. The release kinetics of neomycin from the hydrogel was influenced by the pore size in the hydrogel network. The antibacterial activity of the hydrogels against *S. aureus* incubated over 24 h at 37°C revealed a small inhibition zone for the unloaded hydrogel and a significant inhibition zone of 22 ± 1 mm antibacterial activities of neomycin-loaded hydrogels.

3.2 Sponges

Biomedical sponges have interconnected porous structures and are flexible soft kind of wound dressings. Their porous structure offers several advantages such as high swelling capacity, high hemostatic ability that enables the prevention of exudate accumulation, and high water absorption capability to provide a wet environment while protecting the wound bed from external infection of bacteria [50,51]. Sponges with pore sizes within the ranges of 10–100's of microns and interconnected structures have the capability to support cell proliferation [52]. Sponges prepared from polymers such as chitosan (CS), sodium alginate (SA), graphene oxide (GO), and PVA exhibit bacterial inhibition effects. Chitosan-based sponges exhibit antimicrobial properties, intrinsic hemostatic, biocompatibility, and water transmission rate. However, their improved fluid absorption capability, hydrophilicity, etc. enable them to overcome their selectivity against certain bacterial species, gas permeation, and dehydration [51,53]. PVA has been reported to be useful in improving water solubility of the electrospun CS by alteration of the intermolecular force and promotion of the chain entanglement [54].

Anbazhagan et al. prepared different fungi chitosan sponges loaded with tetracycline hydrochloride (TCH) and the combination of TCH and aloe vera (Aloe). The antibacterial activity of these sponges was tested against *K. pneumonia, E. coli, S. aureus,* and *Bacillus subtilis* and was found to exhibit significant inhibition against both gram-negative and gram-positive strains [55]. Plain CS sponge exhibited reduced inhibition against both bacterial strains signifying that the encapsulation of TCH and Aloe enhanced the antibacterial activity of the sponge. CS-Aloe-TCH sponge exhibited the best inhibition effect against all the tested bacterial species in terms of inhibition zone when compared to CH and CH-TCH sponges. The drug-release studies CH-Aloe-TCH was a Fickian diffusion over a period of 6 h. CH-TCH sponges were best fitted with Higuchi's model. The antibacterial activity of CS sponge was influenced by the antibiotic loaded in them. The structural pore size of the sponges influenced their water absorption water capability and their drug release via a diffusion process.

Ye and coworkers prepared bacterial cellulose/gelatin (BG) sponges loaded with ampicillin. They tested six different sponges from BG-1 to BG-5 with increasing concentrations including the plain BG sponge against three bacterial microorganisms, *C. albicans, E. coli,* and *S. aureus.* The plain BG sponges showed no inhibition effect on the bacteria while BG-5 exhibited a greater inhibition zone in vitro when compared to all the tested sponges against the three bacterial strains. An increase in the concentration of the drug in the sponges increased the antibacterial activity of the sponges. BG-5 presented a great inhibition zone against *E. coli* and *S. aureus* and a partial inhibition against *C. aureus* signifying poor antibacterial activity of ampicillin against *C. aureus.* The mode of action for BG-5 on the bacteria was by the destruction of the bacterial cell membrane leading to the cytoplasm leakage as observed by SEM analysis in which the images of the bacterial cells of *S. aureus, C. albicans,* and *E. coli* revealed a loss of the integrity of cell membranes. The drug release from the hydrogel fitted into a Korsmeyer–Peppas model revealed a non-Fickian diffusion. The release studies showed an initial burst in the first 12 h [56].

Pawar and coworkers prepared chitosan sponges loaded with different antibiotics vancomycin (VAN), CIP, and cefuroxime (CEF). These sponges exhibited significant inhibition on the tested bacterial strains of *S. aureus* for 6, 12, and 24 h, respectively, confirming the antibacterial activity of these antibiotics when incorporated with the chitosan matrix. The antibacterial inhibition increased with an antibiotic concentration in due time of drug released. CS-VAN sponge was reported to exhibit a greater inhibition zone of 6 mm, followed by CS-CEF sponge with 5 mm and CS-CIP with 4.7 mm in the first 30 min. CS-VAN sponges showed

a short-term biological activity as it maintained the inhibition zone of 13 mm for only 2 days. The fast diffusion of VAN is due to its hydrophilic properties in an aqueous environment from the CS matrix. CS-CIP lasted for 13 days with an inhibition zone of 9.5 mm and CS-CEF with 11 mm inhibition zone on the 25th day presenting a sustained antibacterial activity for an extended time [57]. They further reported the sustained CS-CIP sponge activity to be linked with a slow release of the CIP due to its aqueous environment solubility and its hydrophobic properties leading to low diffusion of the drug throughout the scaffold. The three sponges' cell viability was 80% meaning that they were less toxic.

Ma et al. prepared sponges from a combination of sodium alginate, GO, and polyvinyl alcohol and referred to them as SPG. The focus of the research was on the influence of GO in the sponge SP (0 wt%), SPG1 (0.5 wt%), SPG2 (1 wt%), and SPG3 (2 wt%) content of GO on norfloxacin (NFX) encapsulated sponges. All four sponges presented a sustained NFX release with an initial burst within the first 12 h, with SPG1 exhibiting 87% release followed by SPG2 both showing greater drug release. SP exhibited a greater drug release than SPG3 but significantly lower release rates than SPG1 and SPG2 suggesting that the content of GO influenced the release kinetics of the sponges. From their finding, it is notable that a lower GO content has the ability to improve release kinetics of the sponges whereas a larger content has the opposite influence. In vitro antibacterial activity of the four sponges was tested against *E. coli* gram and *S. aureus,* respectively. SPG1 exhibited significant inhibition zones on both bacterial strains when compared to other sponges with inhibition zones of 3.27 cm and 3.04 cm, respectively. SPG3 showed a minimal inhibition zone of 2.82 cm and 2.83 cm, respectively. In vivo healing studies using gauze, SPG2 antibiotic-free and SPG2-NFX loaded with norfloxacin revealed excellent healing of SPG2-NFX that was visible with wound closure when compared to gauze and SPG. The percentage of closure after 2 weeks was 65% for gauze, 80% for SPG, and 90% for SPG2-NFX-treated wounds with smooth wound area and no significant scar formation [58].

3.3 Films

Film wound dressings are made up of adherent and transparent polyurethane that permits the diffusion of carbon dioxide, oxygen, and water vapor from the injury, and it also offers autolytic removal of damaged tissues from the wounds [1,15]. Previously, film dressings were composed of derivatives of nylon with adherent polyethylene frames as the way to make them be occlusive [59]. The nylon films were not utilized for the wounds have high exudates because their absorption capacity is limited and result to wound maceration. The properties of the currently used film dressings are high flexibility, elasticity resulting in its capability to be tailored to any shape, and do not need additional tapping [60]. The wound healing process can be inspected without the removal of dressings due to their transparency. Hence, film wound dressings are preferred for shallow wound, superficial wound, and epithelizing wound with low exudates [61].

Marco Contardi et al. formulated antibiotic-based films by encapsulating ciprofloxacin as an antibiotic into the polyvinylpyrrolidone (PVP) matrix by utilizing acetic acid in water [62]. The antibacterial activity of film dressings was evaluated using *bacillus subtilis* and *Escherichia coli*. The free PVP did not exhibit any antibacterial activity on the bacterial growth of *E. coli*, whereby PVP films concentrated with 1 wt% and 30 wt% acetic acid exhibited inhibition of bacterial growth of 0.5% and 30%, respectively. On the other hand, PVP films loaded with 1 wt% and 30 wt% of ciprofloxacin displayed a broader region of inhibition when compared to films bounded only with acetic acid. Films incorporated with ciprofloxacin and acetic acid demonstrated increased inhibitory efficacy against the *E. coli* growth because of the synergistic effect of ciprofloxacin and acetic acid. Furthermore, no antibacterial activity was exhibited by films bounded only with acetic acid against *B. subtilis* that can be caused by high pH resistance of this bacterial strain. The films loaded with ciprofloxacin demonstrated high inhibition growth of *B. subtilis* strain [62].

The in vitro drug-release studies showed that only 20% of ciprofloxacin was released from the films within 2.5 min. The increase in the volume % of acetic acid in the preparation of PVP films induced a high release of ciprofloxacin up to 100% when compared to the films with low volume % of acetic acid. In addition, these films displayed promising wound resorption characteristics on full-thickness skin wound in vivo on mice model. The absorption of wound exudates by the films was influenced by the concentration of the acetic acid. Furthermore, these films exhibited good antiseptic activity [62].

Thawatchai Phaechamud and coworkers formulated natural porous rubber film dressings encapsulated with gentamicin sulfate, and their properties such as drug-release kinetics, morphology, water erosion and sorption capability, adhesive property, mechanical property, oxygen permeation, water vapor transmission rate, and antimicrobial activity were evaluated [63]. The water

sorption capacity of films was improved by the presence of xanthan gum. The developed morphology of the films with a dense-top layer displayed the low adhesive property, oxygen permeability, and water vapor. In vitro drug-release profiles of porous rubber films demonstrated sustained and controlled gentamicin sulfate-release mechanisms from the films over a period of 7 days. The antimicrobial studies of films loaded with gentamicin exhibited good antibacterial efficacy against *Pseudomonas aeruginosa* and *Staphylococcus aureus* [63].

García and coworkers designed bioadhesive film dressings using dendronized PVP-cross-linked chitosan incorporated with ciprofloxacin [64]. The in vitro antimicrobial assessments of the films loaded with ciprofloxacin showed good antibacterial efficacy by displaying a potential growth inhibition of *S. aureus* and *P. aeruginosa*, which are bacterial species that cause wound infections. The in vitro drug-release mechanism of the films was sustained with extended ciprofloxacin release at pH 5.4 and 7.4, a physiological pH of the skin and plasmatic area, respectively. Furthermore, adhesion and biocompatibility into the conjunctival sacs of the rabbits indicate that these polymeric films have good features to be utilized on the skin wounds for topical applications, favoring a reduced frequent administration and enhancing the residence period of the films [64].

Pawar et al. synthesized wound dressing films from polyox/sodium alginate (POL-SA) and polyox/carrageenan (POL-CAR), loaded with streptomycin and diclofenac [65]. These films were screened for antibacterial activity against *Escherichia coli, staphylococcus aureus,* and *Pseudomonas aeruginosa*. The activity of the films was compared with the commercial silver dressings. Minimum inhibitory concentration exhibited higher values for diclofenac as compared to streptomycin because of the nonpredictable antibacterial efficacy of diclofecan. These wound dressings loaded with streptomycin and diclofenac displayed high antibacterial activity against all the three bacterial species whereby POL-SA formulations entrapped with antibiotics were more active than POL-CAR formulations. These films displayed higher antibacterial activity when compared to the marketed silver dressings and wafer dressings. The combination of streptomycin with diclofenac in films resulted in a synergistic effect for the destruction of the bacteria and the prevention of the development of drug resistance [65].

3.4 Foams

Foam wound dressings are composed of hydrophilic and hydrophobic foam with adhesive boundaries [1].

The outer layer with a hydrophobic nature protects the wound from the liquid but permits water vapor and gaseous exchange. Foam dressings have the ability to absorb variable amounts of wound exudates depending on the wound thickness. The nonadhesive and adhesive foams are available for wound dressing. Foam wound dressings are preferred for application on lower leg ulcers with moderate-to-high exudate wounds. Secondary dressings are not needed for foams because of their moisture vapor permeability and high absorbency [66]. The limitation of foam dressings is that they are not suitable for wound with low exudates, dry scar, and dry wound because they require exudates for their healing mechanism and require frequent dressing [67].

Michailidou et al. formulated polymer-based foams for topical wound transport of chloramphenicol (CHL), a broad-spectrum antibiotic, employing polymer based on the combinations of chitosan and its derivatives. The physicochemical properties of formulated foams were evaluated using FTIR, XRD, SEM, and TGA that revealed their amorphous nature, porous microstructure, and sponge-like structure with a significant increased thermal stability influenced by the concentration of chitosan. The swelling ability studies of chitosan foams displayed a higher degree of swelling capacity for chitosan that combined with its derivatives when compared to free chitosan [68].

The in vitro drug-release profiles of the foams using Korsmeyer—Peppas model displayed an exponential factor of less than 0.5 indicating that the CHL diffuses through the networks with a quasi-Fickian diffusion mode further exhibiting controlled and sustained CHL release. Furthermore, the in vitro antibacterial activity of free chitosan and CHL-loaded foam was evaluated using Mueller Hinton agar plates against *E. coli*, *P. aeruginosa*, and *S. aureus*. The chitosan foams exhibited bactericidal activity against all the bacteria species whereby the CHL-loaded chitosan displayed high growth inhibition of sixfold and eightfold when compared to free chitosan against *E. coli* and *S. aureus*, respectively [68].

Santos and coworkers prepared foam wound dressings loaded with usnic acid using polyurethane (PU)/ and polyaniline (PA) as polymers, and their antibacterial activity was evaluated against gram-positive *Staphylococcus aureus* and gram-negative *Escherichia coli*. The SEM images displayed the porous structures of PU that provide the adhesion site of the antibacterial agents. These images further showed the presence of crystallites confirming the successful interaction of usnic acid with the polymers. The Raman and UV—Vis spectrums exhibited the differences between the chemical structure of the usnic acid loaded and unloaded polyurethane (PU)/and PA foams confirming the successful preparation of them. The antibiofilm studies showed that the inhibition of biofilm formation on both species was 10%, 20%, and 75% for the free PU, PU combined with PA, and PU combined with PA loaded with usnic acid, respectively [69]. The in vitro antibacterial activity studies on the foams showed that free PU foam displayed no bactericidal efficacy against the bacteria species while free PA foam did. These studies further displayed the good bactericidal activity of usnic loaded foams against gram-positive and gram-negative bacteria [69].

McGann et al. designed multifunctional polyHIPE (high internal phase emulsion) foams loaded with ciprofloxacin and/tetracycline using poly(ethylene glycol) diacrylate, sodium/calcium polyacrylate, and PNIPAM as the polymers. The foams displayed increased loading modulus and strain degree, whereby some of the foams demonstrated great stiffness. The cytocompatibility studies of the foams on the viability of the cells revealed nontoxicity. Furthermore, the polymeric foams loaded with antibiotics showed high bacterial growth inhibition against *S. aureus* whereby the blank foams showed no growth inhibition [70].

3.5 Wafers

Wafers are wound dressings that are highly porous and are lyophilized polymers of determined structure and they have been employed as a solid transport for the suspension of matrix metalloproteinase inhibitor, which is insoluble, for the management of chronic wounds [71]. These dressings were originally formulated to absorb fluid of the wound at a predetermined degree forming a physical gel or a viscous solution from which the entrapped molecule can diffuse straight into the aimed tissue whereby the release mechanism is controlled and sustained by swelling wafer network. There are certain polymers, such as xanthan gum and sodium alginate, which are utilized for the fabrication of wafers [71]. The properties of wafers include their capability as topical drug delivery systems, mucoadhesive nature, their prolonged residence on the wound, and their ability to be entrapped with both insoluble and soluble antimicrobial drugs. These wound dressings are used for the treatment of chronic wounds (e.g., diabetic foot ulcers, third-level burn wounds, and mucosal inflammation) that are infected with bacteria [72].

Labovitiadi and coworkers formulated freeze-dried wafers incorporated with chlorhexidine digluconate (CHD) utilizing polysaccharides such as guar gum (GG), xanthan gum (XG), karaya gum (KAG), and SA as the carriers. These polysaccharide wafers were able to transform to highly viscous gels in situ for a long time followed by a controlled and sustained release of the encapsulated antimicrobial. The rheological studies demonstrated that the control gels of SA (5.0%) and KAG (3.0%) did not show apparent yield stress indicating that no important gel network was present with no induced pseudoplastic flow. Incorporating CHD decreased the stability of KAG by 65% but increased the stability of SA by 131% showing the important interaction between CHD and SA. The synergistic combination of KAG and SA incorporated CHD exhibited an increase in both yield stress and stability when compared to free polysaccharide solutions [73].

The drug-release kinetics were performed using the diffusion cell, and the CHD release from the lyophilized gels was faster when compared to lyophilized wafers. The release of CHD from the wafer was in the order: KAG > XG > GG > SA and SA−KAG, KGA wafers showing a concentration of 60 µg/mL at 24 h when compared to SA−KAG that gave up a concentration of approximately 12 µg/mL. The wafers displayed significant antimicrobial activity against *P. aeruginosa*, especially CHD-KAG wafers [73]. Furthermore, Labovitiadi and coworkers prepared wafers using natural vegetable gum, karaya gels, loaded with selected antibiotics and their bactericidal activity was evaluated in vitro against methicillin-resistant *S. aureus*. These dressings exhibited good swelling properties in the presence of simulated wound fluid and released the loaded antibiotics. The wafers incorporated with chlorhexidine and povidone-iodine displayed good bactericidal activity in protein-free buffer whereby neomycin-incorporated wafers activity was improved by the presence of bovine serum albumin [72].

Shiow-Fern Ng et al. formulated wafers using four different polymers: methylcellulose (MC), sodium carboxymethyl cellulose (NaCMC), xanthan gum, and sodium alginate, incorporated with neomycin trisulfate, silver nitrate, and sulfacetamide sodium, but only MC and NaCMC were selected for further examination. Among the preparations, neomycin trisulfate-NaCMC wafer showed the most significant wound dressing features (i.e., sponginess, flexibility, high content drug uniformity, uniform wafer texture) with the controlled and sustained in vitro drug-release profile and the highest inhibition of bacteria growth against both *E. coli* and *S. aureus* bacteria when compared to free neomycin [74].

Gowda and coworkers successfully designed and developed nonfriable, cohesive, disk shape, porous, and antimicrobial wafers for the treatment of chronic wounds using guar gum and sodium alginate as the carriers. These neomycin-based wafers exhibited a drug content of 99.62%, the tensile strength of 0.25 N/m², and viscosity of 12.89 Pa s. The water uptake profile of these wafers displayed 509.7% for drug-loaded wafers and 715.6% for control. Hydration evaluation for antimicrobial wafers exhibited 97.91% and also displayed the most significant characteristics of a wound dressing with the highest drug release in vitro. They also showed good antibacterial activity against *E. coli* and *S. aureus* bacteria with better healing mechanism when compared to the neomycin cream. The in vivo studies of wafers using wounded Wister rats displayed good skin regeneration by exhibiting the highest decrease in the diameter of the wound when compared to Neomycin cream [75].

Pawar et al. formulated wafers of weight ratios of polyox with sodium alginate (50/50) or carrageen (75/25) loaded with diclofecan and streptomycin, for chronic wound healing. These wafers were flexible, soft, nonbrittle, improved porous nature, and elegant in appearance. The mechanical evaluation showed that they were able to withstand normal stress and they were also flexible to protect newly produced skin tissue from the damage. Free wafers displayed relatively high adhesion and swelling when compared to antibiotic-loaded wafers. The release-mechanism profiles of antibiotics from the wafers displayed sustained and controlled drug release. The antibiotic-loaded wafers exhibited good bactericidal activity on *E. coli* and have the capability to reduce pain and swelling related to injury due to the diclofecan antiinflammatory activity [76].

3.6 Fibers

Fibers are regarded as ideal wound dressing materials for chronic wounds due to their drug-delivery characteristics [77−79]. Fibers provide physical protection to the wound site for a long period of time, and they can be incorporated with a high amount of drugs, up to 40%, where there is an adjustment of the release by changing types and compositions of the materials in the fibers [80]. Nanofibers mimic the extracellular matrix, thereby enhancing the proliferation of epithelial cells and the formation of new tissue at the wounded site [81,82]. Their nanometer diameter and nanofibrous meshes promote hemostasis of injured tissues, enhance fluid absorption, promote dermal drug delivery, cell respiration, and high-gas permeation, thereby preventing bacterial infections [81,82]. The electrospinning method makes it difficult to control the pore structure of the nanofibers. The preparation of the nanofibers using a self-assembly method results in nonuniform shapes. When the fiber mesh is utilized for the drug-loading process, the drug loading into the nanofibers becomes low [28,83]. Incorporation of antibiotics into nanofibers improves the efficacy of nanofibers against disease-resistant bacteria (Table 7.2).

Seham et al. reported the application of electrospun fibers as drug and biological agent delivery systems for wound healing. Nanofibers produced through electrospinning have become a promising strategy in the treatment of wounds. They are produced from natural or synthetic polymers. Nanofibers exhibit high surface area, nanoporosity, and can encapsulate potent drugs or enzymes. Therefore nanofibers are excellent candidates for wound treatment and management [84].

Wongkanya et al. prepared fibers blended with soy protein isolate and sodium alginate, encapsulated with vancomycin antibiotic. The fibers were fabricated using the electrospinning method with the assistance of poly(-ethylene oxide) [85]. The results revealed that the release characteristic of vancomycin-loaded sodium alginate/poly(ethylene oxide)/soy protein isolated fibers displayed a ruptured release followed by a controllable release after 2 days when immersed in a phosphate-buffered saline. The drug-loaded fibers were effective against *S. aureus* after 24 h of incubation and the fibers also revealed nontoxicity and biocompatibility characteristics. This suggests that the vancomycin-blended fibers are promising nanomaterials for drug-delivery systems and tissue engineering [85].

Dzikovski et al. prepared an in-jet sprayed nanofibrillar scaffold incorporated with ciprofloxacin antibiotic with biodegradable characteristics. The antibiotic-loaded wound dressing was prepared to avert wound infection [86]. The blends of poly-(ε-caprolactone) (PCL) and poly-D,L-(lactic acid) (PDLLA) were used for the preparation of the matrices. The incorporation of ciprofloxacin leads to a slight increase in the fiber diameter and the mechanical properties of different polymer blends were not modified. Sustained release was observed after more than 23 days. However, the incorporation of antibiotics was efficient in inhibiting bacterial growth of *E. coli* and *B. subtilus* [86].

3.7 Topical Formulation

Topical wound dressings are formulations applied directly to the skin. Topical formulation exists in the form of ointment, gels, creams, powders, pastes, and shake lotions [87]. The type of formulation selected depends on several factors: the nature of the patients' skin, characteristics of the skin problem, the drug allergies, and the previous medication that was taken. The formulation has some advantages, which include the delivery of drugs more selectively to the targeted site, patient compliance, and also overcomes fluctuation in the level of drug [87]. The skin is a complex combination of tissues that perform functions necessary for human survival. Our skin helps to maintain body temperature, receives stimuli from the environment, and stores chemical compounds. The dermal and the epidermal of the skin act as a barrier that prevents drug transport; therefore, it is important to select the right excipient to reduce the immunogenic effect, surface adsorption, and overcome degradation [88]. The application of topical formulation for wounds is limited by reduced penetration in open wounds, systematic absorption in large wounds, risk of hypersensitivity reactions, contamination of the formulation from day-to-day bodily contact because of frequent applications, and alteration of normal cutaneous flora [88].

Some researchers have encapsulated topical formulation with antibiotics to prevent bacterial infection. Toussaint et al. formulated a topical antibiotic ointment for second-degree burns in swine comparing it with silver-containing foam. The silver-containing dressing and the antibiotic ointment were compared to determine which formulation enhanced reepithelialization with less scaring. Deep partial-thickness burns were created on the flanks of three anesthetized female domestic pigs (20–25 kg) using a 150-g aluminum bar preheated in an 80°C water bath and applied to the skin for 20 s using a force of 2 kg. The burn eschars were excised 48 h later with an electric dermatome set at a depth of 0.75 mm. The wound beds were treated with a thin layer of triple-antibiotic petrolatum-based ointment (changed three times weekly) or a silver-containing foam dressing (changed once weekly). Full-thickness punch biopsies were obtained at 9, 11, 14, 16, 18, and 21 days for the determination of percentage complete wound reepithelialization and at 28 days for the measurement of scar depth. Results revealed that a triple-antibiotic ointment enhanced reepithelialization and reduced scar depth and contraction compared with a silver-based foam dressing. This triple-antibiotic ointment should be considered as a control for studies evaluating novel topical burn therapy [89].

Khazri et al. prepared a topical gel composed of rosemary essential oil loaded into nanostructured lipid carriers for in vitro antibacterial activity and in vivo infected wound healing. The gel was applied on two full-thickness wounds, each about 6 mm, at the back of a mouse and each was contaminated with a solution of 10^7 CFU. *Staphylococcus aureus* and *Pseudomonas aeruginosa*. Two animals were treated with the gel containing the rosemary essential oil and nanostructured

TABLE 7.2
Different Types of Wound Dressing Materials.

Wound Dressing	Material	Antibiotic	Therapeutic Outcome	Bacterial Species	Reference
Hydrogels	Chitosan	CIP	Inhibit bacterial growth and has high antibacterial activity. Partially effective	*K. pneumonia, E. coli, S. aureus Streptococcus mutans*	[47]
	PVA	CIP	Inhibit bacterial growth at a greater concentration	*E. coli, S. aureus*	[48]
	Chitosan	CEF	Inhibit bacterial growth on increased concentration levels	*S. aureus*	[40]
	—	Gentamicin Moxifloxacin	Effective Effective	*S. aureus*	[49]
Sponges	Chitosan-aloe vera	TCH	Exhibited significant antibacterial inhibition	*K. pneumonia, E. coli, S. aureus, Bacillus subtilis*	[55]
	Bacterial cellulose/gelatin	Ampicillin	Partially effective. Higher drug concentration had greater antibacterial activity	*C. albicans, E. coli and S. aureus*	[56]
	Sodium alginate, (GO) and PVA	NFX	Effective Effective	*E. coli, S. aureus,*	[58]
	Chitosan	VAN	More effective but suffered from half-life span. Relatively active	*S. aureus*	
		CEF CIP	Less active and last longer at the wound area	*S. aureus*	[57]
Fibers	Sodium alginate	Vancomycin	Effective against bacterial infection, drug delivery, and tissue engineering	*Staphylococcus aureus*	[85]
	PCL PDLLA	Ciprofloxacin	Effective against bacterial growth	*Escherichia coli Bacillus subtilis*	[86]
Topical formulation	NLC	REO	Reduces wound size and bacterial colonization	*Staphylococcus aureus Pseudomonas aeruginosa*	[90]
Topical formulation	Ag gel	Methicillin	Effective against biofilm infection caused by gram-positive and gram-negative bacteria	*Staphylococcus aureus Pseudomonas aeruginosa Acinetobacter baumanii*	[92]
	PVP HA	Ciprofloxacin	Self-adhering strength to human skin. Sustains release of antibiotics within 5 days. Effective against bacterial strains causing infections	*Staphylococcus aureus Escherichia coli Pseudomonas aeruginosa*	[93]
	PVP HA	Ciprofloxacin	Self-adhering strength to human skin. Sustains release of antibiotics within 5 days. Reduces bacterial infection	*Staphylococcus aureus Escherichia coli Pseudomonas aeruginosa*	[94]
Films	Polyvinylpyrrolidone (PVP)	Ciprofloxacin		*B. subtilis and E. coli*	[62]

	Natural porous rubber	Gentamicin	High rate of bacterial growth inhibition and good antiseptic efficacy	P. aeruginosa and S. aureus	[63]
	PVP	Ciprofloxacin	Controlled and sustained drug release rate, and good antibacterial efficacy	P. aeruginosa and S. aureus	[64]
	Polyox/sodium alginate (POL–SA) and polyox/carrageenan (POL–CAR)	Streptomycin and diclofenac	Good bactericidal activity and good adhesion and biocompatibility. Synergistic effect resulting in good antibacterial activity	Staphylococcus aureus and Pseudomonas aeruginosa	[65]
Foams	Chitosan and its derivatives	Chloramphenicol	High degree of swelling capacity and high antibacterial activity	E. coli, Ps. aeruginosa or S. aureus	[68]
	Polyurethane and polyaniline	Usnic acid	Good in vitro antibacterial activity and antibiofilm	S. aureus and E. coli	[69]
	Poly(ethylene glycol) diacrylate, sodium/calcium polyacrylate, and PNIPAM	Ciprofloxacin and tetracycline	High bacteria high growth inhibition	E. coli, and Ps. aeruginosa	[70]
Wafers	Polysaccharides	Chlorhexidine digluconate (CHD)	Sustained stability, and controlled and sustained drug release	P. aeruginosa	[73]
	Natural vegetable gum Methylcellulose (MC), sodium carboxymethyl cellulose (NaCMC), xanthan gum, and sodium alginate	Chlorhexidine	Good swelling properties	S. aureus	[72]
		Neomycin	Good wound dressing features and controlled drug release	E. coli and S. aureus	[74]
	Guar gum and sodium alginate	Neomycin	High drug loading capacity and good skin regeneration	E. coli and S. aureus	[75]
	Polyox, sodium alginate, and carrageen	Diclofecan and streptomycin	Good wound dressing properties and bactericidal activity	E. coli	[76]

lipid carriers: one with Mupirocin and a control. The size, morphology, and antibacterial activity of the gels was characterized. The results revealed the antibacterial activity of the gel against *Staphylococcus aureus, Pseudomonas aeruginosa, Staphylococcus epidermidis, Escherichia coli,* and *Listeria monocytogenes*. However, the gel was capable of reducing the rate of tissue bacterial colonization and the size of the wound. The gel accelerated the wound healing process and is suitable for the treatment of infected wounds [90].

Zilberman et al. prepared a hybrid wound dressing containing a drug-loaded porous poly(di-lactic-co-glycolic acid) with a spongy collagen layer. The layer was incorporated with gentamicin to prevent bacterial infection at the wound site of a burn pig model and compared to neutral nonadherent dressing material Aquacel Ag (ConvaTec) and Melolin (Smith & Nephew). Results revealed a slow gentamicin release (28%), with the highest degree of wound healing compared to previously studied gentamicin-release kinetics. The wound dressing also was suitable for treating large infected burns [91].

Tran et al. formulated a topical silver gel wound dressing for the inhibition of bacterial growth. The formulation was tested in the mouse wound model and tested at different bacterial strains. The gel was analyzed in vivo and in vitro around the infected area. The silver gel was effective against *Staphylococcus aureus, Acinetobacter baumannii,* and *Pseudomonas aeruginosa* for the Ag-gel dressing when compared to the control dressing, using CFU essays, confirmed by in vitro and in vivo results. However, the silver gel was useful in the prevention of biofilm infections caused by both gram-positive and gram-negative bacteria [92].

Marco et al. prepared a transparent bilayer for the release and delivery of antibiotics and cutaneous antiseptic. Polyvinylpyrrolidone was the first layer formulated, containing Neomercurocromo antiseptic. The formulation was tested on full-thickness excisional wound healing mice model. Hyaluronic acid encapsulated with polyvinylpyrrolidone and ciprofloxacin was the second layer. The bilayer film revealed a satisfactory self-adhering strength to the human skin. The polyvinylpyrrolidone and hyaluronic acid formulation interacted via hydrogen bonds with a sustained release of antibiotics within a period of 5 days. The biocompatibility was also observed on the human foreskin fibroblast. The gel was effective against *Escherichia coli, Staphylococcus aureus,* and *Pseudomonas aeruginosa*. The formulations revealed that they are good wound care products for various skin wounds, including skin infections [93].

Shukur and Metwally formulated a topical gel containing rosemary, lemongrass, basil, and thyme oils for the effectiveness against methicillin-resistant *Staphylococcus aureus* skin infections. The formulation was prepared using hydroxypropylmethylcellulose, Carbopol 940, and sodium carboxymethyl cellulose, with lemongrass oil. The spreadability, pH, physical appearance, rheological properties, and antibacterial activity were evaluated against methicillin-resistant *Staphylococcus aureus*. Results revealed that the minimum inhibitory concentration of the thyme oils and lemongrass was 4 and 30 μL/mL. However, the gel containing Carbopol 940 revealed good physical characteristics, including rheological studies and spreadability. The gel containing Carbopol 940 with thyme oil and lemongrass exhibited good antibacterial activity against methicillin-resistant *Staphylococcus aureus* when incorporated into the skin. The gels did not exhibit any kind of swelling or redness on the skin [94].

4 CONCLUSION

Wound dressings scaffolds not encapsulated with antibacterial agents do not exhibit significant antibacterial activity when compared with wound dressings encapsulated with antibiotics. The encapsulation of antibiotics in these scaffolds played a significant role in wound dressing by enhancing the wound healing process. The concentration of the antibiotics and the release mechanism of the wound dressings also influenced the wound healing process. Most of the scaffolds exhibited enhanced reepithelialization with reduced scar depth and contraction. The effectiveness of these scaffolds was proven based on in vitro and in vivo experimental studies. However, preclinical studies are needed to confirm the safety of the scaffolds. An ideal wound dressing enhances the healing process, is permeable for gaseous exchange, is nontoxic, and prevents bacterial infections. Based on the findings reported by several researchers, the incorporation of antibiotics into wound dressing scaffolds is a potential approach for the development of wound dressings for the management of chronic wounds.

ACKNOWLEDGMENTS

The financial assistance of the Medical Research Council (Self-Initiated Research) (MRC) and National Research Foundation (NRF), South Africa toward this research is hereby acknowledged. The views and opinions expressed in this manuscript are those of the authors and not of MRC or NRF.

REFERENCES

[1] S. Dhivya, V. Vijaya, E. Santhini, Review article Wound dressings — a review, Biomedicine 4 (5) (2015) 24—28.

[2] M.C. Robson, D.L. Steed, M.G. Franz, Wound healing: biologic feature and approaches to maximize healing trajectories, Current Problems in Surgery 2 (38) (2001) 72—139.

[3] J.G. Powers, C. Higham, K. Broussard, T.J. Phillips, Chronic wound care and management, Journal of the American Academy of Dermatology 4 (74) (2016) 607—625.

[4] S. Schreml, R. Szeimies, L. Prantl, S. Karrer, M. Landthaler, P. Babilas, Oxygen in acute and chronic wound healing, British Journal of Dermatology 2 (163) (2010) 257—268.

[5] V. Bartels, Handbook of Medical Textiles, Woodhead Publ., 2011. Ser. Text. Number 100 (100).

[6] M.T.L. Chandan, K. Sen, G.M. Gordillo, S. Roy, R. Kirsner, L. Lambert, T.K. Hunt, F. Gottrup, G.C. Gurtner, Human skin wounds: a major and snowballing threat to public health and the economy, Wound Repair and Regeneration (6) (2009) 763—771.

[7] B. Fpg, A. Koehler, Hydrogel wound dressings for bioactive treatment of acute and chronic wounds, European Polymer Journal 100 (2018) 1—11.

[8] M. Flanagan, The physiology of wound healing, Journal of Wound Care 6 (9) (2000) 299—300.

[9] A.E. Rivera, J.M. Spencer, Clinical aspects of full-thickness wound healing, Clinics in Dermatology (25) (2007) 39—48.

[10] M. Strecker-McGraw, T. Jones, D. Baer, Soft tissue wounds and principles of healing, Emergency Medicine Clinics of North America 1 (25) (2007) 1—22.

[11] M. Ramos-e-silva, M. Cristina, R.D.E. Castro, New dressings, including tissue-engineered living skin, Clinics in Dermatology 20 (2002) 715—723.

[12] J. Boateng, K. Matthews, H. Steven, G. Eccleston, Wound healing dressings and drug delivery systems: a review, Journal of Pharmaceutical Sciences 8 (97) (2008) 2892—2923.

[13] S. Dhivya, V.V. Padma, E. Santhini, Wound dressings, Biomedicine (5) (2015) 24—28.

[14] L. Martin, et al., The release of model macromolecules may be controlled by the hydrophobicity of palmitoyl glycol chitosan hydrogels, Journal of Controlled Release 80 (2002) 87—100.

[15] I. Negut, V. Grumezescu, A.M. Grumezescu, Treatment strategies for infected wounds, Molecules 23 (2018) 1—23.

[16] M. Rai, K. Kon, A. Gade, A. Ingle, D. Nagaonkar, P. Paralikar, Antibiotic Resistance. Mechanisms and New Antimicrobial Approaches, New York, USA, 2016.

[17] M. Liu, H. Zhang, D. Min, Dual layered wound dressing with simultaneous temperature & antibacterial regulation properties, Materials Science and Engineering: C 94 (2019) 1077—1082.

[18] R. Han, G. Ceilley, Chronic wound healing: a review of current management and treatments, Advances in Therapy 34 (2017) 599—610.

[19] N.J. Braund, R. Hook, S. Medlicott, The role of topical growth factors in chronic wounds, Current Drug Delivery 4 (2007) 195—204.

[20] Biodermis, The Science of Skin, 2018.

[21] J. Frykberg, R.G. Banks, Challenges in the treatment of chronic wounds, Advanced Wound Care (New Rochelle) 4 (2015) 560—582.

[22] L.A. Guo, S.A. DiPietro, Factors affecting wound healing, Journal of Dental Research 89 (2010) 219—229.

[23] B.A. Aderibigbe, B. Buyana, Alginate in wound dressings, Pharmaceutics 10 (2018) 1—19.

[24] S. Sharma, A. Dua, A. Malik, Third generation materials for wound dressings, International Journal of Pharmaceutical Sciences and Research 6 (5) (2014) 2113—2124.

[25] J. Koehler, F.P. Brandl, A.M. Goepferich, Hydrogel wound dressings for bioactive treatment of acute and chronic wounds, European Polymer Journal 100 (2018) 1—11.

[26] H.P. Felgueiras, M.T.P. Amorim, Functionalization of electrospun polymeric wound dressings with antimicrobial peptides, Colloids and Surfaces B: Biointerfaces 156 (2017) 133—148.

[27] M. Mir, M. Najabat, A. Afifa, B. Ayesha, G. Munam, A. Shizza, M. Asad, Synthetic polymeric biomaterials for wound healing: a review, Progress in Biomaterials 1 (7) (2018) 1—21.

[28] V. Andreu, G. Mendoza, M. Arruebo, S. Irusta, Smart dressings based on nanostructured fibers containing natural origin antimicrobial, anti-inflammatory, and regenerative compounds, Materials 8 (2015) 5154—5193.

[29] R.S. Ambekar, B. Kandasubramanian, Advancements in nano fibers for wound dressing: a review, European Polymer Journal 117 (2019) 304—336.

[30] S. Fahimirad, F. Ajalloueian, Naturally-derived electrospun wound dressings for target delivery of bio- active agents, International Journal of Pharmacuetics 566 (2019) 307—328.

[31] M. Naseri-nosar, Z. Maria, Wound dressings from naturally-occurring polymers: a review on homopolysaccharide-based composites, Carbohydrate Polymers 189 (2018) 379—398.

[32] A. Ajovalasit, M.A. Sabatinoa, S. Todaroa, S. Alessia, D. Giacomazzab, P. Piconec, M.D. Carloc, C. Dispenza, Xyloglucan-based hydrogel films for wound dressing: structure-property relationships, Carbohydrate Polymers 179 (2018) 262—272.

[33] H. Namazi, R. Rakhshaei, H. Hamishehkar, H. Samadi, Antibiotic loaded carboxymethylcellulose/MCM-41 nanocomposite hydrogel films as potential wound dressing, International Journal of Biological Macromolecules 85 (2016) 327—334.

[34] E.A. Kamoun, E.S. Kenawy, X. Chen, A review on polymeric hydrogel membranes for wound dressing applications: PVA-based hydrogel dressings, Journal of Advanced Research 8 (2017) 217—233.

[35] S. Li, S. Dong, W. Xu, S. Tu, L. Yan, C. Zhao, J. Ding, X. Chen, Antibacterial hydrogels, Advanced Science (2018) 1—17 (1700527).

[36] A.T. Zafalon, V.J. dos Santos, F. Esposito, N. Lincopan, V. Rangari, A.B. Lugão, D.F. Parra, Synthesis of polymeric hydrogel loaded with antibiotic drug for wound healing

applications, The Minerals, Metals and Materials Society (2019) 165—176.

[37] N. Wang, W. Xiao, B. Niu, W. Duan, L. Zhou, Y. Zheng, Highly efficient adsorption of fluoroquinolone antibiotics using chitosan derived granular hydrogel with 3D structure, Journal of Molecular Liquids 281 (2019) 307—314.

[38] W. Mozalewska, R. Czechowska-biskup, A.K. Olejnik, R.A. Wacha, P. Ulańskia, J.M. Rosiak, Chitosan-containing hydrogel wound dressings prepared by radiation technique, Radiation Physics and Chemistry 134 (2017) 1—7.

[39] S. Anjum, A. Arora, M.S. Alam, B. Gupta, Development of antimicrobial and scar preventive chitosan hydrogel wound dressings, International Journal of Pharmacuetics 508 (2016) 92—101.

[40] V. Pawar, M. Dhanka, R. Srivastava, Cefuroxime conjugated chitosan hydrogel for treatment of wound infections, Colloids and Surfaces B: Biointerfaces 173 (2019) 776—787.

[41] H. Hamedi, S. Moradi, S.M. Hudson, A.E. Tonelli, Chitosan based hydrogels and their applications for drug delivery in wound dressings: a review, Carbohydrate Polymers 199 (2018) 445—460.

[42] B.T. Hima, M. Vidyavathi, K. Kavitha, T.P. Sastry, Preparation and evaluation of ciprofloxacin loaded chitosan-gelatin composite films for wound healing activity, International Journal of Drug Delivery (2) (2010) 173—182.

[43] C. Flores, M. Lopeza, N. Tabaryb, C. Neutc, F. Chaia, D. Betbederc, C. Herktc, F. Cazauxb, V. Gaucherb, B. Martelb, N. Blanchemain, Preparation and characterization of novel chitosan and ß-cyclodextrin polymer sponges for wound dressing applications, Carbohydrate Polymers 173 (2017) 535—546.

[44] K. Yang, Q. Han, B. Chen, Y. Zheng, K. Zhang, Q. Li, J. Wang, Antimicrobial hydrogels: promising materials for medical application, International Journal of Nanomedicine 13 (2018) 2217—2263.

[45] S.K. Gade, N. Shivshetty, N. Sharma, S. Bhatnagar, V.V.K. Venuganti, S. Pilani, Effect of mucoadhesive polymeric formulation, Journal of Ocular Pharmacology and Therapeutics 34 (2018) 1—9.

[46] D.A. Alves, D. Machado, A. Melo, R.F.C. Pereira, P. Severino, L.M. de Hollanda, D.R. Araújo, M. Lancellotti, Preparation of thermosensitive gel for controlled release of levofloxacin and their application in the treatment of multidrug-resistant bacteria, Biomed Research International 2016 (2016) 1—10.

[47] D.H. Hanna, G.R. Saad, Encapsulation of ciprofloxacin within modified xanthan gum-chitosan based hydrogel for drug delivery, Bioorganic Chemistry 84 (2019) 115—124.

[48] C. Choipang, P. Chuysinuan, O. Suwantong, P. Ekabutr, Hydrogel wound dressings loaded with PLGA/ciprofloxacin hydrochloride nanoparticles for use on pressure ulcers, Journal of Drug Delivery Science and Technology 47 (2018) 106—114.

[49] S. Mohammed, G. Chouhan, O. Anuforo, M. Cook, A. Walsh, P. Morgan-Warren, M. Jenkins, F. Cogan, Thermosensitive hydrogel as an in situ gelling antimicrobial ocular dressing, Materials Science and Engineering: C 78 (2017) 203—209.

[50] Y. Xie, Z. Yi, J. Wang, T. Hou, Q. Jiang, Carboxymethyl konjac glucomannan— crosslinked chitosan sponges for wound dressing, International Journal of Biological Macromolecules 112 (2018) 1225—1233.

[51] X. Yang, W. Liu, G. Xi, M. Wang, B. Liang, Y. Shi, Y. Fenga, X. Ren, C. Shi, Fabricating antimicrobial peptide-immobilized starch sponges for hemorrhage control and antibacterial treatment, Carbohydrate Polymers 222 (2019) 115012.

[52] Y. Feng, X. Li, Q. Zhang, S. Yan, Y. Guo, M. Li, R. You, Mechanically robust and flexible silk protein/polysaccharide composite sponges for wound dressing, Carbohydrate Polymers 216 (2019) 17—24.

[53] C. Chen, L. Liu, T. Huang, Q. Wang, Y. Fang, International Journal of Biological Macromolecules Bubble template fabrication of chitosan/poly (vinyl alcohol) sponges for wound dressing applications, International Journal of Biological Macromolecules 62 (2013) 188—193.

[54] K. Zhang, X. Bai, Z. Yuan, X. Cao, X. Jiao, Y. Li, Y. Qinb, Y. Wena, X. Zhang, Biomaterials Layered nano fiber sponge with an improved capacity for promoting blood coagulation and wound healing, Biomaterials 204 (2019) 70—79.

[55] S. Anbazhagan, K. Puthupalayam, Application of tetracycline hydrochloride loaded-fungal chitosan and Aloe vera extract based composite sponges for wound dressing, Journal of Advanced Research 14 (2018) 63—71.

[56] S. Ye, L. Jiang, C. Su, Z. Zhu, Y. Wen, W. Shao, Development of gelatin/bacterial cellulose composite sponges as potential natural wound dressings, International Journal of Biological Macromolecules 133 (2019) 148—155.

[57] V. Pawar, U. Bulbake, W. Khan, R. Srivastava, Chitosan sponges as a sustained release carrier system for the prophylaxis of orthopedic implant-associated infections, International Journal of Biological Macromolecules 134 (2019) 100—112.

[58] R. Ma, Y. Wang, H. Qi, C. Shi, G. Wei, L. Xiao, Z. Huang, S. Liu, H. Yu, C. Teng, H. Liu, V. Murugadoss, J. Zhang, Y. Wang, Z. Guo, Nanocomposite sponges of sodium alginate/graphene oxide/polyvinyl alcohol as potential wound dressing: in vitro and in vivo evaluation, Composites Part B 167 (2019) 396—405.

[59] G. Han, R. Ceilley, "Chronic wound healing: a review of current management and treatments, Advances in Therapy 34 (3) (2017) 599—610.

[60] B. Gupta, R. Agarwal, M. Alam, Textile-based smart wound dressings, Indian Journal of Fibre and Textile Research 35 (2) (2010) 174—184.

[61] N. Gauze, Dressing Materials, 103—117.

[62] M. Contardi, J.A. Heredia-Guerreroa, G. Perottoa, P. Valentinib, P.P. Pompab, R. Spanòc, L. Goldonic, R. Bertorellic, A. Athanassioua, I.S. Bayer, Transparent ciprofloxacin-povidone antibiotic films and nano fiber mats as

potential skin and wound care dressings, European Journal of Pharmaceutical Sciences 104 (2017) 133–144.

[63] T. Phaechamud, P. Issarayungyuen, W. Pichayakorn, Gentamicin sulfate-loaded porous natural rubber films for wound dressing, International Journal of Biological Macromolecules 85 (2016) 634–644.

[64] M.C. García, A.A. Aldanab, L.I. Tártaraa, F. Aloveroa, M.C. Strumiac, R.H. Manzoa, M. Martinellic, A.F. Jimenez-Kairuz, Bioadhesive and biocompatible films as wound dressing materials based on a novel dendronized chitosan loaded with ciprofloxacin, Carbohydrate Polymers 175 (2017) 75–86.

[65] H.V. Pawar, J. Tetteh, P. Debrah, J.S. Boateng, Comparison of in vitro antibacterial activity of streptomycin-diclofenac loaded composite biomaterial dressings with commercial silver based antimicrobial wound dressings, International Journal of Biological Macromolecules 121 (2019) 191–199.

[66] D. Morgan, Wounds- what should a dressing formulary include? Hospital Pharmacy 9 (2002) 216–261.

[67] M. Ramos-E-Silva, M. Cristina, R.D.E. Castro, New dressings, including tissue-enginnered living skin, Clinics in Dermatology 6 (20) (2002) 715–723.

[68] G. Michailidou, E. Christodoulou, S. Nanaki, P. Barmpalexis, E. Karavasc, S. Vergkizi-Nikolakakid, D.N. Bikiaris, Super-hydrophilic and high strength polymeric foam dressings of modified chitosan blends for topical wound delivery of chloramphenicol, Carbohydrate Polymers 208 (2019) 1–13.

[69] M.R. Santos, J.J. Alcaraz-espinoza, M.M. Costa, H.P. De Oliveira, Usnic acid-loaded polyaniline/polyurethane foam wound dressing: preparation and bactericidal activity, Materials Science & Engineering: C 89 (2018) 33–40.

[70] C.L. Mcgann, B.C. Streifel, J.G. Lundin, J.H. Wynne, Multifunctional polyHIPE wound dressings for the treatment of severe limb trauma, Polymer 126 (2017) 408–418.

[71] M. KH, S. HNE, A. AD, H. MJ, E. GM, Formulation, stability and thermal analysis of lyophilised wound healing wafers containing an insoluble MMP-3 inhibitor and non-ionic surfactant, International Journal of Pharmaceutics 356 (2006) 110–120.

[72] O. Labovitiadi, A.J. Lamb, K.H. Matthews, In vitro efficacy of antimicrobial wafers against methicillin-resistant *Staphylococcus aureus*, Therapeutic Delivery 4 (3) (2012) 443–455.

[73] O. Labovitiadi, A.J. Lamb, K.H. Matthews, Lyophilised wafers as vehicles for the topical release of chlorhexidine digluconate - release kinetics and efficacy against *Pseudomonas aeruginosa*, International Journal of Pharmaceutics 1–2 (439) (2012) 157–164.

[74] S. Ng, N. Jumaat, Carboxymethyl cellulose wafers containing antimicrobials: a modern drug delivery system for wound infections, European Journal of Pharmaceutical Sciences 51 (2014) 173–179.

[75] D.V. Gowda, S. Fredric, A. Srivastava, A.S. A. R, R.A.M. Osmani, Design and development of antimicrobial wafers for chronic wound healing, Der Pharmacia Lettre 7 (8) (2016) 70–79.

[76] H. V Pawar, J.S. Boateng, I. Ayensu, J. Tetteh, Multifunctional medicated lyophilised wafer dressing for effective chronic wound healing, Journal of Pharmaceutical Sciences 103 (6) (2014) 1720–1733.

[77] M. W, J. Wang, Functional electrospun fibers for the treatment of human skin wounds, European Journal of Pharmaceutics and Biopharmaceutics 119 (2017) 283–299.

[78] M. Khezri, A.S. Majd, M.R. Khorasgani, S.J. Moshtaghian, A. Talebi, Application of Chitosan/PVA Nano fiber as a potential wound dressing for streptozotocin-induced diabetic rats, International Journal of Biological Macromolecules 92 (2016) 1162–1168.

[79] M. Liu, X.P. Duan, Y.M. Li, D.P. Yang, Y.Z. L, Electrospun nanofibers for wound healing, Materials Science and Engineering: C 76 (2017) 1413–1423.

[80] K.A. Woodrow, S.F. Chou, Relationships between mechanical properties and drug release from electrospun fibers of PCL and PLGA blends, Journal of the Mechanical Behavior of Biomedical Materials 65 (2017) 724–733.

[81] M. Abrigo, S.L. McArthur, P. Kingshott, Electrospun nanofibers as dressings for chronic wound care: advances, challenges, and future prospects, Macromolecular Bioscience 14 (2014) 772–792.

[82] Y. Zhang, C.T. Lim, S. Ramakrishna, Z.M. Huang, Recent development of polymer nanofibers for biomedical and biotechnological applications, Journal of Materials Science: Materials in Medicine 16 (2005) 933–946.

[83] R.L. Dahlin, F.K. Kasper, A.G. Mikos, Polymeric nanofibers in tissue engineering, Tissue Engineering Part B Reviews 17 (2011) 349–364.

[84] S. Abdelhady, K.M. Honsy, Electro spun- nanofibrous mats: a modern wound dressing matrix with a potential of drug delivery and therapeutics, Journal of Engineered Fibers and Fabrics 10 (2015) 179–193.

[85] C. Pengsuk, R. Wongkanya, P. Chuysinuan, P. Nooeaid, S. Techasakul, K. Lirdprapamongkol, J. Svasti, Electrospinning of alginate/soy protein isolated nanofibers and their release characteristics for biomedical applications, Journal of Science: Advanced Materials and Devices (2017) 309–316.

[86] J.S.M. Dzikovski, N. Castanie, A. Guedon, B. Verrier, C. Primard, Antibiotic incorporation in jet-sprayed nanofibrillar biodegradable scaffolds for wound healing, International Journal of Pharmaceutics 532 (2017) 802–812.

[87] Topical Semi-Solid Dosage Forms. Available online: https://www.malvern.com/en/industry-applications/sample-type-form/topicals-creams-and-gels. (Accessed on 12 December 2017).

[88] Topical Delivery, The Importance of the Right Formulation in Topical Drug Development. Drug Development and Delivery, August 1, 2015. Available online: http://www.drug-dev.com/Main/Back- Issues/TOPICAL-DELIVERY-The-Importance-of-the-Right-Formu-833.

[89] A.J.S.J. Toussaint, W.T. Chung, N. Osman, S.A. Mc Clain, V. Raut, Topical antibiotic ointment versus silvercontaining foam dressing for second-degree burns in swine, Academic Emergency Medicine 22 (2015) 927–933.

[90] K. Khezri, M.R. Farahpour, Accelerated infected wound healing by topical application of encapsulated Rosemary essential oil into nanostructured lipid carriers, Artificial Cells, Nanomedicine, and Biotechnology 1 (47) (2019) 980–988.

[91] M. Zilberman, D. Egozi, M. Shemesh, A. Keren, E. Mazor, M. Baranes-Zeevi, N. Goldstein, I. Berdicevsky, A. Gilhar, Y. Ullmann, Hybrid wound dressings with controlled release of antibiotics: structure-release profile effects and in vivo study in a Guinea pig burn model, Acta Biomaterialia 22 (2015) 155–163.

[92] T. Reid, P.L. Tran, E. Huynh, A.N. Hamood, A. de Souza, D. Mehta, K.W. Moeller, C.D. Moeller, M. Morgan, The ability of a colloidal silver gel wound dressing to kill bacteria in vitro and in vivo, Journal of Wound Care 26 (2017) S16–S24.

[93] M. Contardi, D. Russo, G. Suarato, J.A. Heredia-Guerrero, L. Ceseracciu, I. Penna, N. Margaroli, M. Summa, R. Spanò, G. Tassistro, L. Vezzulli, T. Bandiera, R. Bertorelli, A. Athanassiou, I.S. Bayer, Polyvinylpyrrolidone/hyaluronic acid-based bilayer constructs for sequential delivery of cutaneous antiseptic and antibiotic, Chemical Engineering Journal 358 (2019) 912–923.

[94] M.H. S, G.F. Metwally, Evaluation of topical gel bases formulated with various essential oils for antibacterial activity against methicillin- resistant Staphylococcus aureus, Tropical Journal of Pharmaceutical Research 6 (12) (2013) 877–884.

[95] E.B. Denkbas, E. Ozturk, N. Ozdemir, K. Kececi, Norfloxacin-loaded chitosan sponges as wound dressing material, Journal of Biomaterials Applications 18 (2004) 291–303.

[96] N.S. Alharbi, J.M. Khaled, S. Kadaikunnan, A.S. Alobaidi, A.H. Sharafaddin, S.A. Alyahya, T.N. Almanaa, M.A. Alsughayier, M.R. Shehu, Prevalence of *Escherichia coli* strains resistance to antibiotics in wound infections and raw milk, Saudi Journal of Biological Sciences (2018) 1–6.

[97] L.R. Boles, R. Awais, K.E. Beenken, M.S. Smeltzer, O. Haggard, J.A. Jennings, Local delivery of amikacin and vancomycin from chitosan sponges prevent polymicrobial implant-associated bio film, Military Medicine 183 (2018) 459–465.

[98] B. Singh, S. Sharma, A. Dhiman, Design of antibiotic containing hydrogel wound dressings: biomedical properties and histological study of wound healing, International Journal of Pharmaceutics 457 (2013) 82–91.

[99] M. Varga, B. Sixta, R. Bem, I. Matia, A. Jirkovska, M. Adamec, Application of gentamicin-collagen sponge shortened wound healing time after minor amputations in diabetic patients – a prospective, randomised trial, Archives of Medical Science 2 (10) (2014) 283–287.

CHAPTER 8

Recent Progress on Antibiotic Polymer/Metal Nanocomposites for Health Applications

T.S. MOTSOENENG • M.J. MOCHANE • T.C. MOKHENA • EMMANUEL ROTIMI SADIKU

1 INTRODUCTION

An urgent need to treat chronic ailments pertaining to the indwelling microorganisms that pose a major threat to the healthcare delivery and facilities in medical practices has been an important aspect of research exploration. The elemental additives possess antimicrobial properties, through which the lysis of pathogens leading to infectious deceases can be modulated to afford a complete eradication of colonizing microbes for the healing of the affected area. Innate metals and chemically ionized metal-oxide components provide the toxicity against the deadly microbes via an antibiotic activity. The fundamental characteristics of appropriate metals/metal oxides, including physicochemical properties, uniform dispersion of the nanoparticle aggregates (reduced conglomeration) and biocompatibility within the mammalian cells, account for the improvement of antibiotic efficacy, needed to combat bacterial infections [1−4]. However, the utilization of metals/metal oxides may provide excessive cytotoxicity on the phagocytes, thereby leading to detrimental healthcare issues by alleviating the earnest mechanistic antibiotic activities. Nonetheless, biocidal polymers are usually used in the preparation of metal/metal oxides polymer nanocomposites with well-defined antibiotic effectiveness in which the reduced cytotoxicity, minimal environmental hazards, improved resistance, and extended duration of the release of ions can be observed. Thus metals/metal oxides polymer hybrids are often prepared by using various synthetic pathways to regulate the bactericidal properties, control nanoparticle aggregation, and manipulate the structure−property relationship as well as their biocompatibility for facile use in in vivo applications [5−7]. A plethora of different polymers with a variety of structural architectures, topology, functionality, and number average molar masses have been the subject of interest for quite a number of research groups, both in academia and industry [8−10]. This chapter will outline the use of different elemental metals and their ionization treatments to form metal oxides with enhanced binding strength for biocompatible properties. The influence of the polymer architecture, methods of preparation, and the structure−property relationships of the metal/metal oxide polymer hybrids will be covered as well. The biocidal mechanisms between the microorganisms (cell membrane) and the functional groups of the antibiotic agents, which hamper or disrupt the metabolic proliferation of microbes, are concisely discussed.

2 STRUCTURAL COMPONENTS OF METALLIC ELEMENTS AND METAL OXIDES

Metallic elements, due to their intrinsic structures and earthly compositions, account for the bactericidal activity, which can be employed, primarily, as antibiotic agents to conflict the metabolic replication of microbes. Moreover, the metal oxides are able to offer a pivotal role as vehicles for drug-delivery systems even though there might be a competitive mechanism between antimicrobial resistance and antibiotic agents. Native elements can be chemically and mechanically fabricated to regulate the dimensions for the enhancement of physicochemical or photochemical activity to afford even more biocidal efficacy against the proliferation of detrimental microorganisms. The control of antibiotic efficiency, the properties, and variables, including

Antibiotic Materials in Healthcare. https://doi.org/10.1016/B978-0-12-820054-4.00008-2

chemical stability, low toxicity toward animal cells, low opaque, appropriate hydrophilicity, sufficient oxidizing ability, cost-effective, and efficient resilience during disinfection of pathogens, depending on the inherent compositional generation of the metals [2,11]. The combination of these aspects provides the much-required antibiotic activity, which can overcome more than 70% of biocidal resistance granted by the antibiotic agents. The well-known native properties, potencies, and spectra activity, which have been used for the past decades, instigate the antibiotic efficacy of the most commonly utilized metal elements, inter alia: gold (Au), zinc (Zn), copper (Cu), silver (Ag), and titanium (Ti). In addition, oxidized metals, such as calcium oxide (CaO), cesium oxide (CeO_2), copper oxide (CuO), auric oxide (Au_2O_3), magnesium oxide (MgO), silicon (SiO_2), silver oxide (Ag_2O), and titanium oxide (TiO_2), among others, offer a prominent potency of biocidal aptitude in comparison with metallic elements in free form [3,12−15]. However, their modification techniques and the reduced surface-volume ratio of metals/metal oxide elements are indispensable features that exhibit better antibiotic activity when compared to untreated metals. Thus, the preparation of nanoparticles through the downsizing phenomenon of bulky metals or synthesis methodologies of metal atoms, clusters, and smaller particles afford different prominent properties, which are advantageous for use as biocidal agents in medical facilities. It is a fact that the nanostructures of particles possess enhanced properties, such as catalytic activity, thermal treatment, transportation of mass, and solubility parameters, all of which are due to a discernible increase of fraction surface. The combination of different metals produces a bimetallic alloy through the synergistic upshots, which culminate in the improved properties when compared to monometallic components [2,16−19].

In general, a mechanistic binding of metal ions with oppositely charged ions of the cell membrane initiates an effective antibiotic activity through the production of reactive oxygen species (ROS) in conjunction with less toxicity of the metallic ions. Numerous investigations have been conducted on the utilization of disparate metals/metal oxides as biocidal agents to overcome the plight of antimicrobial resistance against antibiotics. For example, Argueta-Figueroa and coworkers [17] ascertained the use of copper, nickel, and their bimetallic alloy as the antibiotic agents by assessing the minimum inhibitory concentration (MIC) for the determination of the biocidal activity of metal elements. It was observed that all metal nanoparticles provide antibiotic effect (Fig. 8.1), with copper nanoparticles exhibiting the most powerful potency against *S. aureus*, *E. coli*, and *S. mutans*, whereas nickel and bimetal nanoparticles displayed a mere bacteriostatic capacity.

The fabrication of the metal elements for use in the preparation of metal composites promotes the biocidal activity by manifesting a significant hampering of colonization of bacterial species. In the work conducted by Chen and coworkers [11], ZnO and SiO_2 composites were tested for dental caries in the inhibition of cariogenic plague. It was observed that the dual metallic composite denoted by Z_7S_{63} demonstrated an outstanding biocidal activity through the liberation of Zn^{2+} ions, which provided an uncontested effect of demolishing the bacterial cell membrane. The authors deduced that a superb antimicrobial activity was observed in $ZnO-SiO_2$ composite, by manifesting a 99.9% demise of *S. mutans* after 24 h, which was attributed to the convenient conduit given by the SiO_2 mesoporous structure, which provides an efficient movement of Zn^{2+} ions. The conventional synthesis methods of metal/metal oxide nanocomposites that are employed in the surgical treatment of wounds can pose a threat to vital cells of the living organisms and also be hazardous to the environment, due to its potentially high toxicity. Thus the preparation of metal/metal oxide nanocomposites through the green synthesis techniques is necessary for the mitigation of toxic materials that are disposed into the environment to attain eco-friendly nanocomposites at a cost-efficient fashion and facile procedures. Recently, studies on the biosynthesis of metal elements and metal oxides have been undertaken, extensively, to produce nanocomposites with well-controlled size and shape that can be utilized as bactericides in the medical fields [15,16,20]. Tea extracts are known to contain flavonoid compounds, which possess antioxidative properties and metal nanoparticles that can be prepared, conveniently, by using biosynthesis methods and could be employed in combating of multidrug resistant bacterial infections. For instance, Onitsuka and coworkers [21] prepared silver and gold nanoparticles from tea extracts via cost-effective biosynthesis procedure for use as antibiotic agents in gram-positive and gram-negative microbes. The nanoparticles prepared were of uniform size and they were found to possess an efficiently high antimicrobial activity due to the biocidal character that exists in the functionality of polyphenols of the tea leaves.

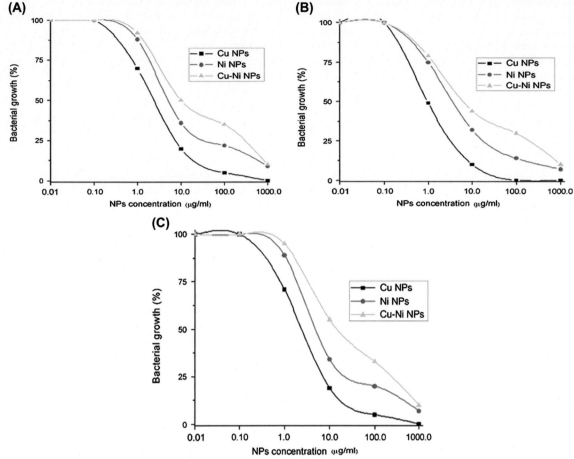

FIG. 8.1 The antimicrobial potency of Cu, Ni, and CU–Ni nanoparticles on the gram-positive bacteria **(A)** *S. aureus*, **(B)** *E. coli*, and **(C)** *S. mutans* [17] Open access.

3 POLYMER MATERIALS AND ARCHITECTURES FOR ANTIBIOTIC ACTIVITY

Polymers are indispensable machinery in the development of antibiotic agents due to their capacity to exterminate or inhibit the metabolic proliferation of microbes, lest the contamination of the surrounding areas of the affected environment. However, the macrostructural configuration, functionality of moieties, the degree of hydrophobicity, and the density of counterions toward bacterial cells are of paramount importance in the device of effective antibiotic polymeric materials with excellent aptitude [22]. Both natural [23,24] and synthetic [25,26] polymers with different structures display a wide range of properties, such as biocompatibility, balanced physicomechanical, and are able to undergo the postpolymerization chain modification, all of

which can be exploited in the potential use of multidrug resistance of evolving microbial colony [9]. The capacity of polymers with antibiotic activity is quantified by the MIC, which is a measure of the polymeric material to hamper the bacterial growth. Despite the quantification tool, for example, MIC, used for the assessment of the capability of polymers against bacterial evolution, the antibiotic agents should be used at appropriate concentrations (fairly lower than the hemolytic concentration (HC_{50})) to avoid a lethal exposure of erythrocytes. Xu and coworkers [27] prepared a conjugated polymer material with a cationic backbone for use in the assessment of antimicrobial potency against gram-negative bacteria (Ampr *E. coli*). It was reported that the positively charged polymer backbone provides an excellent antibiotic activity in comparison with a neutral backbone because of the rapid biocidal activity rate. The development of conjugated polymer materials comprising a superb biocidal activity and viable cytotoxicity has been, of recent, a massive exploration in the academic world. For instance, Lou and coworkers [28] demonstrated the synthesis of new polymer candidates that bear ionic backbones and their utility in the antimicrobial evaluation for the indwelling lung infectious *Klebsiella pneumoniae*. These polyionenes (Fig. 8.2) displayed a powerful in vivo biocidal characteristic, marginal toxicity toward certain organs and enabled the alleviation of resistance toward the evolution of bacteria. Notwithstanding, the biocidal potential of unnatural polymers, the combination of natural—synthetic or synthetic—synthetic polymers for the preparation of more complex polymer architectures provides phenomenal antibiotic characteristics, due to the synergistic effect of the different collective properties emanating from both polymer entities [29—33].

In recent studies, numerous amount of research has been performed on the syntheses of new functionalized, quaternized, and complex polymer materials with outstanding antibiotic activity for the eradication of different bacterial membranes. On the account of feasible cytotoxicity for in vivo applications, these polymers play an intriguing role in daily life as they can be employed in food packaging and textile facilities [34,35]. The research studies conducted for some antibiotic polymers are collated in Table 8.1, which presents the polymer structure, MIC, and the type of targeted bacterial species.

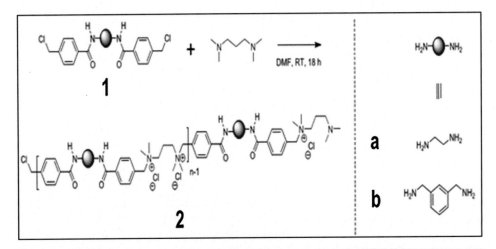

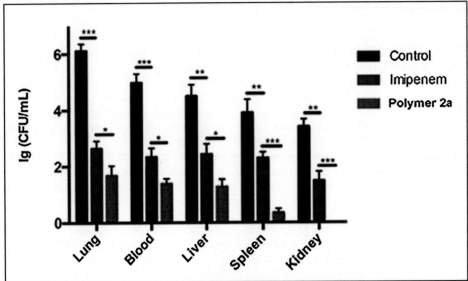

FIG. 8.2 Synthesis of polyionenes containing rigid amide motifs from bis-halide monomers and reduction of viable bacterial colonies in the lung, blood, liver, spleen, and kidney specimens taken at day 5 postinfection [28].

TABLE 8.1
Structure of Synthetic Polymer Materials and Their Antimicrobial Minimal Inhibition Concentration (MIC) Against Various Bacterial Strains.

Name of a Polymer	Synthetic Polymer Structure	Bacterial Species	MIC (mg mL^{-1} or µg mL^{-1} or µmol mL^{-1})		Reference
Poly(quaternary ammonium-2-acrylamido-2-methylpropanesulfonic acid); P[Qa][AMPS]		S. aureus E. coli	24.9 88.32		[36]
Poly(acrylonitrile-co- 2-(((2-(4-methylthiazol-5-yl) ethoxy)carbonyl)oxy)ethyl methacrylate monomer); P(AN-co-MTZ)	R = Me, Bu	S. aureus S. epidermidis P. aeruginosa E. coli C. parapsilosis	Me 128 32 64 128 8	Bu 128 32 64 32 8	[37]
Poly(2-dimethylamino)ethyl methacrylate) containing quaternary ammonium salts P(DMAEMA-BC)		E. coli S. albus C. albicans R. solani Foc4	320 320 320 40 20		[38]
Poly(arbutin-co-arbutin-C8$_{10}$); P(arb)-co-P(arb)-C8$_{10}$		E. coli S. aureus	111 111		[39]
Poly(2-oxazoline) homopolymers, statistical copolymers, and block copolymers		E. coli S. aureus	313 256–512		[40]

Continued

TABLE 8.1
Structure of Synthetic Polymer Materials and Their Antimicrobial Minimal Inhibition Concentration (MIC) Against Various Bacterial Strains.—cont'd

Name of a Polymer	Synthetic Polymer Structure	Bacterial Species	MIC (mg mL^{-1} or µg mL^{-1} or µmol mL^{-1})	Reference
Polyrhodanine		E. coli S. aureus	128–256 256	[41]
Poly(bromooxanorbornene (4-(3-bromopropyl)-10-oxa-4-zatricyclo[5.2.1.02,6] dec8-ene-3,5-dione))		S. aureus E. coli	>62.5 >7.8	[42]
Poly(4-[[4-(dimethylamino) benzylidene]amino]phenol); P(4-DBAP)		S. lutea E. aerogenes E. coli E. feacalis K. pneumoniae B. subtilis C. albicans S. cerevisiae	Not reported	[43]
Polypropylene-polyethyleneimine-styrene maleic anhydride-polyethyleneimine; (PP-PEI-SMA$_{mw}$-PEI)		Escherichia coli O157:H7 Listeria monocytogenes Pseudomonas fluorescens	Not reported	[35]
pPr-DiBocGua		K. pneumoniae A. baumannii P. aeruginosa E. faecium E. coli MRSA	64 16 32 16 32 16	[44]

4 SYNTHESES OF BIOCIDAL METAL/METAL OXIDE POLYMER NANOCOMPOSITES

It is now well-known that the polymer materials and metals/metal oxides are able to, independently, contribute toward the biocidal activity of the resultant metal—polymer nanocomposites, due to their inherent properties and their possession of different characteristics [45]. Thus the preparation of metalized polymer nanocomposites is extremely crucial because it provides an added advantage to the resulting degree of antibiotic capacity, which modulates the resistance of bacterial development. Metals/metal oxides have sufficient durability such that they can be prepared or fabricated in a wide range of chemical conditions, on account of their physicochemical properties, such as toughness, thermal stability, and robust chemical structures (strong van der Waal forces), which exert the antimicrobial character and subsequent performance [1,46,47]. The structural modification of metal oxides plays a significant role in the envisaged biocidal activity of the resultant metal—polymer nanomaterials. Roy and coworkers [7] conducted a study on the structural fabrication of Ag, Cu, and ZnO nanoparticles, treated with montmorillonite clay under UV irradiation for silver and copper metals and in basic sodium hydroxide medium for zinc oxide. These reinforced nanoparticles, Ag-montmorillonite (MMT), Cu-MMT, and ZnO-MMT, were subsequently incorporated in high-density polyethylene for the preparation of high-density polyethylene (HDPE)/metal-MMT nanocomposites by mechanical blending, using twin-screw extruder. The nanocomposites prepared were used in antimicrobial testing and they showed a significant inhibitory effect against gram (−) and gram (+) bacteria. The authors suggested that the degree of antibiotic activity is strongly dependent on the cytoplasmic solubility and redox potential of the treated nanoparticles. In another study, Thokala and coworkers [48] prepared polyamide 11/copper composites by injection molding at 40°C under a 400 bar pressure; the sample showed to be active against E. coli with a 99.9% demise in 24 h due to the effective ion release during prolonged immersion time.

Apart from the mechanical blending techniques for the synthesis of metal—polymer nanocomposites, there is also an in situ methodology, which is greatly employed. This is because it accounts for the optimal distribution of metal nanoparticles within the polymer matrix and could have an immense benefit in providing enhanced biocidal properties [46]. The solution casting method is the most preferred technique for the preparation of antibiotic metal—polymer nanocomposites through an in situ methodology. For example, Mujeeb and coworkers [49] investigated the antibiotic activity of chitosan/ZnO nanocomposites, prepared by solution casting; thus, chitosan was dissolved in aqueous acetic acid under ultrasonication procedure, followed by the addition of zinc acetate and the solution was homogenized and transferred into a glass dish for the formation of uniform films. The results revealed that the colonization of E. coli and S. aureus was vehemently inhibited by chitosan/ZnO nanocomposites, whereas pure chitosan manifested undesirable combat against bacterial development. It was deduced that the presence of ZnO in the hybrid resulted in an excellent antibiotic performance due to the Zn^{2+} ion release in which the bout on the oppositely charged cell membrane of biocides is imminent leading to the destruction or demise of bacteria. Moreover, Jing Xie and Yen-Con Hung [50] incorporated different concentrations of TiO_2 nanoparticle into biodegradable polymer matrices, namely cellulose acetate (CA), polycaprolactone (PCL), and polylactic acid (PLA) via the solution casting technique using chloroform to dissolve PCL and PLA, but utilized acetone for a complete dissolution of CA under vigorous stirring at 25°C. The TiO_2 nanoparticle was dispersed by ultrasonication to attain an optimally homogeneous distribution of the metal oxide nanofiller within the polymers. The antibiotic activity results revealed that the CA/TiO_2 nanocomposite at 5 wt% of TiO_2 nanofiller exhibits the highest biocidal properties amid other prepared TiO_2 polymer nanocomposites against E. coli O157: H7. Conversely, it was discovered that the PLA/TiO_2 and PCL/TiO_2 hybrids showed low antibiotic efficacy toward bacterial development due to photocatalytic properties, notwithstanding the porous structure particularly for PCL matrix through which the penetration of bacteria is looming through the holes and cracks of the films. The authors concluded that CA/TiO_2 nanocomposite possesses a potential application as an antibiotic agent for food packaging facilities.

An in situ preparation methodology is quite versatile, on the account that the synthesis of metal—polymer nanocomposites can be attained either by the dissolution of polymer matrix by using a pertinent solvent system with matching solubility parameters to those of the macromolecule or by polymerization reaction to which the monomer units are treated to form the polymer material in the presence of metals and/or metal oxides. There are various polymerization techniques, in which the metal—polymer nanocomposites can be prepared to afford metal/polymer hybrids with outstanding biocidal properties for use in bioactive applications, including wound healing and inhibition of bacterial development. For example, Khan and coworkers [51] reported the preparation of poly(o-toluidine)V(III) molybdate nanocomposite (POT-V(III)MoO$_4$) by in situ polymerization method, through the sol–gel approach for use as an antibiotic agent against the *Enterococcus faecalis* and *Serratia marcescens* evolution. It was observed that the distribution of V(III)MoO$_4$ nanoparticles were optimally uniform and in an orderly fashion throughout the POT matrix. POT-V(III)MoO$_4$ nanocomposite exhibited excellent antibiotic activity, by showing the minimum inhibitory concentration and minimum bactericidal concentration of 5 and 2.5 µg mL^{-1} for both *Enterococcus faecalis* and *Serratia marcescens*, respectively. A phenomenal advantage for employing an in situ polymerization method entails among other factors: the control of number average molar mass, molar mass distribution, well-defined structural composition of the macromolecules and a high degree of metal nanoparticle distribution within the polymer matrix [5,39,52]. Pavoski and coworkers [52] ascertained the inhibition activity of Si–Ag/polyethylene nanocomposites, prepared via an in situ polymerization reaction of ethylene in the presence of silver metal, treated by the encapsulation of silica nanoparticles. The water contact results demonstrated that the resultant PE/SiAgE nanocomposites possess active inhibitory capacity against *Staphylococcus aureus* colonization.

5 BIOMEDICAL POSTULATIONS FOR AANTIBIOTIC EFFICACY (MECHANISTIC OVERVIEW)

Combat against the development of microbes in living organisms is fundamentally, based on the structural facets of the antibiotic agent, such that antagonistic mechanism can be established between the microbial cell membrane and antibiotic agent, which instigates the demise of bacteria strains. In particular, for ideal biocidal properties, the cytotoxicity of bactericides must appropriately, be acceptable, within the range at which there will not be the annihilation of erythrocytes to avoid the toxicity of human cells. The aptitude of antibiotic agent needed to bout the bacterial replication is assessed by the binding strength or a hefty attachment on the bacterial cell membrane, which results in the inhibition of metabolic proliferation of bacteria. However, certain antibiotics with photocatalytic properties utilize visible light for an efficient biocidal activity to hamper the bacterial colonization [2]. The ions released, emanating from the metal–polymer nanocomposites, due to their inconceivable ability to permeate, have the liberty to enter through the

microbes' cell membrane by endocytosis, especially when the size of the metal is in nanodimensions. Nonetheless, the efficiency of antibiotics depends on the type of metal element or metal oxide in use. Despite the existence of phospholipids, the net charge density of the cell membrane of bacteria is predominantly negative, yet the ions of instable metal nanoparticles composed of the oppositely charged ions (positive net charge density) and thus, an inevitable electrostatic cohesion is possible leading to the oligodynamic mechanistic effect. Notwithstanding the intrinsic nature and properties of the metal, which control the rate of metal ions release, the cations create holes on the microbes' cell membrane, hence leading to subsequent leaching of cytoplasmic fluid. Moreover, the indwelling of the cations within the cell membrane promotes the formation of ROS as the by-products of the Fenton reaction after which the hydrogen peroxide (H$_2$O$_2$) produces neutral metal from its oxidized state via the reduction process. By apoptosis, the ROS radicals then bout the DNA double-helical arrangement to abolish the bacterial strain. The final concentration of ROS radicals in the Fenton reaction is directly proportional to the standard electrode potential of the metal. In addition, the solubility parameters account for the control of the conveyance rate of the cytoplasmic fluid [7,53,54]. In contrast with the role that the polymeric material plays on the attack against bacterial strains, Kajiwara and coworkers [39] hypothesized a mechanistic effect of nontoxic polyarbutin, which was prepared by the polymerization of arbutin for the decolonization of *S. aureus* and *E. coli*. It was postulated that, due to the presence of lipopolysaccharides, polyarbutin undergoes a coordination process with the bacterial cell membrane and subsequent abolition of the strains, as manifested in the following schematic representation (Fig. 8.3).

6 FUTURE PERSPECTIVES OF ANTIBIOTIC POLYMER/METAL NANOCOMPOSITES IN HEALTHCARE

The combat of the development of microbes in living organisms is a thriving aspect in modern society to sustain a healthy living environment for people. Metals and metal oxides reveal a potential characteristic in providing the antibiotic activity through which the demise of harmful microorganisms can be attained. However, due to earthly compositional constituencies of the innate metals, there are detrimental hazards posed to erythrocytes, and hence disfavors the earnest cause of sallying the evolution of unwanted bacterial species without aftermath. Recent studies showed that

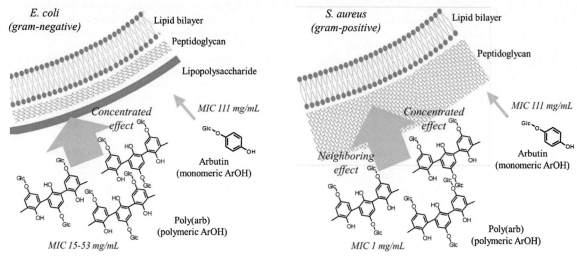

FIG. 8.3 Hypothesis of a possible mechanism occurring between polyarbutin and *S. aureus* and *E. coli* bacterial strains [39].

the incorporation of these untreated and fabricated metals into polymer matrices results in an enhanced antibiotic activity of the resultant polymer/metal nanocomposites. The syntheses of more fascinating polymer materials with intriguing structural compositions, control of molar mass, molar mass distribution, and functionality, particularly the utilization of complex functionalized polymers, would enact the outstanding antibiotic capability for the polymer/metal hybrids against bacteria. Moreover, the pretreatment of metals using pertinent chemical reactions before the formation of polymer/metal nanocomposites should be endorsed to ensure the utmost distribution of the nanostructured metal particles within the domain of polymer matrix for optimal enhancement.

7 CONCLUSION

There are a substantial number of bacteria, which can result in consequential risk infections. It is very important that most of our materials have an additional antimicrobial property. Polymer materials (both synthesized and commercially received) with metal/metal oxides to form polymer—metal composites contribute toward the antimicrobial activity due to their inherent properties. The preparation methods of the polymer—metal composite play a significant role in the degree of antibiotic capacity, which regulates the resistance of bacterial development. Different methods, that is, melt-mixing, solution casting, and injection molding, have been used for the fabrication of polymer—metal/metal oxide composite for antibacterial activity. In

situ polymerization is the preferred method; this is because it accounts for the optimal distribution of metal nanoparticles within the polymer matrix and could have an immense benefit in providing enhanced biocidal properties.

REFERENCES

[1] D. Yoo, D. Lee, J. Park, J. Ahn, S.H. Kim, D. Lee, Porosity control of nanoporous CuO by polymer confinement effect, Scripta Materialia 162 (2019) 58—62.

[2] C. Chambers, S.B. Stewart, B. Su, H.F. Jenkinson, J.R. Sandy, A.J. Ireland, Silver doped titanium dioxide nanoparticles as antimicrobial additives to dental polymers, Dental Materials 33 (2017) e115—e123.

[3] V. Bartunek, B. Vokatá, K. Kolárová, P. Ulbrich, Preparation of amorphous nano-selenium-PEG composite network with selective antimicrobial activity, Materials Letters 238 (2019) 51—53.

[4] M.D. Solmaz, L. Farzaneh, B.-J. Mohammad, H.Z. Mohammad, A. Khosro, Antimicrobial activity of the metals and metal oxide nanoparticles, Materials Science and Engineering: C 44 (2014) 278—284.

[5] T. Yamamotoa, K. Arakawa, Y. Takahashi, M. Sumiyoshi, Antimicrobial activities of low molecular weight polymers synthesized through soap-free emulsion polymerization, European Polymer Journal 109 (2018) 532—536.

[6] S.J. Lam, E.H.H. Wong, C. Boyer, G.G. Qiao, Antimicrobial polymeric nanoparticles, Progress in Polymer Science 76 (2018) 40—64.

[7] A. Roy, M. Joshi, B.S. Butola, Antimicrobial performance of polyethylene nanocomposite monofilaments reinforced with metal nanoparticles decorated montmorillonite, Colloids and Surfaces B: Biointerfaces 178 (2019) 87—93.

[8] H. Xing, M. Lu, T. Yang, H. Liu, Y. Sun, X. Zhao, H. Xu, L. Yang, P. Ding, Structure-function relationships of nonviral gene vectors: lessons from antimicrobial polymers, Acta Biomaterialia 86 (2019) 15−40.

[9] M. Charnley, M. Textor, C. Acikgoz, Designed polymer structures with antifouling−antimicrobial properties, Reactive and Functional Polymers 71 (2011) 329−334.

[10] A. Jain, L.S. Duvvuri, S. Farah, N. Beyth, A.J. Domb, W. Khan, Antimicrobial polymers, Advanced Healthcare Material 3 (12) (2014) 1969−1985.

[11] H. Chen, R. Wang, J. Zhang, H. Hua, M. Zhu, Synthesis of core-shell structured ZnO@m-SiO$_2$ with excellent reinforcing effect and antimicrobial activity for dental resin composites, Dental Materials 34 (2018) 1846−1855.

[12] S.S. Rad, A.M. Sani, S. Mohseni, Biosynthesis, characterization and antimicrobial activities of zinc oxide nanoparticles from leaf extract of Mentha pulegium (L.), Microbial Pathogenesis 131 (2019) 239−245.

[13] S.T. Dubas, S. Wacharanad, P. Potiyara, Tunning of the antimicrobial activity of surgical sutures coated with silver nanoparticles, Colloids and Surfaces A: Physicochemical and Engineering Aspects 380 (2011) 25−28.

[14] A. Akbar, M.B. Sadiq, I. Ali, N. Muhammad, Z. Rehman, M.N. Khan, J. Muhammad, S.A. Khan, F.U. Rehman, A.K. Anal, Synthesis and antimicrobial activity of zinc oxide nanoparticles against foodborne pathogens *Salmonella typhimurium* and *Staphylococcus aureus*, Biocatalysis and Agricultural Biotechnology 17 (2019) 36−42.

[15] S. Sebastiammal, A. Mariappan, K. Neyvasagam, A.L. Fathima, A. Muricata, Inspired synthesis of CeO$_2$ nanoparticles and their antimicrobial activity materials, Today's Office: Proceedings 9 (2019) 627−632.

[16] M. Khatami, R.S. Varma, N. Zafarnia, H. Yaghoobi, M. Sarani, V.G. Kumar, Applications of green synthesized Ag, ZnO and Ag/ZnO nanoparticles for making clinical antimicrobial wound-healing bandages, Sustainable Chemistry and Pharmacy 10 (2018) 9−15.

[17] L. Argueta-Figueroa, R.A. Morales-Luckie, R.J. Scougall-Vilchis, O.F. Olea-Mejía, Synthesis, characterization and antibacterial activity of copper, nickel and bimetallic Cu−Ni nanoparticles for potential use in dental materials, Progress in Natural Science: Materials International 24 (2014) 321−328.

[18] W. Salem, D.R. Leitner, F.G. Zingl, G. Schratter, R. Prassl, W. Goessler, J. Reidl, S. Schild, Antibacterial activity of silver and zinc nanoparticles against *Vibrio cholerae* and enterotoxic *Escherichia coli*, International Journal of Medical Microbiology 305 (2015) 85−95.

[19] G. Sharma, A. Kumar, S. Sharma, Mu. Naushad, R. Prakash Dwivedi, Z.A. Alothman, G.T. Mola, Novel development of nanoparticles to bimetallic nanoparticles and their composites: a review, Journal of King Saud University - Science 31 (2019) 257−269.

[20] S. Andra, S. Balu, R. Ramoorthy, M. Muthalagu, M. Vidyavathy, *Terminalia bellerica* fruit extract mediated synthesis of silver nanoparticles and their antimicrobial activity, Materials Today: Proceedings 9 (2019) 639−644.

[21] S. Onitsuka, T. Hamada, H. Okamura, Preparation of antimicrobial gold and silver nanoparticles from tea leaf extracts, Colloids and Surfaces B: Biointerfaces 173 (2019) 242−248.

[22] T. Calliess, I. Bartsch, M. Haupt, M. Reebmann, M. Schwarze, M. Stiesch, C. Pfaffenroth, M. Sluszniak, W. Dempwolf, H. Menzel, F. Witte, E. Willbold, In vivo comparative study of tissue reaction to bare and antimicrobial polymer coated transcutaneous implants, Materials Science and Engineering: C 61 (2016) 712−719.

[23] G. Zhang, X. Li, X. Xu, K. Tang, V.H. Vu, P. Gao, H. Chen, Y.L. Xiong, Q. Sun, Antimicrobial activities of irradiation-degraded chitosan fragments, Food Bioscience 29 (2019) 94−101.

[24] D.M.S.A. Salem, M.A.E. Sallam, T.N.M.A. Youssef, Synthesis of compounds having antimicrobial activity from alginate, Bioorganic Chemistry 87 (2019) 103−111.

[25] S. Gazzotti, S.A. Todisco, C. Picozzi, M.A. Ortenzia, H. Farina, G. Lesma, A. Silvani, Eugenol-grafted aliphatic polyesters: towards inherently antimicrobial PLA based materials exploiting OCAs chemistry, European Polymer Journal 114 (2019) 369−379.

[26] A. Palantoken, M.S. Yilmaz, N.A. Unubol, E. Yenigul, S. Pişkin, T. Eren, Synthesis and characterization of a ROMP-based polycationic antimicrobial hydrogel, European Polymer Journal 112 (2019) 365−375.

[27] Q. Xu, P. He, J. Wang, H. Chen, F. Lv, L. Liu, S. Wang, J. Yoon, Antimicrobial activity of a conjugated polymer with cationic backbone, Dyes and Pigments 160 (2019) 519−523.

[28] W. Lou, S. Venkataraman, G. Zhong, B. Ding, J.P.K. Tan, L. Xu, W. Fan, Y.Y. Yang, Antimicrobial polymers as therapeutics for treatment of multidrugresistant *Klebsiella pneumoniae* lung infection, Acta Biomaterialia 78 (2018) 78−88.

[29] K. Wojciechowski, M. Kaczorowski, J. Mierzejewska, P. Parzuchowski, Antimicrobial dispersions and films from positively charged styrene and acrylic copolymers, Colloids and Surfaces B: Biointerfaces 172 (2018) 532−540.

[30] D. Demircan, B. Zhang, Facile synthesis of novel soluble cellulose-grafted hyperbranched polymers as potential natural antimicrobial materials, Carbohydrate Polymers 157 (2017) 1913−1921.

[31] M. Benkocká, S. Lupínková, T. Knapová, K. Kolářová, J. Matoušek, P. Slepička, V. Švorčík, Z. Kolská, Antimicrobial and photophysical properties of chemically grafted ultrahigh-molecular-weight polyethylene, Materials Science and Engineering: C 96 (2019) 479−486.

[32] L.J. Bastarrachea, J.M. Goddard, Self-healing antimicrobial polymer coating with efficacy in the presence of organic matter, Applied Surface Science 378 (2016) 479−488.

[33] M.T. Abate, A. Ferri, J. Guan, G. Chen, J.A. Ferreira, V. Nierstrasz, Single-step disperse dyeing and antimicrobial functionalization of polyester fabric with chitosan and derivative in supercritical carbon dioxide, The Journal of Supercritical Fluids 147 (2019) 231−240.

[34] J. Su, J. Noro, S. Silva, J. Fu, Q. Wang, A. Ribeiro, C. Silva, A. Cavaco-Paulo, Antimicrobial coating of textiles by laccase in situ polymerization of catechol and p-phenylenediamine, Reactive and Functional Polymers 136 (2019) 25–33.

[35] Y.-T. Hung, L.A. McLandsborough, J.M. Goddard, L.J. Bastarrachea, Antimicrobial polymer coatings with efficacy against pathogenic and spoilage microorganisms, Lebensmittel-Wissenschaft und -Technologie— Food Science and Technology 97 (2018) 546–554.

[36] Y. Qian, H. Cui, R. Shi, J. Guo, B. Wang, Y. Xu, Y. Ding, H. Mao, F. Yan, Antimicrobial anionic polymers: the effect of cations, European Polymer Journal 107 (2018) 181–188.

[37] R. Cuervo-Rodríguez, A. Muñoz-Bonilla, J. Araujo, C. Echeverría, M. Fernández-García, Influence of side chain structure on the thermal and antimicrobial properties of cationic methacrylic polymers, European Polymer Journal 117 (2019) 86–93.

[38] W. Zhong, C. Dong, R. Liuyang, Q. Guo, H. Zeng, Y. Lin, A. Zhang, Controllable synthesis and antimicrobial activities of acrylate polymers containing quaternary ammonium salts, Reactive and Functional Polymers 121 (2017) 110–118.

[39] R. Kajiwara, A. Seto, H. Kofujit, Y. Shiba, Y. Oishi, Y. Shibasaki, Enhanced antimicrobial activities of polymerized arbutin and its derivatives prepared by oxidative polymerization of arbutin, Reactive and Functional Polymers 138 (2019) 39–45.

[40] C. Krumm, M. Hijazi, S. Trump, S. Saal, L. Richter, G.G.F.K. Noschmann, T.-D. Nguyen, K. Preslikoska, T. Moll, J.C. Tiller, Highly active and selective telechelic antimicrobial poly(2-oxazoline) copolymers, Polymer 118 (2017) 107–115.

[41] J.-W. Kook, S. Kim, J.-Y. Lee, J.-H. Kim, Synthesis of curcumin/polyrhodanine nanocapsules with antimicrobial properties by oxidative polymerization using the Fenton reaction, Reactive and Functional Polymers 109 (2016) 125–130.

[42] A.P. Kaymaz, İ. Acaroğlu-Degitz, M.A. Yapaöz, A.D. Sezer, S. Malta, B. Aksu, T. Eren, Synthesis of 1,4-diazabicyclo [2.2.2]octane and pyridinium based cationic polymers via ROMP technique and examination of their antibacterial activity and cytotoxicity, Materialia 5 (2019) 100246.

[43] N.Y. Baran, M. Saçak, Preparation of highly thermally stable and conductive Schiff base polymer: molecular weight monitoring and investigation of antimicrobial properties, Journal of Molecular Structure 1163 (2018) 22–32.

[44] C. Yang, W. Lou, G. Zhong, A. Lee, J. Leong, W. Chin, B. Ding, C. Bao, J.P.K. Tan, Q. Pu, S. Gao, L. Xu, L.Y. Hsu, M. Wu, J.L. Hedrick, W. Fan, Y.Y. Yang, Degradable antimicrobial polycarbonates with unexpected activity and selectivity for treating multidrug-resistant Klebsiella pneumoniae lung infection in mice, Acta Biomaterialia 94 (2019) 268–280.

[45] G. Pavoski, T. Maraschin, M.A. Milani, D.S. Azambuja, R. Quijada, C.S. Moura, N. de Sousa Basso, G.B. Galland, Polyethylene/reduced graphite oxide nanocomposites with improved morphology and conductivity, Polymer 81 (2015) 79–86.

[46] M. Dhanasekar, V. Jenefer, R.B. Nambiar, S. Ganesh Babu, S. Periyar Selvam, B. Neppolian, S. Venkataprasad Bhat, Ambient light antimicrobial activity of reduced graphene oxide supported metal doped TiO_2 nanoparticles and their PVA based polymer nanocomposite films, Materials Research Bulletin 97 (2018) 238–243.

[47] A. Rauf, J. Ye, S. Zhang, L. Shi, M.A. Akram, G. Ning, Synthesis, structure and antibacterial activity of a copper(II) coordination polymer based on thiophene-2,5-dicarboxylate ligand, Polyhedron 166 (2019) 130–136.

[48] N. Thokala, C. Kealey, J. Kennedy, D.B. Brady, J.B. Farrell, Characterisation of polyamide 11/copper antimicrobial composites for medical device applications, Materials Science and Engineering: C 78 (2017) 1179–1186.

[49] P.M. Rahman, V.M. Abdul Mujeeb, K. Muraleedharan, S.K. Thomas, Chitosan/nano ZnO composite films: enhanced mechanical, antimicrobial and dielectric properties, Arabian Journal of Chemistry 11 (2018) 120–127.

[50] J. Xie, Y.-C. Hung, UV-A activated TiO_2 embedded biodegradable polymer film for antimicrobial food packaging application, Lebensmittel-Wissenschaft und -Technologie— Food Science and Technology 96 (2018) 307–314.

[51] A. Khana, A. Aslam, P. Khana, I. Khan, M. Oves, S. Khan, A.M. Asiria, N. Azuma, L.A. Taib, Y.M. Al Angari, A. Facchetti, Facial synthesis of highly active polymer vanadium molybdate nanocomposite: improved thermoelectric and antimicrobial studies, Journal of Physics and Chemistry of Solids 131 (2019) 148–155.

[52] T.P.T. Nguyen, N. Barroca-Aubry, D. Dragoe, S. Mazerat, F. Brisset, J.-M. Herry, P. Roger, Facile and efficient Cu(0)-mediated radical polymerisation of pentafloro-phenyl methacrylate grafting from poly(ethylene terephthalate) film, European Polymer Journal 116 (2019) 497–507.

[53] S. Demirci, Z. Ustaoğlu, G.A. Yılmazer, F. Sahin, N. Baç, Antimicrobial properties of zeolite-X and zeolite-A ion-exchanged with silver, copper, and zinc against a broad range of microorganisms, Applied Biochemistry and Biotechnology 172 (2014) 1652–1662.

[54] D. Guldiren, S. Aydın, Antimicrobial property of silver, silver-zinc and silver-copper incorporated soda lime glass prepared by ion exchange, Materials Science and Engineering: C 78 (2017) 826–832.

CHAPTER 9

Antibiotic 3D Printed Materials for Healthcare Applications

T.C. MOKHENA • M.J. JOHN • M.J. MOCHANE • EMMANUEL ROTIMI SADIKU •
T.S. MOTSOENENG • MTIBE • P.C. TSIPA

1 INTRODUCTION

Three-dimensional (3D) printing, also called as rapid prototyping, additive layer manufacturing, additive fabrication, or solid-free-form fabrication, attracted unprecedented interest since its introduction in 1980 [1–4]. It is recognized that this technology offered a new platform to manufacture or fabricate various structures with different architectural designs to afford their applications in several emerging technologies, such as biomaterials, photonics, smart textiles, and energy harvesting devices [3–5]. It has been a major breakthrough for design and manufacturing that influenced all stakeholders from economic, geopolitical, and environmental viewpoints.

3D printing due to its capability of simplifying complex structures has opened doors for its application in healthcare (e.g., dental molds, crowns and implants, craniofacial implants, prosthetic parts, medical equipment, surgical models, scaffolds for tissue engineering, and organ printing) [6,7]. It simplifies the designing and fabrication of various constructs obtained from the imaging systems used in healthcare institutions, including computed tomography (CT) and magnetic resonance imaging (MRI) (see Fig. 9.1) [7]. It renders an opportunity to fabricate such constructs related to the patient-specifics that are not possible by conventionally employed techniques.

Besides the promising potential of 3D printing as new technology to overcome some limitations associated with conventional manufacturing technology, there are few aspects that have to be addressed to recognize this technology as a common manufacturing technology in healthcare [7]. One of the primary issues involves the difficulties associated with the printers, such as processing speed, printing speed, and resolution. Despite the improvements and advancements

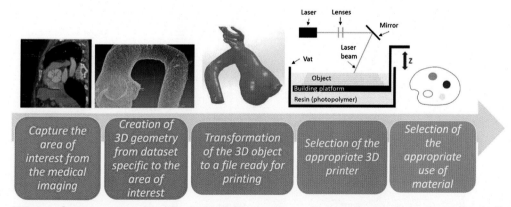

FIG. 9.1 Common stages of the 3D printed model in healthcare. (Reprinted by permission from J. Garcia, Z. Yang, R. Mongrain, R.L. Leask, K. Lachapelle. 3D printing materials and their use in medical education: a review of current technology and trends for the future. BMJ Simulation and Technology Enhanced Learning 4(1) (2018) 27.)

Antibiotic Materials in Healthcare. https://doi.org/10.1016/B978-0-12-820054-4.00009-4

that have been made over the past years, there is a mismatch regarding the printing speed and the resulting constructs resolution. Furthermore, the availability of the inks is limited to most of the materials that are not applicable in healthcare. It is recognized that the 3D printing inks must have some properties to qualify for biomedical applications, such as printability, biocompatibility, appropriate mechanical, and degradation kinetics. In addition, they must have safe degradation by-products and be capable of biomimicking the targeted tissue, meanwhile eliminating the possibility of contamination of the printed object by microorganisms that result in biomaterial-associated infection (BAI) [7,8]. Research has escalated on the use of other means to modify the printing inks to overcome some of these limitations. In this chapter, the recent progress of 3D printing and challenges related to the antibiotic materials and manufacturing process for biomedical applications are discussed.

2 3D PRINTING TECHNOLOGIES

3D printing, also called as rapid prototyping, additive layer manufacturing, additive fabrication, or solid-free-form fabrication, begins from the design of the 3D model, which is then transferred to the 3D printer [1,2]. The model is sliced into several layers that are interpreted by the printer to allow printing of the desired object starting from the base. In the case of medical application, the image obtained from CT and MRI can be translated into a design of a 3D model, which can be transferred to the printer as shown in Fig. 9.1 [6].

It was first patented in 1980 by Hideo Kodama from Nagoya Municipal Industrial Research Institute, Japan. After the patent expired, he published a series of papers from his experiments conducted on 3D models employing UV rays and a photosensitive resin, which later is called stereolithography (SLA). This served as a new era of manufacturing processes in which different well-defined objects can be produced in a short space of time with ease. There are several types of 3D printing techniques ranging from well-established industrial applied technologies to research laboratory settings for more specific applications. These technologies include stereolithography, inkjet printing, selective laser sintering, fused deposition modeling, and laminating object manufacturing (LOM) [1,9].

2.1 Stereolithography

The follow on the Hideo Kodama work by Charles Hull resulted in the development of the stereolithography process using liquid polymers to fabricate plastic devices [1,2]. This led to Hull's first patent in 1986 followed by the first SLA printer being manufactured and commercialized in 1988. This process relies on the directed laser beam polymerizing a liquid resin. It has several approaches that include direct/laser writing (Fig. 9.2A) and mask-based writing (Fig. 9.2B). These approaches are often categorized based on the surface and laser configuration, viz. a free surface (bottom-

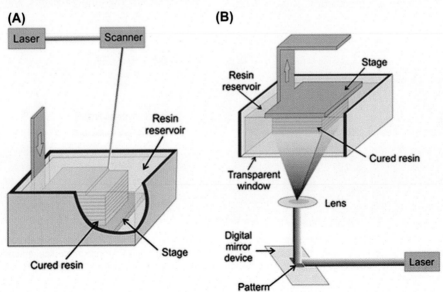

FIG. 9.2 Schematic presentation of SLA: **(A)** bath configuration and **(B)** layer configuration. (Reprinted with permission from B.C. Gross, J.L. Erkal, S.Y. Lockwood, C. Chen, D.M. Spence. Evaluation of 3D printing and its potential impact on biotechnology and the chemical sciences. Analytical Chemistry 86(7) (2014) 3240–3253.)

up) and constrained surface (top-down). The basic components of the direct/laser techniques are movable base, a resin reservoir, UV light beam, and a computer interface; whereas, mask-based writing has a movable platform, resin reservoir, computer, UV beam, and a mask as a digital mirror device that enables the curing of one layer at once.

In the case of bath configuration, the UV beam raster over the resin surface to form a 2D layer with thickness depends on the time of exposure, scan speed, and intensity of the power source. Afterward, the base lowers at a programmed distance to allow polymerization of the next layer on top of the first layer, and the process is repeated until the desired 3D object is completed. The disadvantage of this process includes intense cleaning, a lot of resin waste, and restricted height of the desired object by the vat size.

Yet another approach has similar components as bath configuration is known as layer configuration (Fig. 9.2B). The movable platform is, however, suspended above the resin as opposed to bath configuration where it is submerged. The light source is also placed below the resin reservoir (having optically clear bottom), which overcomes the height restriction for the 3D object. Basically, after the curing of the first layer, the platform is raised to allow the viscous resin to fill the space below the initially cured layer. The step is repeated until the desired 3D item is produced. The advantages of this process include less cleaning, low volumes of resins used, and no height restriction when compared to bath configuration.

It is, however, recognized that posttreatment of the printed items for both approaches is required to ensure that all the resin is completely cured to strengthen bonding between the layers of the printed item. Besides producing well-defined 3D printed objects with high resolution, direct/laser writing is time consuming when compared to the mask-based process in which the employed digital micromirror device (DMD) mirrors can be controlled such that an entire layer is completely cured at once. The efficiency of the layer curing is influenced by several parameters such as the intensity of the light source (E), the critical energy of the resin (E_c), and the depth of light penetration (D_p) as expressed in Eq. (9.1):

$$C_D = D_p ln(E / E_C) \quad (9.1)$$

where (C_D) is the thickness of the cured layer that has to be optimized to overcome the limitations that influence the curing efficiency. This equation is of significant importance to provide the processor with a guidance regarding parameters and resin to choose considering

the object to be printed and intended application. For instance, the lateral resolution has been reported to be directly dependent upon the UV beam diameter (80–250 µm), while the vertical resolution depends on the cured layer thickness ranged from a few submicrometers to few microns.

Despite not being limited, the most commonly used UV light source for SLA is HeCd laser and xenon lamp. Moreover, more than one light source can be utilized to produce high resolution and well-defined 3D objects. On the other hand, resins are limited to the expensive acrylic and epoxy-based polymers that are often entitled to embrittlement and shrinkage upon polymerization. In addition, during printing, only one resin can be employed. The SLA printers are, however, expensive despite their capability of producing high resolution printed items coupled with printing efficiency (i.e., 1.5 cm/h). In addition, it has been limited by its harsh conditions involves (viz., UV cross-linking, inadequate mechanical strength, laborious postprocessing, trapped uncured resin within the completed structure) toward biomedical applications. Nevertheless, the recent advancements in natural and synthetic biodegradable, UV cross-linkable polymers, as well as developments on the technology open door for its future utilization in healthcare.

2.2 Selective Laser Sintering

In the mid-80s, an undergraduate student, Carl Deckard, from the University of Texas developed the selective laser sintering (SLS) process, which is based on the selective solidification of powder using a laser beam to fabricate a desired 3D item [1]. SLS relies on the use of a high-power laser (e.g., CO_2 and Nd:YaG) for sintering polymer powders to generate a 3D model. Basically, the laser is scanned over evenly distributed powder (by a traversing roller) on the stage to increase its temperature to just below the melting temperature such that it promotes the fusion of the powder into a solid form (Fig. 9.3). After each scan is completed, the stage is lowered and in the second layer a polymer powder is added, and then sintered on the desired areas; and the processes are repeated until the 3D model is generated. The unsintered powder serves as a supporting material during sintering and can easily be removed after the printing process. The benefits associated with this technique include a number of various materials that can be utilized as inks, such as polycarbonate, acrylonitrile butadiene styrene (ABS), polyester, nylon, metals, and ceramic. However, there are several aspects that play a major role in the resolution of the printed model, namely laser power, size of the powder, and laser focus.

(A)   **(B)**

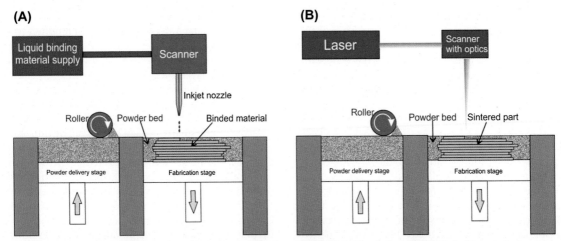

FIG. 9.3 **(A)** Schematic presentation of inkjet printing and **(B)** SLS 3D printing. (Reprinted with permission from B.C. Gross, J.L. Erkal, S.Y. Lockwood, C. Chen, D.M. Spence. Evaluation of 3D printing and its potential impact on biotechnology and the chemical sciences. Analytical Chemistry 86(7) (2014) 3240–3253.)

The limitations of these techniques include the shrinkage of the printed items due to heating and cooling processes, which is one of the most investigated topics to come up with a novel route to overcome this limitation.

2.3 Inkjet

Inkjet was also employed to print different materials such as ceramics, polymers, metals, nucleic acids, and gels over the past decades. It can be categorized into two types, viz. continuous and drop-on-demand (DOD) [1]. The continuous inkjet printing as conceptualized by Sweet from Stanford University in the mid-60s, relies on the printer head having electrostatic plates to control the ink droplets onto the paper during printing or into the waste compartment where it is recycled and reused. The space and size of the ink droplet are directly controlled by the applied pressure wave pattern (Fig. 9.3). In the case of DOD, the ink droplet is directed by voltage and pressure pulse. DOD was developed by Zoltan, and Kyser and Sears. Basically, 3D printing is based on the powdered material having particles of the size 50–100 μm being bound together by droplets of liquids from the inkjet printer head. A layer of powdered material is evenly distributed on the platform to afford the inkjet head to apply the binding liquid material at the desired areas. After the completion of the first layer, the stage lowers and the second layer of powdered material is evenly distributed and again bound together by a binding liquid material. This

process is repeated until the desired 3D item is generated, which is further exposed to posttreatment (e.g., heat) to allow binding of the powder at selected areas. Similar to SLS, the unbounded powder serves as supporting materials.

The benefits of this technique include the use of different materials (e.g., ceramics, polymeric materials, and glass) as long as the liquid binding materials are available for such powdered material. This opened doors for its application from designing simple items to industrial models. On the other hand, the properties of the printed item such as physical and chemical solely rely on the binding material that can limit their applications in certain fields. For instance, the most commonly used binding liquid materials that limit their application in health-related fields are toxic. Moreover, the unbound powder can cause rough surfaces and porous structures that can be either advantageous or disadvantageous depending on the intended application. Due to technology enhancements, there are some nonpowder-based inkjets in which a photopolymer can be used to generate 3D objects (e.g., PolyJet technology).

2.4 Fused Deposition Modeling

Fused deposition modeling (FDM) that relies on the plastic filament or metal wire is heated in a nozzle and then extruded to generate a 3D model that was pioneered by S. Scott and Lisa Crump and patented in 1989. It serves as one of the most used and available printing technologies to date [1,7]. In FDM, a

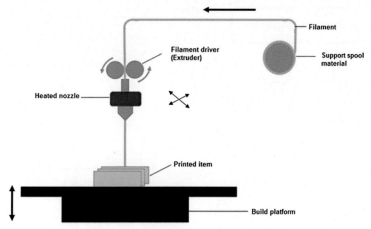

FIG. 9.4 Schematic presentation of FDM.

semimelting plastic filament or metal wire is deposited onto a stage in a layer-by-layer fashion to fabricate the 3D model while guided by a computer as shown in Fig. 9.4. After each deposition, the stage lowers to allow the deposition of the other layer. The process is then repeated to afford the generation of the 3D model in a layer-by-layer fashion. The advantages of this technique include the variety of materials that can be utilized as inks, is cheaper, and offers different inks to be used at the same time by changing the filament. The ink materials that are often employed in the case of antibacterial active products toward healthcare include polymers and nanocomposites. The resolution levels of the FDM techniques for nozzle location of 25 μm in the x,y plane can be afforded with individual line dimensions and layer thickness of 200−500 μm, which directly depends on the nozzle diameter. The disadvantage of this technique includes the lack of capability of fabricating long, unsupported sections or sharp hangs due to lack of structural strength of the extruded filaments upon printing. Nonetheless, the inclusion of fillers has been proposed as an alternative route to overcome this issue.

2.5 Direct Ink Writing
Similar to FDM, direct ink writing (DIW) on the ink is being extruded/pushed through a nozzle as a viscous solution to form single lines that solidify on the stage [7,10,11]. In DIW, also known as solution-cast direct ink writing, as the material is being extruded the nozzle follows the command from the computer model to fabricate a 3D construct layer by layer [10]. The polymer solutions, which are often employed as inks in DIW, are dissolved in the cheaper volatile solvent (e.g.,

dichloromethane (DCM), dimethyl sulfoxide, tetrahydrofuran) that can evaporate during extrusion, leaving behind dry polymeric material. The material used has to be viscous enough to maintain the structure during printing. In addition, different gels having sufficient viscosity to maintain structural stability during printing are also used as inks for the DIW process [10]. In most cases, posttreatment can be employed to maintain the structural stability of the 3D printed constructs. In addition, this technology offers the opportunity to blend different materials together when preparing inks; hence, it has been successfully used in tissue engineering [10]. Depending on the nozzle diameter, the resolution levels with line dimensions from 200 to 500 μm can be achieved. A very recent study by Therriault and coworkers developed a multimaterial DIW method to produce complicated 3D structures by creating a removable support printed from a lower melting temperature metal (i.e., copper) or a ceramic (i.e., alumina) [11]. The authors used a two-step process: (1) room-temperature DIW of metal−polymer composite structures with metal−polymer (ceramic−polymer) composite supports, and (2) a postdeposition thermal treatment to afford fabrication of complex metal structures as shown in Fig. 9.5. In this regard, the inks comprised concentrated steel, copper, and alumina microparticle suspensions dispersed in a polymer solution viz. PLAs dissolved in dichloromethane (PLA/DCM) were used for printing. The supporting materials were built using a secondary material (i.e., copper or alumina ink, which is later removed through postdeposition thermal treatment), while the building structure is printed with the primary material (i.e., steel ink). The lower temperature metallic support (copper)

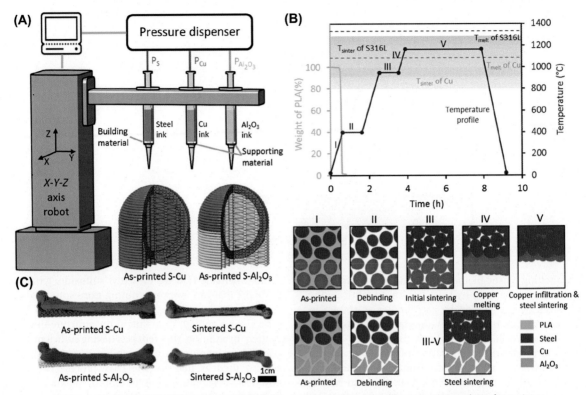

FIG. 9.5 Multimaterial DIW for complex 3D metallic structures with removable supports consists of two steps: **(A)** room-temperature DIW of metal–polymer composite structures with metal–polymer (or ceramic–polymer) composite supports and **(B)** postdeposition thermal treatment turning the as-printed metal–polymer composites structures to metal structures, and a schematic shows the two-phase interface microstructure variations of the as-printed structures at five different stages during the thermal treatment. **(C)** Optical images of as-printed and sintered S–Cu and S–Al$_2$O$_3$ structures printed as human thigh bones at a scale of 1:7. (Reprinted with permission from C. Xu, B. Quinn, L.L. Lebel, D. Therriault, G. L'Espérance. Multi-material direct ink writing (DIW) for complex 3D metallic structures with removable supports. ACS Applied Materials and Interfaces 11(8) (2019) 8499–8506.)

completely melted and filtrated into the pores within the filament of the sintered structure to produce hybrid multifunctional structure. On the other hand, the ceramic support (alumina) survived the thermal treatment cycle and hence was easily removed without affecting the metallic structure (Fig. 9.5). This demonstrates the capability of DIW to possibly fabricate multifunctional complex structures that can afford the usage in medical implants, sensors, and energy storage applications.

2.6 Laminated Object Manufacturing

LOM, which was developed by Helysis from Torrance and shipped in 1991, relies on the basis of stacking the layers of sheets (e.g., papers, plastics, fabrics, composites, and metal) to generate a 3D model (Fig. 9.6)

[9]. The first layer is placed onto the stage and cut into the desired structure using a razor or laser, and then the excess is being removed as shown in Fig. 9.6. The second sheet is then put on top of the previous sheet to be cut into a next defined structure, and then the sheet layers are welded together. These steps are repeated again and again until the 3D object is generated according to the model. The advantages include limited defects as compared to printing techniques such as FDM. However, the limitations involve insufficient lamination between the sheets that may lead to structural collapsing if the parameters are not optimized. Moreover, the choice of the materials used depends on their capability of being processed into sheets as well as how good they are to be welded together.

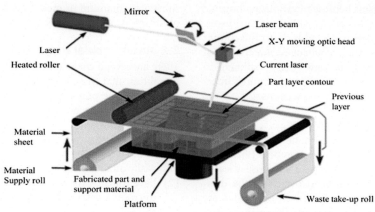

FIG. 9.6 Schematic presentation of laminated object manufacturing (LOM). (Reprinted with permission from D. Ahn, J-H. Kweon, J. Choi, S. Lee. Quantification of surface roughness of parts processed by laminated object manufacturing. Journal of Materials Processing Technology 212(2) (2012) 339–346.)

2.7 Other Novel 3D Technologies

There have been new developments in the above-mentioned 3D printing technologies to overcome some of their limitations, including the lag between the printing process and the resolution of the printed constructs [12]. In 2017, Malinauskas and colleagues from Lithuania managed to develop an ultrafast laser lithography methodology by employing femtosecond laser to create lenses with high clarity and resilience [12]. They reported that pyrolysis performed resulted in the removal of organic constituents leaving behind glass—ceramic structure with homogeneous shrinkage by 35%—40% without compromising the resolution of the printed constructs. In the same year, Chang et al. demonstrated the use of electrohydrodynamic (EHD) printing techniques to produce aligned-fiber antibiotic patches [13]. In this regard, the authors combined the electrospinning technology—involving the application of the electric current onto the needle to produce nanofibers—with 3D technology to enable the control over the deposition of the fibers by layer-to-layer fashion to fabricate 3D constructs. Yet other advanced lithography technology known as "3D interference printing" involves the single-step manufacturing of multilayered ultralong nanochannel arrays with high density, and large-area uniformity was reported by Jeon and colleagues in the year 2015 [14]. The authors fabricated these structures from optical interference phenomenon with restricted near-field diffraction originating from conformal binary gratings. Conformal gratings serve as a floor plan to effectively control the geometry of the printed items such that they possess high resolution within a short period. Similar technology involving single-step manufacturing of multilayered materials was established in 2018 by Dolinski et al. [15]. The approach termed "solution mask liquid lithography (SMaLL)," involves photochromic molecules (solution masks) to render rapid building rates, large depths of cure, good feature resolution, and 3D constructs without layering defects. In the utilization of tunable and visible wavelength photoswitches as well as orthogonal cross-linking reactions, a single mixture can be employed for facile and single-step printing of complex bioinspired constructs.

Zhang et al. [16] pioneered a 3D printing method that affords the fabrication of a portable array chip for target detection. In this regard, the technique known as "3D ice printing" is integrated with photopolymerization sealing in order to avoid complicated preparations commonly employed in wet chemistry-based methods. This technology offers the printability of ice structures having different geometries, sizes, and height as well as multilayered ones specifically for multitarget analysis to be achieved. A 3D printing innovative technology based on a high current plasma focused ion beam system was recently developed and applied for fabrication of the customized MEMS sensor by Watanabe et al. [17]. In comparison to the classic lithography process commonly employed for the fabrication of customized MEMS sensors, the developed 3D method reduced the overall manufacturing time by ~80%.

3 ANTIBACTERIAL MATERIALS

The human health sector has been under a lot of pressure to come up with novel strategies to eradicate harmful microorganisms because of various pathogenic infections resulting from contact with these organisms. Rapid antibiotic resistance also adds up to the current burden. To overcome these challenges, materials that have antibacterial property have been explored in biomedical sectors to reduce healthcare-acquired infections as discussed in the following sections.

3.1 Antibacterial Peptides

Antibacterial peptides (AMPs)—a new class of antibiotic materials available in large numbers (viz. more than 800 sequences)—have attracted immense research interest as a suitable solution to overcome the difficulties related to BAI [18–20]. These molecules are recognized by their antibacterial activity against bacteria, fungi, and viruses. AMPs are low molecular weight substances that show antibacterial efficacy against the common bacteria resistant to antibiotics. The mechanism for the antibacterial activity of the AMPs follows the interaction with specific constituents of the bacterial membrane that leads to depolarization, destabilization, and membrane disruption resulting in bacterial cell death [21]. Owing to the combination of these processes at the same time, there has been hardly any reports on the risk cases associated with bacterial-resistant developments [19]. Most of the AMPs are amphipathic, cationic peptides and are listed in the accessible database for research and support to design novel molecules that are efficient against biofilms [21]. The list offers the information based on the AMPs used and the microbial species and overall conditions at which it was tested [21].

Research has escalated in the immobilization of the AMPs on the surface of medical devices to afford their antibacterial activity. The natural (e.g., magainin, daptomycin, melittin, tritripticin, etc.) and synthetic (melamine, 9-merpeptides from magainin, E14LKK (magainin-class peptide), etc.) AMPs have been covalently bonded on the surface of various materials (carbon nanotubes, gold nanoparticles, titanium surfaces, and many others) to render long-term stability as compared to incorporation release-based systems [19]. Recent studies conducted on the introduction of the AMPs on the surface of the gold nanoparticles demonstrated that these techniques can be applicable for 3D printing [18,19]. It was shown that the coating based on cecropin-melittin peptide maintained good antibacterial activity against gram-positive and negative bacteria as well as multidrug resistance bacteria in the presence of serum and had low cytotoxicity toward human cells; hence, it can be translated on to other substrates (titanium or medical devices) for prevention of device-associated microbial infection [19].

3.2 Antimicrobial Polymers

There are several characteristics, viz. structure and functionalities, that play a major role in the antibacterial property of the polymers. In the case of structure and functionalities, the polymers should at least have the following properties as proposed by Matsuzaki [20,22] and Kenawy et al. [22,23] to exhibit efficient antibacterial activity:

(1) easily attainable at a reasonable cost
(2) have functionalities that enable attachment to the microorganisms and integration with the inside of the membrane
(3) selectively eradicate microorganisms in the presence of mammalian cells
(4) stable for the desired application, viz. for short- or long-term usage as well as storage
(5) no decomposition or release toxic products
(6) can be regenerated, and still maintain its activity
(7) if required eradicate different microorganisms at specified/intended time frame

From these proposed properties, it can clearly be concluded that the functionalities of the polymers play a crucial role in the antibacterial efficiency. It is recognized that the positively charged functional groups, such as quaternary ammonium and phosphonium groups, are commonly used to synthesize cationic polymers that are highly antimicrobial [24,25]. Several authors reported on the utilization of ammonium functionalized polymers to afford good antibacterial activity property [24,25]. The ammonium is protonated in aqueous medium depending on the pH that results in highly antibacterial material [23]. It is worth mentioning that the polymer properties such as molecular weight and distribution of the functional groups are of significant importance toward their antibacterial activity [23]. Ikeda and colleagues found that the synthesized polycationic antibiotic agents with pendent phosphonium salts in comparison with other monomers had a high activity that increased in the order of molecular weight [26–28]. 3D printed constructs having quaternary ammonium groups were reported by Yue et al. [29]. Using monomers containing positively charged quaternary ammonium groups were either directly copolymerized with photocurable diurethane-dimethacrylate/glycerol dimethacrylate resins by photo-curing or prepolymerized by light-induced polymerization to afford dental 3D printed constructs using stereolithography process. The results indicated that the printed parts were able to eradicate the bacteria on contact when the quaternary ammonium was added into the photocurable resin components. It was pointed out that the antibacterial dental constructs with different geometries can be realized with mechanical properties almost identical to photocured polymeric materials. The authors suggested that these types of antibacterial products can also be employed in other applications, such as food packaging, water purification, and entertainment products.

The modification of a commercially available resin, that is, MiiCraft to afford an antibacterial property of the 3D printed constructs was demonstrated by Guo et al. [30]. In this case of 3D printed items using the stereolithography process, posttreatment was developed to grow functional polymer brushes via atomic transfer radical polymerization (ATRP) on fabricated parts to achieve antibacterial properties. In this regard, 3-sulfopropyl methacrylate potassium salt was grown on the printed items via surface-initiated ATRP (SI-ATRP) to induce the antibacterial property against both gram (*E. coli*) and positive (*B. subtilis*) bacteria. The results have shown that both bacteria had a high affinity toward unfunctionalized printed items; however, the functionalization led to a significant adhesion reduction. Moreover, the functionalized printed objects show growth inhibition of the bacteria. The development of the posttreatment on 3D printed constructs using FDM to afford the fabrication of the porous surface items was recently reported by Vargas-Alfredo et al. [31]. The authors functionalized the 3D printed objects by simultaneously introducing pores on the surface such that it would be of interest for functionalized printed objects such that the bacteria can enter the pores and be killed upon contact. This process termed "Breath Figures" approach was used in which a solution of polymer was used to simultaneously coat (functionalization) and introduce pores on the surface of the printed objects. This process renders the formation of porous surfaces by evaporation of the coated polymer solution under moist conditions. Using a similar approach, 3D constructs from high-impact polystyrene fabricated by FDM were chemically modified using the blend solutions of polystyrene (PS) and polystyrene-*b*-poly(acrylic acid) (PS$_{23}$-*b*-PAA$_{18}$) and a quaternized PS-*b*-poly(dimethylaminoethyl methacrylate) (PS$_{42}$-*b*-PDMAEMA$_{17}$) blends to afford antibacterial efficacy toward *S. aureus* in Ref. [32]. The adhesion of the bacteria was similar for all investigated samples as shown in Fig. 9.7A and B. The obtained fluorescence images Fig. 9.7A show that the unmodified 3D printed were permanent green indicating no antibacterial activity over the entire period of exploration (viz. 0, 24, 48 h); whereas, in the chemically modified, there was transition from green surfaces associated with alive bacteria to clear red colored surfaces. The results were quantified using Image J and the values obtained are shown in Fig. 9.7B. The printed objects based on PS-*b*-PAA diblock copolymer show 100% eradication of the adhered bacteria after 24 h, and similar observations were reported for PS-*b*-PDMAEMA-based objects. The antibacterial activity was attributed to the presence of the carboxylic acid groups or charge quaternary ammonium groups and the presence of pores that allows the bacteria to adhere to the printed objects and being killed upon contact.

Other functional groups with excellent antibacterial efficacy include sulfonium, zwitterionic polymers, *N*-halamine groups, and nitric oxide-containing polymers [22,26,33]. *N*- halamines received tremendous interest due to the unique features, such as low cost, easily regenerated, and friendlier to humans and environment. They can be incorporated into polymers using different strategies such as polymerization (i.e., homopolymerization and heteropolymerization), grafting or coating on various substrates, and many others [22,33]. In the case of zwitterionic polymers, they offer distinctive properties such as regeneration, reusability, and excellent biocidal property [34–36]. These polymers possess both positively charged and negatively charged groups along the chain making them neutral while biocidal to gram-negative and -positive bacteria. Polymeric sulfonium compounds have been reported to be lethal to microorganisms [37,38]. The sulfonium compounds are reported to be more effective against gram-positive bacteria than gram-negative bacteria [37]. One of the major limitations of the sulfonium is low thermal stability [38].

3.3 Natural Polymers

Naturally occurring polymers with an antibacterial property are of interest regarding their unique features, such as inexpensiveness, renewability, biodegradability, and biocompatibility [39,40]. In this case, cellulose-based materials have been the most studied natural polymer toward healthcare properties because of its abundant availability as compared to other natural polymers [40]. Cellulose and its derivative have been employed in different fields, including pharmaceuticals, construction, packaging, and many more making them suitable for additive manufacturing [4,41]. Cellulose acetate dissolved in acetone was 3D printed using Printrbot Simple Metal printer followed by conversion to cellulose by treating the printed items with sodium

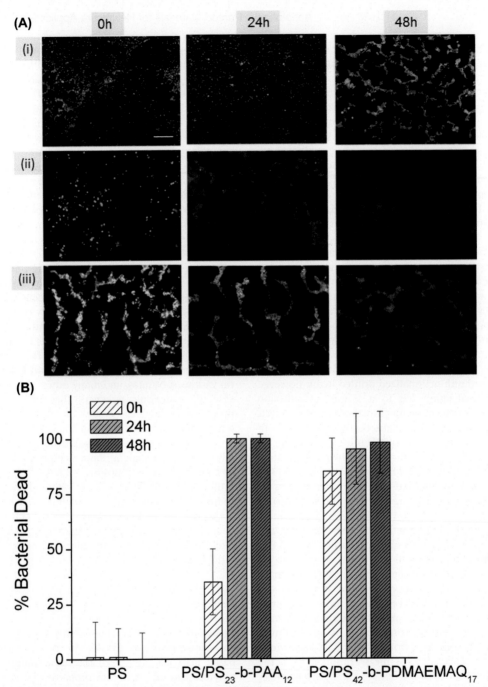

FIG. 9.7 **(A)** Evolution of the bacterial viability at different times, 0, 24, and 48 h, on porous surfaces prepared using (i) only polystyrene, (ii) 20% PS_{23}-b-PAA_{18}/80% PS, and (iii) 20% PS_{42}-b-$PDMAEMAQ_{17}$/80% PS. The fluorescence microscopy images show the polymeric surfaces incubated with *S. aureus* labeled with a green fluorescent protein and stained with propidium iodide (red). Images were acquired using the green and red channels at 63× magnification. For each polymeric surface, overlay images of the two channels are generated using ImageJ. The scale bar corresponds to 20 μm. **(B)** Quantification of the percent of bacteria dead at different times, 0, 24, and 48 h, on the different porous surfaces tested. (Reprinted with permission from N. Vargas-Alfredo, A. Dorronsoro, A.L. Cortajarena, J. Rodríguez-Hernández. Antimicrobial 3D porous scaffolds prepared by additive manufacturing and breath figures. ACS Applied Materials and Interfaces 9(42) (2017) 37454−37462.)

hydroxide by Pattinson and Hart [39]. The mechanical properties of the cellulose acetate printed items (i.e., Young's modulus $(E) = 2-2.4$ GPa, tensile strength $(\sigma) = 42-47$ MPa) were comparable to the commonly used thermoplastic printed through either extrusion or laser sintering, including ABS $(E = 1.6$ GPa, $\sigma = 22-34$ MPa); PLA $(E = 1.5$ GPa, $\sigma = 39$ MPa), and nylon $(E = 0.2-2.6$ GPa, $\sigma = 50-58$ MPa). A statistical 95% reduction in the viable count of bacteria (*Escherichia coli*) demonstrated the possibility of manufacturing sterilized medical or surgical devices.

Chitosan, the derivative of chitin, is also recognized by its antibacterial and antifungal activity as a result of the protonation of NH_2^- on the backbone [42–45]. Chitosan features good biocompatibility and nontoxicity as well as solubility in acid aqueous media at pH values of 5–6 at ambient conditions [42]. Recently, Bergonzi et al. [46] reported an innovative 3D printing method based on chitosan hydrogels for tissue scaffolds. The authors used three different gelation media, that is, KOH 1.5M, Na_2CO_3 1.5M, and ammonia vapors to retain the 3D printed structure. It was found that all used media were suitable to retain the 3D printed structure with the resulting scaffolds being able to enable and drive cell growth up to 21 days. These results indicated that the obtained scaffold could be appropriate for therapeutic implications, considering the fact that scaffolds provided efficient cell growth essential for tissue regeneration. Elsewhere, porous 3D printed chitosan scaffolds for tissue regeneration were conducted in terms of toxicity toward human fibroblasts (Nhdf) and keratinocytes (HaCaT) [47]. It was reported that the best results for cell growth were obtained after 35 days when the Nhdf and HaCaT are seeded together. The results indicated that antibacterial natural polymer can be printed and offer a biomimetic environment than traditional 2D cultures.

3.4 Nanoparticles

A wide variety of nanoparticles (e.g., silver, copper, etc.) are most commonly employed for antimicrobial agents in biomedical applications. Metal nanoparticles feature high antibacterial and antifungal properties at fairly low concentrations; hence, they have been applied in water disinfection and food preservation since ancient times [43,44,48]. Besides the fact that the biocidal properties of the metal nanoparticles cannot fully be explained, several mechanisms have been proposed depending on the metal under investigation [43,44,48,49]. A generally accepted possible mechanism is schematically shown in Fig. 9.8 [48].

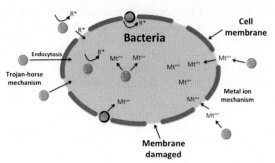

: Metal nanoparticle   R*: Free radicals

: Anchored nanoparticle Mt^{n+}: Metal ions

FIG. 9.8 Schematic presentation on the summary of the mechanisms associated with the antimicrobial behavior of metal nanoparticles: (1) "Trojan-horse effect" due to endocytosis processes; (2) attachment to the membrane surface; (3) catalyzed radical formation; and (4) release of metal ions. (Reprinted with permission from H. Palza. Antimicrobial polymers with metal nanoparticles. International Journal of Molecular Sciences 16(1) (2015) 2099–2116.)

Silver nanoparticles have been the most trusted antimicrobial agent since ancient times [50,51]. It has been often employed in implants [50]. In a recent study, direct metal printing (DMP) was employed to fabricate volume-porous titanium implants coated with silver nanoparticles/chitosan using a plasma electrolytic oxidation process [51]. The results indicated that Ag-coated printed implants had no antibacterial activity toward *Staphylococcus aureus* and impaired normal functioning of neutrophils, the most valuable phagocytic cells for protection against implant-associated infections. It was suggested that caution must be taken when it comes to Ag coatings using a plasma electrolytic oxidation (PEO) process, as this can have a detrimental impact on the innate immune response resulting in persistent infection. Elsewhere, silver nanoparticles were incorporated into the titanium oxide layer of 3D printed titanium implants [52]. The authors employed selective laser melting to produce porous titanium implants having interconnected pores that resulted in 3.75 times larger surface area than solid wire. The silver nanoparticles were embedded into the titanium oxide layer using a PEO. In comparison with the solid implants, the 3D printed implants demonstrated higher Ag release rate (*viz.*, 4.35 times higher) over a period of 28 days and they have strong antibacterial activity against MRSA, that is, two times larger inhibition zones, and an order of magnitude of reduction in the number of colony-forming units (CFU) in the ex vivo antibacterial assay.

3.5 Polymer Composites

Polymers with antibacterial activity are often of interest to print prostheses as they feature unique properties, such as lightweight, easy processability, and availability at fairly low cost [49]. These polymers are often recognized by the presence of the functional groups that usually interact with the microorganism membrane to eradicate bacteria via different mechanisms. On the other hand, the incorporation of the nanoparticles to enhance the antibacterial properties of the resulting polymeric products has been reported in the literature [48,49,53].

A wide variety of nanoparticles (e.g., silver, copper, and zinc) have been incorporated into polymers to produce filaments for FDM technology to fabricate different structures toward healthcare. Muwaffak et al. [54] added zinc, copper, and silver into polycaprolactone (PCL) to produce filaments for 3D printing. They scanned 3D templates for wound dressing and managed to print nose dressing material. It was indicated that the incorporation of the metal ions improved the overall printing of the filaments when compared to other materials (e.g., salicylic acid) regarding complex shapes that can be printed, that is, nose and ears. The authors also demonstrated that the ability to control the release of Cu, Ag, and Zn is attainable as the polymer can act as a barrier to slow down the release of the metals from the system. This is of significant importance considering the fact that the polymeric matrix allows desirable doses to be released during the treatment for a prolonged time such that the dressing stays active for a long time to avoid complications associated with changing the dressing, which can be a painful process. PLACTIVE, as one of the antibacterial polymers containing copper nanoparticles (1%), has recently been printed using FDM to afford its application as finger prosthesis [53]. The antibacterial analysis using ISO 22196 indicated that the printed prosthesis had good antibacterial activity against *Staphylococcus aureus* and *Escherichia coli*, that is, up to 99.99%. Besides the fact that the longevity of the antibacterial activity was not tested, the study showed the capability of printing functional finger prostheses as shown in Fig. 9.9. Moreover, the patient satisfaction results conducted using the Quebec User Evaluation of Satisfaction (QUEST 2.0) clearly indicated that the obtained printed items were quite satisfactory suggesting that the prosthesis was easy to use and effective.

SLA technology was used to fabricate antibacterial active 3D printed dental prosthesis from poly(methyl methacrylate) (PMMA)/titanium oxide (TiO_2) nanocomposites by Totu et al. [55]. The presence of the TiO_2 was depicted by EDX, and these nanoparticles were found to agglomerate at higher loadings. It was reported that 0.4 wt% TiO_2 was sufficient to fabricate dental prosthesis with antibacterial efficacy toward *Candida scotti* using the stereolithographic technique.

3.6 Antibiotics

Antibiotics can also be incorporated into polymers depending on the intended biomedical application. The selection of the antibiotic drug, however, is directly dependent on the following: (1) causative organism; (2) severity of the disease/morbidity; (3) antibiotic resistance; and (4) the involved overall cost [8,56]. On the other hand, it is recognized that the release rate of the antibiotic drug plays a major role in the rehabilitation/therapy of the affected area. The use of the polymers as the host for antibiotic drugs has been amplified due to the benefits associated with this process [8]. These include the choice of polymer depending on the properties required and control over the release rate of the drug over periods of time.

PMMA is one of the most commonly used polymers in reconstructive structure, such as dental implants, reconstruction of craniofacial defects, and bone cement [56]. As PMMA has no antimicrobial property, antibiotic drugs are often incorporated into the system. There are, however, few drawbacks associated with this biocompatible alloplastic material that includes the high exothermic temperatures generated during polymerization, and the limited choice of antibiotic drugs to be employed (e.g., Rifampin, the commonly used antibiotic drug, adversely affect polymerization process by acting as free radical scavengers) [56]. 3D printing also renders the opportunity to design customized patient-specific antibiotic treatment with ease. Moreover, it overcomes the limitations associated with PMMA by avoiding the polymerization step by mixing already polymerized PMMA with the antibiotic drug. Elsewhere, PMMA doped with antibacterial drugs (viz. gentamicin sulfate, tobramycin, and nitrofurantoin) was printed using FDM to afford the development of antibacterial 3D printed medical devices (disks, filaments, and beads) that can be used as a reservoir for localized single drug or suite of drugs delivery, thereby offering local therapeutic levels without the systemic toxicity [56]. It was demonstrated that the printed items had good antibacterial efficacy in the presence of the antibacterial drugs, which indicated that different

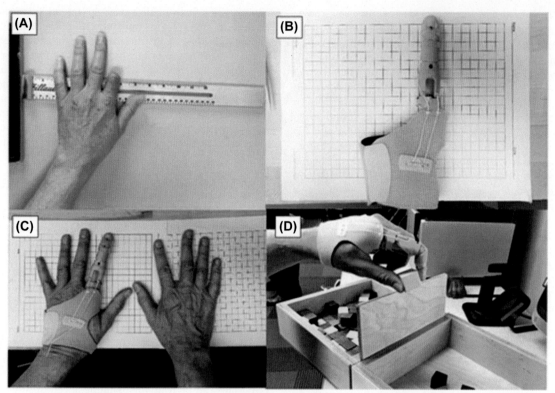

FIG. 9.9 **(A)** Research participant with index finger amputation at the proximal phalanx of the left hand. **(B)** 3D printed finger prosthesis using PLACTIVE antibacterial 3D printing filament. **(C)** Patient using the antibacterial 3D printed finger prosthesis. **(D)** Patient performing the Box and Block Test. (Reprinted with permission from J.M. Zuniga. 3D printed antibacterial prostheses. Applied Sciences 8(9) (2018) 1651.)

shapes and drug loading to target specific clinical condition or patient needs can be met.

Sandler and coworkers [57] demonstrated the ability to incorporate antibacterial agents (nitrofurantoin, NF) into a biodegradable polymer carrier (polylactic acid) to afford 3D printing of model structures having good inhibition of biofilm colonization. The author used *S. aureus* as a model of bacteria to envisage the antibacterial activity of the printed products. The printed samples with NF displayed more than 85% inhibition of biofilm formation was detected regardless of the surface structure of the samples, viz. 89.6% for the smooth and 85.1% for rough surfaces. Elsewhere, PLA doped with different antibiotic agents (i.e., ampicillin sodium salt and vancomycin hydrochloride) was developed using 3D printing technology to establish the in vitro drug-release profiles and the antibacterial activity of the PLA disk samples [58]. The antibiotic agents were incorporated onto disk by using two different surface coating methods, viz. (1) dipping 3D printed disk into a solution of antibiotic agent/poly-lactic-co-glycolic acid (PLGA), and (2) PLA disk immersed into an aqueous solution of antibiotic agents. In the case of the first method of dip coating with PLGA/antibiotic agent, the release of ampicillin continued until 22 days; and the concentrations of ampicillin in buffer solutions at 10−12 days were lower than the value of the minimum inhibition concentration (MIC) for *E. coli* but still higher than the value of MIC 90 for *S. aureus*. In the case of the second method of dipping PLA directly into antibiotic agent solution for an hour (for trial), the stable release of ampicillin in buffer solution continued to about 22 days. The second method was then taken further by increasing the time from an hour to 5 days, because the method was cheaper when compared to the cost associated with PLGA. It was found that the antibiotic agent (i.e., ampicillin) was readily absorbed onto PLA surfaces reaching absorption percentage values of more than 90%, whereas vancomycin reached values higher than 80%.

3.7 Hybrid Antibacterial Materials

The combination of two or more constituents into polymeric material has also been reported to afford multifunctional antibacterial 3D printed objects for biomedical applications [59]. Water et al. [59] prepared PLA containing antibacterial agents (i.e., nitrofurantoin 10%, 20%, 30%) with 5% of hydroxyapatite (HA) in polylactic strands. The drug release was found to be directly dependent on the drug loading in the 3D printed disks with a high level of release being achieved from disks with higher loading. Moreover, the higher loadings (30%) resulted in the prevention of adhesion and planktonic growth of Staphylococcus aureus over a period of 7 days. On the other hand, at low loading (10%), the disks did not inhibit the planktonic growth but prevented adhesion of the bacteria on the surface. The authors indicated that the obtained results in their study demonstrated the possibility of using custom-made, drug-loaded feedstock materials for 3D printing of pharmaceutical products for their controlled release. Blend of PCL and polyvinyl pyrrolidone loaded with tetracycline hydrochloride (TE-HCL) was 3D printed using the EHD printing technique to afford antibacterial active objects in Ref. [13]. It was pointed out that the maximum speed was found to be 80 mm/s after which irregular pore sizes during deposition were obtained. Well-aligned rectangular 3D printed shapes with dimensions $500 \times 500 \ \mu m^2$ were obtained at the solution, working distance, applied voltage, flow rate, and collector speed of 24 w/v%, 2 mm, 2 kV, 0.4 mL/h, and 80 mm/s, respectively; and reducing speed to 50 mm/s afforded the decrease in grid size to $200 \times 200 \ \mu m^2$. Moreover, the composite displayed high encapsulation efficiency of $98.4 \pm 0.9\%$ and a controlled release rate of the antibacterial drug (TE-HCL) over the investigated period (120 h). The possibility of printing complex structures having different sizes, dimension, pore volume, drug loading, and thickness makes this technique ideal for the development of various items with different drug dosages for a variety of anatomical sites and age groups.

Hybrid 3D printed constructs comprising antibiotic agents, nanoparticles, and polymeric materials using ink-jet were reported by Lee and colleagues [60]. It was demonstrated that the potential of utilizing the antibiotic agent and calcium eluting bioresorbable micropatterns as a novel pathway to modify orthopedic implant surfaces for both wound healing and secondary infections. Biphasic calcium phosphate (BCP) nanoparticles and rifampicin (RFP) nanoparticles dispersed in a biodegradable PLGA were used as model materials for 3D printing to demonstrate that the release of the RFP nanoparticles can provide a steady release to kill bacteria near the Ti6Al4V alloy implant surface. The RFP containing 3D printed constructs were able to completely kill S. epidermidis preventing biofilm colony formation. Osteoblasts preferentially adhered to and spread on the Ti alloy surface over the PLGA/BCP pattern surfaces, and the presence of BCP significantly promoted osteogenic differentiation due to its ability to supply calcium ions to the medium with respect to BCP high surface to volume ratio and higher solubility of its components.

4 ANTIBACTERIAL MATERIALS FOR BIOMEDICAL APPLICATIONS AND FUTURE TRENDS

The capability of the 3D technologies to print various items and/or designs having different architectural structures from a wide variety of materials offers the opportunity to afford their use in various applications such as tissue engineering, bone fractures treatment, devices, and wound dressing as summarized in Table 9.1 [8,13,54–57,61–64]. They have improved the manufacturing landscape such that the most complicated structures can be fabricated with ease. For instance, the scans of the patient from MRI or CT scans can be loaded and printed beforehand such that the physicians can be ready by planning ahead of surgery, thereby saving time and avoid complications during surgery. It is, however, recognized that the products/materials employed in other applications such as healthcare require some specifications for them to qualify with the most important one being antibacterial efficiency with low cytotoxicity. The inclusion of the antibacterial agent into the ink material resulted in improving the overall performance of the medical device while reducing secondary infections from bacteria. For instance, bone infection, also known as osteomyelitis, is basically a bacterial infection or rarely fungal infection [56]. It is often treated systemically and locally by delivery of a high dose of the antibiotic drug to the affected site. The antibacterial agents are often incorporated into polymeric materials to control the release rate such that the cell proliferation and differentiation cannot be compromised [60]. Hybrid materials comprised antibacterial agents, and nanoparticles dispersed in the polymeric matrix is one of the promising systems because this type of system affords the inclusion of different components toward various functions. On the other hand, hybrid materials based on the polymers or fillers having different functionalities especially polarity serve as suitable alternatives regarding interaction with the cells [61].

TABLE 9.1
Applications of 3D Printing Technologies in Healthcare.

Composition	3D Technology	Highlights	Intended Application	Refs.
PCL with Cu, Zn, and Ag ions	FDM—MakerBot Replicator 2X Desktop 3D printer	Ag-PCL and Cu-PCL displayed high antibacterial activity toward *S. aureus*, which demonstrates the capability of producing customizable wound dressing designed according to the needs of the patient	Personalized wound dressing	[54]
70/30 poly(butylene-co-adipate terephthalate) (PBAT)/polylactic acid (PLA)/eggshells/silver	Direct ink writing—Hyrel 30 M printer	In vitro antibacterial assessment against *L. monocytogenes* and *S. Enteritidis* revealed that the films were bacteriostatic	Antibacterial coatings	[61]
Thermoplastic polyurethane (TPU-E)/oxacillin sodic	FDM—Sharebot-Q, z-axis resolution of 50 µm	Using the agar disc diffusion method, it was found that no bacterial growth was observed on TPU-containing antibiotic agents after 24 h. The tensile properties were superior for TPU-containing antibiotic agents when compared to neat polymer	Heart valve support	[62]
Polyvinyl pyrrolidone (PVP)-polycaprolactone (PCL) blend loaded with tetracycline hydrochloride	Electrodynamic (EHD) jet printing system	The technology offers the opportunity to tailor dosage forms in a single step	Wound dressing	[13]
Poly(methyl methacrylate) (PMMA)/TiO$_2$	SLA—EnvisonTEC Perfactory 3D printer	3D printed objects containing 0.4, 1 and 2.5 wt % TiO$_2$ inhibited the growth of *Candida scotti*	Dental prosthesis	[55]
Polylactic acid/antibiotics (kanamycin, tetracycline and polymyxin B)	Inkjet—RepRap 3D printer	Different antibiotic agents can be incorporated toward multidrug-resistant bacteria	Diagnostic units for multiplex drug screening	[63]
Polylactic acid (PLA)/nitrofurantion	FDM UP! plus printer	3D printed displayed inhibition of biofilm colonization	Urinary tract infections	[57]
PLA/gentamicin (GS)	FDM—MakerBot 2X 3D printer	PLA containing 1 and 2.5 wt % GS additive showed no growth in broth cultures and a zone inhibition that averaged 12.9 ± 2.56 mm and 21.35 ± 1.0 mm, respectively.	Patient-specific treatment implants to target infection	[64]
Poly(D,L-lactic-co-glycolic) acid (PLGA)/rifampicin (RFP)/biphasic calcium phosphate (BCP) nanoparticles	Inkjet printing—Dimatix Materials Printer, DMP2800, FujiFilm Dimatix, Santa Clara, CA	The RFP-containing 3D printed constructs efficiently prevented the formation of biofilm colonies because of their capability to eradicate bacteria before forming colonies	Orthopedic implants	[60]

Continued

TABLE 9.1
Applications of 3D Printing Technologies in Healthcare.—cont'd

Composition	3D Technology	Highlights	Intended Application	Refs.
PMMA/gentamicin sulfate or tobramycin or nitrofurantoin PLA/gentamicin sulfate or tobramycin or nitrofurantoin	FDM—ExtrusionBot filament extruder	Antibiotic-doped 3D printed beads, disks, and filaments were successfully printed and demonstrated the efficiently inhibited bacterial growth (*E. coli* and *S. aureus*)	Antibiotic deliver on implant sites	[56]

5 CONCLUSION

In this chapter, we discussed the antibacterial activity of the 3D printed constructs fabricated by additive manufacturing toward healthcare applications. The principle of the current additive technologies and new developments were also discussed. Antibacterial materials that can be used as part of 3D printing inks were also elaborated especially regarding the responsible functional groups that play a major role in the antibacterial property of these materials. The capability of additive manufacturing to fabricate different patient-specific objects makes this technology exciting for future healthcare applications. It gives an opportunity to incorporate suitable antibacterial dosages and its controlled release required for various anatomical sites and age groups that can afford their application for medical devices. Furthermore, the ability to incorporate more than two constituents into polymeric matrix/host material renders an opportunity to fabricate multifunctional products for various patient-specific needs. The inclusion of the antibacterial nanoparticles offers additional properties such as good mechanical stability for structural support of the 3D printed constructs while serving as a complement for the antibacterial efficacy of the system under investigation.

ACKNOWLEDGMENTS

Funding from Department of Science and Technology (DST, South Africa)-Biorefinery Program is gratefully acknowledged by the authors.

REFERENCES

[1] P. Parandoush, D. Lin, A review on additive manufacturing of polymer-fiber composites, Composite Structures 182 (2017) 36–53.

[2] B.C. Gross, J.L. Erkal, S.Y. Lockwood, C. Chen, D.M. Spence, Evaluation of 3D printing and its potential impact on biotechnology and the chemical sciences, Analytical Chemistry 86 (7) (2014) 3240–3253.

[3] J.A. Lewis, G.M. Gratson, Direct writing in three dimensions, Materials Today 7 (7) (2004) 32–39.

[4] J. Liu, L. Sun, W. Xu, Q. Wang, S. Yu, J. Sun, Current advances and future perspectives of 3D printing natural-derived biopolymers, Carbohydrate Polymers 207 (2019) 297–316.

[5] T.-M. Tenhunen, O. Moslemian, K. Kammiovirta, A. Harlin, P. Kääriäinen, M. Österberg, et al., Surface tailoring and design-driven prototyping of fabrics with 3D-printing: an all-cellulose approach, Materials and Design 140 (2018) 409–419.

[6] J. Garcia, Z. Yang, R. Mongrain, R.L. Leask, K. Lachapelle, 3D printing materials and their use in medical education: a review of current technology and trends for the future, BMJ Simulation and Technology Enhanced Learning 4 (1) (2018) 27.

[7] M. Guvendiren, J. Molde, R.M.D. Soares, J. Kohn, Designing biomaterials for 3D printing, ACS Biomaterials Science and Engineering 2 (10) (2016) 1679–1693.

[8] S.E. Moulton, G.G. Wallace, 3-dimensional (3D) fabricated polymer based drug delivery systems, Journal of Controlled Release 193 (2014) 27–34.

[9] D. Ahn, J.-H. Kweon, J. Choi, S. Lee, Quantification of surface roughness of parts processed by laminated object manufacturing, Journal of Materials Processing Technology 212 (2) (2012) 339–346.

[10] L. Li, Q. Lin, M. Tang, A.J.E. Duncan, C. Ke, Advanced polymer designs for direct-ink-write 3D printing, Chemistry – A European Journal. 25 (46) (2019) 10768–10781.

[11] C. Xu, B. Quinn, L.L. Lebel, D. Therriault, G. L'Espérance, Multi-material direct ink writing (DIW) for complex 3D metallic structures with removable supports, ACS Applied Materials and Interfaces 11 (8) (2019) 8499–8506.

[12] L. Jonušauskas, D. Gailevičius, L. Mikoliūnaitė, D. Sakalauskas, S. Šakirzanovas, S. Juodkazis, et al., Optically clear and resilient free-form µ-optics 3D-printed via ultrafast laser lithography, Materials (Basel, Switzerland) 10 (1) (2017) 12.

[13] J.-C. Wang, H. Zheng, M.-W. Chang, Z. Ahmad, J.-S. Li, Preparation of active 3D film patches via aligned fiber

electrohydrodynamic (EHD) printing, Scientific Reports 7 (2017) 43924.

[14] J. Park, K.-I. Kim, K. Kim, D.-C. Kim, D. Cho, J.H. Lee, et al., Rapid, high-resolution 3D interference printing of multilevel ultralong nanochannel arrays for high-throughput nanofluidic transport, Advanced Materials 27 (48) (2015) 8000−8006.

[15] N.D. Dolinski, Z.A. Page, E.B. Callaway, F. Eisenreich, R.V. Garcia, R. Chavez, et al., Solution mask liquid lithography (SMALL) for one-step, multimaterial 3D printing, Advanced Materials 30 (31) (2018) 1800364.

[16] H.-Z. Zhang, F.-T. Zhang, X.-H. Zhang, D. Huang, Y.-L. Zhou, Z.-H. Li, et al., Portable, easy-to-operate, and antifouling microcapsule array chips fabricated by 3D ice printing for visual target detection, Analytical Chemistry 87 (12) (2015) 6397−6402.

[17] Plasma ion-beam 3D printing: a novel method for rapid fabrication of customized MEMS sensors, in: K. Watanabe, M. Kinoshita, T. Mine, M. Morishita, K. Fujisaki, R. Matsui, et al. (Eds.), 2018 IEEE Micro Electro Mechanical Systems (MEMS), January 21−25, 2018.

[18] A. Rai, S. Pinto, T.R. Velho, A.F. Ferreira, C. Moita, U. Trivedi, et al., One-step synthesis of high-density peptide-conjugated gold nanoparticles with antimicrobial efficacy in a systemic infection model, Biomaterials 85 (2016) 99−110.

[19] A. Rai, S. Pinto, M.B. Evangelista, H. Gil, S. Kallip, M.G.S. Ferreira, et al., High-density antimicrobial peptide coating with broad activity and low cytotoxicity against human cells, Acta Biomaterialia 33 (2016) 64−77.

[20] K. Matsuzaki, Control of cell selectivity of antimicrobial peptides, Biochimica et Biophysica Acta (BBA) − Biomembranes 1788 (8) (2009) 1687−1692.

[21] M. Di Luca, G. Maccari, G. Maisetta, G. Batoni, BaAMPs: the database of biofilm-active antimicrobial peptides, Biofouling 31 (2) (2015) 193−199.

[22] M.C. González-Henríquez, A.M. Sarabia-Vallejos, J. Rodríguez Hernandez, Antimicrobial polymers for additive manufacturing, International Journal of Molecular Sciences 20 (5) (2019).

[23] E.-R. Kenawy, S.D. Worley, R. Broughton, The chemistry and applications of antimicrobial polymers: A State-of-the-Art review, Biomacromolecules 8 (5) (2007) 1359−1384.

[24] B. Dizman, M.O. Elasri, L.J. Mathias, Synthesis and antimicrobial activities of new water-soluble bis-quaternary ammonium methacrylate polymers, Journal of Applied Polymer Science 94 (2) (2004) 635−642.

[25] C.Z. Chen, N.C. Beck-Tan, P. Dhurjati, T.K. van Dyk, R.A. LaRossa, S.L. Cooper, Quaternary ammonium functionalized poly(propylene imine) dendrimers as effective antimicrobials: structure−activity studies, Biomacromolecules 1 (3) (2000) 473−480.

[26] A. Kanazawa, T. Ikeda, T. Endo, Antibacterial activity of polymeric sulfonium salts, Journal of Polymer Science Part A: Polymer Chemistry 31 (11) (1993) 2873−2876.

[27] T. Ikeda, H. Yamaguchi, S. Tazuke, New polymeric biocides: synthesis and antibacterial activities of polycations

with pendant biguanide groups, Antimicrobial Agents and Chemotherapy 26 (2) (1984) 139−144.

[28] A. Kanazawa, T. Ikeda, T. Endo, Novel polycationic biocides: synthesis and antibacterial activity of polymeric phosphonium salts, Journal of Polymer Science Part A: Polymer Chemistry 31 (2) (1993) 335−343.

[29] J. Yue, P. Zhao, J.Y. Gerasimov, M. van de Lagemaat, A. Grotenhuis, M. Rustema-Abbing, et al., 3D-Printable Antimicrobial composite resins, Advanced Functional Materials 25 (43) (2015) 6756−6767.

[30] Q. Guo, X. Cai, X. Wang, J. Yang, "Paintable" 3D printed structures via a post-ATRP process with antimicrobial function for biomedical applications, Journal of Materials Chemistry B 1 (48) (2013) 6644−6649.

[31] N. Vargas-Alfredo, H. Reinecke, A. Gallardo, A. del Campo, J. Rodríguez-Hernández, Fabrication of 3D printed objects with controlled surface chemistry and topography, European Polymer Journal 98 (2018) 21−27.

[32] N. Vargas-Alfredo, A. Dorronsoro, A.L. Cortajarena, J. Rodríguez-Hernández, Antimicrobial 3D porous scaffolds prepared by additive manufacturing and breath figures, ACS Applied Materials and Interfaces 9 (42) (2017) 37454−37462.

[33] Y. Sun, G. Sun, Novel refreshable N-halamine polymeric biocides: N-chlorination of aromatic polyamides, Industrial and Engineering Chemistry Research 43 (17) (2004) 5015−5020.

[34] Z. Cao, N. Brault, H. Xue, A. Keefe, S. Jiang, Manipulating sticky and non-sticky properties in a single material, Angewandte Chemie International Edition 50 (27) (2011) 6102−6104.

[35] L. Mi, S. Jiang, Integrated antimicrobial and nonfouling zwitterionic polymers, Angewandte Chemie International Edition 53 (7) (2014) 1746−1754.

[36] Z. Cao, L. Mi, J. Mendiola, J.-R. Ella-Menye, L. Zhang, H. Xue, et al., Reversibly switching the function of a surface between attacking and defending against bacteria, Angewandte Chemie International Edition 51 (11) (2012) 2602−2605.

[37] M. Hirayama, The antimicrobial activity, hydrophobicity and toxicity of sulfonium compounds, and their relationship, Biocontrol Science 16 (1) (2011) 23−31.

[38] M.S. Ganewatta, C. Tang, Controlling macromolecular structures towards effective antimicrobial polymers, Polymer 63 (2015) A1−A29.

[39] S.W. Pattinson, A.J. Hart, Additive manufacturing of cellulosic materials with robust mechanics and antimicrobial functionality, Advanced Materials Technologies 2 (4) (2017) 1600084.

[40] C. Thibaut, A. Denneulin, S. Rolland du Roscoat, D. Beneventi, L. Orgéas, D. Chaussy, A fibrous cellulose paste formulation to manufacture structural parts using 3D printing by extrusion, Carbohydrate Polymers 212 (2019) 119−128.

[41] V. Klar, J. Pere, T. Turpeinen, P. Kärki, H. Orelma, P. Kuosmanen, Shape fidelity and structure of 3D printed

high consistency nanocellulose, Scientific Reports 9 (1) (2019) 3822.

[42] T.C. Mokhena, M.J. Mochane, Transforming fishery waste into chitin and chitin based materials, "Waste-to-Profit"? (W-t-P): Value Added Products to Generate Wealth for a Sustainable Economy 1 (2018) 227–249.

[43] T. Mokhena, V. Jacobs, A. Luyt, A Review on Electrospun Bio-Based Polymers for Water Treatment, 2015.

[44] T. Mokhena, A. Luyt, Electrospun alginate nanofibres impregnated with silver nanoparticles: preparation, morphology and antibacterial properties, Carbohydrate Polymers 165 (2017) 304–312.

[45] T.C. Mokhena, A.S. Luyt, Development of multifunctional nano/ultrafiltration membrane based on a chitosan thin film on alginate electrospun nanofibres, Journal of Cleaner Production 156 (2017) 470–479.

[46] C. Bergonzi, A. Di Natale, F. Zimetti, C. Marchi, A. Bianchera, F. Bernini, et al., Study of 3D-printed chitosan scaffold features after different post-printing gelation processes, Scientific Reports 9 (1) (2019) 362.

[47] C. Intini, L. Elviri, J. Cabral, S. Mros, C. Bergonzi, A. Bianchera, et al., 3D-printed chitosan-based scaffolds: an in vitro study of human skin cell growth and an in-vivo wound healing evaluation in experimental diabetes in rats, Carbohydrate Polymers 199 (2018) 593–602.

[48] H. Palza, Antimicrobial polymers with metal nanoparticles, International Journal of Molecular Sciences 16 (1) (2015) 2099–2116.

[49] T.C. Mokhena, M.J. Mochane, T.H. Mokhothu, A. Mtibe, C.A. Tshifularo, T.S. Motsoeneng, Preparation and characterization of antibacterial sustainable nanocomposites, in: Inamuddin, S. Thomas, R. Kumar Mishra, A.M. Asiri (Eds.), Sustainable Polymer Composites and Nanocomposites, Springer International Publishing, Cham, 2019, pp. 215–244.

[50] B.S. Necula, J.P.T.M. van Leeuwen, L.E. Fratila-Apachitei, S.A.J. Zaat, I. Apachitei, J. Duszczyk, In vitro cytotoxicity evaluation of porous TiO$_2$–Ag antibacterial coatings for human fetal osteoblasts, Acta Biomaterialia 8 (11) (2012) 4191–4197.

[51] M. Croes, S. Bakhshandeh, I.A.J. van Hengel, K. Lietaert, K.P.M. van Kessel, B. Pouran, et al., Antibacterial and immunogenic behavior of silver coatings on additively manufactured porous titanium, Acta Biomaterialia 81 (2018) 315–327.

[52] I.A.J. van Hengel, M. Riool, L.E. Fratila-Apachitei, J. Witte-Bouma, E. Farrell, A.A. Zadpoor, et al., Selective laser melting porous metallic implants with immobilized silver nanoparticles kill and prevent biofilm formation by methicillin-resistant *Staphylococcus aureus*, Biomaterials 140 (2017) 1–15.

[53] J.M. Zuniga, 3D printed antibacterial prostheses, Applied Sciences 8 (9) (2018) 1651.

[54] Z. Muwaffak, A. Goyanes, V. Clark, A.W. Basit, S.T. Hilton, S. Gaisford, Patient-specific 3D scanned and 3D printed antimicrobial polycaprolactone wound dressings, International Journal of Pharmaceutics 527 (1) (2017) 161–170.

[55] E.E. Totu, A.C. Nechifor, G. Nechifor, H.Y. Aboul-Enein, C.M. Cristache, Poly(methyl methacrylate) with TiO$_2$ nanoparticles inclusion for stereolithographic complete denture manufacturing – the fututre in dental care for elderly edentulous patients? Journal of Dentistry 59 (2017) 68–77.

[56] D.K. Mills, U. Jammalamadaka, K. Tappa, J. Weisman, Studies on the cytocompatibility, mechanical and antimicrobial properties of 3D printed poly(methyl methacrylate) beads, Bioactive Materials 3 (2) (2018) 157–166.

[57] N. Sandler, I. Salmela, A. Fallarero, A. Rosling, M. Khajeheian, R. Kolakovic, et al., Towards fabrication of 3D printed medical devices to prevent biofilm formation, International Journal of Pharmaceutics 459 (1) (2014) 62–64.

[58] C.-H. Chen, Y.-Y. Yao, H.-C. Tang, T.-Y. Lin, D.W. Chen, K.-W. Cheng, Long-term antibacterial performances of biodegradable polylactic acid materials with direct absorption of antibiotic agents, RSC Advances 8 (29) (2018) 16223–16231.

[59] J.J. Water, A. Bohr, J. Boetker, J. Aho, N. Sandler, H.M. Nielsen, et al., Three-dimensional printing of drug-eluting implants: preparation of an antimicrobial polylactide feedstock material, Journal of Pharmaceutical Sciences 104 (3) (2015) 1099–1107.

[60] Y. Gu, X. Chen, J.-H. Lee, D.A. Monteiro, H. Wang, W.Y. Lee, Inkjet printed antibiotic- and calcium-eluting bioresorbable nanocomposite micropatterns for orthopedic implants, Acta Biomaterialia 8 (1) (2012) 424–431.

[61] B.J. Tiimob, G. Mwinyelle, W. Abdela, T. Samuel, S. Jeelani, V.K. Rangari, Nanoengineered eggshell-silver tailored copolyester polymer blend film with antimicrobial properties, Journal of Agricultural and Food Chemistry 65 (9) (2017) 1967–1976.

[62] E. Gasparotti, E. Vignali, P. Losi, M. Scatto, B.M. Fanni, G. Soldani, et al., A 3D printed melt-compounded antibiotic loaded thermoplastic polyurethane heart valve ring design: an integrated framework of experimental material tests and numerical simulations, International Journal of Polymeric Materials and Polymeric Biomaterials 68 (1–3) (2019) 1–10.

[63] S. Glatzel, M. Hezwani, J. Kitson Philip, S. Gromski Piotr, S. Schürer, L. Cronin, A portable 3D printer system for the diagnosis and treatment of multidrug-resistant bacteria, Inside Cosmetics 1 (3) (2016) 494–504.

[64] J.A. Weisman, J.C. Nicholson, K. Tappa, U. Jammalamadaka, C.G. Wilson, D.K. Mills, Antibiotic and chemotherapeutic enhanced three-dimensional printer filaments and constructs for biomedical applications, International Journal of Nanomedicine 10 (2015) 357.

CHAPTER 10

Inhibition of Bacterial Growth and Removal of Antibiotic-Resistant Bacteria From Wastewater

KEHINDE WILLIAMS KUPOLATI • A.A. BUSARI • EMMANUEL ROTIMI SADIKU •
A. FRATTARI • A.A. ADEBOJE • C. KAMBOLE • K.S. MOJAPELO • M.R. MAITE •
N. MOTSILANYANE • W. BEZUIDENHOUT • A.A. EZE • IDOWU DAVID IBRAHIM •
O.O. AYELERU • T.A. ADEGBOLA • J. SNYMAN • R.J. MOLOISANE • M.M. MOKAE •
J.M. NDAMBUKI • O. AGBOOLA • VICTORIA OLUWASEUN FASIKU •
OLUYEMI OJA DARAMOLA • M.S. ONYANGO • P.A. OLUBAMBI • A.M. BERHE •
M. NDLOVU • L. DE VILLIERS • D.A. BRANGA-PEICU • R.W. SALIM

1 INTRODUCTION

Globally, water is a necessity for human, animal, and plant existence. The demand for water over the years has increased based on the increase in population. Population increase has also led to an increase in wastewater. The wastewater is discharged into sewers in the form of solids or liquids. They can also be in the form of dissolved and suspended organic solids that are often biologically unanalyzable. The treatment method of this waste includes both the chemical and biological methods due to the presence of organic matter, inorganic matter, and metallic materials. The report of the European Union on Urban Waste Water Treatment Council Directives, the effective management of effluent collection, and processing should be in line with the legislative measures and the emission limit [1]. This is because polluted water has myriad effects on both society and the environment. This can be in the form of impurity from industrial plants. The negative effect can affect the reproductive endocrine system, immune system disorders, and so on. It is worthy to note that this also significantly affects the aquatic system [2–5].

The result of the research of [5] showed that wastewater contains all the pathogens found in excreta and antibiotics, and this is not only limited to the underdeveloped and developing nations. This antibiotic and antibiotic-resistant bacterium (ARB) found in wastewater is not usually treated efficiently with the conventional wastewater treatment method. Wastewater contains a high percentage of organic or inorganic disease-causing pathogen including ARB that can alter both plant and animal growth. Generally, two classes of wastewater exist, which are the domestic and industrial waste. The process of removal, modification and removal of this impurity, is called as wastewater treatment [6].

Another broad classification of wastewater, according to the source, is as follows:
1. Domestic waste discharged from residences and industrial establishments and similar facilities
2. Industrial waste
3. Infiltration or inflow waste
4. Stormwater or waste from run-off or floods.

There are two broad treatment methods for wastewater treatment, which are the primary and secondary methods. The two methods involve the removal of biochemical oxygen demand (BOD) and suspended solids from the wastewater. The research of [6] asserted that this method of wastewater treatment is now insufficient as most of the water is not totally free from pathogens including ARB. It has been established that wastewater affects human health and the environment at large. To buttress this point, the health impact report by Ref. [7] revealed that wastewater exposure can increase the prevalence of communicable diseases like diarrhea

Antibiotic Materials in Healthcare. https://doi.org/10.1016/B978-0-12-820054-4.00010-0

that, in turn, affects the immunity of humans. Additionally, this increases the risk of sickness, and diseases can also increase by consuming polluted water. In a bid to proffer a lasting solution to this problem, there are many developments in the field of wastewater treatment with different alternatives for pathogen removal including bacteria and antibiotics. To this end, advanced waste treatment became a focus.

Advance waste treatment methods involve technologies that deal with the reduction of waste and waste utilization initiatives. This provides hope as the inevitable loss of useable water is prevented.

This technology involves the use of microalgae reactors, which are also called photobioreactors. This is one of the sustainable alternatives to wastewater treatment. Furthermore, these microalgae reduced nitrogen and phosphorus compounds as pollutants. Some of this new technology of wastewater treatment may not be capable of removing or reducing the presence of antibiotics in wastewater treatment. Due to the myriad effect of this harmful ARB, this review focused on the inhibition of bacterial growth and the removal of ARB from wastewater.

2 REVIEW OF LITERATURE
2.1 The Need for Clean Water
Sustainable Development Goal 6 of the United Nations, adopted in 2015, focused on equitable access to safe, affordable drinking water for all by 2030. The research of [9] avowed that the essential need for the existence of humans is water. Over the last 2 decades, several achievements have been recorded in the field of water and wastewater improvement [8].

IDW water refers to portable water that is well protected from external contaminants, especially fecal matter and pathogens [9]. The research of [10] reported that a significant improvement was noticed in access to IDW globally. According to the author, 2.6 billion individuals gained access to potable water globally. Despite this high percentage of people with access to IDW, a high percentage of individuals still do not have such access. This can be buttressed with the fact that so many people in the developing and underdeveloped countries rely heavily on untreated water from boreholes and unprotected burrow well, streams, ponds, rainwater, and so on [8].

The report of the post-millennium development goal (MDG) debate focused on the need to reduce the inequalities associated with access to IDW. This includes the reduction of ethnicity and socioeconomic inequalities that exist in several countries of the world [9,11−14].

It is worthy to note that globally, a large percentage of people in developing and underdeveloped nations still consume untreated water as portable water. Report of [15] claims that 159 million people consume this water daily. The author avowed that access to safe potable water is an efficient tool for reducing poverty.

Gloomily, 70% of the population that depends on the use of untreated running or surface water for drinking is from Sub-Saharan Africa [16]. From the result of the same research, 723 million users are from the eastern part of Asia. This showed a decline from 43% to 33% as of 2015.

The effect of this is that 1 million (1,000,000) direct or indirect death is caused by the consumption of this wastewater. Moreover, 80% of this death rate is from developing nations. [17−19] studied the effect of wastewater on the community by assessing secondary data while [20−22] assessed primary data, and the result confirmed similar results.

2.2 Municipal Wastewater Treatment
EPA in 2004 reported that early efforts in pollution management prevented human waste and domestic waste from water supplies. Household waste includes types of wastes from the sinks, faucets, toilets, showers, dishwashers, floors clean up, and laundry. Most of this pathogenic waste is traced to the sewer. Two types of sewers exist, which are separate and combined systems. Although the secondary sewer system deals with the collection of stormwater and domestic waste, the primary sewer system deals with the collection of domestic waste alone [23,24].

The municipal waste treatment method involves the accumulation of discharge of treated effluents in surface water: This could be in the form of the following:
1. Organic material
2. Grease and oils
3. Industrial wastes
4. Nutrients
5. Metals—tiny amounts of metals, that is, iron, copper, and Zn, area unit naturally gift in human wastes. Others like lead, metal.
vi. Chemicals—this is as a result of domestic cleaning (dish laundry detergents and shampoos).

2.2.1 Conventional wastewater treatment
a) Preliminary waste treatment
Wastewater screening involves the removal of solid wastes like coarse aggregates, rags, and so on that can affect the mechanical instrumentation in the preliminary

treatment of waste. Preliminary waste treatment involves the screening and the removal of grits. Grit removal separates significant, inorganic, sandy solids that might settle in the mechanical channels and interfere with treatment processes. Solid waste removal is done to protect the machinery, especially the pumping equipment, and to forestall blockages in smaller pipes and channels. In this method of treatment also, organic waste and BOD are also removed. The discharge of effluents from the deposit tank occurs within the primary stage. It also includes a trickling filter, which is just a bed of stones from 3 to 6 feet deep through which biodegradable pollution passes.
b) Secondary municipal waste treatment

The secondary wastewater treatment involved the use of biological treatment methods in removing microorganisms from the activated sludge
c) Tertiary or advanced waste treatment

This treatment method is used for high-quality effluents. In this method disinfection control for pathogenic microorganisms and viruses is done.

2.2.2 Stages in wastewater treatment

2.2.2.1 Primary waste treatment. In this treatment method, gravity sedimentation of the screened degritted waste to remove solid waste. A large portion of the suspended solids in the form of BOD is being removed [25].

2.2.2.2 Secondary treatment. The secondary treatment involves the removal of about 855 of the organic matter in the sewage by using the bacteria in it. The technology used involved the trickling filter and the activated sludge process. At the stage, the sludge is activated with billions of microorganisms including bacteria by returning into the aeration tank for mixing. From the aeration tank, the partially treated sewage then flows to the deposit tank for the removal of the excess bacteria. To end this process, the sedimentary tank is disinfected before discharged into the receiving waters [26,27].

To finish secondary treatment, effluent from the sedimentary tank is usually disinfected with halogen before being discharged into receiving waters [23,26,27].

2.2.2.3 Other treatment options. Pollution problem has exerted a lot of pressures on the waste treatment systems. The presence of heavy metals, chemical compounds, and toxic substances is very difficult to get rid of in wastewater, and hence other methods of wastewater treatment are always adopted [25].

2.3 The Use of Antibiotics in Wastewater Treatment

2.3.1 Source of antibiotics in wastewater treatment

Antibiotics are either naturally occurring or human-made compounds. Antibiotics have a wide application

in the improvement of human, animal, and plant health. The usage is also adopted in the treatment of infections caused by pathogens [28−31]. Antibiotics inhibit the growth of different types of microorganisms [32]. The extensive usage had led to an increase in the consumption of antibiotics [33,34]. The first antibiotics are from the synthesis of antimicrobial artificial compounds, and this first set of antibiotics occurs naturally and was discovered in 1940. This antibiotic was coincidentally found in plants. Other developments involve antibiotics from soil bacterium, and currently, a lot of antibiotics are synthesized antibiotics [35].

They are divided into different categories, such as β-lactams, quinolones, tetracyclines, macrolides, sulfonamides, aminoglycosides, carbapenems, and cephalosporins (Kummerer). The presence of antibiotics usually affects the ecological function and the structure of the ecosystem and water system [34,36,37]. This negative effect was attributed to the remnants of antibiotics in the aquatic system [38−40].

The increase in the use of antibiotics can be traced to the response of bacterium in the form of antibiotic resistance. Over the years, bacteria have adapted to the increased level of antibiotics usage. Consequently, several antibiotics varieties have been manufactured [41−44]. In 1996, 10,200 tons of antibiotics were used in the European Union alone; 50% of these antibiotics were used in veterinary medicine as growth promoters in animal development, European Federation of Animal Health [45]. The same organization in 1999 reported that 13,280 tons of antibiotics were used in the same part of the world including Switzerland: 65% of this was adopted in human medicine while 45% was used in veterinary medicine [46].

Antibiotics, a form of chemical compound, are usually unconsciously discharged into wastewater from different sources. Some of these sources are as follows (Table 10.1):

2.4 Antibiotics in Wastewater

There are different sources of antibiotics in wastewater. Based on the outcome of the research of [57], the human being is the largest consumer of antibiotics. To this end, they are the major culprit in the discharge of antibiotics into wastewater.

The bacteria in this compound spread their genes into water-indigenous microbes, which also contain resistance genes, thereby altering the microbial activity of the ecosystem [58].

Additionally, the used therapeutic medicine is disposed of into the sewage facilities. If this waste is not properly managed in the soil or the different environmental compartments, they find their way into the surface or subsurface water. This can also be in the form of unmetabolized antibiotic substances.

TABLE 10.1
Sources of Antibiotics in Wastewater.

	Source	Authors
1	Flushing down unused medications Hospital effluent	[47,48]
2	Drug manufacturers could be another source of antibiotics in wastewaters	[49]
3	Wastewaters generated specifically from the activities of agriculture, and aquaculture	[48]
4	Pharmaceuticals in groundwater	[50]
5	Surface water	[49,51]
6	Streams, soil, and sediment samples	[52–54]
7	Pharmaceuticals in sewage	[55,56]

In addition, the waste in the form of excreting from animals contains these antibiotics, which are oftentimes used as manure for plant agricultural purposes [59]. For instance, ciprofloxacin was found in concentrations of between 0.7 and 124.5 µg/L in hospital effluent. In the same manner, Ampicillin was found between 20 and 80 µg/L within the effluent of a medical treatment center [60].

Furthermore, antibiotics are also detected in the µg/L range in municipal biodegradable pollution, within the effluent of biodegradable pollution treatment plants, in surface water and groundwater [52,61–63]. These include quinolones such as ciprofloxacin, sulfon-amides, roxithromycin, dehydrated antibiotic, and so on, whereas others are found within the sediment beneath fish farms as avowed by Ref. [59]. Determination of persistent antibacterial residues in soil fertilized with liquid manure by high liquid performance chromatography using electrospray ionization tandem mass spectrometry.

2.5 Removal of Antibacteria in Wastewater

The waste treatment method determines the reduction efficiencies of antibiotics and bacteria resistant organisms [64,65]. The research of [66–68] affirmed that environmental conditions positively affect gene transfer in most water treatment tanks. Variations in treatment plant design and their operation might influence the fate of ARB and ARGs in waste [32,69,70].

The use of ultraviolet radiation for the removal of ARM was adopted [71–73], while the use of conventional and advanced treatment methods was adopted by Ref. [74].

This was achieved using the method of membrane filtration with pore sizes of 0.2 or 0.45 µm for the removal of ARB and ARG [75,76]. Culture-based methods or molecular-based DNA extraction mostly polymer extractions was used by Refs. [77,78]. The method of authors involves the transfer of bacteria recovered from the membrane filters in the form of selective culture media using standard procedures (Table 10.2).

2.6 Removal of Antibiotics in Wastewater

Municipal waste treatment plants (WWTPs) are known as sources of nutrients, inorganic and organic pollutants as well as ARB and ARGs [77, 82, 83].

Some of these bacteria may be removed by using the conventional treatment method as avowed by Refs. [64,84]. Nevertheless, some percentage of these bacteria survive in the effluents [83,85] and get back to the environment [54,79,86].

Waste management plants are often not designed to cater for the removal of pharmaceutical wastes. Most waste treatment plants have a variety of biological treatment. The removal of antimicrobial bacteria is essential for the prevention of the discharge of ARB and ARG into the receiving waters.

The research of [42,87] showed that ozone can be used to reduce chemical micropollutants. This may be in the

TABLE 10.2
Classification of Antibiotics Found in Wastewater Treatment.

1	Broad-spectrum antibiotics	Veterinary and for human ambulant therapy	Sulfamethoxazole (sulfonamide) concentrations	[79]
2	Antibiotics of last resort and emergency antibiotics	Clinical human and in veterinary therapy	Cephalosporin and quinolone	[80]
3	Narrow spectrum antibiotics	Used in clinical and ambulant human therapy and veterinary therapy	Oxytetracycline	[81]

form of ultraviolet radiation. None the less, the use of this technology can damage the nucleic acids in the cells of the microorganisms [87]. This invariably can reduce the presence and abundance of ARB in wastewater [65,88].

Additionally, this method is considered adequate to inactivate bacteria via the production of highly reactive radicals [89−93].

2.6.1 Physical treatment Method for the Removal of antibiotic-resistance bacteria

Recent studies showed that operational characteristics affect the treatment of wastewater. Bed contact time, filter loading rate, filter media, grain size, and backwash are examples of such characteristics [92,94]. Furthermore, the germicidal effects on ultraviolent lights also induce DNA, proteins, and RNA. This affects the wavelength [95]. This method is known to accelerate gene transfer using mobile genetic elements [96].

2.7 The Use of Nanotechnology in ARM

Several nanomaterials have been used effectively in the removal of ARB and ARG in wastewater. Two prominent methods can be adopted. A functionalized nanomaterial is combined with antibiotics, and nanomaterial enters inside ARB and then releases considerable amounts of toxic ions [108]. Based on the antimicrobial activities of nanomaterials, it is considered as a defense against multiple drug resistance bacteria [109−115].

To properly understand the antibacterial properties of nanomaterials, it is worthy to note that some metals, such as zinc, silver, and copper, exhibit antibacterial mechanisms in their bulk form. However, iron oxide does not show such characteristics in nanoparticulate for [116,117] they also play essential roles in increasing the abundance of certain ARGs.

2.8 Bacteria Inhibition

Water treatment systems contain many various parts such as demineralization units, reverse diffusion, ultrafiltration, and nanofiltration systems of varied pumping units with unit area interconnected. There are a few varieties of treatment that consists of facultative ponds, anaerobic treatment, and aerobic method involving bacterium and protozoa; however, its role is inessential. The research of [124] showed that algae and bacteria live together in phycosphere and interact in complex relationships. The physical and chemical compositions of algae and bacteria affect the treatment and capabilities of the photobioreactor. A photobioreactor utilizing wastewater as a medium of growth contains a wide range of microbial diversity [125]. The use of genetic

evolution using antibiotic resistance is now used in wastewater treatment [126]. There are four major places where the genetic evolution occurs frequently, and antibiotic resistance evolves as shown in Tables 10.3 and 10.4.

TABLE 10.3
Methods of Treating ARB in Wastewater.

Methods	Authors	Research Gap
Granular media filtration and GAC adsorption		
MR in sand and GAC filtration	Plants [92,94, 96−98]	GAC filtration showed a lower enhancement compared to sand filtration
The use of filtration as a tertiary treatment process	[92]	Research required on the fate and transport of ARGs during granular media filtration as a function of operational conditions
Filtration as posttreatment after ozonation	[79,86,92]	
Membrane treatment	[99,100]	Water quality, fouling propensities, and flux has not been studied yet. Further studies required on the underlying mechanisms and to confirm ARGs removal efficiencies as a function of critical operational parameters
Removal of AMR by UV irradiation	[82,95, 101−104]	Specific proteins that are responsible for tetracycline resistance, may absorb UV light, and protect the bacteria. Some positive bacteria are less sensitive to UV exposure than gram-negative bacteria

Continued

TABLE 10.3
Methods of Treating ARB in Wastewater.—cont'd

Methods	Authors	Research Gap
Ozonation	[86]	Germicidal effects of ozone depend on microbial community type
Chlorination	[15,88, 105—107]	Chlorine exposure affects disinfection efficiency. With less removal efficiency for ARBs

TABLE 10.4
Nano Materials Used in Wastewater Treatment.

Nanomaterials	Authors
Silver	[118]
Nitric oxide-releasing nanoparticles	[119]
Silver nanoparticles with NOM and iron	[120]
Superparamagnetic iron oxide Nanoparticles (conjugation r) of iron, zinc, and silver	[121]
Nanoalumina	[122]
Gold nanoparticles with vancomycin	[123]
Iron oxide nanoparticles	[93]

The main venue for the spread of ARB.

	Waste Type	Authors
1	Agriculturally influenced regions	[112,127—129]
2	Constructed a wetland	[130,131]
3	Healthcare centers and hospitals are the most important facilities	[126,132,133]
4	Aquatic and terrestrial environment like soil and surface waters	[85,87,114,115]
5	Aquatic ecosystems	[129,134—136]

The presence of vancomycin-resistant *Enterococcus*, multidrug-resistant *Mycobacterium tuberculosis* and other ARB in wastewater causes life-threatening sickness. This can also be traced to the prevalence of antibiotic-resistant gonorrhea with serious medical complications that is threatening to human life [14,137]. The report [8] avowed that the concentration of ARG in wastewater far supersedes that of natural water and hence the need for the proper treatment of this wastewater before consumption and this will also prevent the dissemination of ARGs in the environment.

3 CONCLUSION

This chapter assessed the inhibition of bacterial and removal of ARB from wastewater. Wastewater has a myriad effect on both society and the environment. This can be in the form of impurity from industrial plants. The negative effect can affect the reproductive endocrine system, immune system disorders, and so on. It is worthy to note that this also significantly affects the aquatic system. It can be concluded as follows:

1. Access to safe potable water is an efficient tool for reducing poverty.
2. The human being is the largest consumer of antibiotics; the extensive usage had led to an increase in the consumption of antibiotics.
3. Waste management plants are often not designed to cater for the removal of pharmaceutical wastes like antibiotics.
4. Environmental condition positively affects gene transfer in most water treatment tanks.
5. Variations in treatment plant design and their operation might influence the removal of ARB and ARGs in waste.
6. The presence of heavy metals in the environment can cause coselection of antibiotic and heavy metal resistance.
7. Typical heavy metals such as Cu and Zn are used widely in industry and play important roles in increasing the abundance of certain ARGs.
8. Ultraviolet radiation as a treatment method for the removal of ARB can damage the nucleic acids in the cells of microorganisms, thereby reducing ARB abundance in the wastewater.
9. Ozonation is considered an adequate method to inactivate bacteria via the production of highly reactive radicals.
10. Some metals such as zinc, silver, and copper exhibit antibacterial mechanisms in their bulk form and this can cause coselection of antibiotics and heavy metal resistance.
11. Heavy metals such as Cu and Zn are used widely in industry and play important roles in increasing the abundance of certain ARGs.

3.1 Recommendation

3.1.1 Recommendation for policymakers

1. Government and policymakers should regulate the discharge of pharmaceutical waste into the

environment as this is the main source of antibiotics in wastewater.

2. As part of the management of clinical biomedical waste, antibiotics and therapeutic drugs should not be flushed.

3. There should be an orientation for medical practitioners, pharmacists, veterinary doctors, and agriculturalists on the adverse effect of releasing antibiotics into the environment.

3.1.2 Recommendations for future research

1. Future research should work on the Diverse effect of ARGs in reducing the susceptibility of pathogens to different antibiotics like sulfonamide (*sul*), tetracycline (*tet*), etc.

2. Nanomaterials and biochar combined with other treatment methods and coagulation process are very recent strategies regarding ARB and ARGs removal and need more investigation and research.

3. Further studies are recommended to understand the underlying mechanisms and to confirm ARGs removal efficiencies as a function of key operational parameters.

4. Why specific proteins that are responsible for tetracycline resistance may absorb UV light and protect the bacteria.

5. Ineffectiveness of UV disinfection and ozonation on ARG removal and the sensitivity of positive bacteria on UV exposure.

REFERENCES

[1] L.-K. Lee, J.-H. Kim, J. Park, J. Kim, Water quality at water treatment plants classified by type, Toxicology and Environmental Health Sciences 67 (6) (2004) 1405–1407.

[2] B. Austin, The effects of pollution on fish health, Journal of Applied Microbiology 85 (1998) 234–242.

[3] G. Gellert, Effect of age on the susceptibility of Zebrafish eggs to industrial wastewater, Water Resources 35 (2007) 3754–3757.

[4] A.M. Vajda, Demasculinization of male fish by wastewater treatment plant effluent, Aquatic Toxicology 103 (2001) 213–221.

[5] K.E. Liney, Health effects in fish of long-term exposure to effluents from wastewater treatment work, Environmental Health Perspectives 114 (2006) 81–89.

[6] S. Amit, R. Ghate, Developments in wastewater treatment methods, Desalination 167 (15) (2004) 55–63.

[7] J.K. Seo, Effect of the sewage and wastewater plant effluent on the algal growth potential in the Nakdong river basin, Algae 18 (2003) 157–167.

[8] World Health Organization, Global Burden of Disease, WHO Press, UNICEF, Geneva, Switzerland, 2015.

[9] Abubakar, Dano, Socioeconomic Challenges and Opportunities of Urbanization in Nigeria Urbanization and its Impacts on Socioeconomic Growth in Developing Regions, IGI Global, Pennsylvania, 2018, pp. 219–240. Google Scholar.

[10] R.C. Mkwate, R.C.G. Chidya, E.M.M. Wanda, Assessment of drinking water quality and rural household water treatment in Balaka District, Malawi, Physics and Chemistry of the Earth 100 (2018) 353–362.

[11] G. Hutton, C. Chase, International Journal of Environmental Research and Public Health 13 (6) (2016), 536-440. CrossRefView Record in ScopusGoogle Scholar.

[12] S. Irianti, P. Prasetyoputra, T.P. Sasimartoyo, Determinants of household drinking-water source in Indonesia: an analysis of the 2007 Indonesian family life survey, Cogent Medicine 3 (2016) 115–143. Google Scholar.

[13] V. Mulenga, B.B. Bwalya, K. Kaliba-Chishimba, Determinants and inequalities in access to improved water sources and sanitation among the Zambian households, International Journal of Development and Sustainability 6 (8) (2017) 746–762. View Record in ScopusGoogle Scholar.

[14] UNICEF/WHO, Progress on Drinking Water, Sanitation and Hygiene: 2017 Update and SDG Baselines, 2017.

[15] D. Yang, H. Mao, Y. Zhou, LuoPrevalence and Fate of Carbapenemase Genes in a Wastewater Treatment Plant in Northern China PLoS One, 2016, pp. 1–14.

[16] World Health Organization, United Nations Children's Fund (UNICEF), Progress on Sanitation and Drinking Water e 2015 Update and MDG Assessment, 2015.

[17] E.A. Adams, G.O. Boateng, J.A. Amoyaw, Socioeconomic and demographic predictors of potable water and sanitation access in Ghana, Social Indicators Research 126 (2) (2016) 673–687.

[18] S. Irianti, P. Prasetyoputra, T.P. Sasimartoyo, Determinants of household drinking-water source in Indonesia: an analysis of the 2007 Indonesian family life survey, Cogent Medicine 3 (2016) 115–143. Google Scholar.

[19] E.S. Osabuohie, U.R. Efobi, C.M. Gitau, Environmental challenge and water access in Africa: empirical evidences based on Nigeria's households survey, in: Berlin Conference of the Human Dimensions of Global Environmental Change, Berlin, October 2012, pp. 5–6.

[20] S. Arouna, Dabbert, Determinants of domestic water use by rural households without access to private improved water sources in Benin: a seemingly unrelated tobit approach, Water Resources Management 24 (2010) 1381–1398.

[21] E.C. Koskei, R.C. Koskei, M.C. Koske, H.K. Koech, Effect of socio-economic factors on access to improved water sources and basic sanitation in Bomet Municipality, Kenya, Research Journal of Environmental and Earth Sciences 5 (12) (2013) 714–719.

[22] S. Madanat, F. Humplick, A model of household choice of water supply systems in developing countries, Water Resources Research 29 (5) (2019) 1353–1358.

[23] Metcalf, Eddy, Wastewater Engineering: Treatment Disposal and Reuse, McGraw-Hill, New York, 1991.

[24] EPA, Wastewater Treatment/Disposal for Small Communities, 1992. EPA 625/R92-005. U.S., Cincinnati, OH.

[25] EPA, Primer for Municipal Wastewater Treatment Systems, September 2004. EPA 832-R-04-001.

[26] C. Lue-Hing, D.R. Zenz, R. Kuchenrither, Municipal Sewage Sludge Management. Processing, Utilization and Disposal Water Quality Management Library, vol. 4, Technomic Publishing, Lancaster, PA, 1992.

[27] A.I. Mytelka, J.S. Czachor, W.B. Guggino, H. Golub, Heavy metals in wastewater and treatment plant effluents, Journal (Water Pollution Control Federation) 45 (1973) 1859−1864.

[28] R.S. Singer, R. Finch, H.C. Wegener, R. Bywater, J. Walters, M. Lipsitch, Antibiotic resistance − the interplay between antibiotic use in animals and human beings, The Lancet Infectious Diseases 3 (2003) 47−51.

[29] F.C. Cabello, Heavy use of prophylactic antibiotics in aquaculture: a growing problem for human and animal health and the environment, Environmental Microbiology 8 (2006) 1137−1144.

[30] K. Kummerer, Antibiotics in the aquatic environment a review part I, Chemosphere 75 (2009a) 417−434.

[31] J.A. Karlowsky, D.C. Draghi, M.E. Jones, C. Thornsberry, I.R. Friedland, D.F. Sahm, Surveillance for antimicrobial susceptibility among clinical isolates of Pseudomonas aeruginosa and Acinetobacter baumannii from hospitalised patients in the United States, 1998 to 2001, Antimicrobial Agents and Chemotherapy 47 (2003) 1681−1688.

[32] R. Lindberg, Environmental Chemistry, Department of Chemistry, Umeå University, Umea, Sweden, 2006. Determination of Antibiotics in the Swedish Environment with Emphasis on Sewage Treatment Plants.

[33] G.D. Wright, The antibiotic resistome: the nexus of chemical and genetic diversity, Nature Reviews Microbiology 5 (2007) 175−186.

[34] R.I. Aminov, The role of antibiotics and antibiotic resistance in nature, Environmental Microbiology 11 (2009) 2970−2988.

[35] J. Clardy, M. Fischbach, C. Currie, The natural history of antibiotics, Current Biology 19 (11) (2009) R437−R441, https://doi.org/10.1016/j. cub.2009 .04.001.

[36] S. Thiele-Bruhn, I.C. Beck, Effects of sulfonamide and tetracycline antibiotics on soil microbial activity and microbial biomass, Chemosphere 59 (2005) 457−465.

[37] A. Kotzerke, S. Sharma, K. Schauss, H. Heuer, S. Thiele-Bruhn, K. Smalla, B.M. Wilke, M. Schloter, Alterations in soil microbial activity and N-transformation processes due to sulfadiazine loads in pig-manure, Environmental Pollution 153 (2008) 315−322.

[38] A.K. Sarmah, M.T. Meyer, A.B.A. Boxall, A global perspective on the use, sales, exposure pathways, occurrence, fate and effects of veterinary antibiotics (VAs) in the environment, Chemosphere 65 (2006) 725−759.

[39] G.D. Wright, The antibiotic resistome: the nexus of chemical and genetic diversity, Nature Reviews Microbiology 5 (2007) 175−186.

[40] N. Kemper, Veterinary antibiotics in the aquatic and terrestrial environment, Ecological Indicators 8 (2008) 1−13.

[41] R.E. Jenkins, R. Cooper, The synergy between oxacillin and Manuka honey sensitises methicillin-resistant, Staphylococcus aureus to Oxacillin 4 (12) (2009).

[42] D.G. Lee, J.M. Urbach, G. Wu, N.T. Liberati, R.L. Feinbaum, S. Miyata, Genomic analysis reveals that Pseudomonas aeruginosa virulence is combinatorial, Genome Biology 7 (2006) R90, https://doi.org/10.1186/gb-2006-7-10-r90 [PMC free article] [PubMed] [CrossRef].

[43] H.J. Morrill, J.M. Pogue, K.S. Kaye, K.L. La Plante, Treatment options for carbapenem-resistant Enterobacteriaceae infections, Open Forum Infectious Diseases (2015) 1−15. Google Scholar.

[44] S.H. Gillespie, Evolution of drug resistance in mycobacterium tuberculosis: clinical and molecular perspective, Antimicrobial Agents and Chemotherapy 46 (2012) 267−274.

[45] European Federation of Animal Health (FEDESA), Antibiotics and Animals, FEDESA/FEFANA Press release, Brussels, Belgium, September 8, 1997.

[46] European Federation of Animal Health (FEDESA), Antibiotic Use in Farm Animals Does Not Threaten Human Health, FEDESA/FEFANA Press release, Brussels, Belgium, July 13, 2001.

[47] Harvard Health Letter. Drugs in the Water. Harvard Health Publications, Harvard

[48] K. Kummerer, Antibiotics in the aquatic environment a review, part I, Chemosphere 75 (2009a) 417−434.

[49] A. Gulkowska, Y. He, M.K. So, L.W.Y. Yeung, H.W. Leung, J.P. Giesy, P.K.S. Lam, M. Martin, B.J. Richardson, The occurrence of selected antibiotics in Hong Kong coastal waters, Marine Pollution Bulletin 54 (8) (2007) 1287−1293.

[50] M.E. Lindsey, M. Meyer, E.M. Thurman, Analysis of trace levels of sulfonamide and tetracycline antimicrobials in groundwater and surface water using solid-phase extraction and liquid chromatography/mass spectrometry, Analytical Chemistry 73 (19) (2001) 4640−4646.

[51] W. Xu, G. Zhang, S. Zou, X. Li, Y. Liu, Determination of selected antibiotics in the Victoria Harbour and the Pearl River, South China using high-performance liquid chromatography−electrospray ionization tandem mass spectrometry, Environmental Pollution 145 (3) (2007) 672−679.

[52] D. Kolpin, E.T. Furlong, M.T. Meyer, Pharmaceuticals, hormones, and other organic wastewater contaminants in U.S. streams, 1999−2000: a national reconnaissance, Environmental Science and Technology 36 (2002) 1202−1211.

[53] M.S. Díaz-Cruz, D. Barceló, LC/MS2 trace analysis of antimicrobials in water, sediment and soil, Trends in Analytical Chemistry 24 (7) (2005) 645−657.

[54] S. Kim, Potential ecological and human health impacts of antibiotics and antibiotic-resistant bacteria from

wastewater treatment plants, Journal of Toxicology and Environmental Health B 10 (2007) 559−573.

[55] C.G. Daughton, T.A. Ternes, Pharmaceuticals and personal care products in the environment: agents of subtle change? Environmental Health Perspectives 107 (6) (1999) 907−938.

[56] D. Ashton, M. Hilton, K.V. Thomas, Investigating the environmental transport of human pharmaceuticals to streams in the United Kingdom, The Science of the Total Environment 333 (3) (2004) 167−184.

[57] European Federation of Animal Health (FEDESA). Antibiotic Use in Farm Animals Does Not Threaten Human Health, FEDESA/FEFANA Press release, Brussels, Belgium, July 13.

[58] A. Alonso, P. Sanchez, J.L. Martinez, Environmental selection of antibiotic resistance genes, Environmental Microbiology 3 (2001) 1−9.

[59] R. Wise, Antimicrobial resistance: priorities for action, Journal of Antimicrobial Chemotherapy 49 (2002) 585−601.

[60] K. Kummerer, Antibiotics in the aquatic environment— a review—part I, Chemosphere 75 (2009) 417−434.

[61] E. Zuccato, D. Calamari, M. Natangelo, Presence of therapeutic drugs in the environment, Lancet 335 (2000) 1789−1790.

[62] E.M. Golet, A.C. Alder, A. Hartmann, Trace determination of fluoroquinolone antibacterial agents in urban wastewater by solid-phase extraction and liquid chromatography with fluorescence detection, Analytical Chemistry 73 (2001) 3632−3638.

[63] F. Sacher, H.-J. Brauch, F.T. Lange, The occurrence of antibiotics in groundwater in Baden-Württemberg, Germany—results of a comprehensive monitoring program, in: Abstracts of the 11th Annual Meeting of SETAC Europe (Society of Environmental Toxicology and Chemistry), Madrid, Spain, 2001. Abstract M/EH056, SETAC Europe, Brussels, Belgium, 2001, p. 112.

[64] L. Guardabassi, A. Dalsgaard, Occurrence and Fate of Antibiotic-Resistant Bacteria in Sewage, 2002. Danish EPA Environmental Project Report 722.

[65] J.J. Huang, H.Y. Hu, F. Tang, Y. Li, S.Q. Lu, Y. Lu, Inactivation and reactivation of antibiotic-resistant bacteria by chlorination in secondary effluents of a municipal wastewater treatment plant, Water Research 45 (2001) 2775−2781.

[66] H. Kruse, H. Sorum, Transfer of multiple drug resistance plasmids between bacteria of diverse origins in natural microenvironments, Applied and Environmental Microbiology 60 (1994) 4015−4021.

[67] J. Poté, M.T. Ceccherini, V.T. Van, W. Rosseli, W. Wildi, P. Simonet, T.M. Vogel, Fate and transport of antibiotic resistance genes in saturated soil columns, European Journal of Soil Biology 39 (2003) 65−71.

[68] J. Davies, Sanitation: sewage recycles antibiotic resistance, Nature 487 (2012) 302.

[69] T. Iwane, T. Urase, K. Yamamoto, The possible impact of treated wastewater discharge on incidence of antibiotic-resistant bacteria in river water, Water Science and Technology 43 (2001) 91−99.

[70] L. Guardabassi, D.M. Lo Fo Wong, A. Dalsgaard, The effects of tertiary wastewater treatment on the prevalence of antimicrobial-resistant bacteria, Water Research 36 (2002) 1955−1964, https://doi.org/10.1016/S0043-1354(01)00429-8 [PubMed] [CrossRef] [Google Scholar].

[71] M. Umar, F. Roddick, L. Fan, Moving from the traditional paradigm of pathogen inactivation to controlling antibiotic resistance in water − the role of ultraviolet irradiation, Science Total Environment 662 (2016) 923−939.

[72] Michael-Kordatou, P. Karaloia, D. Fatta-Kassinos, The role of operating parameters and oxidative damage mechanisms of advanced chemical oxidation processes in the combat against antibiotic-resistant bacteria and resistance genes present in urban wastewater, Water Research 129 (2018) 208−230.

[73] W. Yan, Y. Xiao, W. Yan, R. Ding, S. Wang, F. Zhao, The effect of bio-electrochemical systems on antibiotics removal and antibiotic resistance genes: a review, Chemical Engineering Journal 358 (2019) 1421−1437.

[74] P. Krzeminski, M.C. Tomei, P. Karaolia, A. Langenhoff, C.M.R. Almeida, E. Felis, F. Gritten, H.R. Andersen, T. Fernandes, C.M. Manaia, L. Rizzo, D. Fatta-Kassinos, Performance of secondary wastewater treatment methods for the removal of contaminants of emerging concern implicated in crop uptake and antibiotic resistance spread: a review, The Science of the Total Environment 648 (2004) 1052−1081.

[75] EPA, Method 1600: Enterococci in Water by Membrane Filtration Using Membrane-Enterococcus Indoxyl-B-D-Glucoside Agar (mEI), 2002.

[76] K. Wong, M. Munir, I. Xagoraraki, Release of -antibiotic-resistant bacteria and genes in the effluent and biosolids of five wastewater utilities in Michigan, Water Research 45 (2001) 681−693, https://doi.org/10.1016/j.watres.2010.08.033 [PubMed] [CrossRef].

[77] L. Rizzo, C. Manaia, C. Merlin, T. Schwartz, C. Dagot, M.C. Ploy, Urban wastewater treatment plants as hotspots for antibiotic resistant bacteria and genes spread into the environment: a review, The Science of the Total Environment 447 (2013) 345−360, https://doi.org/10.1016/j.scitotenv.2013.01.032 [PubMed] [CrossRef] [Google Scholar].

[78] Tom CurtisBacterial Pathogen Removal in the Wastewater Treatment Plant, December 2003, https://doi.org/10.1016/B978-012470100-7/50031-5. Handbook of Water and Wastewater Microbiology.

[79] J. Alexander, A. Bollmann, W. Seitz, T. Schwartz, Microbiological characterization of aquatic microbiomes targeting taxonomical marker genes and antibiotic resistance genes of opportunistic Bacteria, The Science of the Total Environment 5 (2) (2015) 23−28.

[80] s. Galvin, F. Boyle, P. Hickey, A. Vellinga, D. Morris, M. Cormican, Enumeration and characterization of antimicrobial-resistant *Escherichia coli* bacteria in effluent from municipal, hospital, and secondary

treatment facility sources, Applied and Environmental Microbiology 3 (2010) 4772−4779.

[81] S.C. Kim, K. Carlson, Temporal and spatial trends in the occurrence of human and veterinary antibiotics in aqueous and river sediment matrices, Environmental Science and Technology 41 (1) (2007) 50−57.

[82] M. Guo, J. Huang, H. Hu, W. Liu, J. Yang, UV inactivation and characteristics after photoreactivation of *Escherichia coli* with plasmid: health safety concern about UV disinfection, Water Research 46 (2012) 4031−4036.

[83] N. Hembach, F. Schmid, J. Alexander, C. Hiller, E.T. Rogall, T. Schwartz, The occurrence of the mcr-1 colistin resistance gene and other clinically relevant antibiotic resistance genes in microbial populations at different municipal wastewater treatment plants in Germany, Frontiers in Microbiology 8 (2017) 1282, https://doi.org/10.3389/fmicb.2017.01282 [PMC free article] [PubMed] [CrossRef] [Google Scholar].

[84] P.M. Da Costa, P. Vaz-Pires, F. Bernardo, Antimicrobial resistance in Enterococcus spp. isolated in inflow, effluent and sludge from municipal sewage water treatment plants, Water Research 40 (2006) 1735−1740, https://doi.org/10.1016/j.watres.2006.02.025 [PubMed] [CrossRef] [Google Scholar].

[85] A. Pruden, R. Pei, H. Storteboom, K.H. Carlson, Antibiotic resistance genes as emerging contaminants: studies in northern Colorado, Environmental Science and Technology 40 (2006) 7445−7450, https://doi.org/10.1021/es060413l [PubMed] [CrossRef] [Google Scholar].

[86] N. Czekalski, T. Berthold, S. Caucci, A. Egli, H. Bürgmann, Increased levels of multiresistant bacteria and resistance genes after wastewater treatment and their dissemination into Lake Geneva, Frontiers in Microbiology 3 (106) (2012) 1−18.

[87] C.W. McKinney, A. Pruden, Ultraviolet disinfection of antibiotic-resistant bacteria and their antibiotic resistance genes in water and wastewater, Environmental Science and Technology 46 (2012) 13393−13400, https://doi.org/10.1021/es303652q [PubMed] [CrossRef] [Google Scholar].

[88] M. Munir, I. Xagoraraki, Levels of antibiotic resistance genes in manure, biosolids, and fertilized soil, Journal of Environmental Quality 40 (2011) 248−255.

[89] J. Hollender, S.G. Zimmermann, S. Koepke, M. Krauss, C.S. McArdell, C. Ort, Elimination of organic micropollutants in a municipal wastewater treatment plant upgraded with a full-scale post-ozonation followed by sand filtration, Environmental Science and Technology 43 (2009) 7862−7869, https://doi.org/10.1021/es9014629 [PubMed] [CrossRef].

[90] S.G. Zimmermann, M. Wittenwiler, J. Hollender, M. Krauss, C. Ort, H. Siegrist, Kinetic assessment and modelling of an ozonation step for full-scale municipal wastewater treatment: micropollutant oxidation, by-product formation and disinfection, Water Research 45 (2011) 605−617, https://doi.org/10.1016/j.watres.2010.07.080 [PubMed] [CrossRef].

[91] M.C. Dodd, Potential impacts of disinfection processes on elimination and deactivation of antibiotic resistance genes during water and wastewater treatment, Journal of Environmental Monitoring 14 (2012) 1754−1771, https://doi.org/10.1039/c2em00006g [PubMed] [CrossRef].

[92] F. Lüddeke, S. Heß, C. Gallert, J. Winter, H. Guede, H. Loeffler, Removal of total and antibiotic-resistant bacteria in advanced wastewater treatment by ozonation in combination with different filtering techniques, Water Research 69 (2015) 243−251, https://doi.org/10.1016/j.watres.2014.11.018 [PubMed].

[93] N. Tran, P. Drogui, J.F. Blais, G. Mercier, Phosphorus removal from spiked municipal wastewater using either electrochemical coagulation or chemical coagulation as tertiary treatment, Separation and Purification Technology 95 (2012) 16−25, https://doi.org/10.1016/j.seppur.2012.04.014 [CrossRef] [Google Scholar].

[94] H.T. El-Zanfaly, Antibiotic resistant bacteria: a factor to be considered in safe drinking water, Journal of Environmental Protection and Sustainable Development 1 (3) (2014) 134−143. View Record in ScopusGoogle Scholar.

[95] C. Jungfer, T. Schwartz, U. Obst, UV-induced dark repair mechanisms in bacteria associated with drinking water, Water Research 41 (2007) 188−196, https://doi.org/10.1016/j.watres.2006.09.001 [PubMed] [CrossRef] [Google Scholar].

[96] Y. Chao, L. Ma, Y. Yang, F. Ju, X.X. Zhang, W.M. Wu, Metagenomic analysis reveals significant changes in microbial compositions and protective functions during drinking water treatment, Scientific Reports 3 (2013) 3550, https://doi.org/10.1038/srep03550 [PMC free article] [PubMed] [CrossRef] [Google Scholar].

[97] J. Xu, Y. Xu, H. Wang, C. Guo, H. Qiu, Y. He, Y. Zhang, X. Li, W. Meng, Occurrence of antibiotics and antibiotic resistance genes in a sewage treatment plant and its effluent-receiving river, Chemosphere 119 (2014) 1379−1385.

[98] S. Zhang, W. Lin, X. Yu, Effects of full-scale advanced water treatment on antibiotic resistance genes in the Yangtze Delta area in China FEMS Microbiol, The Ecologist 92 (5) (2012) 1−9.

[99] U. Böckelmann, H.-H. Dörries, M. Neus, Ayuso-Gabella, M.S. de Marcay, V. Tandoi, C. Levantesi, C. Masciopinto, E. Van Houtte, U. Szewzyk, T. Wintgens, E. Grohmann, Quantitative PCR monitoring of antibiotic resistance genes and bacterial pathogens in three European artificial groundwater recharge systems, Applied and Environmental Microbiology 75 (2009) 154−163.

[100] M.R. Breazeal, J.T. Novak, P.J. Vikesland, A. Pruden, Effect of wastewater colloids on membrane removal of antibiotic resistance genes, Water Research 47 (2013) 130−140.

[101] Tchobanoglous, B. Bernados, K. Bourgeous, K.G. Linden, A. Salveson, F. Soroushian, Ultraviolet Disinfection Guidelines for Drinking Water and Water Reuse, third ed., National Water Research Institute in collaboration with Water Research Foundation, 2012.

[102] Bourrouet, J. Garcia, R. Mujeriego, G. Penuelas, Faecal bacteria and bacteriophage inactivation in a full-scale UV disinfection system used for wastewater reclamation, Water Science and Technology 43 (10) (2001) 187–194.

[103] J.-L. Martínez, R. Canto, Antibiotics and antibiotic resistance in water environments, Current Opinion in Biotechnology Journal 19 (2009) 260–265.

[104] W.A.M. Hijnen, E.F. Beerendonk, G.J. Medema, Inactivation credit of UV radiation for viruses, bacteria and protozoan (oo) cysts in water: a review, Water Research 40 (2006) 3–22.

[105] Y. Zhuang, H. Ren, J. Geng, Y. Zhang, Y. Zhang, L. Ding, Inactivation of antibiotic resistance genes in municipal wastewater by chlorination, ultraviolet, and ozonation disinfection, Environmental Science and Pollution Research International 22 (2015) 7037–7044, https://doi.org/10.1007/s11356-014-3919-z [PubMed] [CrossRef] [Google Scholar].

[106] J. Oh, D. Salcedo, C.A. Medriano, S. Kim, Comparison of different disinfection processes in the effective removal of antibiotic-resistant bacteria and genes, Journal of Environmental Sciences 26 (2014) 1238–1242.

[107] N. Fahrenfeld, Y. Ma, A. O'Brien, PrudenReclaimed water as a reservoir of antibiotic resistance genes: distribution system and irrigation implications, Frontiers in Microbiology 4 (130) (2014) 1–10.

[108] D.M. Aruguete, B. Kim, M.F. Hochella Jr., Y. Ma, Y. Cheng, A. Hoegh, Antimicrobial nanotechnology: its potential for the effective management of microbial drug resistance and implications for research needs in microbial nanotoxicology, Environmental Science: Processes and Impacts 15 (2013) 93–102, https://doi.org/10.1039/c2em30692a [PubMed] [CrossRef] [Google Scholar].

[109] A.R. Shahverdi, A. Fakhimi, H.R. Shahverdi, S. Minaian, Synthesis and effect of silver nanoparticles on the antibacterial activity of different antibiotics against *Staphylococcus aureus* and *Escherichia coli*, Nanomedicine 3 (2007) 168–171, https://doi.org/10.1016/j.nano.2007.02.001 [PubMed] [CrossRef] [Google Scholar].

[110] D.R. Monteiro, L.F. Gorup, A.S. Takamiya, A.C. Ruvollo-Filho, E.R. de Camargo, D.B. Barbosa, The growing importance of materials that prevent microbial adhesion: antimicrobial effect of medical devices containing silver, International Journal of Antimicrobial Agents 34 (2009) 103–110, https://doi.org/10.1016/j.ijantimicag.2009.01.017 [PubMed] [CrossRef] [Google Scholar].

[111] S.J. Lam, N.M. O'Brien-Simpson, N. Pantarat, A. Sulistio, E.H. Wong, Y.Y. Chen, Combating multidrug-resistant gram-negative bacteria with structurally nanoengineered antimicrobial peptide polymers, Nature Microbiology 1 (2016) 16162, https://doi.org/10.1038/nmicrobiol.2016.162 [PubMed] [CrossRef] [Google Scholar].

[112] W. Yu, S. Zhan, Z. Shen, Q. Zhou, D. Yang, Efficient removal mechanism for antibiotic resistance genes from aquatic environments by graphene oxide nanosheet, Chemical Engineering Journal 313 (2016) 836–846, https://doi.org/10.1016/j.cej.2016.10.107 [CrossRef] [Google Scholar].

[113] M.J. Hajipour, K.M. Fromm, A.A. Ashkarran, D. Jimenez de Aberasturi, I.R. de Larramendi, T. Rojo, Antibacterial properties of nanoparticles, Trends in Biotechnology 30 (2012) 499–511, https://doi.org/10.1016/j.tibtech.2012.06.004 [PubMed] [CrossRef] [Google Scholar].

[114] V.K. Sharma, K.M. Siskova, R. Zboril, J.L. Gardea-Torresdey, Organic-coated silver nanoparticles in biological and environmental conditions: fate, stability and toxicity, Advances in Colloid and Interface Science 204 (2014) 15–34, https://doi.org/10.1016/j.cis.2013.12.002 [PubMed] [CrossRef] [Google Scholar].

[115] J.H. Miller, J.T. Novak, W.R. Knocke, K. Young, Y. Hong, P. Vikesland, Effect of silver nanoparticles and antibiotics on antibiotic resistance genes in anaerobic digestion, Water Environment Research 85 (2013) 411–421, https://doi.org/10.2175/106143012X13373575831394 [PubMed] [CrossRef] [Google Scholar].

[116] B. Riquelme, V. Maria, J.T. Novak, P.J. Vikesland, A. Pruden, Effect of wastewater colloids on membrane removal of antibiotic resistance genes, Water Research 47 (2013) 130–140, https://doi.org/10.1016/j.watres.2012.09.044 [PubMed] [CrossRef] [Google Scholar].

[117] H. Li, M. Duan, J. Gu, Y. Zhang, X. Qian, J. Ma, Effects of bamboo charcoal on antibiotic resistance genes during chicken manure composting, Ecotoxicology and Environmental Safety 140 (2017) 1–6, https://doi.org/10.1016/j.ecoenv.2017.01.007 [PubMed] [CrossRef] [Google Scholar].

[118] R. Singh, M.S. Smitha, S.P. Singh, The role of nanotechnology in combating multi-drug resistant bacteria, Journal of Nanoscience and Nanotechnology 14 (2014) 4745–4756, https://doi.org/10.1166/jnn.2014.9527 [PubMed] [CrossRef] [Google Scholar].

[119] V.K. Sharma, N. Johnson, L. Cizmas, T.J. McDonald, H. Kim, A review of the influence of treatment strategies on antibiotic resistant bacteria and antibiotic resistance genes, Chemosphere 150 (2016) 702–714, https://doi.org/10.1016/j.chemosphere.2015.12.084 [PubMed] [CrossRef] [Google Scholar].

[120] M. Pavithra, R. Menezes, J. Suarez, S. Santra, Antimicrobial properties of copper and silver loaded silica nanomaterials, Acta Materialia 46 (2015) 17–26, https://doi.org/10.1002/9781118217511.ch6 [CrossRef] [Google Scholar].

[121] E.N. Taylor, K.M. Kummer, N.G. Durmus, K. Leuba, K.M. Tarquinio, T.J. Webster, Superparamagnetic iron oxide nanoparticles (SPION) for the treatment of antibiotic-resistant biofilms, Small 8 (2012) 3016–3027, https://doi.org/10.1002/smll.201200575 [PubMed] [CrossRef] [Google Scholar].

[122] Z. Qiu, Y. Yu, Z. Chen, M. Jin, D. Yang, Z. Zhao, Nano-alumina promotes the horizontal transfer of multi-resistance genes mediated by plasmids across genera,

Proceedings of the National Academy of Sciences of the United States of America 109 (2012) 4944–4949, https://doi.org/10.1073/pnas.1107254109 [PMC free article] [PubMed] [CrossRef] [Google Scholar].

[123] A. Mohammed Fayaz, M. Girilal, R. Mashihur, R. Venkatesan, P.T. Kalaichelvan, Biosynthesis of silver and gold nanoparticles using thermophilic bacterium Geobacillus stearothermophilus, Process Biochemistry 46 (2011) 1958–1962, https://doi.org/10.1016/j.procbio.2011.07.003 [CrossRef] [Google Scholar].

[124] R.I. Aminov, R.I. Mackie, Evolution and ecology of antibiotic resistance genes, FEMS Microbiology Letters 271 (2007) 147–161.

[125] F. Ju, F. Guo, L. Ye, Y. Xia, T. Zhang, Metagenomic analysis on seasonal microbial variations of activated sludge from a full-scale wastewater treatment plant over 4 years, Environmental Microbiology Reports 6 (2014) 80–89.

[126] F. Baquero, J.L. Martínez, R. Cantón, Antibiotics and antibiotic resistance in water environments, Current Opinion in Biotechnology 19 (2008) 260–265, https://doi.org/10.1016/j.copbio.2008.05.006 [PubMed] [CrossRef] [Google Scholar].

[127] L.Y. He, Y.S. Liu, H.C. Su, J.L. Zhao, S.S. Liu, J. Chen, Dissemination of Antibiotic resistance genes in representative broiler feedlots environments: identification of indicator ARGs and correlations with environmental variables, Environmental Science and Technology 48 (2014) 13120–13129, https://doi.org/10.1021/es5041267 [PubMed] [CrossRef] [Google Scholar].

[128] H.Y. Done, R.U. Halden, Reconnaissance of 47 antibiotics and associated microbial risks in seafood sold in the United States, Journal of Hazardous Materials 282 (2015) 10–17, https://doi.org/10.1016/j.hazmat.2014.08.075 [PMC free article] [PubMed] [CrossRef] [Google Scholar].

[129] Y.G. Zhu, T.A. Johnson, J.Q. Su, M. Qiao, G.X. Guo, R.D. Stedtfeld, Diverse and abundant antibiotic resistance genes in Chinese swine farms, Proceedings of the National Academy of Sciences of the United States of America 110 (2013) 3435–3440, https://doi.org/10.1073/pnas.1222743110 [PMC free article] [PubMed] [CrossRef] [Google Scholar].

[130] H. Fang, Q. Zhang, X. Nie, B. Chen, Y. Xiao, Q. Zhou, Occurrence and elimination of antibiotic resistance genes in a long-term operation integrated surface flow constructed wetland, Chemosphere 173 (2017) 99–106, https://doi.org/10.1016/j.chemosphere.2017.01.027 [PubMed] [CrossRef] [Google Scholar].

[131] J. Chen, G.G. Ying, X.D. Wei, Y.S. Liu, S.S. Liu, L.X. Hu, Removal of antibiotics and antibiotic resistance genes from domestic sewage by constructed wetlands: effect of flow configuration and plant species, The Science of the Total Environment 571 (2016) 974–982, https://doi.org/10.1016/j.scitotenv.2016.07.085 [PubMed] [CrossRef] [Google Scholar].

[132] N. Devarajan, A. Laffite, N.D. Graham, M. Meijer, K. Prabakar, J.I. Mubedi, Accumulation of clinically relevant antibiotic-resistance genes, bacterial load, and metals in freshwater lake sediments in Central Europe, Environmental Science and Technology 49 (2015) 6528–6537, https://doi.org/10.1021/acs.est.5b01031 [PubMed] [CrossRef] [Google Scholar].

[133] W.P.M. Rowe, C. Baker-Austin, D.W. Verner-Jeffreys, J.J. Ryan, C. Micallef, D.J. Maskell, Overexpression of antibiotic resistance genes in hospital effluents over time, Journal of Antimicrobial Chemotherapy 72 (2017) 1617–1623, https://doi.org/10.1093/jac/dkx017 [PMC free article] [PubMed] [CrossRef] [Google Scholar].

[134] J. Du, J. Geng, H. Ren, L. Ding, K. Xu, Y. Zhang, Variation of antibiotic resistance genes in municipal wastewater treatment plant with A2O-MBR system, Environmental Science and Pollution Research 22 (2015) 3715–3726, https://doi.org/10.1007/s11356-014-3552-x [PubMed] [CrossRef] [Google Scholar].

[135] Y.B. Xu, M.Y. Hou, Y.F. Li, L. Huang, J.J. Ruan, L. Zheng, Distribution of tetracycline resistance genes and AmpC β-lactamase genes in representative non-urban sewage plants and correlations with treatment processes and heavy metals, Chemosphere 170 (2017) 274–281, https://doi.org/10.1016/j.chemosphere.2016.12.027 [PubMed] [CrossRef] [Google Scholar].

[136] A. Di Cesare, E.M. Eckert, S. D'Urso, R. Bertoni, D.C. Gillan, R. Wattiez, Co-occurrence of integrase 1, antibiotic and heavy metal resistance genes in municipal wastewater treatment plants, Water Research 94 (2016) 208–214, https://doi.org/10.1016/j.watres.2016.02.049 [PubMed] [CrossRef] [Google Scholar].

[137] L.H. Su, C.H. Chiu, C. Chu, J.T. Ou, Antimicrobial resistance in nontyphoid salmonella serotypes: a global challenge, Clinical Infectious Diseases 39 (2004) 546–551, https://doi.org/10.1086/422726 [PubMed] [CrossRef] [Google Scholar].

CHAPTER 11

Nosocomial Bacterial Infection of Orthopedic Implants and Antibiotic Hydroxyapatite/Silver-Coated Halloysite Nanotube With Improved Structural Integrity as Potential Prophylaxis

JIMMY LOLU OLAJIDE • D.A. DESAI • J.O. AJIBOLA • GBOLAHAN JOSEPH ADEKOYA • OLUYEMI OJO DARAMOLA • K.K. ALANEME • VICTORIA OLUWASEUN FASIKU • EMMANUEL ROTIMI SADIKU

1 INTRODUCTION

Above the quadragenarian age, millions of people across the globe experience musculoskeletal complications (MCs). Moreover, this is highly prevalent in developed countries due to the high rate of road/home accidents, orthopedic trauma, and the ever-increasing population of the geriatric community [1–3]. Although the pervasiveness of MCs increases with age, younger people also experience these complications during their peak income-earning years, where women's peak is between ages 35 and 54 while that for men is between 45 and 64 [4]. According to an online Fact Sheets published by World Health Organization (WHO), MCs were the foremost cause of disability in four of the six WHO regions in 2017 wherein the East Mediterranean Region and the African Region ranked second and third, respectively [4]. In light of this given information, a positive synergistic effect of the consolidated research efforts of biomaterials' researchers, biomedical engineers, orthopedic surgeons, and microbiologists in the design, development, and deployment of reliable orthopedic implants (OIs), to deal with MCs is tremendously anticipated.

OIs are generally designed and manufactured by materials scientists and engineers to replace or support damaged/diseased bones and joints in the body. These OIs are inserted by the orthopedic surgeons in the body to treat congenital and accident-induced deformities, normalize faulty body posture, and restore the normal musculoskeletal functions [1–4]. In the context of orthopedics, materials for OIs must have biostable structural integrity and excellent biocompatibility with the host tissue. The past few decades have seen OIs become an integral part of regenerative medicine and reconstructive surgery [1–5].

As of 2010, as reported by Kremers et al., 4.7 million Americans have undergone total knee arthroplasty (TKA) and 2.5 million have undergone total hip arthroplasty (THA) and are living with OIs (Fig. 11.1) [6]. In Europe, as of 2018, data collected from 24 different European joint registries revealed the total numbers of primary THA and TKA to be over 3.1 million and 2.5 million, respectively [7]. Similar details for Asia, Middle East, and Africa are available in the works of Kim et al. [8], Al-Taiar et al. [9], and George and Ofori-Atta [10], respectively. Regardless of the geographical location of the patients, one common denominator in this development is the progressive increase in these arthroplasties over time [6–10]. In comparison to what was obtainable in the past, this is somewhat a milestone achievement [5,11,12]. A brief insight into this dynamic transformation is considered necessary in this chapter. Thus, the next few paragraphs focus on that.

Antibiotic Materials in Healthcare. https://doi.org/10.1016/B978-0-12-820054-4.00011-2

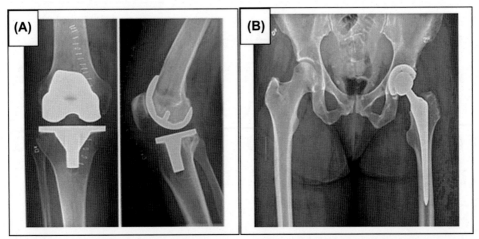

FIG. 11.1 X-ray images showing **(A)** total knee replacement [13] and **(B)** hip bone replacement [14].

Centuries ago, the treatment of bone fractures, bone resorption, osteoarthritis, scoliosis, osteoporosis, damaged tissues, and failing cells constituted a major problem for the orthopedic surgeons and materials scientists. This was as a result of limited materials, financial resources, and technologies available to people then. In consequence, many people who suffered from the previously mentioned orthopedic challenges experienced excruciating pains and high mortality rate [5,11,12].

Interestingly, between the 1860s and 1870s, the introduction of antiseptic surgery by the British surgeon Lord Joseph Lister [15], ether as a surgical anesthetic by the American dentist William T.G. Morton [16], and chloroform as another anesthetic by the Scottish obstetrician James Young Simpson [17] revolutionized the protocols of orthopedic surgery. These revolutionary breakthroughs (although the use of chloroform later halted as a result of its adverse effects on people), coupled with advancement in materials' processing technologies, preeminently encouraged orthopedic surgeons to experiment with artificial materials as implants to treat orthopedic defects in humans and animals [18–21].

The prelisterian materials were wax, decalcified bones, rubber, gold, silver, platinum, and iron [22]. The postlisterian materials comprise a long list. However, cobalt-chromium [23], steels—majorly stainless steels [24], titanium alloys [24,25], and magnesium alloys [26] are on top of the list. Amid all these materials, only titanium alloys have remarkably stood the test of time, especially one of its derivatives by the name of Nitinol (Nickel–Titanium alloy) [27]. Most recently, green biopolymers such as protein, polysaccharides, and green polyesters [22]; synthetic polymers such as ultrahigh molecular weight polyethene and high-density polyethene [22]; and also a class of bioceramics that are bioinert, bioactive, and bioresorbable have made significant inroads into the sphere of orthopedic implant materials (OIMs) [2,3,22,28–30].

It is noteworthy to point out that each class of these materials, that is metals, polymers, and ceramics, has its exclusive advantages and disadvantages when employed as OIM. In consequence, extensive studies have been carried out and are still ongoing on the development of superior composites materials with permissible limitations for OIMs [31–33]. These superior composites are derivable from a systematic combination of different classes of materials [31–33]. In this chapter, the primary focus is on a selected bioactive bioceramic material (BBM). Therefore, the next few subsections cover market evolution, types, properties, orthopedic applications, prevalent drawbacks of BBMs, and suggested solutions.

1.1 Role of Bioceramics in Orthopedics

In 2017, the OIs' marketplace was estimated to be USD 45, 901 million and is hypothesized to reach USD 66,636 million by 2025, registering a compound annual growth rate (CAGR) of 4.7% from 2018 to 2025 [34]. Over the years, the market for OIs has undergone a radical transformation, due to the transition from traditional orthopedic surgeries to the use of state-of-the-art fixations and prostheses. This phenomenon, alongside other factors such as congenital bone defects, workplace injuries, and the rise in the geriatric global population (which increases the susceptibility of people to MCs), is principally responsible for the continuous increase in the demand for OIs [1–4,34–36].

Concerning bioceramic materials, their market should grow at a CAGR of approximately 7% from

2019 to 2024 [37]. Additional factors linked with this projected growth are increasing rate of road accidents in developing countries, more funds for orthopedic research, and the continuous quest to combat chronic diseases such as cancer and arthritis [37]. Nowadays, the availability of many types of bioceramic-based OIs has led to the increase in spinal and ortho-biological surgeries, an increase in dental care, and increasing repair/replacement of diseased/damaged tissues/organs in the body [34,35]. With this development, materials with excellent long-term biocompatibility and service-proven structural integrity in vivo have become attractive candidates for the fabrication of OIs. The BBMs are great examples of such materials [28–30,34–37]. Fig. 11.2 furnishes the schematic visualization of the growth rate of bioceramics' market projected from 2019 to 2024 with North America, Europe, and Asia having the highest projections. In Fig. 11.3, typical bioceramics used as OIs are presented.

The commercial grades of BBMs that are currently available are the calcium-based and silica-based (bioactive glasses) bioceramics. The calcium-based bioceramics include calcium carbonate [$CaCO_3$ in aragonite form], calcium sulfate (Plaster of Paris), [$CaSO_4.2H_2O$], calcium phosphates (hydroxyapatite (HAP), [$Ca_{10}(PO_4)_6$]), beta-tricalcium phosphate (β-TCP), [$Ca_3(PO_4)_2$], and biphasic calcium phosphate [28–30]. The silica-based bioactive glasses include 45S5 Bioglass—NovaBone/Perioglas, S53P4—AbminDent1, and Cerabone [28,39]. For improved performance in most orthopedic applications, these materials are seldom used in their pristine forms; instead, they are either functionalized or combined to form composite materials [40,41]. Intensive research works in the aforementioned area are actively ongoing [41–43].

In this work, considerable emphasis is on HAP, as it is the BBM of interest to the authors of this chapter. Nevertheless, for scholars interested in bioceramics for orthopedics, tissue engineering, and regenerative medicine (OTERM) purposes, the works of Mara and Ruby Celsia [38], Ahmad and Soodeh [28], Zhou et al. [2], Ginebra et al. [44], Garrido et al. [29], and Albulescu et al. [45] contain stimulating and detailed information.

Five decades back in the orthopedic field, BBMs such as bioglass and HAP received minuscule attention of researchers, unlike the bioinert ones such as alumina and zirconia, which do not initiate a reaction or interact when in contact with biological tissues [46]. This paltry interest was as a result of the mundane nature of biomaterial science and biomedical technology back then that was insufficient to cater for the high susceptibility of the hosts to perioperative and postsurgical complications and discomforts associated with the interaction between OIs and the body system [47]. Some of these complications and discomforts include shock, bacterial infection (BI), arthritis, constrained motion, hemorrhage, deep vein thrombosis, and pulmonary embolism [48,49]. Later in this chapter, the BI of OIs will be thoroughly discussed, as it is pivotal to this write-up.

On a brighter note, the topical consolidation of nanotechnology, advanced materials processing technologies [50–53], selective functionalization of

High

Medium

Low

FIG. 11.2 Bioceramics' market growth rate by region, 2019–2024 [37].

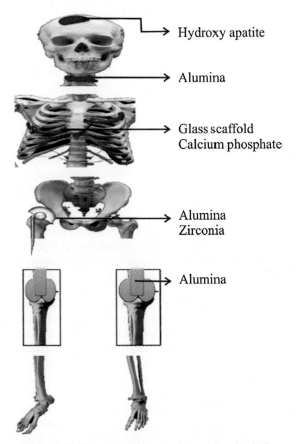

FIG. 11.3 Examples of bioceramics used as OIs [38].

→ Hydroxy apatite

→ Alumina

→ Glass scaffold
Calcium phosphate

→ Alumina
Zirconia

→ Alumina

The outstanding acceptance of BBMs for OTERM purposes lies in their unique biomechanical properties, appreciable biocompatibility, bioactivity, osteoconductivity, hydrophilicity, long-term in vivo mechanical integrity, and bone-like mineral composition [1–3,28–30,59]. These properties allow BBMs to mimic the nature of natural bone in vivo, such as bone self-healing and regeneration. In actuality, this accounts for their continuous and rapid proliferation in the implant materials' market, and by this means, they are quickly replacing the metallic and polymeric ones [1–3,34,35,37,59]. Fig. 11.4 furnishes the fundamental design principle and potential applications of bioceramics (BBMs inclusive) used in OTERM.

Despite the impressive records of the deployment of bioceramics in OTERM, previous studies have also shown that in their pristine forms, they are prone to nosocomial BIs (NBIs) that usually lead to the premature failure of BBM implants [60–65]. In fact, despite the notable accomplishments accrued from new protocols of modern prophylaxis and regimens, the NBI of OIMs remains disturbingly recurrent, particularly in the scope of every so often devastating consequences of septic complications [60–65]. On the one hand, the failure mentioned earlier invariably exposes the patients to postsurgery complications, such as chronic pains, sociofinancial crisis, unanticipated diseases, and even death [60–65]. On the other hand, the ever-increasing use of OIs, especially in patients innately and highly prone to the risk of NBI, has necessitated the clinical significance of infectious complications [60–65]. In light of this, the next few paragraphs will present the readers with an overview of the effects of BI on OIMs/OIs and how to prevent and treat them with antibiotic materials. More importantly, various antibiotic materials have demonstrated the capability to reduce the risk of infectious complications during and after implant surgery [60–63].

1.2 Nosocomial Bacterial Infection of Orthopedic Implants

Since the advent of implant surgery, BI has remained the major complication of OIs. Projections from the actual trend indicate that TKA and THA and their consequent burden are inevitably going to increase tremendously, as time goes on [6–12,60–65].

As for NBI, it is known as one of the major causes of OIMs failures. NBI of OIMs is a very critical issue because of the resultant high morbidity rate and propensity to severe relapses. Preeminently, NBIs are difficult to manage, harm the quality of life of the patient (such as amputation in severe cases), resulting in prolonged admission in hospitals, and constitute unanticipated financial burden [60–65]. The average cost of

biomaterials' surfaces [54–56], in vitro/in vivo characterization of novel biomaterials [57], and substantial financial grants from various health organizations across the globe [58] have tremendously aroused the interests of researchers in BBMs for OTERM applications.

For insight, implanted BBMs can induce bone ingrowth and a linkable interface to support physiological loads in the body [28–30,59]. The observed interaction between BBMs and bone is due to a complex combination of interwoven phenomena that cut across the biology, chemistry, physics, and mechanics of the body. Some of the primary factors that influence these processes are the deformation of organic molecules, the discharge of chemical elements by the BBMs, and the phagocytosis processes of particles released by the implanted materials [28–30,59]. The main factors that determine the in vivo condition of the implants are implantation site, tessutal trauma, structural properties, surface of the material, and disgregation of material at the interface with bone [59].

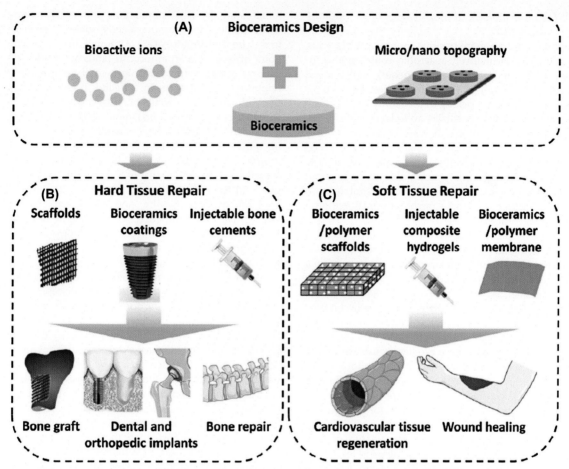

FIG. 11.4 The design protocols and promising applications of bioceramics for OTERM purposes: **(A)** fundamental design principle of bioceramic materials; **(B)** bioceramics in the form of porous scaffolds, injectable bone cements, or composites with metals or polymers for hard tissue regeneration; and **(C)** bioceramics combined with biopolymer materials for soft tissue applications [2].

combined medical and surgical treatment is estimated to be USD 15,000 for an infected implant [65]. Regarding arthroplasties, the financial implication of the associated complication per patient approximately amounts to USD 50,000 and 250 million per year [65]. Taking these problems into account, intensive research to provide the needed solutions is still ongoing, but unfortunately, OIMs are highly prone to microbial colonization and are known to accommodate the inception of NBIs [60–65].

The primary sources of NBIs may include but not limited to the operating room's environment, surgical equipment, apparel worn by medical/paramedical staff, resident bacteria on the patient's skin, and bacteria that are already denizens of the patient's internal body system [66,67].

At this point, it is worthwhile to mention that staphylococci cause the majority of all implant-associated infections [62,68,69]. The two staphylococcal species, *Staphylococcus aureus* (*S. aureus*) and *Staphylococcus epidermidis* (*S. epidermidis*), account for two out of three infection isolates. They represent, in absolute, the main causative pathogenic agents in OTERM [62,68,69]. Besides, researchers have identified them as the leading etiologic agents of OIMs' infection. Staphylococci are responsible for many forms of OIMs' infections. *S. aureus* is a major cause of nosocomial infection of surgical wounds, and *S. epidermidis* causes infections associated with implanted medical devices [62,68–70]. The next few subsections present an overview of the epidemiological significance of both bacteria in OTERM.

1.2.1 Epidemiology of staphylococci in medical implants

Staphylococci are gram-positive (approximately 1 μm in diameter), facultatively anaerobic, commensal components of the mammalian cutaneous, and mucosal microflora that can grow at temperatures between 18 and 40°C [71]. They belong to the family Staphylococcaceae in the order Bacillales and class Bacilli. From microscopic observations, they are nonspore-forming, immotile spherical organisms that divide into several planes forming asymmetrical grape-like clusters. The cell wall peptidoglycan is susceptible to lysostaphin and lysozyme-resistant. The cell envelope contains ribitol wall teichoic acid and a membrane-anchored glycerol lipoteichoic acid [72].

Across the globe, staphylococci are markedly regarded as the etiological agents of several opportunistic human and animal infections with an emphasis on community-associated and nosocomial diseases [62–70]. They have been identified as the foremost causes of many infections of the skin, tissues, and implants. Staphylococci-related infections can range from mild (requiring no treatment) to chronic and potentially terminal infections (if left untreated). Staphylococci have excellent resistance toward dry conditions and high salt concentrations (10% NaCl) and are well suited to their ecological habitat, that is, the skin. Nonetheless, there are claims from microbiologists that they can also survive for long periods in other environments [62–70].

More importantly, the ambidextrous nature of these bacteria has necessitated the exigent demand for a quick and accurate identification of isolates of specific species and strains even to a genotypical level of bacteria. Because, in most cases, the wrong identification of clinical isolates may manifest in lethal upshots in morbidity and mortality [62–70].

On the one hand, coagulase-positive staphylococci, primarily, *S. aureus* and *Staphylococcus intermedius*, are most frequently concerned with pathologic processes. On the other hand, coagulase-negative staphylococci are mostly considered to be fewer common causes of disease in mammals [71,72]. However, their role as nosocomial pathogens in humans is becoming increasingly significant. The next two subsections present a brief overview of the roles of *S. aureus* and *S. epidermidis* as causative agents of NBIs.

1.2.2 S. epidermidis-mediated nosocomial infections of medical implants

S. epidermidis is a gram-positive, catalase-positive, coagulase-negative, immotile, and facultative anaerobe and also a common microbiota of the skin and mucous membranes. Its large population and omnipresence render it one of the most prevalently isolated organisms in the clinical laboratory [73]. Although, from a general point of view, it is an innocuous Janus-faced microorganism, recent claims have deemed it an important opportunistic pathogen. In actuality, it is the one of the leading causative agents of NBIs (mostly associated with foreign body infections), most notably in hospitalized patients with a low level of neutrophils and immunocompromised systems living with intravascular devices and/or OIs [73–75]. Its prevalence is now somewhat equivalent to that of *S. aureus* due to its complex virulence factors, such as the formation of antibiotic-resistant biofilm and fibrinogen binding [76,77]. The group of people who are highly prone to *S. epidermidis-associated* infections includes users of intravenous drugs, catheters, heart valves, shunts, stents, and other artificial appliances [73–77].

Whereas *S. epidermidis*-related infections occasionally manifest into deadly or terminal diseases, their rate of recurrence and the formidable challenge they constitute concerning treatment (antibiotic resistance) represent a demanding task for the public health sector [78,79]. In the United States alone, the financial report of the public health system has revealed that the cost of treating and managing 2 million NBIs associated with *S. epidermidis* is approximately USD 11 billion per annum [78]. In worst-case scenarios, these infections become life-threatening if they find their way into the bloodstream of the host where they can cause nosocomial sepsis, which most often than not results in insufferable financial burden, chronic diseases and high mortality rate [78,79]. Unfortunately, the emergence of the penicillinase-resistant strain of *S. epidermidis* popularly known as methicillin-resistant *S. epidermidis* (MRSE) and its nearly infeasible treatment have demanded immediate and effective therapeutic interventions [80,81]. On a brighter note, research findings have shown that the appropriate administration of antibiotic regimens such as vancomycin (added with rifampin or aminoglycoside), nafcillin, and oxacillin might be efficacious in combating perioperative MRSE infections [80–82].

1.2.3 S. aureus-mediated infections of medical implants

S. aureus is a commensal human microbiota, and at the same time, an opportunistic pathogen with a broad-spectrum of clinical indications, most especially in the context of infections. *S. aureus* infections can be community-acquired or nosocomial. Some of these infections are skin and soft tissue infections (e.g., furuncles, carbuncles, abscesses, boils, cellulitis, folliculitis, impetigo and scalded skin syndrome); pulmonary infections (e.g., pneumonia and empyema); cardiac infections (e.g., endocarditis); and bone and joint infections (BJIs) (e.g., osteomyelitis, septic arthritis, and prosthetic device infections) [82–86]. Invasive infections and/or toxic-mediated diseases caused by *S. aureus* are chiefly associated with the strains involved and the site of infection. *S. aureus* is spread by touching infected blood or body fluids, most often by contaminated hands [84].

To date, *S. aureus* is still one of the most prevalent causes of NBIs and frequently the cause of postsurgical wound infections [82–85]. For example, information retrieved from Wikipedia on *S. aureus* indicates that per annum around 500,000 patients in hospitals of the United States contract a staphylococcal infection, chiefly by *S. aureus* and almost 10% of this population's death is associated with *S. aureus* infections [85]. Unfortunately, treatment of these infections and consequent diseases similarly remain a disturbing challenge to tackle due to the emergence of multidrug-resistant strains such as methicillin-resistant *S. aureus* (MRSA) [82,87]. MRSA causes outbreaks in hospitals that can result in a serious epidemic. Phagocytosis [88], ultraclean laminar airflow, preoperative antibiotic prophylaxis, antibiotic-loaded bone cement, and antimicrobial coating are the most important mechanisms for combatting staphylococcal infections. However, biofilm growth on implants is resistant to phagocytosis [89]. The antibiotic-resistant strains often cause *S. aureus*-related NBIs and only vancomycin can treat these infections [90,91]. Fig. 11.5 presents a clinical case of a patient with bacteria-infected TKA being treated with antibiotic therapy and surgery.

1.2.4 Formation of bacteria biofilms, their mechanisms of antibiotic resistance and implant-associated failures

For a very long time, bacteria biofilm-associated infections (BBAIs) have remained amid the leading public health problems and concerns as they severely affect the health of humans and animals living with and without medical implants/devices. According to data provided by the National Institutes of Health, 65% of microbial infections and 80% of chronic infections are related to biofilm formation [92,93]. This occurrence is preeminently due to their strong resistance to several antibiotics and antibiotic prophylaxis/therapies, which in turn renders BBAIs recrudescent with the possibility of higher severity than before a remission, however, thanks to the unyielding resilience and relentless efforts of researchers who are constantly making pivotal moves to break this barrier

[92–96]. Interestingly, remarkable research progress in this context has immensely facilitated the understanding of bacterial biofilms, most importantly, polymicrobial communities, which are predominantly linked with chronic infections. For instance, community genomics has revealed that a polymicrobial community exhibits passive resistance, metabolic cooperation, quorum-sensing network architectures, and high genetic diversity, among others [92]. Before delving further into this subject, it would be worthwhile to present newbies in this field with the definition of bacterial biofilm.

A biofilm is a complex three-dimensional network of the immobile microbiome with different colonies (genetically alike) of bacteria/fungi or mono cells in a group embedded within a slimy matrix of extracellular polymeric substance (EPS) that it produces to protect itself from external stress such as bactericidal agents [92–96]. Usually, the thickness of the EPS matrix is 0.2–1.0 mm; however, the size of the biofilm does not exceed 10–30 nm. Typically, 5%–35% of the biofilm volume is constituted by the microorganisms, while the remaining volume is extracellular matrix. The main components of the EPS matrix are polysaccharides, proteins, extracellular DNA, RNA, lipids, water, and ions (bound and free). Typically, a biofilm comprises any syntrophic consortium of microbes in which cells permanently adhere to each other and, in most cases, to living and inanimate surfaces alike. Studies have shown that the rate and extent of this irreversible adherence are dependent on the characteristics of both the substratum and the cell surface [94].

Bacterial biofilms are typically opportunistic and pathogenic and are the major causes of NBIs of OIMs, which can primarily be treated by their removal due to the weak activity of antibiotics in eradicating BBAIs and the resistance of bacterial biofilm to host immune response (HIR) [95]. In light of this, this study considers it very paramount to give an overview of the steps involved in the formation of bacterial biofilms and their mechanisms of antibiotic resistance.

1.2.4.1 Steps involved in the formation of bacteria biofilm.
The formation of biofilm is a survival mechanism for microbes in antibiotic environments as this is their means of coexisting with their host, that is, their living habitat [95]. The process of biofilm formation consists of five basic steps that are brought about by cell-to-cell communication among different colonies of microbial species, also known as quorum sensing [96]. The five steps are given as follows [92–100]:
1. Migration and colonization of planktonic bacteria to the surface of a material by four forces, namely

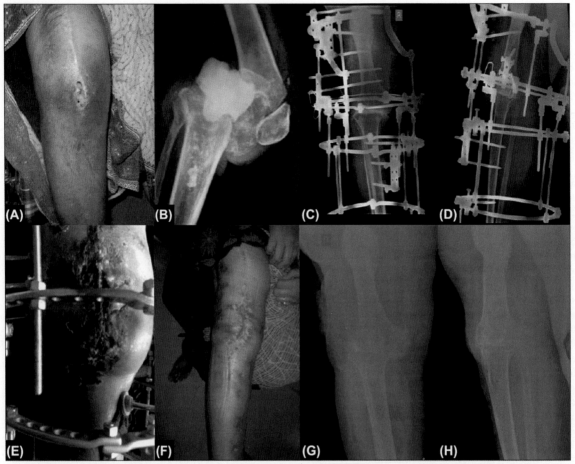

FIG. 11.5 **(A)** clinical picture of a patient with infected TKA of the right side with a discharging sinus over the anterior aspect of the knee, **(B)** X-ray after stage I procedure where an antibiotic cement spacer was placed after thorough debridement. **(C** and **D)** X-ray (anteroposterior and lateral views) of the same patient after stage II surgery. **(E** and **F)** Clinical picture showing marginal wound necrosis after stage II surgery, which successfully healed with the secondary intention on conservative management. **(G** and **H)** X-ray of the right knee (anteroposterior and lateral views) at follow up of 5 years showing solid consolidation [64].

acid–base, hydrophobic, Lifshitz–Van der Waals, and electrostatic interaction forces. This colony is mediated by quorum sensing.

2. Production of the EPS that gives strength to the biofilm and also responsible for shielding the microbes against antimicrobial agents and the HIR by impaired activation of phagocytes and complement system. It equally aids the trapping of nutrients for their survival.

3. Maturation of the EPS matrix for a spontaneous increase of microbes and the formation of channels for nutrient and waste.

4. Maximum densification of bacterial cells with the establishment of three-dimensional, multicellular, and multilayered architecture.

5. Detachment of microcolonies of microbes and the subsequent migration to a new surface, thereby spreading the infections.

For instance, several factors with interrelated functional characteristics mediate the attachment and colonization of *S. epidermidis* biofilm [98]. During the accumulative phase of biofilm formation in *S. epidermidis*, the production of active molecules such as polysaccharide intercellular adhesin (PIA), accumulation associated protein, and the extracellular matrix binding protein function by inducing cell aggregation that results in the assembly of biofilm consortium [95–98]. The appreciation of biofilm formation is very instrumental in thoroughly understanding the virulent nature of *S. aureus* and *S. epidermidis* and their consequent infections of OIMs. More importantly, this will help medical practitioners to make clinical decision to treat these infections. Fig. 11.6 shows a schematic illustration of biofilm formation and a vista of opportunity for effective eradication of bacteria.

1.2.4.2 Virulence of *S. epidermidis* biofilm in OTERM.

S. epidermidis is very virulent due to its ability to form biofilms that are preeminently recalcitrant toward antibiotic therapies. One probable reason for this could be its surface proteins, which readily bind blood and extracellular matrix proteins. This phenomenon produces PIA, which comprises sulfated polysaccharides that allow other bacteria to efficiently bind to the already existing biofilm, thereby creating a multilayer biofilm. Such biofilms are known to reduce the metabolism of bacteria within them. This decreased metabolism, in conjunction with hindered penetration of antibiotics, is responsible for the difficulty of antibiotics to clear *S. epidermidis*-related infections effectively. Correspondingly, most *S. epidermidis* strains are often resistant to several antibiotics such as rifampin, fluoroquinolones, gentamicin, tetracycline, clindamycin, and sulfonamides, among others. MRSE is prevalent, with 75%–90% of hospital isolates resistant to methicillin [96–98].

By its virulence factor, *S. epidermidis* readily colonizes plastic devices and causes biofilms to grow on them. Hence, plastic-based OIMs used intravenously are highly susceptible to *S. epidermidis*-related infections. Such infections include but not limited to endocarditis, necrosis, and sepsis [99]. *S. epidermidis* may as well be involved in prosthetic joint, vascular graft, surgical site, and central nervous system shunt infections. These infections often begin with the introduction of bacteria from the hospital environment or body of the patient during device/implant insertion. In one work, *S. epidermidis* accounted for at least 22% of bloodstream infections in the intensive care unit patients in the United States. Second, only to *S. aureus*, *S. epidermidis* is responsible for approximately 13% of prosthetic valve endocarditis infections, with 38% intracardiac abscesses and 24% mortality [81,100].

1.2.4.3 Virulence of *S. aureus* biofilm in OTERM.

S. aureus biofilm (SBF) is usually found in vivo on implanted medical devices and human tissues. It is commonly found with the opportunistic pathogenic yeast, *Candida albicans*, thereby forming multispecies biofilms. The yeast is presumed to assist *S. aureus* to penetrate human tissue [83,84]. SBF is the predominant cause of implant-related infections with a very high mortality rate linked to its ability to form multispecies biofilms [83–85]. After implantation, the surfaces of these devices/implants become covered with host proteins, which in turn provide a rich surface for an irreversible bacterial attachment and biofilm formation [83–87]. Once the device/implant becomes infected, the only and most effective treatment known to date is to remove it completely because SBFs are very difficult, if not impossible, to destroy regardless of the strength of any current antibiotic treatments whatsoever. The ability of all bacteria to permanently attach to surfaces and form protective biofilm worsens the situation. The difficulty is very complicated because bacteria in biofilms are difficult to detect in routine diagnostics

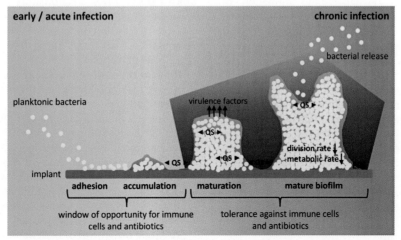

FIG. 11.6 Biofilm formation and window of opportunity for effective clearance of bacteria [100].

and are inherently intractable to HIR and antimicrobial therapy [83–87,97].

SBFs have demonstrated very high resistance to HIR. Although the actual mechanism of resistance is unknown, *S. aureus* biofilms have increased growth under the presence of cytokines produced by the HIR. Host antibodies are not that effective against *SBFs* because of nonhomogenous distribution of antigen, which may make antigen present in some areas of the biofilm, but absent from other areas [98].

Researchers have tried to explain the underlying mysteries surrounding the indistinct mechanism of antibiotic resistance of *SBF*. Two assumptions to this effect are as follows:

1. The SBF's matrix shields the embedded cells by acting as a barrier to prevent antibiotic penetration. Nevertheless, the biofilm matrix is composed of about 97% water channels; hence, the premise of this hypothesis is highly questionable. On a brighter note, SBF's matrix possibly contains antibiotic-degrading enzymes such as β-lactamases, which are capable of preventing antibiotic penetration.
2. The conditions in SBF's matrix favor the formation of persister cells. Persister cells are dormant cells that form instinctively within a biofilm and are highly recalcitrant to antibiotics.

1.2.4.4 Mechanisms of antibiotic resistance of bacterial biofilm.

Several mechanisms of resistance adopted by biofilm against external stress responses include the presence of Capsules or Glycocalyx in the biofilm, the transformation of enzyme-mediated bactericide to a nontoxic form, heterogeneity in metabolism and growth rate among the participating bacteria, quorum sensing, genetic adaptation by plasmids/transposons exchange, and formation of persister cells [4,6,7]. In line with acquired resistance, studies have indicated that the roles of plasmids and transposons can become very active in biofilms under several conditions. Plasmids or transposons are nonchromosomal circles of DNA that predominantly encode antibiotic resistance genes to specific bacteria. This plethora of diverse mechanisms is characterized by a distinct complex profile of biological functions and varies across different bacterial colonies. In summary, there are at least three reasons that have been identified for the intrinsic antimicrobial resistance of biofilms [98–100]. The reasons are given in a chronological order:

1. Antimicrobial agents must penetrate the EPS matrix to contact and deactivate the pathogens embedded within the biofilm.

2. Biofilm-associated pathogens have reduced growth rates, which, in turn, lessens the intake rate of antimicrobial agents by the cell and, therefore, distressing inactivation kinetics.
3. The immediate environment surrounding the cells within a biofilm is capable of providing protective conditions for the pathogen.

Fig. 11.7 shows the potential mechanisms of antibiotic resistance in biofilms' communities.

1.2.4.5 Implant-associated failures due to antibiotic resistance of bacteria biofilm.

The intravenous application of implants is one of the most significant risk factors in their NBIs. NBIs are responsible for approximately 45% of implant-related failures. These infections of implants are almost insusceptible to HIR and antibiotic therapy, and hence they continue to affect the host until removal of the implant (removal is the most effective therapy to date) [76]. However, the prominent downsides to this therapy may include perioperative and postoperative complications such as nerve and tissue damage, vessel injury, fracture, mental illness, and chronic bacterial infections [76–79].

In recent times, mounting evidence now points to the role of bacterial biofilms in orthopedic device-related infection (ODRI) as the primary cause of implant failure. Bacterial infection in prosthetic implants is caused by free-floating bacteria that change when they come into contact with the surface of an implant. These bacteria form an EPS matrix, which eventually binds firmly to the underlying surface. Over time, the biofilm reaches critical mass (maximum population of bacteria) on the contaminated implant, which triggers a host inflammatory reaction and can lead to the total failure of the implant. More importantly, bacteria within biofilms are extremely less prone to antibiotics, HIR, and antiseptics rendering them impervious to treatment [98].

The formation of biofilms on implants' surfaces presents three major problems, and they are the following:

- First, bacterial communities on these surfaces represent a nidus of bacteria that can be transferred into the body, leading to a chronic infection.
- Second, biofilm bacteria are highly insusceptible to the treatment of antibiotics; therefore, once these polymicrobial communities form, they are impervious to eliminate with traditional antimicrobial therapies.
- Finally, HIR and antimicrobial therapies often inactivate the growth of bacteria in a biofilm, which

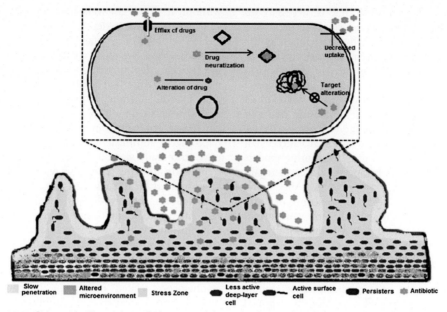

FIG. 11.7 Schematic illustration of the potential mechanisms of antibiotic resistance in the biofilms' communities [95].

activates the HIR and stimulates the generation of inflammatory mediators, which enhances bacterial activity and toxins at the site of implants [93].

If bacterial adhesion occurs before tissue regeneration takes place, HIR often cannot prevent surface colonization for specific bacterial species such as staphylococci (*S. aureus* and *S. epidermidis* in particular) that are capable of forming a protective biofilm layer in hostile environments. Therefore, inhibiting bacterial adhesion is essential to prevent implant-associated infection, as these biofilms are extremely recalcitrant to HIR, antibiotics, and many other biofilm eradication protocols [62].

1.2.5 Mini review of antibiotic therapy of orthopedic implants' infections

The recalcitrant nature of bacterial biofilms can be nightmares for the patients, the healthcare personnel, and the researchers because the infected patient continues to be in pain (in case of chronic infections), keeps spending money on management/treatment, and remains a regular customer of the hospital. In addition, the efforts of both the health personnel and researchers are rendered futile. This occurrence is extremely discouraging and makes one wonder if there is ever going to be a therapeutic protocol with 100% biofilm destruction/inactivation with highly negligible or no side effects at all. Therefore, in this chapter, the

authors considered it paramount to present a brief account of what has been done so far on the treatment of orthopedic infections with popular antibiotic regimens and plant extracts.

1.2.5.1 Rifampin. Rifampin ($C_{43}H_{58}N_4O_{12}$), also known as rifampicin, is an example of an antibiotic that has demonstrated strong antibiotic potency against staphylococci. However, to circumvent resistance to its potency, studies have evinced that rifampin cannot be used as a standalone antibiotic but must be invariably combined with another antimicrobial agent. With this latest development, orthopedic implant-associated infections are likely to be eradicated in up to 80% −90% of patients [62,101]. Leijtens et al. studied the synergistic potency of clindamycin-rifampin combination against staphylococcal prosthetic joint infections (PJIs). They reported a success rate of 86%, with five recurrent/persistent PJIs in 36 treated patients. The cure rate was 78% (14/18) in patients that underwent debridement and retention of the implant (DAIR patients) and 94% (17/18) in the revision group. Five patients (14%) stopped the clindamycin-rifampin regimen because of adverse effects. Of the 31 patients taking the clindamycin-rifampin regimen, 29 patients (94%) were cured. Although they concluded that this protocol is safe, well-tolerated, and effective, they also pointed out that a randomized controlled trial is necessary to establish its full potency against staphylococcal PJIs [102]. Shiels et al. investigated the effectiveness of the local application of rifampin powder, rifampin-vancomycin, and rifampin-daptomycin regimens in treating matured biofilm S. aureus-induced musculoskeletal trauma in vitro and in vivo in 63 mouse models. They found out that using rifampin powder as a standalone antibiotic was more effective in vivo than when combined with either vancomycin or daptomycin. Furthermore, only the murine models that underwent the monotherapy were able to regain the presurgical weight. Nonetheless, the poly-therapeutic effects of rifampin-vancomycin and rifampin-daptomycin supersede that of vancomycin and daptomycin when used mono-therapeutically. On the other hand, the polytherapy exhibited additional antibiofilm resistance in vitro. Despite documenting >75% success rate of antibiofilm resistance of rifampin powder in the mouse models, these authors proposed that the influence of locally applied rifampin powder on the local tissue environment and fracture healing needs an evaluation to establish its efficacy on S. aureus biofilm-induced musculoskeletal infections [103]. Trombetta et al. demonstrated the outstanding therapeutic regimen of a local drug delivery scaffold made of poly(lactic-co-glycolic acid)-coated calcium phosphate loaded with rifampin-sitafloxacin (PLGA-CP-RS) against methicillin-susceptible S. aureus (MSSA). The therapy was administered in combination with a systemic vancomycin treatment to treat S. aureus osteomyelitis in a preclinical murine model. This same procedure was repeated for gentamycin-loaded polymethyl methacrylate/clinical control sample, rifampin-loaded PLGA-CP, and sitafloxacin-loaded PLGA-CP. Of all the samples, the clinical control sample showed the least efficacy after a 3-week and a 10-week inspection, respectively. In addition, the sitafloxacin-loaded PLGA-CP showed significant inactivation of persister cells. Despite these impressive outcomes, the authors encouraged similar research for MRSA-induced orthopedic infections [104]. A long-term (1−2 years), postoperative, follow-up investigation carried out by Jacobs et al. on the therapeutic success rate of rifampin-clindamycin and rifampin-teicoplanin regimens administered to 60 patients with PJIs due to *Propionibacterium* ended with an impressive outcome. They reported an overall success rate of 93% at 1-year follow-up and 86% at 2-year follow-up. However, they found similar results for patients treated with rifampicin combination therapy and those treated

without it. They as well have encouraged further studies for a more in-depth insight into this outcome [105]. To overcome rifampin resistance by MRSA biofilm, John et al. combined rifampin with levofloxacin, daptomycin, vancomycin, and linezolid, respectively. From their findings, rifampin-daptomycin combination therapy displayed the best activity against MRSA with a 67% cure rate and prevented the emergence of rifampin resistance. The dose of daptomycin administered to their specimens (guinea pigs) once-daily was 30 mg/kg, which corresponds to 6 mg/kg in humans. They also reported that the cure rate observed for this regimen was comparable to that observed for the levofloxacin-rifampin combination but higher than the one observed for rifampin-vancomycin combination, which failed to prevent the emergence of rifampin resistance [106]. More detailed information regarding the role of rifampin monotherapy and combination therapy in treating staphylococci-related orthopedic infections in vivo and in vitro is available in a bibliographical review written by Zimmerli and Sendi [107]. One core information from this review is that the best antibiotic agents for combination therapy for rifampin are fluoroquinolones [107].

1.2.5.2 Daptomycin. Daptomycin ($C_{72}H_{101}N_{17}O_{26}$) is a naturally occurring lipopeptide antibiotic that kills susceptible gram-positive bacteria by disrupting their membrane potential [108]. Some researchers have demonstrated its bactericidal potency against staphylococci-related infections of orthopedic implants. Rosslenbroich et al. studied the bactericidal efficacy of polymethyl methacrylate (PMMA)-daptomycin monotherapy and combination therapy (with rifampin and gentamicin) against infection caused by MRSA in human osteoblasts model. The study was conducted in vitro for 20 and 40 h, respectively. Their result indicated no significant difference in the bactericidal activities of daptomycin and gentamicin and their combination therapy. However, both rifampin monotherapy and combination therapy evinced outstanding performances. Even though this is a positive outcome, studies have shown that monotherapeutic administration of rifampin in vivo is liable to rapid resistance induction due to mutations in the *rboB* gene of MRSA. In addition, admixing rifampin alone with PMMA adversely alters its curing process. Owing to this, Rosslenbroich et al. suggested that in place of PMMA/rifampin-based regimen, PMMA/daptomycin-based regimen deficient of the abovementioned shortcomings might be a better alternative to treat osteomyelitis and implant-associated infection in trauma and orthopedic surgery associated with multi-resistant strains. They encouraged further study in vivo to assess the realistic efficacy of their proposed regimen [109]. In a recent cohort study, Sawada et al. compared the effectiveness and associated adverse reactions of administering daptomycin and linezolid therapies for the treatment of gram-positive pathogens-associated PJIs in 82 patients with and without implant removal. According to their findings, there was no marked difference between infection control rates for both therapies. The overall infection control rates documented against gram-positive pathogens in linezolid-treated patients and daptomycin-treated patients were 79% and 77%, respectively. In addition, infection control rates were 94% and 58% in linezolid-treated patients and 75% and 80% in daptomycin-treated patients, without and with implant removal, respectively. Their Fisher's exact analysis result revealed higher clinical success rates and lower adverse event rates in patients treated with daptomycin, including higher red blood cell and platelet counts and lower C-reactive protein levels in comparison with patients treated with linezolid. Side effects observed for patients treated with linezolid include pancytopenia (8 cases) and loss of appetite or nausea (4 cases) while for daptomycin-treated patients, drug eruptions, decreased renal function, and anaphylactic shock occurred in one case each [110]. Some of the limitations that might affect the credibility of this study, as pointed out by the authors are small sample size, lack of randomized control trial, excessive administration of antibiotics due to initiation of antibiotic regimen before identification of the specific causative pathogen, and unexplored combination effect of rifampin [110]. In a previous study, Corona Pérez-Cardona et al. demonstrated the efficacy of using daptomycin therapy (and in some cases with rifampin combination) in treating 20 patients suffering from knee and hip PJIs predominantly caused by staphylococci. Ultimately, they reported a success rate of 78.6% after a median follow-up of 20 months. However, some associated severe side effects include acute renal failure (attributed to rhabdomyolysis and one of eosinophilic pneumonia) and asymptomatic transient creatine phosphokinase (CPK) level elevation. Despite the overall success rate, they expounded that the viability of using a daptomycin regimen in conjunction with an adequate surgical approach to treat stubborn gram-positive PJIs is dependent on close monitoring of

serum CPK level as an unwanted level might result in potentially terrible outcomes [111]. In another cohort study by Malizos et al., on the administration of daptomycin regimen to 6075 patients suffering from osteomyelitis or ODRIs primarily caused by MRSA, they reported a clinical success rate of 82.7% and 81.7% in *S. aureus* and *S. aureus* coagulase-negative staphylococcal infections, respectively. In addition, they found out that the regimen resulted in adverse effects in 6.7% of the patients and serious adverse effects in 1.9% of the patients. In consequence, they discontinued the regimen in 5.5% of the patients (adverse effects-related) and 1.6% of the patients (death-related). Nonetheless, they concluded that this regimen is effective and safe [112]. According to Luengo et al. the polytherapeutic efficacy of daptomycin-fosfomycin regimen in treating an early postsurgical infection of a totally replaced femur in a 79-year old woman caused by *S. epidermidis* was established. This regimen was very successful as the treated patient did not have to remove the infected orthopedic device, and more importantly, she was able to walk with no associated pains whatsoever. Based on these findings, they proposed fosfomycin as a suitable alternative to be combined with daptomycin in place of rifampin (especially when unavailable) for the treatment of infected total femoral replacements and in particular, those that are difficult to treat [113]. In addition, Niska et al. have exposed the indifference between the high dose of daptomycin, vancomycin, and tigecycline prophylaxes in treating surgical site infections caused by MSSA or MRSA in murine models. On the other hand, they were able to demonstrate that low doses of daptomycin and tigecycline are more effective regimens than low vancomycin regimen against the implant infection. They encouraged further investigations in humans in vivo for a quantitative efficacy of these regimens [114]. More detailed information on the efficacy of daptomycin antimicrobial therapy against BJIs and PJIs is available in the systematic review of Telles et al. The major conclusion drawn from this review is that the daptomycin regimen compares relatively well with other evaluated therapies for the treatment of BJIs [115].

1.2.5.3 Gentamicin. Gentamicin ($C_{21}H_{43}N_5O_7$) is an antibiotic active against a wide range of bacterial infections, mostly gram-negative and gram-positive Staphylococcus. It is used in the treatment of blood, bone, and soft tissue infections [116,117]. Recently,

Thompson et al. investigated the bactericidal efficacy of titanium aluminum niobium (TAN) discs coated with clinically approved CP infused with gentamicin (10 wt.%). The coating applied was via intraoperative loading. There results in vitro on the adsorption and release kinetics of gentamicin and in vivo on the bactericidal efficacy of the coated TAN discs proved significantly successful. They reported an 87.5% cure rate in eight infected murine models and proposed their regimen for preventing perioperative orthopedic device-related *S. aureus* infections in vivo [118]. However, they also reported their choice of a 7-day endpoint of study might influence this success rate as *S. aureus* can colonize both murine and human submicron canaliculi (an indication that noninfected screws might get infected after 7 days) [118]. A similar study by Lucke et al. revealed the same efficacy of locally applied poly(D, L-lactide) PDLLA coating infused with gentamicin (10 wt.%) against *S. aureus*-related orthopedic implant-related infection (SAROII) in a rat model [119]. Vesta et al. have also demonstrated the impressive drug release kinetics and antibiotic efficacy of PDDLA-gentamicin (10%) coating on Kirschner wires implanted in murine models with SAROII. More importantly, they found out that the coating process did not have any adverse effect on the drug release kinetics and antibacterial efficacy of the coating in vitro and in vivo [120]. As for Yang et al., they proposed an antibiotic therapy against SAROII based on gentamicin-loaded titanium nanotubes, which proved very efficacious in a murine model. They arrived at this conclusion after comparisons with monolithic titanium nanorod and titanium nanotube without gentamicin addition. Ultimately, they asserted this prophylaxis as an efficient drug-delivery approach to combat the NBIs of OIs [121]. Similarly, Zyaie and coauthors studied the efficacy of local application of gentamicin in combination with a systemic antibiotic regimen in 2 groups of patients (60 in each group) going through open reduction and internal fixation of a single long bone fracture. They evaluated the efficacy of this approach based on the rates of superficial and deep infection rates observed in the patients. They found only 3 cases of superficial infections in the group locally administered gentamicin while up to 12 and 2 superficial and deep infections were observed in the control group, respectively. In addition, 3 of the 12 superficial cases progressed to deep infections. Although the outcome of their experiment evinced that local application of gentamicin in combination with a systemic antibiotic regimen can significantly reduce the rate of infection, the noncompliance of some of the patients disallowed the generalization of this prophylaxis for surgical infections [122]. In a fascinating study, Diefenbeck et al. successfully prepared a bioactive titanium surface with excellent osseointegration and bactericidal properties in vivo as a result of gentamicin-sodium dodecyl sulfate and gentamicin-tannic acid coatings. These prophylactic regimens were able to suppress bacterial growth, prevent biofilm formation, prevent implant contamination and enhanced osseointegration, and also implant fixation. Regarding implant contamination, they reported a 100% success rate for the gentamicin-sodium dodecyl sulfate coating and a 90% success rate for the gentamicin-tannic acid coating, respectively. However, they encouraged further investigations on the influence of sodium dodecyl sulfate and tannic acid on the bactericidal potency of the coatings. They tested their prophylaxis in murine models infected with *S. aureus* chronic osteomyelitis [123]. Similarly, Liu et al. have demonstrated the excellent antibiotic behavior and biocompatibility of a novel implant made of nanotubular anodized titanium coated with gentamicin. Their findings indicated improved prophylactic properties in a rabbit model in vivo, wherein the prevention of implant-related osteomyelitis (caused by *S. aureus*) and enhanced bone biocompatibility occurred significantly. More importantly, this novel implant outperformed gentamicin-coated titanium, monolithic nanotubular anodized titanium, and monolithic titanium in terms of cytocompatibility and antibacterial efficacy [124]. More related information concerning the use of gentamicin-based regimens in the treatment of orthopedic-related infections is available in a review authored by Knaepler [125].

1.2.5.4 The fluoroquinolones. Fluoroquinolones are clinically approved antibiotics and are highly effective against certain bacterial strains. The fluoroquinolone antibiotics include levofloxacin, ciprofloxacin, moxifloxacin, gemifloxacin, and ofloxacin [126]. Some of the pharmacokinetic properties of this class of antibiotics are high oral bioavailability, a large volume of distribution, and broad-spectrum antimicrobial activity, among others. Unfortunately, the extensive use of fluoroquinolones over the years has manifested in resistance by some bacterial strains. Furthermore, fluoroquinolones are associated with the risk of serious adverse effects (*such as* tendinopathy and neuropathy) and have multiple drug–drug interactions. Hence, the administration of fluoroquinolone regimen is typically reserved for cases wherein the benefits outweigh the risks [127]. In addition, this may be responsible for the little interest of researchers in their monotherapies for the prevention and treatment of orthopedic-related infections. Nevertheless, the next few paragraphs give a concise overview of fluoroquinolone-related regimens investigated for the treatment of orthopedic infections.

1.2.5.5 Levofloxacin. Levofloxacin ($C_{18}H_{20}FN_3O_4$) is an antibiotic that helps to fight bacteria in the body. It is used to treat different types of bacterial infections (e.g., respiratory infections, urinary infections, and BJIs, among others) and also people that have been exposed to anthrax or certain types of plague [128]. The administration of levofloxacin can be orally, intravenously, or in eye drop form [128,129]. In line with the use of orally administered levofloxacin regimen for the treatment of BJIs, Asseray et al. conducted a retrospective cohort study involving 230 patients with BJIs (predominantly caused by MSSA) to assess the benefit−risk ratio of this regimen in terms of optimum dosage. They found out that oral administration of levofloxacin at 500 mg once daily has low toxicity risk, well-tolerated by the patients, and potent for the treatment of BJIs. The posttreatment results were impressive with total or partial recovery (including orthopedics aftermath) observed in 89% −93% of the patients. However, the adverse effects observed in some of the patients were insomnia, rash−hepatitis−pancreatitis−neutropenia−myalgia, and rash−hepatitis−acute renal failure [130]. In addition, Muller-Serieys et al. have demonstrated the modified bactericidal behavior of levofloxacin-rifampin combination therapy administered to rabbits with *S. aureus*-induced prosthetic knee infections. Although the bactericidal activity of the combination therapy did not necessarily increase, the observed resistance to rifampin by *S. aureus* in rifampin monotherapy was eradicated [131].

1.2.5.6 Ciprofloxacin. Ciprofloxacin ($C_{17}H_{18}FN_3O_3$) is an antibiotic used to treat several severe bacterial infections such as BJIs, intraabdominal infections, respiratory tract infections, skin infections, and urinary tract infections, among others [132,133]. Some researchers were able to successfully use ciprofloxacin in combination therapy with ceftazidime to cure 9 of 9 patients (100% cure rate) with *Pseudomonas. aeruginosa* (*P. aeruginosa*)-infected osteosynthetic implant and 4 of 5 patients (80% cure rate) with hip and knee

prostheses without removing the implant. Although observed side effects include arthralgia in one patient and rash in another patient, this regimen is considered efficacious and safe for the treatment of *P. aeruginosa*-infected orthopedic implants [134]. In a previous study, Berdal et al. orally administered ciprofloxacin-rifampin combination therapy to 29 patients with early PJIs caused by *S. aureus*, *S. epidermidis*, and *Corynebacterium JK*. With this therapy, they were able to achieve an 83% cure rate without removing the prostheses. Due to the low failure rates associated with this regimen, they ultimately considered it as an effective and safe approach for the management of early manifestation of implant infections [135]. Castro et al. have also demonstrated the potential antibiotic behavior of an orthopedic implant based on polylactic acid (PLA)/phosphates/ciprofloxacin formulations against *S. aureus* in rabbit models. After they found out that the ciprofloxacin concentrations throughout the rabbits' infected femurs were above the minimum inhibitory concentrations of the significant causative bacteria for approximately 4 weeks, and that one of the implant material (PLA/phosphate ratio = 20:80 and ciprofloxacin = 10%) is well tolerated and rapidly induces new bone formation. They suggest that this particular implant could be potentially efficient for treating bone infection. The significant causative bacteria in their study were *S. aureus*, *S. epidermidis*, *P. aeruginosa*, *Proteus mirabilis* (*P. mirabilis*), and *Escherichia coli* (*E. coli*) [136].

1.2.5.7 Moxifloxacin. Moxifloxacin ($C_{21}H_{24}FN_3O_4$) is an antibiotic used to treat many bacterial infections such as pneumonia, conjunctivitis, endocarditis, tuberculosis, sinusitis, and bone infections, among others. Its methods of administration can be oral, ocular, or intravenous [137,138]. Just like other fluoroquinolones, its monotherapy for the treatment of orthopedic infections is not widespread. However, researchers who believe there could be significant contributions to knowledge in the exploration of this regimen have attempted to do so. Suan Juan et al. evaluated the safety and clinical efficacy of moxifloxacin monotherapy in 48 patients with orthopedic implant-related staphylococcal infections. Twenty patients had joint prosthesis infections, and 28 patients had osteosynthesis material infections. The etiologies were MSSA in 33 patients and coagulase-negative staphylococcus (CoNS) in 15 patients. Surgical management was performed for most patients (37/48; 77%), and 21 patients have their implants retained (43.8%). From their results, they observed complete drug compliance in all; however, two patients (4.2%) discontinued the usage of the drug due to adverse events (diarrhea and dizziness). No moxifloxacin-induced arrhythmia occurred. The global cure rate was 38/46 (82.6%), and the cure rate for patients with implant retention was 15/21 (71.4%). The global cure rate for the 32 patients with a minimum follow-up of 2 years was 80%. Out of 8 cases of relapse observed, they obtained microbiological confirmation in 6 cases, and all bacteria recovered were susceptible to quinolone. Conclusively, they affirmed moxifloxacin monotherapy to be an effective, safe, and secure alternative regimen for the long-term treatment of orthopedic implant-related infections caused by quinolone-sensitive staphylococcal strains. As a means to firmly assert the credibility of their findings, they encouraged further studies to compare the efficacy of the use of moxifloxacin monotherapy with that of rifampin and other quinolones combination therapy using a follow-up period longer than 2 years, which was the case in their study [139].

1.2.5.8 Others

1.2.5.8.1 Vancomycin. Vancomycin ($C_{66}H_{75}Cl_2N_9O_{24}$) is an antibiotic used to treat several bacterial infections. It is usually recommended intravenously as a treatment for complicated skin infections, bloodstream infections, endocarditis, meningitis, and BJIs caused by MRSA [140,141]. Inzana and colleagues investigated the treatment of *S. aureus*-related BJIs with tissue debridement, and locally/systemically administered vancomycin in a novel mouse model. They found out that, regardless of the regimen adapted to treat the infected mice, only a reduction in bacterial colonization occurred while the infection remained at the end of the study. Even though localized vancomycin delivery alone tended to decrease the bacterial burden and osteolysis, these effects were only significant when combined with systemic antibiotic therapy. Ultimately, their novel mouse model replicates important features of implant-associated osteomyelitis that render treatment extremely difficult, such as biofilm formation and osteolysis, and emulates the clinical practice of placing an antibiotic-laden spacer after infected tissue debridement. Moreover, the model reveals the drawbacks of current PMMA spacers and could be an invaluable tool for assessing alternative antimicrobial prophylaxis for implant-associated bone infections [142].

1.2.5.8.2 Plant extracts. The search for and phytochemical use of drugs and dietary supplements derived from plants for the treatment of infectious diseases have expanded in recent years. However, this latest development has not made significant inroads into the field of orthopedic infections treatment yet [143]. Very few researchers have explored this avenue, of which ALhlale et al. happens to be among them. In their study, they demonstrated the inhibitory efficacy of methanolic and aqueous extracts of the leaves and stems of *Euphorbia hirta* (*E. hirta*) against surgical infections associated with *E. coli*, *P. aeruginosa*, *Proteus mirabilis*, and *S. aureus*. Their results indicated that the extracts have a significant inhibitory effect on most bacterial species, notably *P. mirabilis* bacteria. In addition, the *E. hirta* methanolic extracts were more efficient than aqueous extracts in the inhibition of tested bacteria at a concentration of 150 μL. However, in comparison with ciprofloxacin, the observed effectiveness did not match up. Nonetheless, they can serve as suitable low-cost alternatives where antibiotics such as ciprofloxacin are inaccessible [143].

1.2.5.9 Summary.

The roles of some conventional antibiotic regimens in the treatment of ODRIs/OIs' infections were briefly reviewed. The predominant causative pathogen of most of these infections is MSSA/MRSA (gram-positive and coagulase-negative strains). The most significant cure rates observed were for combination therapies with rifampin-fluoroquinolones rating highest. Regardless of the therapy type, the failure rates reported by most researchers were generally low. More importantly, the bactericidal efficacy of each regimen is dependent on factors such as antibiotic dosage, combination technique in cases of combination therapy, and mode of therapy administration, among others. Overall, two underlying limitations holding back the credibility of most of the regimens for human specimens are (1) lack of randomized clinical trial and (2) insufficient posttreatment follow-up period mostly due to noncompliance of the patients. For most of the successful therapies in murine and rabbit models in vivo, their clinical trials in human in vivo are encouraged. A concise summary of causative pathogens of ODRIs/OIs' infections, researched ways to treat them in vivo, and associated side effects is presented in Table 11.1.

1.3 Emergence of Silver Nanoparticle-Based Materials as Effective Treatment for Bacteria Biofilm-Induced Infections of Orthopedic Implants

The emerging novel treatments for *S. aureus* biofilm involving nanoparticles, bacteriophages, and plant-

TABLE 11.1
Summary of Causative Pathogens of Orthopedic Infections, Their Antibiotic Regimens and Associated Side Effects: An In Vivo Assessment.

Pathogenic Infection/ Causative Pathogen	Antibiotic Therapy/ Material and Dosage	Duration of Therapy	Experimental Nature and Specimens	Number of Case Studies	Cure Rate (%)	Side Effects
RIFAMPIN MONOTHERAPY AND COMBINATION THERAPY						
Staphylococci PJI (*S. aureus* or coagulase-negative *Staphylococcus*)	Oral clindamycin-rifampin therapy (clindamycin: 600 mg thrice daily and rifampin: 450 mg twice daily)	Minimum of 70 days for DAIR patients and 66 days for REV patients (minimum of 3 months and a follow-up period of up to 2 years)	In vivo (humans)	36	86–94	Fatigue and loss of appetite (1 patient), rash (1 patient), allergy to clindamycin (1 patient), and diarrhea without demonstration of *Clostridium difficile* toxin (2 patients) [102]
Musculoskeletal infection (*S. aureus* osteomyelitis)	Locally applied rifampin powder (dosage = 50 mg)	14 days + (broad-spectrum systemic antibiotic (cefazolin; SQ, 5 mg/kg) for 72 h posttreatment placement twice daily)	In vivo (murine model)	63	≥75	N/A [103]
Osteomyelitis (MRSA)	PLGA-coated rifampin-sitafloxacin-laden- calcium phosphate scaffold (rifampin: ~28 μg and sitafloxacin ~33 μg) + systemic doses of vancomycin (110 mg/kg subcutaneously twice daily) for 21 days postrevision.	3 weeks (short-term assessment) 10 weeks (long-term assessment)	In vivo (murine model)	8	100	N/A [104]
PJI (*Propionibacterium*)	Oral or intravenous rifampin-clindamycin therapy (rifampin: 450 mg twice a day and clindamycin: 600 mg thrice a day) (Cefazoline (1000 mg 3 times a day) for 5 days before the rifampin combination therapy)	3 months with a follow-up period of 2 years	In vivo (humans)	33	93 (after 1 year) 86 (after 2 years)	Gastrointestinal symptoms (16 patients), Diarrhea with demonstration of *Clostridium difficile* (1 patient) and skin allergy (3 patients maybe or maybe not from this therapy) [105]

Continued

TABLE 11.1
Summary of Causative Pathogens of Orthopedic Infections, Their Antibiotic Regimens and Associated Side Effects: An In Vivo Assessment.—cont'd

Pathogenic Infection/Causative Pathogen	Antibiotic Therapy/Material and Dosage	Duration of Therapy	Experimental Nature and Specimens	Number of Case Studies	Cure Rate (%)	Side Effects
DAPTOMYCIN MONOTHERAPY AND COMBINATION THERAPY						
PJI (gram-positive pathogens)	Daptomycin: daily dose of 6.0 mg/kg/day (3.1—10.6 mg/kg/day)	Mean duration of 24 days (1—106 days) with a follow-up period of 22 months (range: 1—48 months)	In vivo (humans)	30	77 (80 and 75 with and without implant removal, respectively)	Drug eruptions (1 case), decreased renal function (1 case), and anaphylactic shock (1 case) [110]
PJI (methicillin-resistant coagulase-negative Staphylococcus (40%))	Daptomycin: salvage therapy at a median dose of 6.6 mg/kg (range: 4—9 mg/kg/day)	Mean duration: 44.9 days (14—69 days) with a median follow-up period of 20 months (median global time of therapy: 57.5 days with range 14—90 days)	In vivo (humans)	20 (14 patients evaluable)	78.6	Acute renal failure due to massive rhabdomyolysis (1 case), eosinophilic pneumonia (1 case) and asymptomatic transient creatine phosphokinase (CPK) elevation (2 cases) [111]
Osteomyelitis and ODRIs (*S. aureus was the most isolated pathogen with >50% being MRSA*)	Daptomycin: 6.0 mg/kg/day and >6.0 mg/kg/day was used in one-third of the total number of patients Range: >4 to >10 mg/kg/day	Median duration: 20 days (range: 1—246 days) with a follow-up period of up to 2 years	In vivo (humans)	638	82.7 for *S. aureus*-related infections and 81.7 for coagulase-negative staphylococcal infections	Rhabdomyolysis (3 patients), myositis (1 patient), myalgia (1 patient), elevated CPK (11 patients), and deaths (10 patients) [112]
Total femoral replacement infection (*S. aureus*-related)	Daptomycin (700 mg qd ≈ 10 mg/kg/qd)—intravenous fosfomycin (2 g qid) Patient continued with chronic suppressive oral amoxicillin (1 g bid)	42 days with a follow-up period of up to 2 years	In vivo (human)	1	≈100	N/A [113]
GENTAMICIN MONOTHERAPY AND COMBINATION THERAPY						
ODRI infection (*S. aureus*-related)	CaP-coated titanium aluminum niobium screw dipped into 40 mg/mL for a period of (1 s—60 min)	7 days	In vivo (murine model)	8	87.5	N/A [118]

Infection	Treatment	Duration	Model	n	Cure rate (%)	Outcome [Ref]
ODRI/osteomyelitis infection (*S. aureus-related*)	Titanium Kirschner wires coated with PDLLA +10% gentamicin	42 days	In vivo (murine model)	10	30 (completely sterile)	N/A [119]
ODRI/osteomyelitis infection (*S. aureus-related and S. epidermidis-related*)	Titanium Kirschner wires coated with PDLLA +10% gentamicin (systemic single-shot application)	42 days	In vivo (murine model)	10	30 (completely sterile against *S. aureus*) In addition, effective against *S. epidermidis (cure rate not given)*	N/A [120]
Intramedullary implant-related infection (*S. aureus-related*)	Gentamicin-loaded TNT (gentamicin PBS solution: 100 mg/mL)	42 days	In vivo (murine model)	9	Effective against *S. aureus, however an additional systemic antibiotic administration might be needed for 100% cure rate*	N/A [121]
Implant-related osteomyelitis infection (*S. aureus-related*)	Bioactive titanium alloy (TiAl6V4) coated with gentamicin + sodium dodecyl sulfate (average coat deposited: 0.590 mg ± 0.036) Bioactive TiAl6V4 coated with gentamicin + tannic acid average coat deposited: 1.11 mg ± 0.061)	28 days	In vivo (murine model)	10 / 10	100 / 90	N/A [123]
Implant-related osteomyelitis infection (*S. aureus-related*)	Anodized TNT-coated with gentamicin	42 days	In vivo (rabbits)	8	≈ 96 in comparison with uncoated anodized TNT	N/A [124]
FLUOROQUINOLONES' MONOTHERAPY AND COMBINATION THERAPY						
BJIs (*predominantly Staphylococci-related*)	Orally administered levofloxacin (500 mg once daily in 74 cases) (750 mg once daily in 4 cases of obesity) (250 mg once daily in 1 case of chronic renal failure) combined with other regimens Before this therapy, intravenous antibiotic combination of β-lactams and glycopeptides was administered	Median interquartile range: 13 (6–13) weeks with a follow-up period of up to 1 year Median interquartile range for follow-up period: 12 (4–18) months	In vivo (humans)	79	83–94 (total or partial recovery, including orthopedic aftermath)	Insomnia (Grade A); rash—hepatitis —pancreatitis —neutropenia —myalgia (Grade C) and rash—hepatitis —acute renal failure (Grade C) observed for 3 cases [130]

Continued

TABLE 11.1
Summary of Causative Pathogens of Orthopedic Infections, Their Antibiotic Regimens and Associated Side Effects: An In Vivo Assessment.—cont'd

Pathogenic Infection/Causative Pathogen	Antibiotic Therapy/Material and Dosage	Duration of Therapy	Experimental Nature and Specimens	Number of Case Studies	Cure Rate (%)	Side Effects
Prosthetic knee infection (MSSA-related)	Levofloxacin—Rifampin combination therapy Levofloxacin: 25 mg/kg twice daily (intravenous administration) Rifampin: 10 mg/kg twice daily (intramuscular administration)	17 Days (the regimen was administered for 7 days and rabbit killed on the 17th day)	In vivo (rabbits)	12	50	N/A [131]
Orthopedic prostheses infections (P. aeruginosa-related and Staphylococci-related)	Ciprofloxacin—Ceftazidime combination therapy Ciprofloxacin: 500 mg thrice daily (oral administration) Ceftazidime: 1500 mg twice daily (intravenous or intramuscular administration) In addition, 1500 mg of ciprofloxacin alone was administered until the end of therapy	6 weeks with a mean follow-up period of 21 months (range: 6 –60 months)	In vivo (humans)	14	100 (9/9)/for patients with P. aeruginosa-infected osteosynthetic materials and 80 (4/5) for patients with infected hip or knee prostheses without removal of the implant	Arthralgia (1 patient) and rash (1 patient) [134]
PJIs (S. aureus-related, S. epidermidis-related, and Corynebacterium JK-related)	Ciprofloxacin—Rifampin Ciprofloxacin: 500 mg twice daily (oral administration) Rifampin: 450 mg twice daily (oral administration) In addition, Dicloxacillin 1000 mg 4 times daily was started at the surgery if the infection was suspected and was continued until inclusion, 3–7 days postsurgery	3 months with a median observation of 674 days	In vivo (humans)	29	≈83 (time to failure: 47 days (range: 47 –217 days))	N/A [135]
PJI and osteosynthesis materials infections (methicillin-sensitive S. aureus-related and coagulase-negative staphylococcus)	Moxifloxacin monotherapy: 400 mg thrice a day	3 months with a median follow-up period of 716 days (range: 102 –1613 days)	In vivo (humans)	48	82.6 (global cure rate)	Diarrhea and dizziness (2 patients) [139]

derived antibiotic agents are being currently studied [72]. These protocols have shown inhibitory effects against *S. aureus* embedded in biofilms [85–87]. Alternative ways to combat antibiotic-resistant biofilms are also furnished in Fig. 11.8. The nanoparticle of interest to this present work is AgNP.

The ability of bacteria such as *S. aureus* and *S. epidermidis*, which are otherwise virtually avirulent, to develop resistance to host defenses and antibiotic therapy, has necessitated the design and development of alternative methods such as infection-resistant materials [76]. This increasing tendency of microbial infections coupled with the rapid emergence of drug resistance to different antibiotics and quick evolution through mutation necessitates the development or modification of antimicrobial compounds/antibiotic materials and alternative treatments [94]. Thus, it has become imperative to provide immediate and practical solutions to these life-endangering problems. In light of this, research into furnishing these materials with antibiotic agents is actively ongoing. One of the antibiotic agents that have notably gained the attention of researchers is silver (Ag).

In general, Ag is one of the most potent antimicrobial agents that have continuously stimulated the interest of researchers owing to its broad array of antibiotic features such as potency against antibiotic-resistant bacteria, zero cytotoxicity (at clinically recommended doses), acceptable stability in vivo, and a very slim possibility to develop resistant strains, [144]. The essential derivatives of antibiotic Ag include Ag powder, AgNPs, Ag nanofilms, and Ag oxide (Ag_2O). Amid the above-mentioned forms of Ag, AgNPs are the most prevalently investigated and commercialized form because of their ease of production, excellent antimicrobial properties, wound healing capabilities, and emerging green routes of synthesis [145–151].

1.3.1 Antibacterial activities of biosynthesized AgNPs

In our daily lives, antibacterial agents find extensive applications, and they effectively protect global public health. As time goes by, antibacterial properties of nanomaterials, such as AgNPs, have increasingly aroused the interest of researchers across the globe. AgNPs have received much attention due to their potent

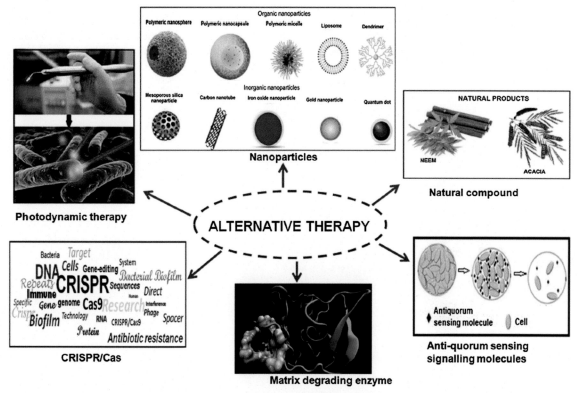

FIG. 11.8 Schematic illustration of the alternative approaches to combat antibiotic-resistant biofilms [95].

antibiotic behavior toward a broad spectrum of bacterial and fungal species, including antibiotic-resistant strains, low toxicity (in the right dosage), high thermal stability, and low volatility. There are claims that AgNPs are more reactive than bulk metallic forms due to their more active sites that resulted from high specific surfaces [145]. Behravan et al. studied the antibacterial activity of biosynthesized AgNPs from *Berberis vulgaris* leaf and root aqueous extract on *E. coli* and *S. aureus* bacteria. They reported the excellent bactericidal effect of the AgNPs on the investigated bacteria [146]. Anandan et al. studied the inhibitory effect of AgNPs synthesized from *Dodonaea viscosa* leaves on *Streptococcus pyogenes*. They found out that the AgNPs effectively inhibited the growth of A549 NSCLC cells [147]. According to Manikandan et al., AgNPs synthesized from epilithic green seaweed *Caulerpa scalpelliformis* demonstrated good antiproliferative, proapoptotic, and wound healing activities. More importantly, the AgNPs proved to be cytotoxic toward the breast cancer cell line, MCF-7 [148]. In another investigation carried out by Manjunath and Chandrashekhar, AgNPs biogenically synthesized from endophytic fungus derived from *Sophora wightii* showed effective inhibition against the growth of *E. coli* (MTCC 118), *S. aureus* (MTCC 7443), *Bacillus subtilis* (MTCC 441), *S. epidermidis* (MTCC 435), and *Candida albicans* (MTCC 183) [149]. A comprehensive review of AgNPs synthesized from plant extracts for antimicrobial applications is available in the work of Ahmed et al. [150]. Similarly, an extensive review with emphasis on the biogenic synthesis of AgNPs and their broad-spectrum applications is available in the work of Rafique et al. [151].

With these attractive properties of AgNPs, researchers have considered it medically and biomedically necessary to improve the antibacterial, disinfectant, and wound healing properties of nanotubes used as standalone biomaterials and reinforcements for biocomposite biomaterials with AgNPs [152–170]. Typical examples of these nanotubes with improved antibiotic properties are carbon nanotubes (CNTs), titania nanotubes (TNTs), boron nitride nanotubes (BNNTs), and halloysite nanotubes (HNTs) functionalized/decorated with Ag-based materials [152–180]. Herein, the bactericidal mechanism and biocompatibility of these materials are discussed with great emphasis on the antibiotic properties of AgNPs-functionalized/decorated HNTs and AgNP-HAP-based materials in general [152–180]. Fig. 11.9 furnishes conventional biogenic routes to synthesize AgNP.

1.3.2 Antibacterial activities of AgNPs-decorated CNTs

CNTs are allotropes of carbon that consist of graphite in cylindrical geometries characterized by a high aspect ratio. They have a broad spectrum of attractive and unique properties (structural, mechanical, thermal, and electronic properties) due to their relatively small size and mass, their strong mechanical integrity, and their excellent electrical and thermal conductivity. Owing to their

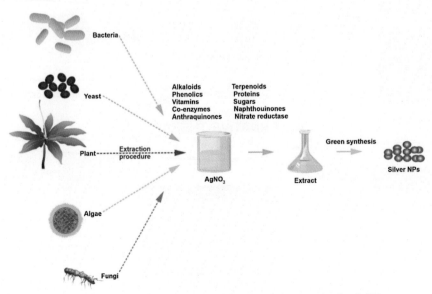

FIG. 11.9 Schematic illustration of biogenic routes to synthesize AgNP.

high surface area, they possess the capability to adsorb or conjugate with several therapeutic and diagnostic agents that account for their trending biomedical/medical applications [152,153]. They are very widespread in today's world of bioresearch, most notably in the fields of efficient drug delivery and biosensing methods for disease treatment and health monitoring. Recently, they are being functionalized by antibiotic agents for improved antibiotic behavior against specific bacterial strains [154—157]. Dinh et al. demonstrated the bactericidal efficacy of AgNPs-decorated multiwalled CNTs (MWCNTs) against gram-negative *E. coli* and gram-positive *S. aureus*. They explained that the bactericidal mechanism of this material is physical interaction with the cell membrane of the bacteria, which resulted in the formation of large aggregates of cell-AgNPs-MWCNTs, which in turn destroyed the cell membrane and disrupted its function [154]. In addition, Haider et al. studied the influence of exposure time and concentration of Ag ions on the bactericidal efficacy of AgNPs-functionalized MWCNTs against *E. coli* and *S. aureus*. From their results, they concluded that the number of bacterial colonies decreased with increasing exposure time and increasing concentration of Ag ions. Similarly, their observed bactericidal mechanism was based on physical interaction and substantial contact of the AgNPs-functionalized MWCNTs with the cytoplasmic cell membrane and also excellent adhesion to the bacteria [155]. Seo and co-researchers on the antibacterial activity and cytotoxicity of AgNPs-decorated MWCNTs showed that 30 μg/mL of this material is effective in inhibiting the growth and proliferation of *Methylobacterium* spp and *Sphingomonas* spp while maintaining biocompatibility with mammalian cells [156]. However, despite these improved antibiotic properties of AgNPs-CNTs materials, their biocompatibility in vivo is still under discussion [157]. In light of this, research into TNTs, HNTs, and BNNTs as suitable alternatives have emerged [158—180].

1.3.3 Antibacterial activities of AgNP-decorated TNTs

Titanium oxide nanotubes or Titania nanotubes (TNTs) are nano-sized versions of titanium oxide with morphology and properties analogous to that of CNTs. They are one-dimensional structures that exist as multicell or bundled tubes characterized by a high aspect ratio (diameter range of 30—80 nm and a tube length range of 10—220 μm) [158,159]. As analogs of CNT, they are used in a broad array of applications, which include white/opaque pigments, antifogging coatings, and coatings/sunscreens (for UV radiation shielding). They also find applications as ceramic filtration membranes,

photocatalysis, photovoltaic cells, drug delivery systems, carriers of proteins, antibiotic coatings, and orthopedic implants, among others [160]. Of primary interest to the present work is their antibiotic performance as orthopedic implants, especially when they are filled/coated/decorated/functionalized with AgNP/Ag$_2$O-NP.

Studies have shown that TNT arrays, in particular, can significantly improve the functions of osteoblasts and repress the activity of osteoclasts; thus, having promising applications in orthopedics, orthodontics, pharmacology, and biomechanics [144,158—160]. Moreover, in comparison with pristine titania layers, they are characterized by higher surface roughness, enhanced wettability, ability to functionalize surfaces, with a diverse range of peptides, and better transport properties, among others [161]. Similarly, relative to pristine titanium (Ti), TNTs have demonstrated superior resistance to corrosion and cytocompatibility/biocompatibility with human and animal cells in vivo and in vitro. In addition, Young's modulus of TNTs is much closer to that of natural bone than that of pristine/pure Ti that somewhat solves the problem of elastic mismatch between implants and natural bone [161]. Despite these attractive properties of TNTs, just like most other implant materials, they are susceptible to perioperative bacterial infections of implants [144,160,161]. This drawback has prompted researchers to investigate different approaches to improving the antibiotic potency of TNTs. Zhao and co-authors developed antibacterial TNT-AgNP coatings and tested the bactericidal potency of the coatings against planktonic and adherent *S. aureus* bacteria. Their result indicated 100% bactericidal efficiency of nearly all the coatings against the planktonic bacteria in the first few days of incubation and about 80% efficiency against adherent bacteria for 30 days, which is the total incubation period. According to them, 30 days is enough for the perioperative prevention of adherent bacterial infection of orthopedic implants. However, a little downside associated with their research is the cytotoxicity observed for the coatings with higher concentrations (1.5 and 2.0 M) of AgNP, but on a brighter note, they concluded that this could be taken care of by controlling the Ag ion (Ag$^+$) release rate [144]. By employing magnetron sputtering and anodization, Gao et al. were able to successfully fabricate TNT arrays incorporated with Ag$_2$O-NP (TNT-Ag$_2$O-NP) having a unique structure. The materials demonstrated potent antibacterial activity against *S. aureus* and *E. coli* characterized by 100% efficiency on the first day of 28 days incubation period. Although the observed efficiency diminished slightly with time (particularly for the material with least concentration of Ag_2O), up to 97% efficiency was still maintained for 28 days, which translates to a long-term antibacterial activity of these materials. More importantly, no cytotoxicity was observed against osteoblast cells regardless of Ag_2O concentration, which the researchers attributed to the controlled release rate of Ag^+ as a function of the materials' structure. In addition, Ag_2O incorporation favored cell proliferation [160]. In a quest to overcome excessive Ag deposition on/in TNTs associated with chemical fabrication routes, Mei et al. used a dual process containing anodization and Ag plasma immersion ion implantation (Ag PIII) to fabricate TNT-embedded Ag materials. Against *Porphyromonas gingivalis* (Pg, ATCC33277) and *Actinobacillus actinomycetemcomitans* (Aa, ATCC29523), the materials fabricated were able to inactivate these bacterial strains. On the one hand, it was found that less sustained antimicrobial activity and impaired cytocompatibility with epithelial cells and fibroblasts in vitro and in vivo were more pronounced with TNTs with a large amount of surface Ag. On the other hand, TNTs with a lower amount of surface Ag demonstrated improved sustained antibacterial activity for up to 7 days and equally evinced biocompatibility with epithelial cells and fibroblasts in vitro and less inflammatory responses in vivo. According to these researchers, their adapted processing technique for TNT-Ag development facilitated the easy control of the amount of surface Ag [162]. Zhang et al. reported a novel strategy to develop surface-functionalized implant coating with hierarchical antibacterial property based on TNT, AgNP, and bio-inspired Polydopamine (PDA) layer formulation. In comparison with pristine titanium, TNT, and TNT/PDA, the TNT-PDA-grafted AgNP coating elicited superior bactericidal activity against *S. aureus* in vitro for 5 days. Furthermore, although the PDA film significantly facilitated the hindrance of bacteria proliferation at the early stage of incubation, this was short-lived as the bactericidal potency of the film decreased with time. However, the most significant function of the PDA was assisting to graft AgNP in situ onto the surface of TNT in a hierarchical manner. In terms of cytotoxicity, the TNT-PDA-grafted AgNP coating demonstrated moderate cytotoxicity toward osteoblast cells in vitro. Ultimately, this coating elicited enhanced antibacterial properties (powerfully potent for 5 days) and minimized cellular toxicity [163]. In an attempt to control the amount of Ag release to mammalian cells from their novel TNT-AgNP implant (to improve biocompatibility), Gunputh et al. coated the implant with nano HAP (nHAP). They

found out that the coating did not, in any way, debilitate the bactericidal potency of the implant against *S. aureus* >80% biocidal activity was observed for the implant accompanied by slow release of dissolved Ag through the nHAP coating [164]. Another in vivo study by Zeng et al. showed that TNT-Ag_2O could effectively deactivate *S. aureus* and relieve inflammatory response from prosthetic joint replacements in animal models [165].

1.3.4 Antibacterial activities of AgNP-decorated HNTs

The particles of halloysite can be found naturally in a variety of morphologies, such as short tubular, spheroidal, platy clays-kaolin, montmorillonite, and the most predominant being a dioctahedral 1:1 clay mineral of the kaolin group that contains octahedral gibbsite $Al(OH)_3$ and tetrahedral SiO_4 sheets, and consists of tubular cylinders formed by multiple rolled layers separated by a monomer of water molecules (Fig. 11.10) [166–169]. In a more specific context, HNT is a double-layered aluminosilicate clay mineral with a micronano-sized tubular morphology and an external diameter of 50–200 nm, lumen diameter of 5–3 nm, and length of 0.5–2 μm [166–169]. In terms of chemical structure, HNT is similar to kaolinite [168–170]. Different from other nano-sized tubes such as CNTs and BNNTs, HNTs are readily available (mined from countries like China, New Zealand, America, Brazil, and France), relatively inexpensive, and possess unique structures [166–170]. Due to its tubular shape with an empty lumen, halloysite is considered environmentally-friendly/green containers naturally available for the encapsulation of the active agents that benefit the biomedical, medical, pharmaceutical, and chemical industries [166–170].

Research activities in this area have significantly increased in recent years due to the attractive benefits of HNTs, such as elongated tubular morphology, empty lumen, ease of mixing with a broad range of polymers, biocompatibility, bio-friendliness, and availability [166–171]. They passed as potential nanoscale containers for the encapsulation of biologically active molecules such as biocides and drugs. They also exhibit a high degree of biocompatibility and nontoxicity with human dermal fibroblasts and breast cells. Moreover, halloysite-based nanocomposites are very suitable for vast biomedical applications such as antibacterial and drug delivery applications [166–171]. Zhang et al. studied the potent antibacterial activity of a novel AgNP-HNT nanocomposite powder. The antibacterial test evaluated the antibacterial properties of the novel

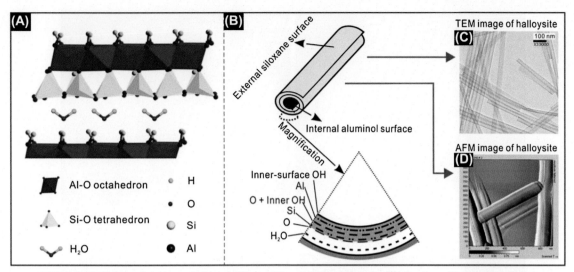

FIG. 11.10 Schematic representation of **(A)** crystalline structure halloysite, **(B)** structure of halloysite particle/HNT, **(C)** transmission electron microscopy (TEM) image of HNT, and **(D)** atomic force microscopy (AFM) image of HNT.

material on gram-negative *E. coli* and gram-positive *S. aureus* bacteria. They found out that the novel material is a potent antibacterial agent against both bacterial strains, which inspired them to suggest the material for bio-adhesive, antibacterial fabrics, and antibacterial membranes [171]. Jana et al. developed HNTs with immobilized AgNPs for antibacterial application. They considered the *E. coli* bacteria in their study in vitro. They concluded that the material developed exhibited bactericidal properties, and more importantly, the bactericidal effect is more potent for the samples of HNTs with AgNPs under blue light illumination, which indicates the role of the plasmonic excitation of AgNPs for their bioactive properties [172]. Yu et al. have also reported the enhanced antibacterial activity of AgNPs/HNTs/Graphene nanocomposites with a sandwich-like structure. They studied the antibacterial properties of the materials against gram-negative bacterial strains, *E. coli* and gram-positive bacterial strains, *S. aureus*.

From their result supported by Fig. 11.11B and D, the bacteriostasis rate nearing 100% against both bacterial strains demonstrated the effectiveness and popularity of the antibacterial properties of these materials. In addition, the morphological observation of the *E. coli* cells after exposure to the AgNPs/HNTs/Graphene material for 16 h revealed ruptured cells, pale along with the release of cytoplasm and a large portion of decomposed cells. Such irreversible cellular damage demonstrates the effectiveness of the antibacterial properties of AgNPs/HNTs/Graphene nanocomposites [173]. Shu et al. investigated the synergistic biocidal effect of AgNP and nanoparticles

of Zinc Oxide (ZnONP) anchored on HNT against *E. coli*. From their results, they revealed that while HNT facilitated the dispersion and stability of ZnONP and brought them in close contact with bacteria, AgNP, on the other hand, promoted the separation of photogenerated electron–hole pairs and improved the antimicrobial potency of ZnONP. The close contact with the cell membrane allowed the nanoparticles to produce the increased concentration of reactive oxygen species and the metal ions to infuse into the cytoplasm, and thus induced rapid death of bacteria [174]. Ultimately, it is evident that not much research has been done in the areas of AgNP-decorated HNTs for antibiotic applications (confined to very few authors), and this further contributes to the significance of this study.

1.3.5 Antibacterial activities of AgNP-decorated BNNT

BNNTs are a type of nanostructured one-dimensional tube composing of hexagonal boron nitride sheet or boron-nitrogen bond networks. They are low-density materials with unique mechanical, physical, chemical, thermal, electrical, and biomedical properties. Although many of these properties are comparable to that of CNTs, some also supersede them [175–177]. The elastic modulus of BNNTs can be as high as 1.3 TPa, and the tensile strength as high as 33 GPa. In comparison with CNTs, BNNTs possess piezoelectric and radiation shielding characteristics together, superior thermal conductivity, radiation absorption,

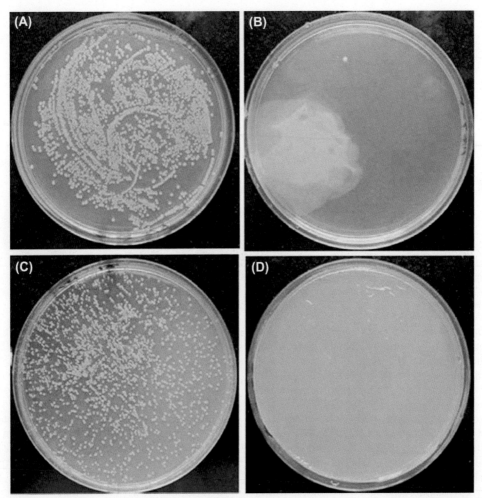

FIG. 11.11 Photographical representation of bacterial colonies formed by **(A)** *E. coli*, **(B)** control groups, **(C)** *S. aureus*, and **(D)** treated with AgNP-HNTs-Graphene [173].

hydrogen storage capacity, oxidation resistance, and excellent thermal and chemical stabilities. For example, BNNTs can survive at up 800°C in air, while CNTs start to oxidize at 400°C. The flexible and elastic nature of BNNTs and their ability to withstand severe deformation could help prevent their failure during high-pressure application and high-temperature processing [175–177].

BNNTs have progressively continued to be useful in different areas of applications due to the uniqueness of their abovementioned properties. They are highly valued for their possible biomedical and pharmaceutical applications, including novel biomaterials development, drug delivery, and neutron capture therapy. Furthermore, an important consideration to use BNNTs for orthopedic applications is as a result of their

biocompatibility and biomechanical integrity. More importantly, previous studies have shown that BNNTs are noncytotoxic to osteoblasts and macrophages, the two foremost cell lineages related to orthopedic applications. BNNTs are also found noncytotoxic to human-derived cell line (SH-SY5Y). No record is available regarding the adverse effect of BNNT on the living cells [175–178]. However, due to their high hydrophobicity, BNNTs can only be used in biological applications after noncovalent or covalent modifications to improve their water dispersibility [175]. In addition, according to Firestein et al., the antibacterial properties of BNNTs have not been thoroughly investigated [179]. In a study by Nithya et al., the bactericidal potency of pristine BNNTs and polyethylenimine-coated BNNTs against *S. aureus* was investigated. It was found that pristine BNNTs exhibit nonbactericidal effects, while polyethylenimine-coated BNNTs exhibit the bactericidal effect. This occurrence was attributed to the intrinsic hydrophobicity of the pristine nanotubes and weak interactions between OH groups and BNNT, which would prevent effective interaction with the bacteria [180]. To date, very few studies are available on antimicrobial properties of BNNT-reinforced composites [175–180]. Therefore, research in this direction is strongly encouraged.

1.3.6 Antibacterial activities, cytocompatibility, drug loading capacity and drug release of AgNP-hydroxyapatite nanomaterials

Calcium phosphate (CP) is a family of minerals that primarily consists of calcium cations and phosphate anions, of which they could be orthophosphates, metaphosphates, or pyrophosphates. HAP ($Ca_{10}(PO_4)_6(OH)_2$ is a naturally occurring form of CP similar to the human hard tissues in morphology and composition [181–184]. Principally, it has a hexagonal structure and a stoichiometric Ca/P ratio of 1.67, which is identical to bone apatite. It is the main component of bone and teeth. Relative to other CP compounds, HAP is the most thermodynamically stable under physiological conditions such as temperature, pH, and composition of the body fluids [181–184]. Thanks to the relentless efforts of materials' scientists and engineers who are continually pushing the boundaries of nanotechnology for advanced materials design and development [184]. In consequence, a remarkable impact on science and technology of nanomaterials design and development has emerged. In recent years, nHAP and its composites have garnered interest as biomaterials/coatings for use in prosthetics and orthopedic applications [185,186]. Some of their attractive properties including but not limited to outstanding biocompatibility, biodegradability, bioactivity, and biomechanical integrity [181–186].

The current and ever-increasing demand for reliable long-term implant materials and bone substitutes with excellent biocompatibility, bioactivity, biomechanical/mechanical, biofunctional, and antibiotic properties deficient of host immune complication is a great challenge for biomaterials' researchers [187]. These implants and bone substitute structures should be specially designed and developed for individual patients with all details meticulously controlled on the micrometer length scale. Similarly, nontoxic, biocompatible targeted drug delivery systems with controlled pharmacokinetics, and antibiotic efficacy are of great interest [187–191]. Remarkable research efforts have been made to develop

novel implant materials and bone substitutes with the properties as mentioned earlier [188–191]. Materials based on nHAP/modified HAP and antibiotic agents' formulations are excellent material for these purposes [192–194]. Pertinent research references to this development are given in the next few paragraphs [195–205].

Zhou et al. studied the cytocompatibility and antibacterial properties of AgNP-doped HAP/alginate nanocomposite microparticles ($\sim$550 nm) prepared by a facile double-emulsion approach. They found out that the developed materials were cytocompatible with A549 cells and exhibited excellent antibacterial activity against gram-negative *E. coli* and gram-positive *S. aureus* even at a low concentration of 100 µg/mL. However, they observed that the cytocompatibility and antibacterial activity of these materials are dependent on the incubation period and concentration of the microparticles [195].

Lazic et al. developed a novel and synthetic approach to investigate the antibacterial activities and nontoxicity of AgNp-decorated 5-aminosalicylic-functionalized HAP against gram-negative bacteria *E. coli*, gram-positive bacteria *S. aureus*, and yeast *C. albicans*. They documented inactivation of the *E. coli* and *S. aureus* bacteria by the developed material at low concentration (1–40 µm/mL) of applied AgNP (see Fig. 11.12). They also observed that the antibacterial activities of the materials remained potent after 1000 min (see Fig. 11.12), which translates to the long-term biomedical application of these materials. However, they observed the toxicity for these materials, which they attributed to the release of Ag^+ as the AgNPs are strongly bonded to the functionalized support. The only case of nontoxicity observed was for the yeast *albicans*. Similarly, they reported the dependency of the bacteria inactivation on time and concentration of the antibacterial agents [196].

Andrade et al. [197] developed a novel eco-friendly method to synthesize HAP-AgNP nanocomposites with excellent antibacterial properties. Their results, as shown in Fig. 11.13, revealed effective bacteriostasis of *E. coli* and *S. aureus* bacterial strains. They also found an excellent antibacterial effect for low concentration (0.024 wt.%) against both bacterial strains. More importantly, this effect was more pronounced at bacterial concentrations of 10^5 CFU/mL (see Fig. 11.13), which is similar to the real infection scenario of an implant, as reported by Poelstra et al. [198].

In an attempt to promote oral hygiene and combat the formation of bacterial-induced white spot lesions prevalent among orthodontic patients, Sodagar et al. fabricated orthodontic adhesives based on AgNP and

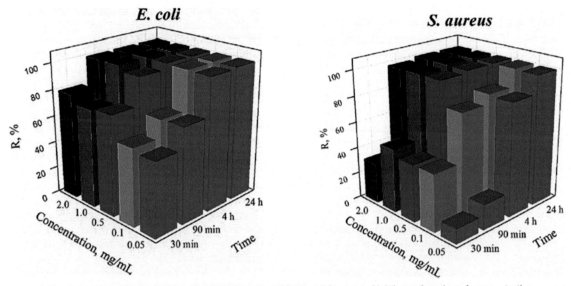

FIG. 11.12 Microbial reduction (R, %) against *E. coli* (left) and *S. aureus* (right) as a function of concentration of AgNP-decorated 5-aminosalicylic-functionalized HAP nanocomposite [196].

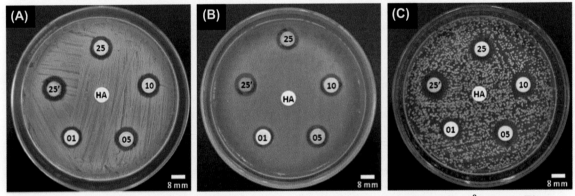

FIG. 11.13 Variation of AgNp concentrations with the inhibition zones for **(A)** *S. aureus* with 10^8 CFU/mL, **(B)** *E. coli* with 10^8 CFU/mL, and **(C)** *S. aureus* with 10^5 CFU/mL [197].

nHAP with reported potent antibacterial properties. They conducted antibacterial and biofilm tests on the adhesives to investigate the potency of their biocidal effects on three bacterial biofilms derived from *Streptococcus mutans*, *Streptococcus sanguinis*, and *Lactobacillus acidophilus*, respectively. They found out that the effect of concentration of the nanoparticles on the antibacterial effect of the adhesives on the biofilms is not significant. Nonetheless, the greatest colony count was observed for the cariogenic *S. mutans* strain at 5% concentration of the nanoparticles [199].

To develop environmentally-friendly HAP nanorods with antibacterial potency, Kumar et al. employed the green synthesis route involving leaf extracts of *Azadirachta indica* and *Coccinia grandis* as a solvent to develop their proposed antibiotic materials. They investigated the inhibiting effect of the antibiotic nanorods against *E. coli* and *S. aureus*. They reported significant inactive bacterial zones around the nanorods with the most significant inactivity observed for *S. aureus*. They somewhat attributed this antimicrobial occurrence to the presence of biomolecules such as tetratriterpenoids, flavonoids, and protein compounds with the potential to inhibit the enzymes required for bacteria growth [200].

To study the effect of combustion on the antibacterial properties and cytocompatibility of HAP, Lamkhao et al. employed two synthesis routes, namely hot plate approach and microwave approach (combustion approach facilitated by hydrogen peroxide), to synthesize HAP nanoparticles. They found out that the microwave-assisted HAP potently inhibited the growth of *S. aureus* and *E. coli* for up to 720 h. They attributed this occurrence to the presence of radicals. Similarly, the microwave-assisted HAP did not have any adverse effect on the cytocompatibility of HAP [201].

Suvannaprak et al. studied the antibacterial and cytotoxicity of HAP-doped with nanosized AgNP-phosphate and impregnated with vancomycin and also gentamicin synthesized by low-temperature phosphorization technique. They found out that the addition of nanosized AgNP-phosphate to vancomycin-impregnated HAP conferred antibacterial properties to the material against gram-negative bacterial strain *P. aerugisona* even though vancomycin is known to be ineffective against gram-negative bacterial strains. In the case of gentamycin-impregnated HAP, the addition of nanosized Ag-phosphate improved the bacterial inhibition efficiency against *P. aerugisona*; however, the duration of inhibition reduced with extraction time in comparison with HAP-AgNP with no gentamycin. Notably, the samples with the highest AgNP-phosphate showed the most prolonged duration of inhibition (15 days). The conducted cytotoxic effect indicated that the cytotoxicity of the materials is dependent on the concentration of the antibiotic loadings and extraction time. Against *S. aureus*, all the materials demonstrated effective bactericidal efficacy toward this

bacterial strain. Similarly, the inhibition zone, duration of inhibition, and cell viability (against osteoblasts) were dependent on the type of preloaded antibiotics and concentration of the AgNP-phosphate. These same factors also influenced the drug loading capacity of the materials. Overall, the synergistic antibacterial effect of the preloaded antibiotics and nanosized AgNP-phosphate is more pronounced against both bacterial strains than that of their singular effect [202].

With the primary objective of effectively combating bacterial infections associated with *E. coli* and *S. aureus*, Xiong et al. have studied the long-term (over 56 days) antibacterial activities of novel materials/papers based on ultralong HAP nanowires (HAPNWs), AgNP, and ciprofloxacin. From their results, they reported high drug loading capacity (447.4 mg/g), good biocompatibility (in vitro), large diameter of inhibition zones, and low minimum inhibitory concentrations of 30 and 40 µg/mL (see Fig. 11.15) [203]. Similar to what was reported by Suvannaprak et al. [202], the most effective bactericidal activities against both bacterial strains were demonstrated by the papers containing both AgNP and ciprofloxacin that translates to a positive synergy between the two materials in combating the previously mentioned bacterial infections as furnished in Figs. 11.14 and 11.15 [203].

Xie et al. successfully conducted interesting and impressive research on how to enhance the osteoconductivity and antibacterial properties of metallic implant surfaces using AgNP, HAP, chitosan (CHS), and bone morphogenetic protein 2(BMP-2)/heparin

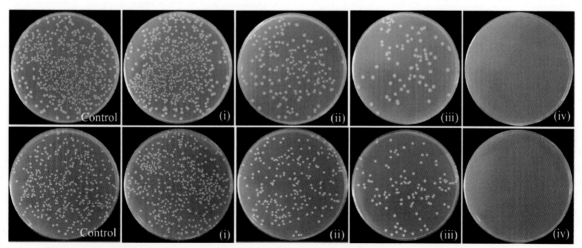

FIG. 11.14 Digital images of recultured *E. coli* (top) and *S. aureus* (bottom) colonies on solid nutrient agar plates after culture with the samples for 24 h: (i) HAPNWs, (ii) HAPNWs-AgNPs, (iii) HAPNWs-ciprofloxacin, and (iv) HAPNWs-AgNPs-ciprofloxacin.

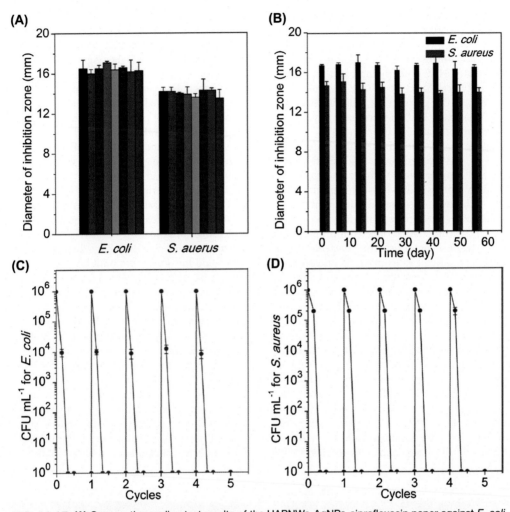

FIG. 11.15 **(A)** Consecutive cycling test results of the HAPNWs-AgNPs-ciprofloxacin paper against *E. coli* and *S. aureus* for eight cycles by the inhibition zone method; **(B)** long-term stability test results of the HAPNWs-AgNPs-ciprofloxacin paper against *E. coli* and *S. aureus* for 2 months; 5 cycles of antibacterial tests of the HAPNWs-AgNPs-ciprofloxacin paper against **(C)** *E. coli*, and **(D)** *S. aureus* by the liquid broth culture method. At each initial point, fresh bacterial fluids were added to the liquid broth and each cycle time was 24 h [203].

solution. Their results of antibacterial tests indicated that the HAP-AgNP-CHS coatings have high antibacterial properties against both *S. epidermidis* and *E. coli*. Osteoblasts culture revealed that the HAP-AgNP-CHS coatings exhibit excellent biocompatibility. The addition of CHS and BMP has no significant effect on the antibacterial activity of HAP-AgNP coatings. However, enhancement in biocompatibility was attributed to CHS addition and sustained release kinetics to BMP addition [204].

Ni et al. developed a simple and green rapid approach for the synthesis of uniform poriferous HAP and poriferous AgNPs-decorated HAP (HAP-AgNP) nanocomposites with excellent antibacterial properties. With this approach, they were able to develop porous HAP-AgNP nanocomposite uniformly decorated with AgNP. Against *S. aureus*, *E. coli*, and *P. aeruginosa*, the composites demonstrated excellent antibacterial properties, especially for *S. aureus*. Interestingly, HAP as a carrier and by its large porosity and specific chemical composition improved the durable bactericidal efficacy of the nanocomposites. From a microstructural morphology perspective as furnished in Fig. 11.16A–E, the original uniform and smooth morphologies of the investigated bacteria before they came in contact with the nanocomposite became significantly

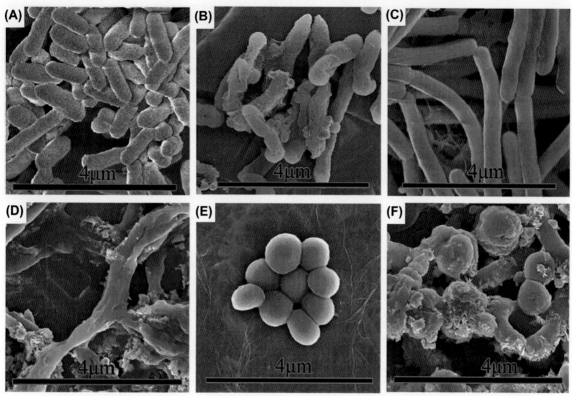

FIG. 11.16 SEM images of the normal and treated bacteria with 100 mg/mL of HAP-AgNP nanocomposites solution for *E. coli* (**A**: control and **B**: treated), *P. aeruginosa* (**C**: control and **D**: treated), and *S. aureus* (**E**: control and **F**: treated) [205].

distorted and roughened with loose cell debris after contact, which translates to an interaction between the bacteria cells and the nanocomposite that resulted in the destruction of bacteria cells and effective inhibition of their growth. According to these authors, dependent on the well-known hypotoxicity of AgNPs and high biocompatibility of HAP, the as-prepared HAP-AgNP nanocomposites are promising antibiotic materials, which can be directly applied in the human body [205].

Judging by the findings of these researchers [195–205], it is seen that antibacterial properties and cytocompatibility of AgNP-HAP-based materials are highly influenced by concentration of Ag, mode of Ag$^+$ release, combination with other antibiotic materials, fabrication technique and time of exposure to bacteria culture (for biocidal potency only).

2 MATERIALS AND METHODS

2.1 Materials

HAP was selected as the matrix phase of the composite, while HNTs and BNNTs coated and uncoated with AgNPs were selected as the reinforcing phases of the composites, respectively. The properties of the materials used for the development of the representative volume element (RVE) models and finite element analysis (FEA) are presented in Table 11.2.

2.2 Methods

2.2.1 The finite element model

The finite element model (FEM) is a specific numerical method for solving partial differential equations in two or three space variables (i.e., some boundary value problems). FEA is simply an alternative technique for solving the governing equations of simple and complex structural problems [215]. In FEA, the structures are usually assumed to be composed of discrete parts (i.e., finite elements), which are then assembled in such a way as to represent the deformation or response of the structure under the specified loads (thermal, mechanical, thermomechanical, and acoustic, among others). Each element has an assumed displacement field, and the researcher must understand how to apply the method to select appropriate elements of the correct

TABLE 11.2
Material Properties of the HAP, HNT, BNNT, and AgNP Used for the FEA.

Materials	Density (kg/m³)	Young modulus (MPa)	Poisson ratio	Aspect ratio
HAP	3080 [206]	104,000 [207]	0.27 [207]	–
HNT	2140 [208]	140,000 [209]	0.4 [210]	40 [211]
BNNT	2100 [212]	1,200,000 [207]	0.21 [207]	40 [207]
AgNP	10,500 [213]	83,000 [213]	0.37 [214]	–

size and distributions (the FE mesh). FEA has been widely used to study composites subjected to different types of loads [215,216].

Any homogeneous and isotropic material can be represented at a specific length scale as a composite consisting of a finite or infinite number of constituent elements or subelements. The behavior of these composite materials or multiphase materials, in general, is difficult to define by their complex microstructures, therefore modeling and analyzing them constitute a significant challenge in the mechanics of materials engineering. However, taking into account the interaction of the individual phases and properties of a multiphase material, the researcher can employ a multiscale modeling technique to model and simulate its mechanical response under an applied load [216].

At present, virtual mechanical testing of heterogeneous materials is done at different length and time scales, dependent on the degree of predictive accuracy and precision required. Multiscale modeling is beneficial for evaluating the stress–strain behavior of multiphase materials and reveals important engineering properties, such as Young's modulus, shear modulus, Von Mises stress, and Poisson's ratio [210,215,216]. The objective of this type of modeling is to predict the interaction between the macroscopic (effective) properties and the scale below it (meso- or in some cases, microscopic properties). The different levels can be combined by averaging methods, that is, using homogenization schemes through an RVE modeling concept. To do this, the properties of the different phases that can be considered individually heterogeneous (physical parameters, global/effective, or macrolevel material constants), the distribution, orientation, and geometry of the inclusions in the matrix must be known [207,210,215,216].

2.3 Development of the RVE Model

In the theory of composite materials, the RVE or "unit cell" is the smallest volume over which a measurement can be made that will produce a value representative of the whole system [210]. To avoid the expensive, huge computational calculations (with associated high computational power and time) of composites in FEA modeling, it is advisable to use the RVE approach [210]. In RVE modeling, numerical homogenization is essential to reconstruct an illustrative simulation of the real composite microstructure. The underlying theory in this method is that the RVE has strain energy stored in it, and the real composite must be approximately the same. Therefore, an RVE should contain an adequate number of inclusion elements with preferred distribution and orientation in the matrix [207,210,216]. Fig. 11.17 gives a schematic illustration of an RVE with two different arrays.

To model the RVE, HAP was used as the matrix and 4 wt.% of the HNTs and BNNTs as the reinforcements. Considering the density ratios between the HNTs and HAP and HAP and BNNTs, the volume fractions of HNTs in the HAP-HNTs and BNNTs in HAP-BNNTs were calculated as 5.5% for each composite [207]. The coordinates and random orientation of the HNTs and BNNTs in the HAP were automatically generated in Digimat multiscale modeling software. By using the coordinates and orientations of the BNNTs, five cubic RVEs sized as L3 with 15, 30, 45, 60, and 75 HNTs and BNNTs were generated [207]. The sizes of the RVEs calculated were proportional to the aspect ratios and numbers of the HNTs and BNNTs, respectively. After that, all sizes of RVEs were set up parametrically and depended on the numbers of HNTs and BNNTs [207]. For each RVE, the minimum relative distance between inclusions was set to 0.25 (to avoid overlapping), the minimum relative volume (relative to elementary inclusion volume) was also set to 0.25, and the maximum number of attempts for random placement was set to 2000 [217].

In this work, the aspect ratio selected for HNTs and BNNTs was 40 [207,211], and for simplicity, the HNTs and BNNTs were assumed to be solid cylindrical bars. In addition, the matrix and the nanotubes were

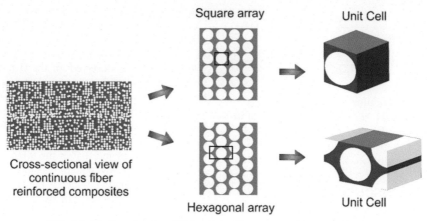

Square array Unit Cell

Cross-sectional view of
continuous fiber
reinforced composites

Hexagonal array Unit Cell

FIG. 11.17 Sample of an RVE model with square and hexagonal arrays.

TABLE 11.3
Validation of the Predictive Effective Elastic Moduli of HAP-HNT and HAP-BNNT With 60 and 45 Number of Inclusions, Respectively.

Material	FEA	Halpin Tsai	Literature	Experimental
HAP	104		104 [207]	—
HAP-HNT_60	105.9	105.9	—	—
HAP-BNNT_45	126.3	132.7	129.1 [207]	205 [177]

considered to be linearly elastic, isotropic, and homogeneous [207,210]. Correspondingly, for simplicity, the AgNP decoration on the nanotubes was assumed as continuous thin films. Digimat software was used for the analysis and computation of their elastic moduli. The matrix and reinforcement elements were meshed with nonconforming voxel mesh to reduce computational time and avoid meshing problems as the geometries are simple ones [218,219]. The total number of elements in each analysis was 125,000. The contact of the matrix and nanotubes was assumed to be perfectly bonded, and 100% load transferring occurred [207]. All necessary data for an elastic analysis were input into the models according to the data in Table 11.3. The predicted elastic moduli of the composites are an average of 10 simulations per RVE due to the random positioning of the inclusions in the matrix [220].

2.4 Periodic Boundary Condition

In general, periodic boundary conditions (PBCs) are a set of boundary conditions that are usually selected for approximating a large (infinite) system by using RVE. They are used in computer simulations and mathematical models. In this study, the mechanical loading (uniaxial) was applied under PBC with a strain rate of 1 and a peak value of 0.03 [217]. PBCs ensure that the nodes and elements that cross or are near the periodic boundaries interact with elements in the neighboring unit cell [221].

2.5 Validation Technique for the FEM Model
2.5.1 The Halpin–Tsai model

For validation of the effective elastic modulus (EEM) predicted by the FEA model, the validating micromechanical model employed was modified Halpin–Tsai. This approach considered the aspect ratio and volume fraction variables, which are effective factors when it comes to assessing the effective elastic properties of composites reinforced with randomly distributed cylindrical inclusions such as the ones considered in this work.

The Halpin–Tsai model has been extensively used for the prediction of elastic modulus of composites filled with different types of inclusion geometries [207,222–224]. The model is as follows:

$$\frac{E_c}{E_m} = \frac{1 + \cap \vartheta V_i}{1 - \cap V_i} \qquad (11.1)$$

$$\cap = \frac{E_i/E_m - 1}{E_i/E_m + \vartheta} \qquad (11.2)$$

where "E_c"—elastic modulus of composite, "E_m"—elastic modulus of matrix, "E_i"—elastic modulus of inclusion, "V_i"—volume fraction of inclusion, and "ϑ" corresponds to the shape factor that is dependent on the geometry of the inclusion.

3 RESULTS AND DISCUSSION

3.1 Effective Elastic Properties of HAP-HNT Nanocomposites

The result of stress–strain distribution profiles (SSDPs) of the monolithic HAP (matrix) is presented in Fig. 11.18. The maximum stress value (MSV) generated in the RVE was 3907.86 MPa, while the maximum principal total strain (MPTS) was approximately 0.06. The obtained values were set as the reference values for comparative studies between HAP and the virtually developed composites.

The results of the stress and strain profiles for the HAP-HNTs as a function of inclusion number are presented in Fig. 11.19A–E. From the results, it was seen that the maximum stress in all the RVEs was on the HNT reinforcement, which is an indication of effective load transfer from the HAP matrix to the HNT

reinforcements [225]. The MSVs were observed to increase progressively from HAP-HNT_15 with an MSV of 4752.75 MPa to HAP-HNT_60 with an MSV value of 4972.19 MPa; however, a sharp drop in this value to 4800.13 MPa was observed for the HAP-HNT_75. This result clearly shows that the critical number of inclusions has been reached at 60 HNTs. Nonetheless, in comparison with the MHAP, all the composites exhibited better load-bearing capacities. In the case of MPTS, no significant difference was observed between the MPTS values of the composites and that of the MHAP. However, the only exception to this pattern was for one instance that seems to be an outlier, which is the HAP-HNT_45 with an anomalous MPTS value of approximately 0.02 while that of others approximates to 0.06. Ultimately, this is a fascinating spectacle that shows HNT can be used to improve the stiffness of HAP without compromising its ductility [226].

With respect to EEM, the number of HNT inclusions does not have any significant effect. The observed variation in the elastic moduli of the HAP-HNT composites as a function of the number of HNTs did not exceed 0.05. Similarly, in comparison with the monolithic HAP, all the composites exhibited better stiffness properties. The elastic modulus of the MHAP increased from 104 to 105.9 GPa when compared with the HAP-HNT_15 and HAP-HNT_60. This slight improvement can be attributed to the close elastic modulus values of MHAP and HNT, which are 104 and 140 GPa, respectively. Interestingly, this also points to the fact that the

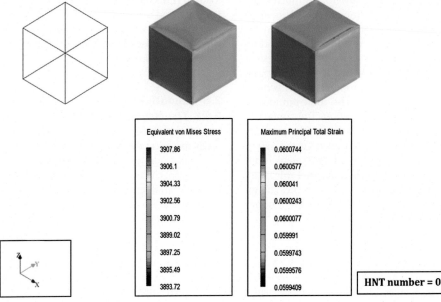

FIG. 11.18 Predicted stress and strain distribution profiles of the monolithic HAP (MHAP).

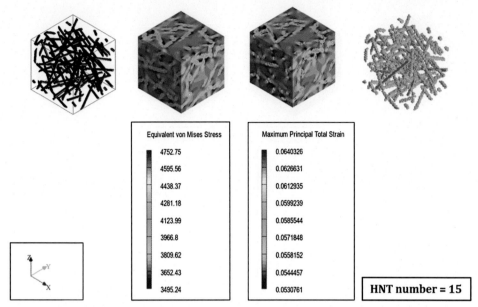

FIG. 11.19A Predicted stress and strain distribution profiles of the HAP-HNT with 15 HNT inclusions.

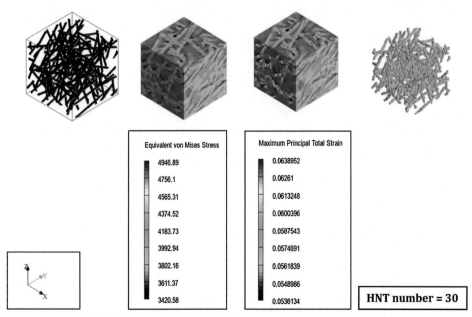

FIG. 11.19B Predicted stress and strain distribution profiles of the HAP-HNT with 30 HNT inclusions.

HAP-HNT composites can be conveniently loaded with antibiotic agents such as AgNP with higher values of elastic modulus that can prevent the scenario of elastic mismatch in the event of using this material as a bone substitute. Some researchers have reported that the mismatch of the elastic modulus between implant materials and bone tissue can result in stress shielding, bone resorption, and poor osseointegration [227]. Thus, the composite with the highest MSV (HAP-HNT_60) was selected as the representable sample of the composites, and its EEM was validated by the results from only the Halpin–Tsai model's prediction as no

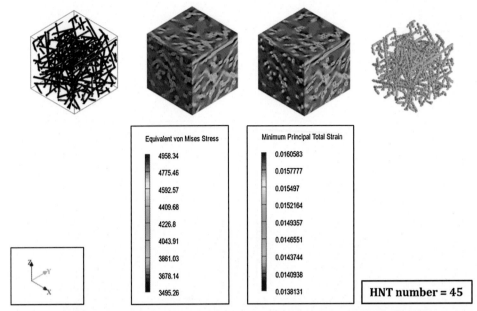

FIG. 11.19C Predicted stress and strain distribution profiles of the HAP-HNT with 45 HNT inclusions.

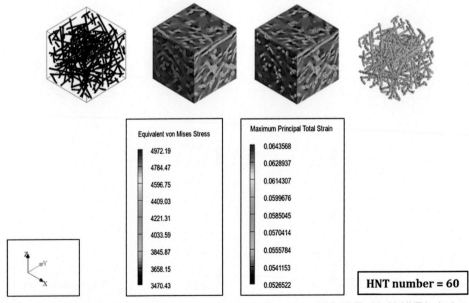

FIG. 11.19D Predicted stress and strain distribution profiles of the HAP-HNT with 60 HNT inclusions.

results are currently available from experiments and open literature though limited to the knowledge of the authors. More importantly, it is envisaged in the event of a physical experiment such as spark plasma sintering of the HNT-HAP composites, the experimental elastic modulus of the composites might be higher than those predicted. A similar scenario was reported by Ref. [207] where high-temperature sintering shrunk the HAP particles and allowed BNNT to assist in a heat treatment process, which resulted in a more compacted composite with high elastic modulus [207].

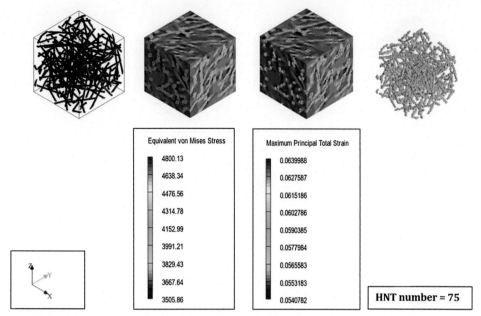

FIG. 11.19E Predicted stress and strain distribution profiles of the HAP-HNT with 75 HNT inclusions.

3.2 Effective Elastic Properties of HAP-BNNT Nanocomposites

Fig. 11.20A—E shows the results of the SSDPs for the HAP-BNNTs as a function of inclusion number. From the results, the maximum stresses in all the RVEs were on the BNNT reinforcements, which translate to efficient stress transfer from the matrix to the nanotubes, as was observed for the HAP-HNT composites [225]. The observed MSVs increased progressively from HAP-BNNT_15 with an MSV of 41,938.10 MPa to HAP-BNNT_45 with an MSV value of 50,511.80 MPa. However, a sharp drop in this value to 37,404.30 MPa was observed for the HAP-BNNT_60 and then to 34,604.40 MPa for HAP-BNNT_75. Correspondingly, this clearly shows that the critical number of inclusions is 45 for the BNNTs. Nonetheless, in comparison with the MHAP, all the composites exhibited better load-bearing capacities. A slight variation was observed in the MPTS values of the HAP-BNNT composites for inclusion numbers. In comparison with that of MHAP, all the HAP-BNNT composites evinced reduced MPTS, which can be attributed to the very high elastic modulus (1.2 TPa) of BNNT used in this study [207].

For EEM, the number of BNNT inclusions similarly has a slight but notable effect on this property. In comparison with the MHAP, all the composites exhibited better stiffness properties. The EEM of the MHAP increased from 104 to 126.31 GPa when compared

with the HAP-BNNT_45. This significant improvement is due to the very high elastic modulus value of BNNT relative to that of HNT. Ali et al. have reported a similar finding [207]. It is believed that this composite can also be laden with antibiotic agents for enhanced antibacterial activities. The composite with the highest MSV (HAP-BNNT_45) was selected as the representable sample of the composites, and its EEM was validated by the results from the Halpin—Tsai model's prediction, experiment and open literature [177,207].

The reason for a large discrepancy between experimental and predicted results for the BNNT-HAP nanocomposites had been linked to the influence of processing parameters such as temperature and pressure on the resultant composites (high temperature may lead to improved densification and higher stiffness in the composites [207]. The predicted EEMs of HAP-HNT and HAP-BNNT as a function of inclusion number is comparatively presented in Fig. 11.21.

3.3 Effect of AgNP Coating Thickness on the Effective Elastic Properties of Silver Nanoparticle-Decorated HAP-HNT and HAP-BNNT Nanocomposites

Fig. 11.22A and B show the effect of AgNP coating thickness on the EEM of HAP-HNT_60 and HAP-BNNT_45. The results indicated that the EEM of the HAP-HNT_60 is not influenced by variation in coating thickness,

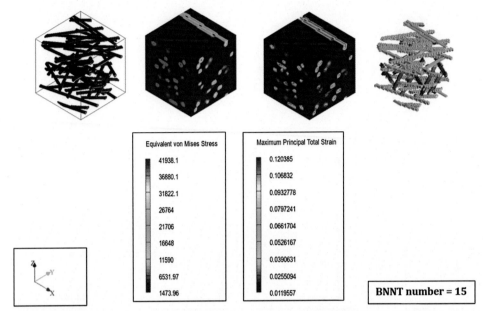

FIG. 11.20A Predicted stress and strain distribution profiles of the HAP-BNNT with 15 BNNT inclusions.

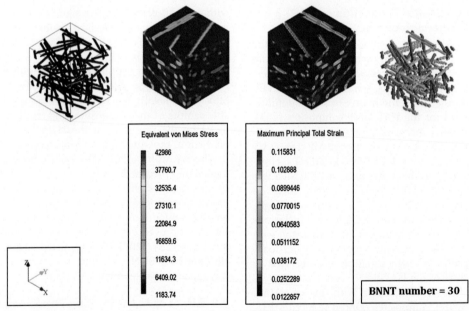

FIG. 11.20B Predicted stress and strain distribution profiles of the HAP-BNNT with 30 BNNT inclusions.

while in the case of HAP-BNNT_45, the least coating thickness of 2 nm reduced its EEM from 126.31 to 122.27 GPa. The explanation for this will be that, for the HAP-HNT_60, the coating is within the optimum thickness range, and there is still sufficient matrix to retain the interfacial bonding energy between the matrix and inclusion effectively, which is supported by its SSPD, MSV, and MPTS. As for the HAP-BNNT_45, it is believed that the optimum coating thickness had been exceeded in the sense that, the least coating thickness of 2 nm immediately reduced the interinclusion distance between the nanotubes, and hence reduced HAP

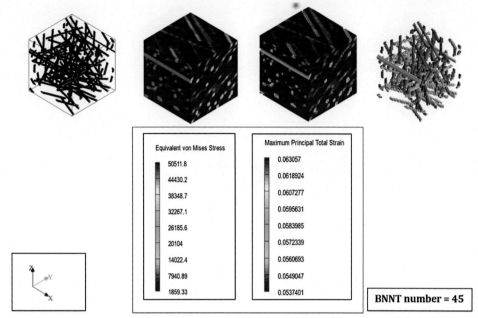

FIG. 11.20C Predicted stress and strain distribution profiles of the HAP-BNNT with 45 BNNT inclusions.

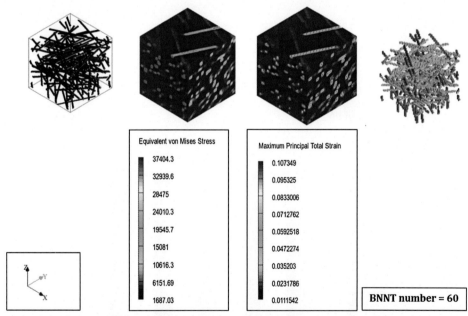

FIG. 11.20D Predicted stress and strain distribution profiles of the HAP-BNNT with 60 BNNT inclusions.

matrix to accommodate the fattened nanotubes that in turn resulted in reduced EEM of this composite. Furthermore, this could be the same reason, while the MSV of this composite dropped beyond 45 BNNTs. Poelma et al. have worked on similar research and linked this occurrence with optimum coating thickness, bucking, and waviness of the nanotubes, which might also account for why there are scarce research findings on the functionalization of BNNTs with AgNP for antibiotic purposes [220].

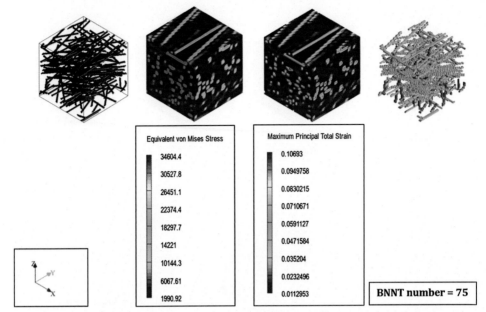

FIG. 11.20E Predicted stress and strain distribution profiles of the HAP-BNNT with 75 BNNT inclusions.

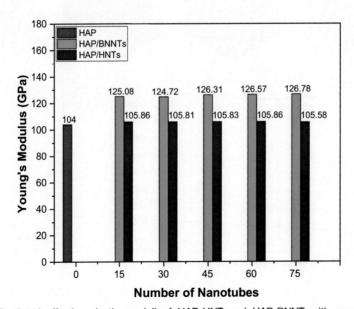

FIG. 11.21 Predicted effective elastic moduli of HAP-HNT and HAP-BNNT with varying number of inclusions.

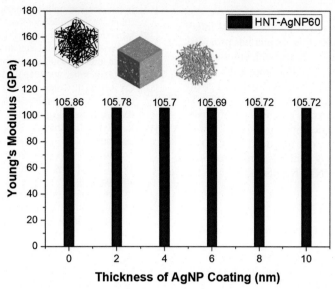

FIG. 11.22A Variation of thickness of AgNP coating with the predicted effective elastic modulus of HAP-HNT_60.

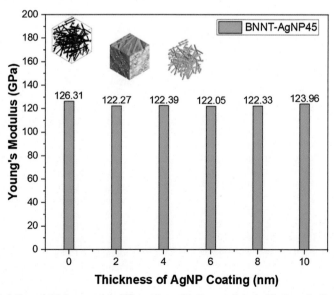

FIG. 11.22B Variation of thickness of AgNP coating with the predicted effective elastic modulus of HAP-BNNT_45.

4 CONCLUSIONS

To address the challenges raised by nosocomial infections of orthopedic implants, this chapter has provided a great insight into this subject matter with emphasis on their causative pathogens, trending, and most efficacious regimens, and finally, a new material with improved structural integrity based on HAP-HNT-AgNP formulation is proposed as potential prophylaxis. This material can be fabricated by spark plasma sintering, and Ag-based materials can be added via CVD or laden in the HNT tube using methods similar to those used for AgNP-decorated CNT/TNT/BNNT.

REFERENCES

[1] H. Ma, C. Feng, J. Chang, C. Wu, 3D-printed bioceramic scaffolds: from bone tissue engineering to tumor therapy, Acta Biomaterialia 79 (October 1, 2018) 37–59.

[2] Y. Zhou, C. Wu, J. Chang, Bioceramics to regulate stem cells and their microenvironment for tissue regeneration, Materials Today 24 (April 1, 2019) 41–56.

[3] M.J. Zafar, D. Zhu, Z. Zhang, 3D printing of bioceramics for bone tissue engineering, Medicine 4 6.

[4] https://www.who.int/news-room/fact-sheets/detail/musculoskeletal-conditions.

[5] K. Markatos, G. Tsoucalas, M. Sgantzos, Hallmarks in the history of orthopaedic implants for trauma and joint replacement, AMHA-Acta Medico-Historica Adriatica 14 (1) (2016) 161–176.

[6] H.M. Kremers, D.R. Larson, C.S. Crowson, W.K. Kremers, R.E. Washington, C.A. Steiner, W.A. Jiranek, D.J. Berry, Prevalence of total hip and knee replacement in the United States, The Journal of Bone and Joint Surgery American 97 (17) (September 2, 2015) 1386.

[7] A. Lübbeke, A.J. Silman, C. Barea, D. Prieto-Alhambra, A.J. Carr, Mapping existing hip and knee replacement registries in Europe, Health Policy 122 (5) (May 1, 2018) 548–557.

[8] H.A. Kim, S. Kim, Y.I. Seo, H.J. Choi, S.C. Seong, Y.W. Song, D. Hunter, Y. Zhang, The epidemiology of total knee replacement in South Korea: national registry data, Rheumatology 47 (1) (January 1, 2008) 88–91.

[9] A. Al-Taiar, R. Al-Sabah, E. Elsalawy, D. Shehab, S. Al-Mahmoud, Attitudes to knee osteoarthritis and total knee replacement in Arab women: a qualitative study, BMC Research Notes 6 (1) (December 2013) 406.

[10] A. George, P. Ofori-Atta, Joint replacement surgery in Ghana (West Africa)—an observational study, International Orthopaedics 43 (5) (May 2, 2019) 1041–1047.

[11] E.K. Song, J.K. Seon, J.Y. Moon, Y. Ji-Hyoun, P. Kinov, The Evolution of Modern Total Knee Prostheses, vol. 10, InTech, Rijeka, Croatia, February 20, 2013, p. 54343.

[12] S.R. Knight, R. Aujla, S.P. Biswas, Total hip arthroplasty-over 100 years of operative history, Orthopedic Reviews 3 (2) (September 6, 2011).

[13] https://www.gponline.com/total-knee-replacement/musculoskeletal-disorders/joint-replacements/article/1406657.

[14] https://en.wikipedia.org/wiki/Hip_replacement.

[15] M. Worboys, Joseph Lister and the performance of antiseptic surgery, Notes and Records of the Royal Society 67 (3) (September 20, 2013) 199–209.

[16] C.Y. Chang, E. Goldstein, N. Agarwal, K.G. Swan, Ether in the developing world: rethinking an abandoned agent, BMC Anesthesiology 15 (1) (December 2015) 149.

[17] R.A. Kyle, M.A. Shampo, James Young Simpson and the introduction of chloroform anesthesia in obstetric practice, Mayo Clinic Proceedings 72 (4) (April 1, 1997) 372 (Mayo Foundation for Medical Education and Research).

[18] I.V. Shishkovsky, M.V. Kuznetsov, Y.G. Morozov, Porous titanium and nitinol implants synthesized by SHS/SLS: microstructural and histomorphological analyses of tissue reactions, International Journal of Self-Propagating High-Temperature Synthesis 19 (2) (June 1, 2010) 157–167.

[19] C.G. Ambrose, B.E. Hartline, T.O. Clanton, W.R. Lowe, W.C. McGarvey, Polymers in orthopaedic surgery, in: Advanced Polymers in Medicine, Springer, Cham, 2015, pp. 129–145.

[20] M.F. Ulum, W. Caesarendra, R. Alavi, H. Hermawan, In-vivo corrosion characterization and assessment of absorbable metal implants, Coatings 9 (5) (May 2019) 282.

[21] M. Sancho-Tello, F. Forriol, J.J. de Llano, C. Antolinos-Turpin, J.A. Gómez-Tejedor, J.L. Ribelles, C. Carda, Biostable scaffolds of polyacrylate polymers implanted in the articular cartilage induce hyaline-like cartilage regeneration in rabbits, The International Journal of Artificial Organs 40 (7) (July 2017) 350–357.

[22] O.O. Daramola, J.L. Olajide, S.C. Agwuncha, M.J. Mochane, E.R. Sadiku, Nanostructured green biopolymer composites for orthopedic application, in: Green Biopolymers and Their Nanocomposites, Springer, Singapore, 2019, pp. 159–190.

[23] A.F. Mavrogenis, P.J. Papagelopoulos, G.C. Babis, Osseointegration of cobalt-chrome alloy implants, Journal of Long-Term Effects of Medical Implants 21 (4) (2011).

[24] B.A. Steel, How to choose between the implant materials steel and titanium in orthopedic trauma surgery: Part 2—biological aspects, Acta Chirurgiae Orthopaedicae et Traumatologiae Cechoslovaca 84 (2) (2017) 85–90.

[25] W. Wang, C.K. Poh, Titanium alloys in orthopaedics, in: Titanium Alloys-Advances in Properties Control, May 15, 2013, pp. 17–43.

[26] S. Kamrani, C. Fleck, Biodegradable magnesium alloys as temporary orthopaedic implants: a review, Biometals 32 (2) (April 1, 2019) 185–193.

[27] M. Bahraminasab, B.B. Sahari, NiTi shape memory alloys, promising materials in orthopedic applications, in: Shape Memory Alloys-Processing, Characterization and Applications, April 3, 2013, pp. 261–278.

[28] O. Ahmad, A. Soodeh, Application of Bioceramics in Orthopedics and Bone Tissue Engineering.

[29] C.A. Garrido, S.E. Lobo, F.M. Turíbio, R.Z. LeGeros, Biphasic calcium phosphate bioceramics for orthopaedic reconstructions: clinical outcomes, International Journal of Biomaterials 2011 (2011).

[30] C. Gao, Y. Deng, P. Feng, Z. Mao, P. Li, B. Yang, J. Deng, Y. Cao, C. Shuai, S. Peng, Current progress in bioactive ceramic scaffolds for bone repair and regeneration, International Journal of Molecular Sciences 15 (3) (March 2014) 4714–4732.

[31] M.S. Scholz, J.P. Blanchfield, L.D. Bloom, B.H. Coburn, M. Elkington, J.D. Fuller, M.E. Gilbert, S.A. Muflahi, M.F. Pernice, S.I. Rae, J.A. Trevarthen, The use of composite materials in modern orthopaedic medicine and prosthetic devices: a review, Composites Science and Technology 71 (16) (November 14, 2011) 1791–1803.

[32] M. Saad, S. Akhtar, S. Srivastava, Composite polymer in orthopedic implants: a review, Materials Today: Proceedings 5 (9) (January 1, 2018) 20224–20231.

[33] R. Sedlacek, Z. Padovec, M. Ruzicka, P. Ruzicka, M. Kral, Reinforced thermoplastic composite materials for use in orthopaedics, in: Front. Bioeng. Biotechnol. Conference Abstract: 10th World Biomaterials Congress, 2016.

[34] https://www.globenewswire.com/news-release/2019/07/24/1887203/0/en/Global-Orthopedic-Implants-Market-Is-Expected-to-Reach-66-63-Billion-by-2025.html.

[35] https://www.polarismarketresearch.com/industry-analysis/orthopedic-implants-market.

[36] M. Gomez-Galan, J. Perez-Alonso, A.J. Callejon-Ferre, J. Lopez-Martinez, Musculoskeletal disorders: OWAS review, Industrial Health 55 (4) (July 31, 2017) 314–337.

[37] https://www.mordorintelligence.com/industry-reports/bioceramics-market.

[38] R. Mala, A.S.R. Celsia, Bioceramics in orthopaedics: a review, in: S. Thomas, P. Balakrishnan, M.S. Sreekala (Eds.), Fundamental Biomaterials: Ceramics, Woodhead Publishing, February 16, 2018.

[39] F. Baino, S. Hamzehlou, S. Kargozar, Bioactive glasses: where are we and where are we going? Journal of Functional Biomaterials 9 (1) (March 2018) 25.

[40] D. Bellucci, R. Salvatori, M. Cannio, M. Luginina, R. Orrù, S. Montinaro, A. Anesi, L. Chiarini, G. Cao, V. Cannillo, Bioglass and bioceramic composites processed by Spark Plasma Sintering (SPS): biological evaluation versus SBF test, Biomedical Glasses 4 (1) (January 2018) 21–31.

[41] D.M. Yunos, O. Bretcanu, A.R. Boccaccini, Polymer-bioceramic composites for tissue engineering scaffolds, Journal of Materials Science 43 (13) (July 1, 2008) 4433–4442.

[42] A. Yelten, S. Yilmaz, A novel approach on the synthesis and characterization of bioceramic composites, Ceramics International 45 (12) (August 15, 2019) 15375–15384.

[43] H. Yu, Y. Chen, M. Mao, D. Liu, J. Ai, W. Leng, PEEK-biphasic bioceramic composites promote mandibular defect repair and upregulate BMP-2 expression in rabbits, Molecular Medicine Reports 17 (6) (June 1, 2018) 8221–8227.

[44] M.P. Ginebra, M. Espanol, Y. Maazouz, V. Bergez, D. Pastorino, Bioceramics and bone healing, EFORT Open Reviews 3 (5) (May 2018) 173–183.

[45] R. Albulescu, A.C. Popa, A.M. Enciu, L. Albulescu, M. Dudau, I.D. Popescu, S. Mihai, E. Codrici, S. Pop, A.R. Lupu, G.E. Stan, Comprehensive in vitro testing of calcium phosphate-based bioceramics with orthopedic and dentistry applications, Materials 12 (22) (January 2019) 3704.

[46] M. Vallet-Regí, Evolution of bioceramics within the field of biomaterials, Comptes Rendus Chimie 13 (1–2) (January 1, 2010) 174–185.

[47] G.I. Im, Biomaterials in orthopaedics: the past and future with immune modulation, Biomaterials Research 24 (1) (December 1, 2020) 7.

[48] D. Ricketts, R.A. Rogers, T. Roper, X. Ge, Recognising and dealing with complications in orthopaedic surgery, Annals of the Royal College of Surgeons of England 99 (3) (March 2017) 185–188.

[49] https://www.marinahospital.com/faq/what-are-the-most-common-complications-and-risks-of-orthopedic-surgery.

[50] R. Garimella, A.E. Eltorai, Nanotechnology in orthopedics, Journal of Orthopaedics 14 (1) (March 1, 2017) 30–33.

[51] M. Sato, T.J. Webster, Nanobiotechnology: implications for the future of nanotechnology in orthopedic applications, Expert Review of Medical Devices 1 (1) (September 1, 2004) 105–114.

[52] L. Yang, Nanotechnology-Enhanced Orthopedic Materials: Fabrications, Applications and Future Trends, Woodhead Publishing, July 28, 2015.

[53] S. Kumar, M. Nehra, D. Kedia, N. Dilbaghi, K. Tankeshwar, K.H. Kim, Nanotechnology-based biomaterials for orthopaedic applications: recent advances and future prospects, Materials Science and Engineering: C (September 2, 2019) 110154.

[54] C. Hu, D. Ashok, D.R. Nisbet, V. Gautam, Bioinspired surface modification of orthopedic implants for bone tissue engineering, Biomaterials (July 15, 2019) 119366.

[55] H. Shahali, A. Jaggessar, P.K. Yarlagadda, Recent advances in manufacturing and surface modification of Titanium orthopaedic applications, Procedia Engineering 174 (2017) 1067–1076.

[56] B. Priyadarshini, M. Rama, Chetan, U. Vijayalakshmi, Bioactive coating as a surface modification technique for biocompatible metallic implants: a review, Journal of Asian Ceramic Societies 7 (4) (October 2, 2019) 397–406.

[57] L. Xia, B. Feng, P. Wang, S. Ding, Z. Liu, J. Zhou, R. Yu, In vitro and in vivo studies of surface-structured implants for bone formation, International Journal of Nanomedicine 7 (2012) 4873.

[58] D. McCoy, S. Chand, D. Sridhar, Global health funding: how much, where it comes from and where it goes, Health Policy and Planning 24 (6) (November 1, 2009) 407–417.

[59] S. Giannini, A. Moroni, G. Coppola, L. Ponziani, A. Ravaglioli, A. Kraiewski, A. Venturini, M. Pigato, D. Zaffe, Bioceramics in orthopaedic surgery: know how status and preliminary results, in: Bioceramics and the Human Body, Springer, Dordrecht, 1992, pp. 295–301.

[60] I. Uçkay, P. Hoffmeyer, D. Lew, D. Pittet, Prevention of surgical site infections in orthopaedic surgery and bone trauma: state-of-the-art update, Journal of Hospital Infection 84 (1) (May 1, 2013) 5–12.

[61] F.A. Al-Mulhim, M.A. Baragbah, M. Sadat-Ali, A.S. Alomran, M.Q. Azam, Prevalence of surgical site infection in orthopedic surgery: a 5-year analysis, International Surgery 99 (3) (May 2014) 264–268.

[62] M. Ribeiro, F.J. Monteiro, M.P. Ferraz, Infection of orthopedic implants with emphasis on bacterial adhesion process and techniques used in studying bacterial-material interactions, Biomatter 2 (4) (October 1, 2012) 176–194.

[63] T.F. Moriarty, R. Kuehl, T. Coenye, W.J. Metsemakers, M. Morgenstern, E.M. Schwarz, M. Riool, S.A. Zaat, N. Khana, S.L. Kates, R.G. Richards, Orthopaedic device-related infection: current and future interventions for improved prevention and treatment, EFORT Open Reviews 1 (4) (April 2016) 89–99.

[64] V.G. Reddy, R.V. Kumar, A.K. Mootha, C. Thayi, P. Kantesaria, D. Reddy, Salvage of infected total knee arthroplasty with Ilizarov external fixator, Indian Journal of Orthopaedics 45 (6) (November 2011) 541.

[65] A.M. Eid, Infected orthopedic implants, Egyptian Orthopaedic Journal 51 (3) (July 1, 2016) 187.

[66] T.T. Gebremariam, M.F. Declaro, Operating theaters as a source of nosocomial infection: a systematic review, Saudi Journal for Health Sciences 3 (1) (January 1, 2014) 5.

[67] Z. Guo, Analysis on risk factors of nosocomial infection in orthopedic patients and research on nursing strategies, Infection International 5 (1) (March 1, 2016) 27–30.

[68] L. Montanaro, P. Speziale, D. Campoccia, S. Ravaioli, I. Cangini, G. Pietrocola, S. Giannini, C.R. Arciola, Scenery of Staphylococcus implant infections in orthopedics, Future Microbiology 6 (11) (November 2011) 1329–1349.

[69] M. Drancourt, A. Stein, J.N. Argenson, A. Zannier, G. Curvale, D. Raoult, Oral rifampin plus ofloxacin for treatment of Staphylococcus-infected orthopedic implants, Antimicrobial Agents and Chemotherapy 37 (6) (June 1, 1993) 1214–1218.

[70] L. Salih, S. Tevell, E. Månsson, Å. Nilsdotter-Augustinsson, B. Hellmark, B. Söderquist, Staphylococcus epidermidis isolates from nares and prosthetic joint infections are mupirocin susceptible, Journal of Bone and Joint Infection 3 (1) (2018) 1.

[71] https://www.msdmanuals.com/professional/infectious-diseases/gram-positive-cocci/staphylococcal-infections.

[72] https://en.wikipedia.org/wiki/Staphylococcus.

[73] K. Becker, C. Heilmann, G. Peters, Coagulase-negative staphylococci, Clinical Microbiology Reviews 27 (4) (October 1, 2014) 870–926.

[74] A.E. Namvar, S. Bastarahang, N. Abbasi, G.S. Ghehi, S. Farhadbakhtiarian, P. Arezi, M. Hosseini, S.Z. Baravati, Z. Jokar, S.G. Chermahin, Clinical characteristics of Staphylococcus epidermidis: a systematic review, GMS Hygiene and Infection Control 9 (3) (2014).

[75] P.H. Kwakman, A.A. te Velde, C.M. Vandenbroucke-Grauls, S.J. Van Deventer, S.A. Zaat, Treatment and prevention of Staphylococcus epidermidis experimental biomaterial-associated infection by bactericidal peptide 2, Antimicrobial Agents and Chemotherapy 50 (12) (December 1, 2006) 3977–3983.

[76] J.M. Schierholz, J. Beuth, Implant infections: a haven for opportunistic bacteria, Journal of Hospital Infection 49 (2) (October 1, 2001) 87–93.

[77] M. Nilsson, L. Frykberg, J.I. Flock, L. Pei, M. Lindberg, B. Guss, A fibrinogen-binding protein of Staphylococcus epidermidis, Infection and Immunity 66 (6) (June 1, 1998), 2666-73.Infection. 2001 Oct 1;49(2):87-93.

[78] M. Otto, Staphylococcus epidermidis—the'accidental'pathogen, Nature Reviews Microbiology 7 (8) (August 2009) 555–567.

[79] T.H. Nguyen, M.D. Park, M. Otto, Host response to Staphylococcus epidermidis colonization and infections, Frontiers in Cellular and Infection Microbiology 7 (March 21, 2017) 90.

[80] M. Widerström, C.A. McCullough, G.W. Coombs, T. Monsen, K.J. Christiansen, A multidrug-resistant Staphylococcus epidermidis clone (ST2) is an ongoing cause of hospital-acquired infection in a Western Australian hospital, Journal of Clinical Microbiology 50 (6) (June 1, 2012) 2147–2151.

[81] R.R. ASSESSMENT, Multidrug-Resistant Staphylococcus epidermidis. https://www.ecdc.europa.eu/sites/default/files/documents/15-10-2018-RRA-Staphylococcus%20epidermidis%2C%20Antimicrobial%20resistance-World_ZCS9CS.pdf.

[82] C. Carbon, MRSA and MRSE: is there an answer? Clinical Microbiology and Infection 6 (August 1, 2000) 17–22.

[83] T.A. Taylor, C.G. Unakal, Staphylococcus aureus, InStatPearls [Internet], StatPearls Publishing, March 27, 2019.

[84] T. Foster, Staphylococcus. InMedical Microbiology, fourth ed., University of Texas Medical Branch at Galveston, 1996.

[85] R. Gutierrez Jauregui, H. Fleige, A. Bubke, M. Rohde, S. Weiss, R. Forster, IL-1b promotes Staphylococcus aureus biofilms on implants in vivo, Frontiers in Immunology 10 (2019) 1082.

[86] https://apic.org/monthly_alerts/staphylococcus-aureus.

[87] C.G. Gemmell, D.I. Edwards, A.P. Fraise, F.K. Gould, G.L. Ridgway, R.E. Warren, Guidelines for the prophylaxis and treatment of methicillin-resistant Staphylococcus aureus (MRSA) infections in the UK, Journal of Antimicrobial Chemotherapy 57 (4) (April 1, 2006) 589–608.

[88] T. Lu, A.R. Porter, A.D. Kennedy, S.D. Kobayashi, F.R. DeLeo, Phagocytosis and killing of *Staphylococcus aureus* by human neutrophils, Journal of Innate Immunity 6 (5) (2014) 639–649.

[89] A.F. Widmer, New Developments in Diagnosis and Treatment of Infection in Orthopedic Implants.

[90] E.J. Choo, H.F. Chambers, Treatment of methicillin-resistant *Staphylococcus aureus* bacteremia, Infection and Chemotherapy 48 (4) (December 1, 2016) 267–273.

[91] J. Tang, J. Hu, L. Kang, Z. Deng, J. Wu, J. Pan, The use of vancomycin in the treatment of adult patients with methicillin-resistant *Staphylococcus aureus* (MRSA) infection: a survey in a tertiary hospital in China, International Journal of Clinical and Experimental Medicine 8 (10) (2015) 19436.

[92] A.D. Verderosa, M. Totsika, K.E. Fairfull-Smith, Bacterial biofilm eradication agents: a current review, Frontiers in Chemistry 7 (2019) 824.

[93] M. Jamal, W. Ahmad, S. Andleeb, F. Jalil, M. Imran, M.A. Nawaz, T. Hussain, M. Ali, M. Rafiq, M.A. Kamil, Bacterial biofilm and associated infections, Journal of the Chinese Medical Association 81 (1) (January 1, 2018) 7–11.

[94] R. Roy, M. Tiwari, G. Donelli, V. Tiwari, Strategies for combating bacterial biofilms: a focus on anti-biofilm agents and their mechanisms of action, Virulence 9 (1) (December 31, 2018) 522–554.

[95] D. Sharma, L. Misba, A.U. Khan, Antibiotics versus biofilm: an emerging battleground in microbial communities, Antimicrobial Resistance and Infection Control 8 (1) (December 2019) 76.

[96] C.W. Hall, T.F. Mah, Molecular mechanisms of biofilm-based antibiotic resistance and tolerance in pathogenic bacteria, FEMS Microbiology Reviews 41 (3) (May 1, 2017) 276–301.

[97] G.G. Anderson, G.A. O'toole, Innate and induced resistance mechanisms of bacterial biofilms, in: Bacterial Biofilms, Springer, Berlin, Heidelberg, 2008, pp. 85–105.

[98] S.T. Bekele, G.K. Abay, B. Gelaw, B. Tessema, Bacterial Biofilms; Links to Pathogenesis and Résistance Mechanism.

[99] H. Büttner, D. Mack, H. Rohde, Structural basis of *Staphylococcus epidermidis* biofilm formation: mechanisms and molecular interactions, Frontiers in Cellular and Infection Microbiology 5 (February 17, 2015) 14.

[100] E. Seebach, K.F. Kubatzky, Chronic implant-related bone infections—can immune modulation be a therapeutic strategy? Frontiers in Immunology 10 (2019) 1724.

[101] https://en.wikipedia.org/wiki/Rifampicin.

[102] B. Leijtens, J.B. Elbers, P.D. Sturm, B.J. Kullberg, B.W. Schreurs, Clindamycin-rifampin combination therapy for staphylococcal periprosthetic joint infections: a retrospective observational study, BMC Infectious Diseases 17 (1) (December 2017) 321.

[103] S.M. Shiels, D.J. Tennent, J.C. Wenke, Topical rifampin powder for orthopedic trauma part I: rifampin powder reduces recalcitrant infection in a delayed treatment musculoskeletal trauma model, Journal of Orthopaedic Research 36 (12) (December 2018) 3136–3141.

[104] R.P. Trombetta, M.J. Ninomiya, I.M. El-Atawneh, E.K. Knapp, K.L. de Mesy Bentley, P.M. Dunman, E.M. Schwarz, S.L. Kates, H.A. Awad, Calcium phosphate spacers for the local delivery of sitafloxacin and rifampin to treat orthopedic infections: efficacy and proof of concept in a mouse model of single-stage revision of device-associated osteomyelitis, Pharmaceutics 11 (2) (February 2019) 94.

[105] A.M. Jacobs, M.L. Van Hooff, J.F. Meis, F. Vos, J.H. Goosen, Treatment of prosthetic joint infections due to Propionibacterium: similar results in 60 patients treated with and without rifampicin, Acta Orthopaedica 87 (1) (January 2, 2016) 60–66.

[106] A.K. John, D. Baldoni, M. Haschke, K. Rentsch, P. Schaerli, W. Zimmerli, A. Trampuz, Efficacy of daptomycin in implant-associated infection due to methicillin-resistant *Staphylococcus aureus*: importance of combination with rifampin, Antimicrobial Agents and Chemotherapy 53 (7) (July 1, 2009) 2719–2724.

[107] W. Zimmerli, P. Sendi, Role of rifampin against staphylococcal biofilm infections in vitro, in animal models, and in orthopedic-device-related infections, Antimicrobial Agents and Chemotherapy 63 (2) (February 1, 2019) e01746-18.

[108] https://www.drugbank.ca/drugs/DB00080.

[109] S.B. Rosslenbroich, M.J. Raschke, C. Kreis, N. Tholema-Hans, A. Uekoetter, R. Reichelt, T.F. Fuchs, Daptomycin: local application in implant-associated infection and complicated osteomyelitis, Science World Journal 2012 (2012).

[110] M. Sawada, K. Oe, M. Hirata, H. Kawamura, N. Ueda, T. Nakamura, H. Iida, T. Saito, Linezolid versus daptomycin treatment for periprosthetic joint infections: a retrospective cohort study, Journal of Orthopaedic Surgery and Research 14 (1) (December 1, 2019) 334.

[111] P.S. Corona Pérez-Cardona, V. Barro Ojeda, D. Rodriguez Pardo, C. Pigrau Serrallach, E. Guerra Farfán, C. Amat Mateu, X. Flores Sanchez, Clinical experience with daptomycin for the treatment of patients with knee and hip periprosthetic joint infections, Journal of Antimicrobial Chemotherapy 67 (7) (July 1, 2012) 1749–1754.

[112] K. Malizos, J. Sarma, R.A. Seaton, M. Militz, F. Menichetti, G. Riccio, J. Gaudias, U. Trostmann, R. Pathan, K. Hamed, Daptomycin for the treatment of osteomyelitis and orthopaedic device infections: real-world clinical experience from a European registry, European Journal of Clinical Microbiology and Infectious Diseases 35 (1) (January 1, 2016) 111–118.

[113] G. Luengo, J. Lora-Tamayo, E. Paredes, I. Muñoz-Gallego, A. Díaz, E. Delgado, Daptomycin plus fosfomycin as salvage therapy in a difficult-to-treat total femoral replacement infection, Journal of Bone and Joint Infection 3 (4) (2018) 207.

[114] J.A. Niska, J.H. Shahbazian, R.I. Ramos, J.R. Pribaz, F. Billi, K.P. Francis, L.S. Miller, Daptomycin and

tigecycline have broader effective dose ranges than vancomycin as prophylaxis against a *Staphylococcus aureus* surgical implant infection in mice, Antimicrobial Agents and Chemotherapy 56 (5) (May 1, 2012) 2590–2597.

[115] J.P. Telles, J. Cieslinski, F.F. Tuon, Daptomycin to bone and joint infections and prosthesis joint infections: a systematic review, Brazilian Journal of Infectious Diseases 23 (3) (June 14, 2019) 191–196.

[116] https://en.wikipedia.org/wiki/Gentamicin.

[117] https://www.webmd.com/drugs/2/drug-1496/gentamicin-injection/details.

[118] K. Thompson, S. Petkov, S. Zeiter, C.M. Sprecher, R.G. Richards, T.F. Moriarty, H. Eijer, Intraoperative loading of calcium phosphate-coated implants with gentamicin prevents experimental *Staphylococcus aureus* infection in vivo, PLoS One 14 (2) (2019).

[119] M. Lucke, G. Schmidmaier, S. Sadoni, B. Wildemann, R. Schiller, N.P. Haas, M. Raschke, Gentamicin coating of metallic implants reduces implant-related osteomyelitis in rats, Bone 32 (5) (May 1, 2003) 521–531.

[120] H. Vester, B. Wildemann, G. Schmidmaier, U. Stöckle, M. Lucke, Gentamycin delivered from a PDLLA coating of metallic implants: in vivo and in vitro characterisation for local prophylaxis of implant-related osteomyelitis, Injury 41 (10) (October 1, 2010) 1053–1059.

[121] Y. Yang, H.Y. AO, S.B. Yang, Y.G. Wang, W.T. Lin, Z.F. Yu, T.T. Tang, In vivo evaluation of the anti-infection potential of gentamicin-loaded nanotubes on titania implants, International Journal of Nanomedicine 11 (2016) 2223.

[122] A. Zyaie, A.R. Saied, Prophylactic topical gentamicin after open reduction and internal fixation of long bone fractures in orthopedics, Journal of Applied Sciences 8 (December 2008) 3753–3756.

[123] M. Diefenbeck, C. Schrader, F. Gras, T. Mückley, J. Schmidt, S. Zankovych, J. Bossert, K.D. Jandt, A. Völpel, B.W. Sigusch, H. Schubert, Gentamicin coating of plasma chemical oxidized titanium alloy prevents implant-related osteomyelitis in rats, Biomaterials 101 (September 1, 2016) 156–164.

[124] D. Liu, C. He, Z. Liu, W. Xu, Gentamicin coating of nanotubular anodized titanium implant reduces implant-related osteomyelitis and enhances bone biocompatibility in rabbits, International Journal of Nanomedicine 12 (2017) 5461–5468.

[125] H. Knaepler, Local application of gentamicin-containing collagen implant in the prophylaxis and treatment of surgical site infection in orthopaedic surgery, International Journal of Surgery 10 (January 1, 2012) S15–S20.

[126] https://www.webmd.com/cold-and-flu/qa/what-are-fluoroquinolones.

[127] https://www.uptodate.com/contents/fluoroquinolones.

[128] https://www.drugs.com/mtm/levofloxacin.html.

[129] https://en.wikipedia.org/wiki/Levofloxacin.

[130] N. Asseray, C. Bourigault, D. Boutoille, L. Happi, S. Touchais, S. Corvec, P. Bemer, D. Navas, Levofloxacin at the usual dosage to treat bone and joint infections: a cohort analysis, International Journal of Antimicrobial Agents 47 (6) (June 1, 2016) 478–481.

[131] C. Muller-Serieys, A.S. Mghir, L. Massias, B. Fantin, Bactericidal activity of the combination of levofloxacin with rifampin in experimental prosthetic knee infection in rabbits due to methicillin-susceptible *Staphylococcus aureus*, Antimicrobial Agents and Chemotherapy 53 (5) (May 1, 2009) 2145–2148.

[132] https://en.wikipedia.org/wiki/Ciprofloxacin.

[133] https://www.drugs.com/ciprofloxacin.html.

[134] P. Brouqui, M.C. Rousseau, A. Stein, M. Drancourt, D. Raoult, Treatment of *Pseudomonas aeruginosa*-infected orthopedic prostheses with ceftazidime-ciprofloxacin antibiotic combination, Antimicrobial Agents and Chemotherapy 39 (11) (November 1, 1995) 2423–2425.

[135] J.E. Berdal, I. Skra, P. Mowinckel, P. Gulbrandsen, J.V. Bjørnholt, Use of rifampicin and ciprofloxacin combination therapy after surgical debridement in the treatment of early manifestation prosthetic joint infections, Clinical Microbiology and Infection 11 (10) (October 1, 2005) 843–845.

[136] C. Castro, E. Sanchez, A. Delgado, I. Soriano, P. Nunez, M. Baro, A. Perera, C. Evora, Ciprofloxacin implants for bone infection. In vitro–in vivo characterization, Journal of Controlled Release 93 (3) (December 12, 2003) 341–354.

[137] https://en.wikipedia.org/wiki/Moxifloxacin.

[138] https://www.drugs.com/mtm/moxifloxacin.html.

[139] R. San Juan, A. Garcia-Reyne, P. Caba, F. Chaves, C. Resines, F. Llanos, F. López-Medrano, M. Lizasoain, J.M. Aguado, Safety and efficacy of moxifloxacin monotherapy for treatment of orthopedic implant-related staphylococcal infections, Antimicrobial Agents and Chemotherapy 54 (12) (December 1, 2010) 5161–5166.

[140] https://en.wikipedia.org/wiki/Vancomycin.

[141] https://reference.medscape.com/drug/firvanq-vancocin-vancomycin-342573.

[142] J.A. Inzana, E.M. Schwarz, S.L. Kates, H.A. Awad, A novel murine model of established Staphylococcal bone infection in the presence of a fracture fixation plate to study therapies utilizing antibiotic-laden spacers after revision surgery, Bone 72 (March 1, 2015) 128–136.

[143] M.F. ALhlale, A.H.A. Saleh, K.S. Alsweedi, W.H.A. Edrees, The inhibitory effect of Euphorbia hirta extracts against some wound bacteria isolated from Yemeni patients, Chronicles of Pharmaceutical Science 3 (2) (2019).

[144] L. Zhao, H. Wang, K. Huo, L. Cui, W. Zhang, H. Ni, Y. Zhang, Z. Wu, P.K. Chu, Antibacterial nanostructured titania coating incorporated with silver nanoparticles, Biomaterials 32 (24) (August 1, 2011) 5706–5716.

[145] X.F. Zhang, Z.G. Liu, W. Shen, S. Gurunathan, Silver nanoparticles: synthesis, characterization, properties, applications, and therapeutic approaches, International

Journal of Molecular Sciences 17 (9) (September 2016) 1534.

[146] M. Behravan, A.H. Panahi, A. Naghizadeh, M. Ziaee, R. Mahdavi, A. Mirzapour, Facile green synthesis of silver nanoparticles using *Berberis vulgaris* leaf and root aqueous extract and its antibacterial activity, International Journal of Biological Macromolecules 124 (March 1, 2019) 148–154.

[147] M. Anandan, G. Poorani, P. Boomi, K. Varunkumar, K. Anand, A.A. Chuturgoon, M. Saravanan, H.G. Prabu, Green synthesis of anisotropic silver nanoparticles from the aqueous leaf extract of *Dodonaea viscosa* with their antibacterial and anticancer activities, Process Biochemistry 80 (May 1, 2019) 80–88.

[148] R. Manikandan, R. Anjali, M. Beulaja, N.M. Prabhu, A. Koodalingam, G. Saiprasad, P. Chitra, M. Arumugam, Synthesis, characterization, antiproliferative and wound healing activities of silver nanoparticles synthesized from *Caulerpa scalpelliformis*, Process Biochemistry 79 (April 1, 2019) 135–141.

[149] M.M. Hulikere, C.G. Joshi, Characterization, antioxidant and antimicrobial activity of silver nanoparticles synthesized using marine endophytic fungus-*Cladosporium cladosporioides*, Process Biochemistry 82 (July 1, 2019) 199–204.

[150] S. Ahmed, M. Ahmad, B.L. Swami, S. Ikram, A review on plants extract mediated synthesis of silver nanoparticles for antimicrobial applications: a green expertise, Journal of Advanced Research 7 (1) (January 1, 2016) 17–28.

[151] M. Rafique, I. Sadaf, M.S. Rafique, M.B. Tahir, A review on green synthesis of silver nanoparticles and their applications, Artificial Cells, Nanomedicine, and Biotechnology 45 (7) (October 3, 2017) 1272–1291.

[152] N. Saifuddin, A.Z. Raziah, A.R. Junizah, Carbon nanotubes: a review on structure and their interaction with proteins, Journal of Chemistry 2013 (September 17, 2012).

[153] D. Janas, Towards monochiral carbon nanotubes: a review of progress in the sorting of single-walled carbon nanotubes, Materials Chemistry Frontiers 2 (1) (2018) 36–63.

[154] N.X. Dinh, N.V. Quy, T.Q. Huy, A.T. Le, Decoration of silver nanoparticles on multiwalled carbon nanotubes: antibacterial mechanism and ultrastructural analysis, Journal of Nanomaterials 2015 (2015).

[155] A.J. Haider, M.R. Mohammed, E.A. Al-Mulla, D.S. Ahmed, Synthesis of silver nanoparticle decorated carbon nanotubes and its antimicrobial activity against growth of bacteria, Rendiconti Lincei 25 (3) (September 1, 2014) 403–407.

[156] Y. Seo, J. Hwang, J. Kim, Y. Jeong, M.P. Hwang, J. Choi, Antibacterial activity and cytotoxicity of multi-walled carbon nanotubes decorated with silver nanoparticles, International Journal of Nanomedicine 9 (2014) 4621.

[157] P. Kumarathasan, D. Breznan, D. Das, M.A. Salam, Y. Siddiqui, C. MacKinnon-Roy, J. Guan, N. de Silva, B. Simard, R. Vincent, Cytotoxicity of carbon nanotube variants: a comparative in vitro exposure study with A549 epithelial and J774 macrophage cells, Nanotoxicology 9 (2) (February 17, 2015) 148–161.

[158] https://www.americanelements.com/titanium-oxide-nanotubes-13463-67-7.

[159] https://www.sigmaaldrich.com/technical-documents/articles/materials-science/titania-nanotubes-synthesis-and-applications.html.

[160] Gao A, Hang R, Huang X, Zhao L, Zhang X, Wang L, Tang B, Ma S, Chu PK. The effects of titania nanotubes with embedded silver oxide nanoparticles on bacteria and osteoblasts. *Biomaterials*, 35 (13), April 1, 2014, 4223-4235.

[161] B. Mi, W. Xiong, N. Xu, H. Guan, Z. Fang, H. Liao, Y. Zhang, B. Gao, X. Xiao, J. Fu, F. Li, Strontium-loaded titania nanotube arrays repress osteoclast differentiation through multiple signalling pathways: in vitro and in vivo studies, Scientific Reports 7 (1) (May 24, 2017) 1–6.

[162] S. Mei, H. Wang, W. Wang, L. Tong, H. Pan, C. Ruan, Q. Ma, M. Liu, H. Yang, L. Zhang, Y. Cheng, Antibacterial effects and biocompatibility of titanium surfaces with graded silver incorporation in titania nanotubes, Biomaterials 35 (14) (May 1, 2014) 4255–4265.

[163] Y. Zhang, C. Dong, S. Yang, T.W. Chiu, J. Wu, K. Xiao, Y. Huang, X. Li, Enhanced silver loaded antibacterial titanium implant coating with novel hierarchical effect, Journal of Biomaterials Applications 32 (9) (April 2018) 1289–1299.

[164] U.F. Gunputh, H. Le, K. Lawton, A. Besinis, C. Tredwin, R.D. Handy, Antibacterial properties of silver nanoparticles grown in situ and anchored to titanium dioxide nanotubes on titanium implant against *Staphylococcus aureus*, Nanotoxicology 14 (1) (January 2, 2020) 97–110.

[165] Z. Zeng, X. He, B. Tan, C. Dai, W. Zheng, Titanium oxide nanotubes embedded with silver dioxide nanoparticles for *Staphylococcus aureus* infections after prosthetic joint replacement in animal models, International Journal of Clinical and Experimental Medicine 11 (7) (January 1, 2018) 7392–7399.

[166] D. Rawtani, Y.K. Agrawal, Multifarious applications of halloysite nanotubes: a review, Reviews on Advanced Materials Science 30 (3) (April 2012) 282–295.

[167] Y. Zhang, A. Tang, H. Yang, J. Ouyang, Applications and interfaces of halloysite nanocomposites, Applied Clay Science 119 (January 1, 2016) 8–17.

[168] X. Li, J. Ouyang, H. Yang, S. Chang, Chitosan modified halloysite nanotubes as emerging porous microspheres for drug carrier, Applied Clay Science 126 (June 1, 2016) 306–312.

[169] P. Yuan, D. Tan, F. Annabi-Bergaya, Properties and applications of halloysite nanotubes: recent research advances and future prospects, Applied Clay Science 112 (August 1, 2015) 75–93.

[170] E. Abdullayev, Y. Lvov, Halloysite for controllable loading and release, in: Developments in Clay Science, vol. 7, Elsevier, January 1, 2016, pp. 554–605.

[171] Y. Zhang, Y. Chen, H. Zhang, B. Zhang, J. Liu, Potent anti-bacterial activity of a novel silver nanoparticle-halloysite nanotube nanocomposite powder, Journal of Inorganic Biochemistry 118 (January 1, 2013) 59—64.

[172] S. Jana, A.V. Kondakova, S.N. Shevchenko, E.V. Sheval, K.A. Gonchar, V.Y. Timoshenko, A.N. Vasiliev, Halloy-site nanotubes with immobilized silver nanoparticles for anti-bacterial application, Colloids and Surfaces B: Biointerfaces 151 (March 1, 2017) 249—254.

[173] L. Yu, Y. Zhang, B. Zhang, J. Liu, Enhanced antibacterial activity of silver nanoparticles/halloysite nanotubes/graphene nanocomposites with sandwich-like structure, Scientific Reports 4 (April 11, 2014) 4551.

[174] Z. Shu, Y. Zhang, Q. Yang, H. Yang, Halloysite nanotubes supported Ag and ZnO nanoparticles with synergistically enhanced antibacterial activity, Nanoscale research letters 12 (1) (December 2017) 1—7.

[175] S. Kalay, Z. Yilmaz, O. Sen, M. Emanet, E. Kazanc, M. Çulha, Synthesis of boron nitride nanotubes and their applications, Beilstein Journal of Nanotechnology 6 (1) (January 8, 2015) 84—102.

[176] J.H. Kim, T.V. Pham, J.H. Hwang, C.S. Kim, M.J. Kim, Boron nitride nanotubes: synthesis and applications, Nano Convergence 5 (1) (December 2018) 1—3.

[177] D. Lahiri, V. Singh, A.P. Benaduce, S. Seal, L. Kos, A. Agarwal, Boron nitride nanotube reinforced hydroxy-apatite composite: mechanical and tribological performance and in-vitro biocompatibility to osteoblasts, Journal of the Mechanical Behavior of Biomedical Materials 4 (1) (2011) 44—56.

[178] A. Merlo, V.R. Mokkapati, S. Pandit, I. Mijakovic, Boron nitride nanomaterials: biocompatibility and bio-applications, Biomaterials Science 6 (9) (2018) 2298—2311.

[179] K.L. Firestein, D.G. Kvashnin, A.N. Sheveyko, I.V. Sukhorukova, A.M. Kovalskii, A.T. Matveev, O.I. Lebedev, P.B. Sorokin, D. Golberg, D.V. Shtansky, Structural analysis and atomic simulation of Ag/BN nanoparticle hybrids obtained by Ag ion implantation, Materials and Design 98 (May 15, 2016) 167—173.

[180] J.S. Nithya, A. Pandurangan, Aqueous dispersion of polymer coated boron nitride nanotubes and their anti-bacterial and cytotoxicity studies, RSC Advances 4 (60) (2014) 32031—32046.

[181] N.M. Pu'ad, P. Koshy, H.Z. Abdullah, M.I. Idris, T.C. Lee, Syntheses of hydroxyapatite from natural sources, Heliyon 5 (5) (May 1, 2019) e01588.

[182] V.S. Gshalaev, A.C. Demirchan, Hydroxyapatite: Synthesis, Properties, and Applications, Nova Science Publishers, 2012.

[183] J. Thirumalai, Introductory chapter: the testament of hydroxyapatite: new prospects in regenerative medicinal treatments. Hydroxyapatite: advances in composite nanomaterials, Biomedical Applications and Its Technological Facets (February 14, 2018) 1.

[184] E.M. Rivera-Muñoz, Hydroxyapatite-based materials: synthesis and characterization, Biomedical Engineering-Frontiers and Challenges (August 1, 2011) 75—98.

[185] T.G. Galindo, Y. Chai, M. Tagaya, Hydroxyapatite nano-particle coating on polymer for constructing effective biointeractive interfaces, Journal of Nanomaterials 2019 (2019).

[186] M.R. Shirdar, N. Farajpour, R. Shahbazian-Yassar, T. Shokuhfar, Nanocomposite materials in orthopedic applications, Frontiers of Chemical Science and Engineering 13 (1) (March 1, 2019) 1—3.

[187] B. Li, T. Webster (Eds.), Orthopedic Biomaterials: Advances and Applications, Springer, March 22, 2018.

[188] M. Bongio, J.J. Van Den Beucken, S.C. Leeuwenburgh, J.A. Jansen, Development of bone substitute materials: from 'biocompatible' to 'instructive', Journal of Materials Chemistry 20 (40) (2010) 8747—8759.

[189] G. Fernandez de Grado, L. Keller, Y. Idoux-Gillet, Q. Wagner, A.M. Musset, N. Benkirane-Jessel, F. Bornert, D. Offner, Bone substitutes: a review of their characteristics, clinical use, and perspectives for large bone defects management, Journal of Tissue Engineering 9 (June 2, 2018), 2041731418776819.

[190] R. Canaparo, F. Foglietta, F. Giuntini, C. Della Pepa, F. Dosio, L. Serpe, Recent developments in antibacterial therapy: focus on stimuli-responsive drug-delivery systems and therapeutic nanoparticles, Molecules 24 (10) (January 2019) 1991.

[191] M.C. Chifiriuc, A.M. Holban, C. Curutiu, L.M. Ditu, G. Mihaescu, A.E. Oprea, A.M. Grumezescu, V. Lazar, Antibiotic drug delivery systems for the intracellular targeting of bacterial pathogens, in: Smart Drug Delivery System, IntechOpen, February 10, 2016.

[192] E. Andronescu, A.M. Grumezescu, M.I. Guşă, A.M. Holban, F.C. Ilie, A. Irimia, I.F. Nicoară, M. Tone, Nano-hydroxyapatite: novel approaches in biomedical applications, in: Nanobiomaterials in Hard Tissue Engineering, William Andrew Publishing, January 1, 2016, pp. 189—213.

[193] K. Szurkowska, A. Laskus, J. Kolmas, Hydroxyapatite-based materials for potential use in bone tissue infections, Hydroxyapatite-Advances in Composite Nanomaterials Biomedical Applications and Its Technological Facets, Intechopen, (February 14, 2018) 109—136.

[194] N. Roveri, M. Iafisco, Evolving application of biomimetic nanostructured hydroxyapatite, Nanotechnology, Science and Applications 3 (2010) 107.

[195] Q. Zhou, T. Wang, C. Wang, Z. Wang, Y. Yang, P. Li, R. Cai, M. Sun, H. Yuan, L. Nie, Synthesis and characterization of silver nanoparticles-doped hydroxyapatite/alginate microparticles with promising cytocompatibility and antibacterial properties, Colloids and Surfaces A: Physicochemical and Engineering Aspects 585 (January 20, 2020) 124081.

[196] V. Lazić, I. Smičiklas, J. Marković, D. Lončarević, J. Dostanić, S.P. Ahrenkiel, J.M. Nedeljković, Antibacterial ability of supported silver nanoparticles by functionalized hydroxyapatite with 5-aminosalicylic acid, Vacuum 148 (February 1, 2018) 62—68.

[197] F.A. Andrade, L.C. de Oliveira Vercik, F.J. Monteiro, E.C. da Silva Rigo, Preparation, characterization and antibacterial properties of silver nanoparticles—hydroxyapatite composites by a simple and eco-friendly method, Ceramics International 42 (2) (February 1, 2016) 2271—2280.

[198] K.A. Poelstra, N.A. Barekzi, D.W. Grainger, A.G. Gristina, T.C. Schuler, A novel spinal implant infection model in rabbits, Spine 25 (4) (February 15, 2000) 406—410.

[199] A. Sodagar, A. Akhavan, E. Hashemi, S. Arab, M. Pourhajibagher, K. Sodagar, M.J. Kharrazifard, A. Bahador, Evaluation of the antibacterial activity of a conventional orthodontic composite containing silver/hydroxyapatite nanoparticles, Progress in Orthodontics 17 (1) (December 2016) 1—7.

[200] G.S. Kumar, S. Rajendran, S. Karthi, R. Govindan, E.K. Girija, G. Karunakaran, D. Kuznetsov, Green synthesis and antibacterial activity of hydroxyapatite nanorods for orthopedic applications, MRS Communications 7 (2) (June 2017) 183—188.

[201] S. Lamkhao, M. Phaya, C. Jansakun, N. Chandet, K. Thongkorn, G. Rujijanagul, P. Bangrak, C. Randorn, Synthesis of hydroxyapatite with antibacterial properties using a microwave-assisted combustion method, Scientific Reports 9 (1) (March 8, 2019) 1—9.

[202] W. Suvannapruk, F. Thammarakcharoen, P. Phanpiriya, J. Suwanprateeb, Development of antibiotics impregnated nanosized silver phosphate-doped hydroxyapatite bone graft, Journal of Nanomaterials 2013 (2013).

[203] Z.C. Xiong, Z.Y. Yang, Y.J. Zhu, F.F. Chen, Y.G. Zhang, R.L. Yang, Ultralong hydroxyapatite nanowires-based paper co-loaded with silver nanoparticles and antibiotic for long-term antibacterial benefit, ACS Applied Materials and Interfaces 9 (27) (July 12, 2017) 22212—22222.

[204] C.M. Xie, X. Lu, K.F. Wang, F.Z. Meng, O. Jiang, H.P. Zhang, W. Zhi, L.M. Fang, Silver nanoparticles and growth factors incorporated hydroxyapatite coatings on metallic implant surfaces for enhancement of osteoinductivity and antibacterial properties, ACS Applied Materials and Interfaces 6 (11) (June 11, 2014) 8580—8589.

[205] Z. Ni, X. Gu, Y. He, Z. Wang, X. Zou, Y. Zhao, L. Sun, Synthesis of silver nanoparticle-decorated hydroxyapatite (HA@ Ag) poriferous nanocomposites and the study of their antibacterial activities, RSC Advances 8 (73) (2018) 41722—41730.

[206] S. Roopalakshmi, R. Ravishankar, S. Belaldavar, R.V. Prasad, A.R. Phani, Investigation of structural and morphological characteristic of hydroxyapatite synthesized by sol-gel process, Materials Today: Proceedings 4 (11) (January 1, 2017) 12026—12031.

[207] D. Ali, S. Sen, Finite element analysis of the effect of boron nitride nanotubes in beta tricalcium phosphate and hydroxyapatite elastic modulus using the RVE model, Composites Part B: Engineering 90 (April 1, 2016) 336—340.

[208] M.J. Saif, H.M. Asif, M. Naveed, Properties and modification methods of halloysite nanotubes: a state-of-the-art review, Journal of the Chilean Chemical Society 63 (3) (2018) 4109—4125.

[209] B. Lecouvet, J. Horion, C. D'haese, C. Bailly, B. Nysten, Elastic modulus of halloysite nanotubes, Nanotechnology 24 (10) (February 20, 2013) 105704.

[210] Y. Rémond, S. Ahzi, M. Baniassadi, H. Garmestani, Applied RVE Reconstruction and Homogenization of Heterogeneous Materials, John Wiley & Sons, June 14, 2016.

[211] K. Roy, S.C. Debnath, A. Pongwisuthiruchte, P. Potiyaraj, Up-to-date review on the development of high performance rubber composites based on halloysite nanotube, Applied Clay Science 183 (December 15, 2019) 105300.

[212] https://www.americanelements.com/boron-nitride-nano tubes-10043-11-5.

[213] https://www.americanelements.com/silver-nanoparticles -7440-22-4.

[214] https://www.americanelements.com/silver-nanorods-74 40-22-4.

[215] M.A. Neto, A. Amaro, L. Roseiro, J. Cirne, R. Leal, Engineering Computation of Structures: The Finite Element Method, Springer, Cham, September 29, 2015.

[216] Y.A. Orban, D.L. Manea, C. Aciu, Study of methods for simulating multiphase construction materials, Procedia Manufacturing 22 (January 1, 2018) 256—261.

[217] MSC Software Company, Digimat Users' Manual, E-xstream Engineering, 2016.

[218] I. Watanabe, A. Yamanaka, Voxel coarsening approach on image-based finite element modeling of representative volume element, International Journal of Mechanical Sciences 150 (January 1, 2019) 314—321.

[219] J. Segurado, R.A. Lebensohn, J. LLorca, Computational homogenization of polycrystals, in: Advances in Applied Mechanics, vol. 51, Elsevier, January 1, 2018, pp. 1—114.

[220] R.H. Poelma, X. Fan, Z.Y. Hu, G. Van Tendeloo, H.W. van Zeijl, G.Q. Zhang, Effects of nanostructure and coating on the mechanics of carbon nanotube arrays, Advanced Functional Materials 26 (8) (February 2016) 1233—1242.

[221] U. Galvanetto, M.H. Aliabadi, Multiscale Modeling in Solid Mechanics: Computational Approaches, World Scientific, 2010.

[222] I. Alfonso, I.A. Figueroa, V. Rodriguez-Iglesias, C. Patino-Carachure, A. Medina-Flores, L. Bejar, L. Perez, Estimation of elastic moduli of particulate-reinforced composites using finite element and modified Halpin—Tsai models, Journal of the Brazilian Society of Mechanical Sciences and Engineering 38 (4) (April 1, 2016) 1317—1324.

[223] M.K. Hassanzadeh-Aghdam, J. Jamali, A new form of a Halpin—Tsai micromechanical model for characterizing the mechanical properties of carbon nanotube-reinforced polymer nanocomposites, Bulletin of Materials Science 42 (3) (June 1, 2019) 117.

[224] Z. Luo, X. Li, J. Shang, H. Zhu, D. Fang, Modified rule of mixtures and Halpin–Tsai model for prediction of tensile strength of micron-sized reinforced composites and Young's modulus of multiscale reinforced composites for direct extrusion fabrication, Advances in Mechanical Engineering 10 (7) (July 2018), 1687814018785286.

[225] B. Chen, S. Li, H. Imai, L. Jia, J. Umeda, M. Takahashi, K. Kondoh, Load transfer strengthening in carbon nanotubes reinforced metal matrix composites via in-situ tensile tests, Composites Science and Technology 113 (June 5, 2015) 1–8.

[226] B. Chen, K. Kondoh, H. Imai, J. Umeda, M. Takahashi, Simultaneously enhancing strength and ductility of carbon nanotube/aluminum composites by improving bonding conditions, Scripta Materialia 113 (March 1, 2016) 158–162.

[227] L. Shi, L. Wang, Y. Duan, W. Lei, Z. Wang, J. Li, X. Fan, X. Li, S. Li, Z. Guo, The improved biological performance of a novel low elastic modulus implant, PLoS One 8 (2) (February 21, 2013) e55015.

CHAPTER 12

Antibiotics as Emerging Pollutants in Water and Its Treatment

MARIEL GODOY • JULIO SÁNCHEZ

1 EMERGING POLLUTANTS

At present, it is essential to find new alternatives to preserve the most valuable resource our planet has, which is the water. There are limitations to the availability of clean water, and it is recognized as a global issue. Population growth and industrial development are the main reasons for the formation of large quantities of pollutants in various states of matter, which usually end up in the atmosphere, soil, as well as in rivers, oceans, groundwater, and drinking water. Each country has a waste management plan, some more comprehensive than others.

Aqueous pollutants can be of various types. We can classify aqueous pollutants into two large groups: inorganic and organic. Within the category of inorganic pollutants, we can find various toxic metals such as lead, zinc, copper, mercury, silver, nickel, chromium, cadmium, and metalloids such as arsenic: waste belonging to industrial wastewater such as metal plating, mining activities, smelting, battery manufacturing, oil refining, paint production, pesticide industries, pigment manufacturing, printing, and photography.

On the other hand, organic pollutants are those that are produced by the oil, food, pharmaceutical, and refinery industries, among others; dyes, surfactants, oils, lubricants, organic solvents, petroleum products, pharmaceutical products (e.g., pharmaceuticals, antibiotics, hormones, etc.) can be found in their wastewater, which is nowadays known as emerging pollutants. These are compounds of different origin and chemical nature, essentially organic compounds or synthetic chemicals [1]. Personal care products, surfactants, plasticizers, industrial additives, as well as pharmaceutically active compounds such as hormones, steroids, lipid regulators, and antibiotics are considered in this classification [2,3]. These compounds are considered emerging because their regulation and entry into the ecosystem have not yet been regulated. These compounds enter the ecosystem in various ways: industrial, hospital, livestock, agricultural, or domestic effluents. Although some of them can be degraded and/or removed from the environment, not all are susceptible to these processes and their constant introduction to the environment has generated alertness in the scientific community [2–4].

2 ANTIBIOTICS: IMPACT ON THE ECOSYSTEM AND HUMAN HEALTH

According to the World Health Organization, there are currently few systematic monitoring programs and comprehensive studies on human exposure to pharmaceutical products present in drinking water. These are not considered in water treatment programs and are not monitored [2,3].

Since the discovery of penicillin in 1929, the incorporation of a varied amount of medicines that preserve the quality and duration of life in living beings has significantly improved their health and longevity, but given the improvement of human health, it was not taken in counts the long-term effects of medications over the years [4].

Today, we find in the literature that the presence of pharmaceutical chemicals in aquatic environments has been recognized and established, and various processes of removal of these are investigated. Depending on the physicochemical properties of the drugs, their metabolites, degradation products, and the characteristics of the soils, these substances can reach groundwater and contaminate the aquifers or be retained in the soil and accumulate and can affect the ecosystem and humans through the food chain.

The main routes of entry of pharmaceutical products into the environment are through industrial production processes, human and animal excretion, hospital effluents, inappropriate disposal of pharmaceutical products, animal and agricultural effluents, as well as irrigation

Antibiotic Materials in Healthcare. https://doi.org/10.1016/B978-0-12-820054-4.00012-4

with contaminated water. A wide range of pharmaceutical products has been detected in surface and groundwater. These pharmaceutical wastes are transported to the water cycle by different routes. Among these, wastewater treatment plants act as a gateway for these products to water streams, as many of these compounds are not really retained in their processes [3,5–8]. The presence of these compounds in the environment goes unnoticed because their distribution is random and their concentrations are low. Environmental studies that detect their presence in surface waters have been conducted, generating questions about their potential environmental impact and the health of living beings [5].

Antibiotics are widely used drugs in the world whose effects against pathogenic microorganisms in animals and humans are known. Its production and consumption have increased, which has favored the increase of large discharges on the bodies of water. There is evidence of the presence of antibiotic residues in the environment and their involvement in the defense mechanisms of living organisms. Antibiotics with the greatest presence in water streams include tetracyclines, aminoglycosides, macrolides, β-lactams, vancomycins, phenols, and fluoroquinolones, among others [9–12]. Some researchers suggest that prolonged exposure to traces of antibiotics leads to the development of antibiotic-resistant bacteria, and also long-term adverse effects on the ecosystem and human health [13].

Scientific research worldwide has focused on how to extract these antibiotics from aqueous effluents. Antibiotics in the β-lactam family or β-lactams have been of great interest in this research for their extensive use in both human and veterinary medicine. β-Lactams are a group of antibiotics of natural or semisynthetic origin that is characterized by possessing in their chemical structure a β-lactam ring. These work by inhibiting the last stage of bacterial cell-wall synthesis. They constitute the largest family of antimicrobials and the most widely used in clinical practice [4,14]. In the case of the possible effects of these species, recent studies on amoxicillin

(AMX), sodium ampicillin (MPI), and sodium cloxacillin (CLX) (see Fig. 12.1) indicate various ecosystem damage. Some authors have found AMX and CLX in wastewater, further demonstrating that these pharmaceutical products can influence the structure and function of algae communities in areas receiving these effluents [15].

Other researchers have quantified the amount of these antibiotics in surface water, groundwater, and final effluents from wastewater treatment plants, in concentrations in the nanogram/microgram per liter range [12].

The use of these pharmaceutically active compounds takes many decades; however, monitoring and remediation of these compounds in water have been done only in the last 10 years.

3 DEGRADATION OF ANTIBIOTICS

Within the treatments applied for the remediation of antibiotics in water, we find the use of conventional treatments (biological processes and filtration), oxidation processes (advanced oxidation and chlorination), adsorption processes, membranes, and combination treatments.

Conventional wastewater treatments are not suitable or efficient for the elimination of large amounts of antibiotics. For example, water treatment based on biological processes has been insufficient to eliminate most pharmaceuticals. Efficient removal of antibiotics can be ensured using processes such as advanced oxidation, but the high cost of equipment and maintenance as well as energy supply are disadvantages, as well as the cost required for peroxide cooling in advanced oxidation processes discards their use for large amounts of aqueous effluents [12,13]. Optimization is always possible, complementing treatments, taking into account that treatments such as advanced oxidation degrade contaminant molecules, and those such as polymeric membranes can remove contaminants from water (see Fig. 12.2).

Below we will define the most commonly used processes applied in antibiotics.

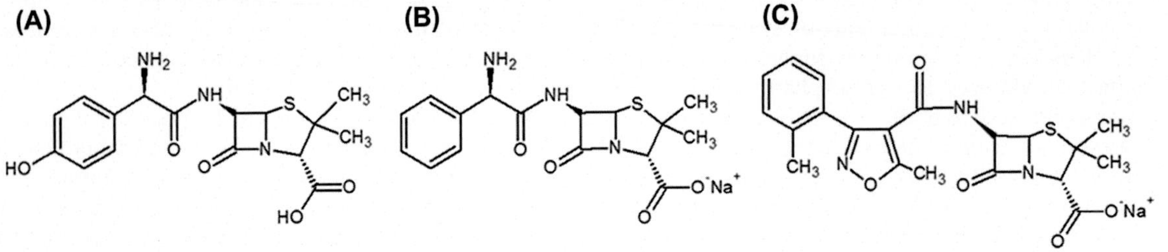

FIG. 12.1 Chemical structure of β-lactam antibiotics: **(A)** amoxicillin, **(B)** sodium ampicillin, and **(C)** sodium cloxacillin.

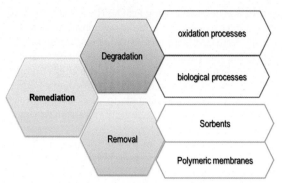

FIG. 12.2 General classification of aqueous remediation of antibiotics.

3.1 Advanced Oxidation

Advanced oxidation processes begin with the use of free radical species (strong oxidizers). These processes serve to degrade contaminants to simple, nontoxic molecules. Free radical species are atoms or molecules containing at least one unpaired electron, such as hydroxyl radical (HO•), superoxide radical (O$_2$•-), hydroperoxyl radical (HO$_2$•), or alkoxyl radical (RO•), the most attractive being the radical HO•, for its features: its nonselective nature, high reactivity, and powerful oxidizing capabilities [16].

According to reports in the literature, advanced oxidation processes (AOPs) commonly applied for pharmaceutical wastewater include three types: photochemical, nonphotochemical processes, and hybrid or combined processes (see Fig. 12.3).

A considerable number of work related to the application of AOPs for the reduction and/or degradation of pharmaceutical products in water has been published. In this section, we will mention those linked to antibiotics, such as ozonation, ozonation/ultrasonic irradiation, Fenton and photo-Fenton, UV/H$_2$O$_2$, sonolysis, and combined processes.

3.2 Ozonation

The use of ozone (O$_3$) is reported as a pretreatment process in waters containing pharmaceuticals. The experiments are conducted on synthetic wastewater, surface water, and effluents from the wastewater treatment plant. The degradation efficiency of four groups of pharmaceutical products (antibiotics, estrogens, acidic and neutral pharmaceuticals) has been studied, and the effect of ozone dose and pH on degradation efficiency is monitored. A Microtox bioassay was used to assess the change in the toxicity of aqueous solutions before and after ozonization. The oxidation efficiency of antibiotics, estrogens, and neutral pharmaceuticals increased as the dose of ozone and pH increased. It was experimentally determined that the optimal ozone input dose corresponds to 188.1, 222.3, and 222.4 mg/h, producing the highest oxidation efficiency for pharmaceuticals studied in synthetic wastewater, surface water, and effluents of the wastewater treatment plant, respectively. A specific mean ozone dose of 2.05 for antibiotics, 1.11 for estrogen, and 1.30 mg O$_3$/mg DOC (degraded organic compounds) for neutral pharmaceuticals significantly reduced the acute toxicity of aqueous solutions and mineralized more than 40%, 33%, and 23%, respectively, of DOC in less than 1 min. The results indicate the effectiveness of ozone-based AOPs for the effective degradation of pharmaceutical products present in water. The results showed that the ozonation process is more effective than other

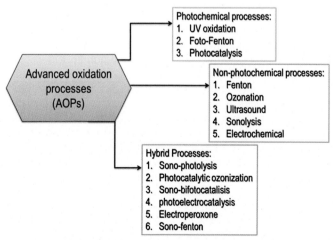

FIG. 12.3 Types of AOPs in pharmaceutical wastewater treatments.

conventional oxidation processes (Cl_2 and ClO_2) to eliminate pharmaceuticals and reduce wastewater toxicity. Specific ozone doses ranging from 0.82 to 2.55 mg O_3/mg DOC resulted in >99.9% elimination for most of the pharmaceutical products studied. Increased toxicity for aqueous solutions of acid pharmaceuticals increased the dose of ozone to 2.24 mg O_3/mg DOC for these, due to the formation of more toxic by-products [17].

3.3 Ozonation and Ultrasound Irradiation

This technique has been used to treat water with the amoxicillin antibiotic by ultrasonic irradiation of medium–high frequency and/or ozonation. The ultrasonic irradiation process was carried out in a batch reactor for amoxicillin aqueous solutions at three different frequencies (575, 861, and 1141 kHz). The applied ultrasonic power was 75 W, and the diffuse power was calculated as 14.6 W/L. The highest elimination was achieved at an ultrasonic frequency of 575 kHz (99%) with the first-order reaction constant of pseudo–first-order greater than 0.04 min^{-1} at pH 10, but the mineralization achieved was about 10%. The presence of alkalinity and humic acid species had a negative effect on elimination efficiency (50% decrease). To improve these results, ozonization was applied with or without ultrasound. Ozone eliminated amoxicillin at a rate 50 times faster than ultrasound. In addition, due to the synergistic effect, ozone coupling and ultrasound resulted in a speed constant of 2.5 min^{-1} (625 times more than ultrasound). In the processes where ozone was used, humic acid showed no significant effect because the velocity constant was so high that ozone has easily overcome the effects of removing natural water constituents. In addition, intermediate compounds, after incomplete oxidation mechanisms, have been analyzed to reveal possible amoxicillin degradation pathways through ultrasonic irradiation and ozonation applications. The results of the intermediate compound and toxicity experiments were investigated to provide a clear explanation of the safety of the resulting solution. The relevance of all results concluded that the hybrid advanced oxidation system is a good choice for amoxicillin removal [18].

3.4 Fenton and Photo-Fenton

A study on the influence of iron species on the degradation of AMX through Fenton and photo-Fenton has been reported. The degradation of AMX was favored in the presence of the potassium ferrioxalate complex (FeO_x) compared to $FeSO_4$. Total oxidation of AMX in the presence of FeO_x was obtained after 5 min, while

15 min was necessary using $FeSO_4$. The results obtained with the *Daphnia magna* bioassays showed that the toxicity decreased from 65% to 5% after 90 min of irradiation in the presence of $FeSO_4$. However, toxicity increased again to a maximum of 100% after 150 min, indicating the generation of toxic intermediates [19].

Similarly, research related to the treatment of fluoroquinolone ciprofloxacin (CIP) has been reported, through the photo-Fenton process in Milli-Q water, to evaluate the degradation of this antibiotic at high and low concentrations. In the presence of different iron sources (iron citrate, iron oxalate, and modified iron nitrate) and at different pHs (2.5, 4.5, and 6.5), the following results are generated. For a high concentration of CIP, degradation was greatly influenced by the source of iron, which resulted in a much lower efficiency with iron nitrate. At pH 4.5, the highest removal of total organic carbon (TOC) (0.87) was achieved in the presence of iron citrate, while similar values of CIP degradation with oxalate and citrate (0.98 after 10 min) were obtained. For the low concentration of CIP, a much higher conversion was observed in the presence of citrate or oxalate relative to iron nitrate up to pH 4.5 [20].

3.5 UV/H_2O_2 Degradation

The effectiveness of UV processes in the elimination of pharmaceutical products depends largely on the UV absorption presented by the pharmaceutical product. The critical parameters that determine the degradation kinetics of direct UV photolysis also comprise the speed constants, kUV, quantum yield, and the molar extinction coefficient. In addition, UV combined with H_2O_2 (UV/H_2O_2) generally provides better removal efficiencies for pharmaceutical products with low UV absorption. UV/H_2O_2 processes are governed by the concentration of H_2O_2, HO radical formation rate, UV light intensity, water components, the chemical structure of the pharmaceutical industry, and also the pH of the solution [16].

The degradation of the antibiotic amoxicillin, AMX, treated with direct photolytic processes UV-C and UV/H_2O_2, was investigated. In addition, the antibacterial activity of the UV/H_2O_2 advanced oxidation treated solution was compared with the ozone-treated AMX solution. The rate of AMX degradation in both processes was adjusted to pseudo–first-order kinetics and rates increased up to six times by increasing the addition of H_2O_2 to 10 mM of H_2O_2 compared to direct photolysis. However, low mineralization was achieved in both processes, showing a maximum of 50% removal of TOC with UV/H_2O_2 after a reaction time of 80 min with the addition of 10 mM H_2O_2. In addition, microbial growth inhibition bioassays were performed to determine any residual antibacterial activity of the possible photoproducts left in the treated solutions. An increase in antibacterial activity was observed in the treated samples compared to the untreated sample in a time-based comparison. However, the UV/H_2O_2 process effectively eliminated any antibacterial activity of AMX and its intermediate photoproducts at 20 min of contact with a 10 mM H_2O_2 dose after the complete elimination of AMX [21].

3.6 Sonolysis

Sonolysis or ultrasonic irradiation is a promising technique for the degradation of persistent organic molecules such as pharmaceuticals into wastewater. In the research cited, sonolytic degradation is used in the antibiotic of the fluoroquinolone family, CIP. During the first experiment at 25°C and 544 kHz, the degradation of a CIP 15 mg/L solution demonstrated a degradation constant of pseudo−first-order k^1 equal to 0.0067 ± 0.0001 min^{-1} ($n = 3$). Experiments with the addition of t-butanol as a radical remover showed that reaction with OH radicals. This is the main CIP degradation path. Obtaining OH• radicals was greater than 544 kHz, this is considered the most favorable frequency for CIP degradation compared to 801 ($k^1 = 0.0055$ min^{-1}) and 1081 kHz ($k^1 = 0.0018$ min^{-1}). The degradation constant also depends to a large extent on the temperature of the solution. The degradation constant increased significantly by increasing the temperature from 0.0055 min^{-1} a 15°C a 0.0105 min^{-1} a 45°C. According to the law of Arrhenius, it was determined that the apparent activation energy was 17.5 kJ mol^{-1}. This suggests that the degradation of CIP is controlled by diffusion [22].

3.7 Hybrid Oxidation Processes

In general, it is observed that these techniques have varied advantages in the degradation of antibiotics having better results, as they are optimized. For example, the degradation of two organic pollutants (amoxicillin and diclofenac) in 0.1 mM aqueous solutions was studied using advanced oxidation processes: ozonation, photolysis, photolytic ozonation, photocatalysis, and photocatalytic ozonation. Diclofenac was rapidly degraded under direct photolysis by artificial light (medium-pressure steam arc, exc > 300 nm), while amoxicillin remained very stable. In the presence of ozone, regardless of the type of process, a complete degradation of both organic pollutants was observed in less than 20 min. Photolysis or ozonization alone led to modest TOC removal values (16% or

41%, respectively, within 180 min), while for photocatalysis (no ozone presence) a significant fraction of nonoxidized compounds remained in the water treatment (−15% after 180 min). In the case of photolytic ozonization, the kinetics of TOC elimination was slow. In contrast, a relatively rapid and complete mineralization of amoxicillin and diclofenac (30 and 120 min, respectively) was achieved by applying the photocatalytic ozonization process (a hybrid process). The absence of toxicity of treated waters was confirmed by growth inhibition tests using two different microorganisms, *Escherichia coli* and *Staphylococcus aureus*. Photocatalytic ozonization was also applied to urban wastewater enriched with amoxicillin and diclofenac. The original contaminants oxidized easily, but the TOC removal was only 68%, mainly due to the persistent presence of oxamic acid in the treated sample hybrid [23].

4 SORBENTS AND MEMBRANES APPLIED IN ANTIBIOTIC REMOVAL

Sorption could serve as a preconcentration treatment of contaminants for its ease of implementation, efficiency, among other qualities, and would have applicability in combined processes such as those mentioned earlier. Antibiotic removal investigations with sorbents are what we will discuss in this section [24−28]. One of the advantages of working with solid−liquid extraction systems with sorbents is the separation of both phases easily through physical processes, favoring the removal of the solid phase if used in batch, or elution in the matrix understudy if it is used in the column. Sorbents such as activated carbon, carbon nanotubes, graphene, biomaterials derived from chitosan, and cellulose, among others, have been reported for the remediation of aqueous matrices with different antibiotics [8−13,15]. For example, granular activated carbon is an applicable adsorbent; however, its regeneration is difficult and the efficiency of activated carbon removal is low for these emerging pollutants.

Preliminary studies indicate that the membrane separation method and the use of biosorbents represent an appropriate alternative for antibiotic removal applications due to their high efficiency and satisfactory separation without using additional chemical compounds.

Below are some cases where various sorbents are used for the removal of antibiotics, both from the β-lactam family, and the fluoroquinolone family.

4.1 β-Lactam Antibiotic Removal

The ionic characteristics of AMX vary with pH (see Fig. 12.4), resulting in a low adsorption capacity of

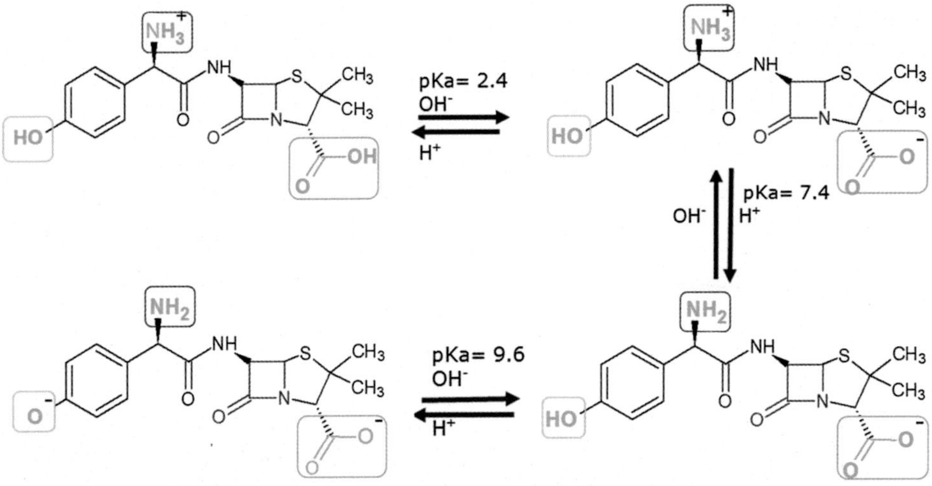

FIG. 12.4 Ionic characteristics of amoxicillin at various pHs.

nonmodified AMX adsorbents. pH is a key factor in establishing, for example, the antibiotic load.

Depending on the pH the AMX has different ionic forms. when that pH is acid, the amino group is protonated, at pH neutral, the AMX behaves like zwitterion; at pH over 7.4 until 9.6, the AMX presents the carboxylic acid group as carboxylate, and on this basic pH, in addition, hydroxyl is deprotonated. The application and functionalization of polymeric or biomaterial materials facilitate the removal of these products.

4.1.1 Biosorbents

As a first example, we have flax noil (FN), which is the main byproduct of the separation of flax fiber by grafting [29]. This is a natural and sustainable material consisting of cellulose and lignin. It is used as a reinforcement material for biodegradable green compounds and in the production of cotton fiber. In the research noted, these fibers are modified with quaternary ammonium groups by grafting; getting quaternary cellulose from flax noil (QCFN), which was used as an adsorbent. QCFN's capacity for AMX adsorption in aqueous solution was investigated. In addition, the characterization of the initial product and quaternized fiber by FTIR spectroscopy, XPS, and XR was performed, confirming the incorporation of the quaternary ammonium group. Batch adsorption studies allow you to determine the retention percentage (efficiency), the optimal retention time using kinetic studies, and maximum sorption capacity [29]. Results show that adsorption capacity increased with increased pH and maximum adsorption capacity reached 183.14 mg/g. The behavior of the AMX adsorption on the adsorbent conforms to the pseudo–second-order kinetic model, the intraparticular diffusion model, and the Langmuir isotherm. The activation energy is 42.25 kJ/mol, which implies that AMX adsorption is a chemically controlled process. The positive value of H (24.9 kJ/mol) and the negative value of G (−4.26, 5.11, 6.19 kJ/mol) showed that adsorption was endothermic and spontaneous. QCFN was shown to be a promising and efficient adsorbent for the removal of AMX from the aqueous solution, having a potential application in the purification of wastewater containing AMX.

4.1.2 Membranes

On the other hand, nanofiltration membranes can be manufactured based on polymers, and many of them are functionalized with the presence of functional groups such as carboxylic or sulfonic acids. The application of membranes has also been reported for the removal of β-lactam antibiotics. As the pore sizes of microfiltration and ultrafiltration membranes are very large for removing antibiotics, only nanofiltration and reverse osmosis membranes can be used for antibiotic removal [12,13].

These polymeric materials can be used because they can act selectively through the electrostatic interaction between ions and functional groups. Various mechanisms have been proposed for these surface processes: steric impediment, electrostatic interaction between charged solutes and polymer-attached groups, the interaction between ions and polarized charge [8,9,15].

In the research described below, pH-sensitive polysulfone (PSf)/polyacrylic acid (PAA) nanofiltration membranes were synthesized for the separation of amoxicillin from pharmaceutical wastewater [30].

Ultrafiltration membranes with different pore sizes were prepared, using polyethylene glycol for this purpose, by phase reversal process. In addition, PSf and PAA were incorporated through a UV grafting process.

The efficiency of amoxicillin separation improved as a result of its pH-sensitive nature, as well as surface activity and pH sensitivity of developed nanofiltration membranes. Finally, the separation of amoxicillin with synthesized nanofiltration membranes reaches its optimal efficiency at pH 10 reaching 91% retention, while the flow is 108.3 L/hm^2, using the mechanism proposed in Fig. 12.5. The proposed mechanism suggests that by adding NaOH they increase negative loads on the membrane surface by increasing the amount of dissociated carboxyl groups, as well as the negative loads of AMX favoring separation with increased pH.

4.2 Removing Fluoroquinolones Antibiotics

Amoxicillin is not the only antibiotic being studied in aqueous removal systems with polymers or biomaterials. In addition, in the literature, the use of polymeric ultrafiltration membranes coupled to water-soluble polymers (polymer-assisted liquid phase retention, LPR) for the removal of the antibiotic from the fluoroquinolone family, ciprofloxacin [31], is reported. This antibiotic, like AMX, can be in dissolution in three ion forms depending on pH: as cation at pH under 5.9, as anion at pH over 8.89, and as zwitterion at neutral pH.

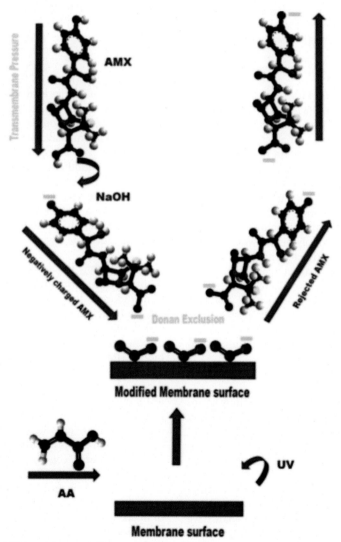

FIG. 12.5 Schematic amoxicillin separation process by using pH-sensitive membranes [30].

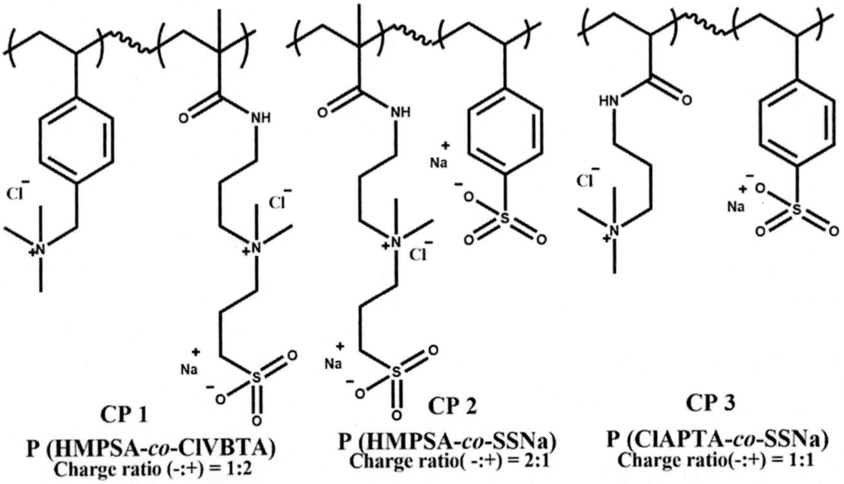

FIG. 12.6 Water-soluble copolymers studied in the removal of the antibiotic ciprofloxacin by the LPR technique [31].

4.2.1 LPR technique

The research reported details of the preparation of three copolymers: CP1, CP2, and CP3 (see Fig. 12.6), and the LPR technique analyses the retention capacity of ciprofloxacin against various variables such as pH, ionic strength, optimal concentration of polymer, and maximum retention capacity. Efficient retention of ciprofloxacin (80%) was reported at the molar polymer concentration: ciprofloxacin 20:1 to pH 5, for CP2 and CP3. The maximum retention capacity of ciprofloxacin was determined by the enrichment method, demonstrating that the C3 copolymer shows the highest retention capacity [31]. In summary, the results showed a retention capacity of ciprofloxacin by the three copolymers delivering retention values of 70% and 80% at pH 5 and remarkably low retention at pH 3 and 9. In addition, the retention of ciprofloxacin decreased as ion strength increased due to competition between the ions present and the antibiotic by the sites loaded in the polymer chains.

4.2.2 Magnetic biochar

Another example of removal of the same family of antibiotics is the use of sorbents obtained by modifying stems and potato leaves (BC) with Fe_3O_4 and MnOx forming a biomaterial called magnetic biochar (MMB), which was successfully characterized and used for elimination of norfloxacin (NOR), CIP, and enrofloxacin (ENR) [32]. The results of the characterization indicate that Fe_3O_4 and MnO_x are incorporated into the formed biomaterial and that these materials have great porosity. Adsorption experiments showed that the maximum MMB adsorption capabilities for NOR, CIP, and NRR were 6.94, 8.37, and 7.19 mg/g,

respectively (see Fig. 12.7). Compared to BC, MMB adsorption capabilities increased 1.2, 1.5, and 1.6 times for NOR, CIP, and ENR, respectively. The kinetic model and the Langmuir model were successfully correlated with the experimental data. Thermodynamic studies revealed that the adsorption processes were spontaneous and endothermic. MMB adsorption capacity decreases by increasing solution pH (between 3.0 and 10.0) and with increased ion strength. In conclusion, the MMB has a high efficiency of elimination of FQs, an easy separation; in addition, these materials presented an adequate regeneration capacity [32].

4.3 Removing Phenicol Antibiotics

In the past year, other literature reports report on sorption studies in the family of phenicol antibiotics. In this study, a structured porous 3D biogel (3D-PBA) is first synthesized, with a large surface area (2607 m^2/g), which was applied in the removal of phenicol antibiotics (PABs). The adsorption capacity of chloramphenicol, florfenicol, and thiamphenicol was determined to be obtained by 786.1, 751.5, and 691.9 mg/g at 298 K, respectively. The Langmuir–Freundlich isothermal model describes the effectiveness of biomaterial where PABs can be completely eliminated in 10 min with 90% removal efficiency, even at an initial concentration as high as 40 mg/L. As a result, the effect of pores, the interaction of the π-π/n-π EDA, and the electrostatic interaction plays a crucial role and is also accompanied by the interaction of hydrogen bonds. In addition, the synergistic action of hydrophobic interaction also acts in the adsorption process. In general, 3D-PBA can act as an efficient candidate for ultrafast removal of PABs [33].

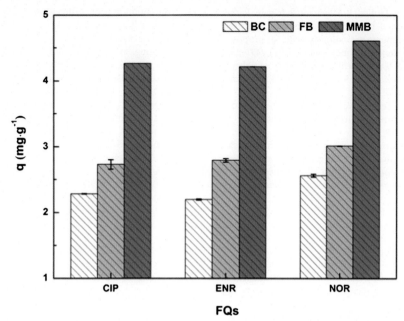

FIG. 12.7 Adsorption capacity of three CF antibiotics (CIP, ENR, and NOR) by BC, FB, and MMB sorbents [32].

5 PERSPECTIVES

It is essential to generate new materials that can be applied in the removal of antibiotics (or any emerging contaminants), but it is also necessary to combine the processes described previously to find a solution to the environmental problems that are generated every day.

It is possible to use sorbents such as hydrogels: polymers insoluble in aqueous solution, with a great capacity to absorb water and adopt the form of the container that contains them, they are also very versatile as they have been applied in various areas of medicine, food industry, and research. These materials can be functionalized according to the contaminant to be removed; therefore, they are possible to apply in the removal of antibiotics from water and are proposed as a new alternative to solve environmental problems.

6 CONCLUSIONS

The effectiveness of remediation depends on several factors. Defining the treatments used in remediation, degradation processes have high remediation rates in aqueous effluents, being so far the most reported in the literature.

In the case of removal techniques, one of the main factors is the variation of pH, and this factor affects the chemical speciation of the antibiotic and may or may not promote removal.

Sorption processes are acquiring more strength in the research area of emerging pollutants removal,

discovering in them a simple alternative of remediation with optimal removal rates. A combination of the two processes would facilitate their implementation on an industrial scale.

ACKNOWLEDGMENTS

The authors thank the FONDECYT Project no. 1191336.

REFERENCES

[1] A. Watkinson, E. Morby, S. Constanzo, Removal of antibiotics in conventional and advanced wastewater treatment: implications for environmental discharge and wastewater recycling, Water Research 41 (2007) 4164–4176.

[2] L.R. Braga, T.O. Carvalho, A.R. Nunes, K.R. Araújo, A.G. Prado, Removal of emergent pollutants (oxicam, nonsteroidal anti-inflammatory drug) from water by chitosan microspheres, Journal of Thermal Analysis and Calorimetry 130 (2017) 1697–1706.

[3] A.A. Basheer, New generation nano-adsorbents for the removal of emerging contaminants in water, Journal of Molecular Liquids 261 (2018) 583–593.

[4] J.M. Beale, J.H. Block (Eds.), Wilson and Gisvold's Textbook of Organic Medicinal and Pharmaceutical Chemistry, 2011.

[5] C. Barrios-Estrada, M. de Jesús Rostro-Alanis, B.D. Muñoz-Gutiérrez, H.M. Iqbal, S. Kannan, R. Parra-Saldívar, Emergent contaminants: endocrine disruptors and their laccase-assisted degradation—a review, The Science of the Total Environment 612 (2018) 1516–1531.

[6] T. Robinson, G. McMullan, R. Marchant, P. Nigam, Remediation of dyes in textile effluent: a critical review on current treatment technologies with a proposed alternative, Bioresource Technology 77 (2001) 247—255.

[7] S. Wang, Y. Peng, Natural zeolites as effective adsorbents in water and wastewater treatment, Chemical Engineering Journal 156 (2010) 11—24.

[8] S. Richardson, T. Ternes, Water analysis: emerging contaminants and current issues, Analytical Chemistry 77 (2005) 3807—3838.

[9] N. Kumar, R. Rajendra, Gold-palladium nanoparticles aided electrochemically reduced graphene oxide sensor for the simultaneous estimation of lomefloxacin and amoxicillin, Sensors and Actuators B: Chemical 243 (2016) 658—668.

[10] M.R. Awual, M.M. Hasan, M.A. Khaleque, M.C. Sheikh, Treatment of copper (II) containing wastewater by a newly developed ligand based facial conjugate materials, Chemical Engineering Journal 288 (2016) 368—376.

[11] S. Pandey, S. Tiwari, Facile approach to synthesize chitosan based composite—characterization and cadmium (II) ion adsorption studies, Carbohydrate Polymers 134 (2015) 646—656.

[12] E. Elmolla, M. Chaudhuri, Degradation of amoxicillin, ampicillin and cloxacillin antibiotics in aqueous solution by the UV/ZnO photocatalytic process, Journal of Hazardous Materials 176 (2010) 445—449.

[13] L. Rizzo, C. Manaia, C. Merlin, T. Schwartz, C. Dagot, M. Ploy, I. Michael, D. Fatta-Kassinos, Urban wastewater treatment plants as hotspots for antibiotic resistant bacteria and genes spread into the environment: a review, The Science of the Total Environment 447 (2013) 345—360.

[14] K. Strømgaard, P. Krogsgaard-Larsen, U. Madsen (Eds.), Textbook of Drug Design and Discovery, CRC press, 2017.

[15] A. Moarefian, H. Golestani, H. Bahmanpour, Removal of amoxicillin from wastewater by self-made polyethersulfone membrane using nanofiltration, Journal of Environmental Health Science and Engineering 12 (2014) 127.

[16] D. Kanakaraju, B.D. Glass, M. Oelgemöller, Advanced oxidation process mediated removal of pharmaceuticals from water: a review, Journal of Environmental Management 219 (2018) 189—207.

[17] F.A. Almomani, M. Shawaqfah, R.R. Bhosale, A. Kumar, Removal of emerging pharmaceuticals from wastewater by ozone-based advanced oxidation processes, Environmental Progress and Sustainable Energy 35 (2016) 982—995.

[18] R. Kıdak, S. Dogan, Medium-high frequency ultrasound and ozone based advanced oxidation for amoxicillin removal in water, Ultrasonics Sonochemistry 40 (2018) 131—139.

[19] A.G. Trovo, R.F.P. Nogueira, A. Aguera, A.R. Fernandez-Alba, S. Malato, Degradation of the antibiotic amoxicillin by photo-Fenton process — chemical and toxicological assessment, Water Research 45 (2011) 1394—1402.

[20] J.A. De Lima Perini, M. Perez-Moya, R.F.P. Nogueira, Photo-Fenton degradation kinetics of low ciprofloxacin concentration using different iron sources and pH, Journal of Photochemistry and Photobiology A: Chemistry 259 (2013) 53—58.

[21] Y.J. Jung, W.G. Kim, Y. Yoon, J.W. Kang, Y.M. Hong, H.W. Kim, Removal of amoxicillin by UV and UV/H$_2$O$_2$ processes, The Science of the Total Environment 420 (2012) 160—167.

[22] E. De Bel, C. Janssen, S. De Smet, H. Van Langenhove, J. Dewulf, Sonolysis of ciprofloxacin in aqueous solution: influence of operational parameters, Ultrasonics Sonochemistry 18 (2011) 184—189.

[23] N.F.F. Moreira, C.A. Orge, A.R. Ribeiro, J.L. Faria, O.C. Nunes, M.F.R. Pereira, A.M.T. Silva, Fast mineralization and detoxification of amoxicillin and diclofenac by photocatalytic ozonation and application to an urban wastewater, Water Research 87 (2015) 87—96.

[24] V. Lozano-Morales, I. Gardi, S. Nir, T. Undabeytia, Removal of pharmaceuticals from water by clay-cationic starch sorbents, Journal of Cleaner Production 190 (2018) 703—711.

[25] Y. Kong, L. Wang, Y. Ge, H. Su, Z. Li, Lignin xanthate resin—bentonite clay composite as a highly effective and low-cost adsorbent for the removal of doxycycline hydrochloride antibiotic and mercury ions in water, Journal of Hazardous Materials 368 (2019) 33—41.

[26] E.C. Lima, Removal of emerging contaminants from the environment by adsorption, Ecotoxicology and Environmental Safety 150 (2018) 1—17.

[27] A. Rossner, S.A. Snyder, D.R. Knappe, Removal of emerging contaminants of concern by alternative adsorbents, Water Research 43 (2009) 3787—3796.

[28] S. Li, X. Zhang, Y. Huang, Zeolitic imidazolate framework-8 derived nanoporous carbon as an effective and recyclable adsorbent for removal of ciprofloxacin antibiotics from water, Journal of Hazardous Materials 321 (2017) 711—719.

[29] D. Hu, L. Wang, Adsorption of amoxicillin onto quaternized cellulose from flax noil: kinetic, equilibrium and thermodynamic study, Journal of the Taiwan Institute of Chemical Engineers 64 (2016) 227—234.

[30] M. Homayoonfal, M. Mehrnia, Amoxicillin separation from pharmaceutical solution by pH sensitive nanofiltration membranes, Separation and Purification Technology 130 (2014) 74—83.

[31] D.A. Palacio, B.L. Rivas, B.F. Urbano, Ultrafiltration membranes with three water-soluble polyelectrolyte copolymers to remove ciprofloxacin from aqueous systems, Chemical Engineering Journal 351 (2018) 85—93.

[32] R. Li, Z. Wang, X. Zhao, X. Li, X. Xie, Magnetic biochar-based manganese oxide composite for enhanced fluoroquinolone antibiotic removal from water, Environmental Science and Pollution Research 25 (2018) 31136—31148.

[33] H. Liu, Y. Wei, J. Luo, T. Li, D. Wang, S. Luo, J.C. Crittenden, 3D hierarchical porous-structured biochar aerogel for rapid and efficient phenicol antibiotics removal from water, Chemical Engineering Journal 368 (2019) 639—648.

CHAPTER 13

Aptamer and Riboswitches: A Novel Tool for the Need of New Antimicrobial Active Compounds

VÍCTOR DÍAZ-GARCÍA • GERARDO RETAMAL-MORALES

1 INTRODUCTION

Since the 1990s, several studies have shown that the indiscriminate use of antibiotics by the community is directly related to the high levels of antibiotic-resistance from microorganisms, affecting public health, veterinary, and the food industry, causing a negative impact in the fight against many infectious diseases, including tuberculosis, pneumonia, and cholera among other highly relevant infections [1,2]. On the other hand, the use and abuse of antibiotics have generated a number of side effects for patients, such as allergic reactions, kidney, hematological, hepatic, gastrointestinal, and neuronal problems [3—5]. This is how, over time, more antibiotic-resistant bacteria rise, with several resistance mechanisms, which lead to the lower effectiveness of antibiotic treatments and increasing morbidity and mortality in the population due to bacterial infections. This makes bacterial resistance one of the most important problems facing humanity today [6—9].

Bacterial resistance mechanisms can vary depending on the antibiotic that is used. However, a common factor is an adaptation at a molecular level, either by preventing the entry of the antibiotic, by modifying the site of action, or directly degrading the antimicrobial compound [7,8]. Thus, the adaptability of bacteria is a problem as relevant that researchers have come to the conclusion that even the novel antibacterial drugs are projected soon to be ineffective [10]. Therefore, not only the development of new drugs but also the search for new antimicrobial targets is necessary for facing the problem of bacterial resistance.

One of the most common strategies to develop new compounds with antimicrobial activity is to enhance the effectiveness of existing antibiotics, using them in the conjugate form with other compounds that present antibacterial activity, or with particles or polymers that allow a controlled accumulation and release of antibiotics at the specific sites of infection [11]. Some examples of this include the production of antibiotics using model organisms [12,13] as plants or other microorganisms as to antibiotic biofactories, the performance of chemical modifications or the combination of preexisting antibiotics with molecules that increase their activity [14], the development of new antibiotic from bacterial natural products [15], the use of nanoparticles, either because of their toxic characteristics or as a vehicle to increase the selective accumulation of the antimicrobial compound [16,17], and the development of new molecules with antibacterial capacity combining the use of engineering with synthetic biology [18]. Regarding this last category, in this chapter, we will discuss a novel strategy that was developed thanks to advances in synthetic biology, the development of aptamers and riboswitches.

2 APTAMERS

Aptamers are short stranded single oligonucleotides (DNA or RNA) or small peptides, with the ability to interact with high specificity and affinity with different target molecules [19,20]. Aptamers were first described in 1990, by three different studies that were performed independently. In March 1990, Robertson and Joyce described an RNA enzyme that showed genetic and catalytic properties (phosphoester-transfer reactions). That was obtained by in vitro techniques for the mutation, selection, and amplification of catalytic RNA, based on the construction of a sequence pool with several structural variants of the tetrahymena ribozyme and the selection of these by its interaction with ribozyme target sequence [21]. In August of the same year, Tuerk and Gold reported a method to obtain an RNA oligonucleotide with a high affinity for the T4 DNA polymerase from a Bacteriophage (gp43) from a pool of over 65,000 different sequences. Using alternated cycles of ligand selection from pools of variant sequences and amplification of the gp43 bound species, authors called this process, Systematic evolution of ligands by exponential enrichment [22]. Finally, Ellington and Szostak reported a method to obtain high-affinity nucleic acid ligands for small organic molecules from a pool of 10^{10} different sequences [23]. These three studies showed in combination the revolutionary aspect of this new nucleic acid-based technology and a common methodology that involve an evolutionary selection process of the nucleic acid sequences. Together, they mark the beginning of the aptamer's era.

In this chapter, we will focus specifically on aptamers constructed using oligonucleotides, not to be confused from those of protein nature, as peptides with antimicrobial activity are addressed in earlier chapters of this book.

3 SYSTEMATIC EVOLUTION OF LIGANDS BY EXPONENTIAL ENRICHMENT

Aptamers are synthesized and selected through a process called systematic evolution of ligands by exponential enrichment (SELEX). SELEX is a ligand selection technique, in which a quadrillion (10^{15}) random sequences of oligonucleotides (DNA or RNA) or small peptides are incubated with a specific target (large, small molecules, ions, whole cells, viruses, etc.). Subsequently, the bound sequence(s), are separated and purified from those that did not interact with the specific target. The selected oligonucleotides are amplified by a polymerase chain reaction (PCR) and are used as refined libraries for the next round of selection. This process is repeated many times until a limited number of molecules that bind specifically to its ligand is obtained [23] (Fig. 13.1).

Finally, the obtained aptamers are characterized to determine their ability to bind to their target molecule and/or therapeutic capacity, depending on the motivation in their construction [24].

4 SELEX STEP BY STEP

4.1 Construction of DNA-RNA Library

The SELEX process begins with the synthesis of partially or completely randomized oligonucleotides. When we talk about partially randomized, it implies the use of

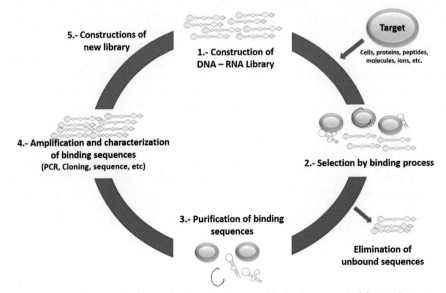

FIG. 13.1 Diagram of general in vitro selection of aptamers. Bind were separated from a large amount of randomized library and are enriched through the selection process where the bound molecules are regenerated by amplification processes for the next selection cycle.

preconceived sequences that have a specific activity within the aptamer, for example, the development of a hybrid aptamer from an oligonucleotide library, containing a known aptamer sequence and random sequence, which will interact with the target.

These random sequences are frequently constructed with primers that flank them, which allow the amplification (by PCR) of the selected sequences (DNA aptamers) or by in vitro transcription of the sequences (RNA aptamers) (Fig. 13.2). In addition, these flanking sequences may contain restriction-enzyme binding sites, to facilitate cloning [24,25].

For this process, construction by chemical synthesis of oligonucleotides is usually used, which allows to obtain a nucleotide library of about 10^{15} sequences, with a random zone of between 15 and 75 nucleotides [26,27]. Currently, the construction of oligonucleotide libraries can be done commercially by several companies.

4.2 Selection by Binding Process

This stage is essential in the SELEX process, as it depends on various parameters that directly influence the ability of the aptamer to interact with its target. In solution, the oligonucleotide chains will form secondary and tertiary structures in a nondeterministic manner. Through weak interactions generated by the nitrogenous bases [28–30], these chains will interact with the target with which they are incubated. This aptamer selection process allows the separation of those chains that, due to their three-dimensional structure, can join the targets.

As this process depends on the three-dimensional conformation that oligonucleotide chains adopt, it is recommended to slowly heat and then cool the oligonucleotide solution, to favor that the oligonucleotide population is mostly linearized (primary structure), seeking to favor interaction with the target and that the oligos adopt the 3D conformation in its presence. Usually, the selection process is performed using the target molecule immobilized to a matrix, this immobilization is carried out to favor the separation of attached aptamers [31]. However, due to their chemical nature, some target molecules do not allow this methodology. For the latter, several processes can make their purification possible,

such as the use of molecular exclusion columns, nitrocellulose filters, among others [27].

A drawback of the use of immobilization of targets is the possibility of unions of aptamers to the immobilization matrix, generating an unwanted selection of nonspecific oligonucleotides. To avoid this, Ellington and Szostak (1992) proposed a method called negative selection [31]. This includes, in the early stages of selection, the first step of incubation of the oligonucleotide library with the matrix alone, without a target. This allows the elimination of those aptamers that interact nonspecifically with the matrix, and then, with this refined library, the SELEX process begins. This study shows that the negative selection increased the affinity of the aptamers for their target up to 10 times, in opposition to a process without negative selection. Additionally, another methodology that increases the affinity for negative selection is the use of nonspecific competitors. These are small molecules with physical properties similar to the oligonucleotides in solution; the function of these is to generate a competitive environment for the interaction of the aptamers to their target, encouraging the selection of those oligonucleotides with higher affinity.

Another parameter to consider when performing a selection by the binding process is the conditions of the buffer used for the incubation of the randomized oligonucleotides and the targets, seeking to increase the possibility of interaction between the chains. Buffers with high saline content generate an ionic environment conducive to the interaction of oligonucleotides with small charged molecules. The current literature usually indicates the use of buffers, commonly named SELEX buffer [32].

Finally, the relative concentration of the target used is a parameter to consider in the selection by binding process. A higher concentration of targets favors the interaction with oligonucleotides by increasing the probability of the sequences to be bound during incubation and to be preserved, but this does not generate an effective selection pressure, so it is very likely that the selected aptamers will present a high affinity with the ligand. Therefore, if the aptamer is desired to have a high affinity, it is convenient that the oligonucleotides are in excess compared to the target, to encourage the competition between sequences for available binding sites [31].

4.3 Purification of Binding Sequences

To separate the oligonucleotide sequences that bound the target from those that did not bind, several methodologies can be used depending on the type of target molecule. Generally, the process is carried out by means of washes to allow the elimination of the bound

FIG. 13.2 Representative diagram of a typical oligonucleotide using for SELEX. In red, the primer zone that allows the amplification. In blue, the random oligonucleotide zone.

sequences from the unbound. Once the linked sequences are separated, they are washed using denaturing conditions to unbind them from the target and allow their characterization and amplification [33].

One of the most commonly used methodologies to favor this process is to bind the target, in a covalent way, to a matrix and immobilize it. Among other methods, the following can be highlighted: (1) Columns of agarose or sepharose that are attached to the target molecule, allowing separation by fractional elution. (2) Magnetic beads attached to the target, allowing separation using magnetism [34]. (3) Bits attached to the target, allowing separation using centrifugation, among others.

If the target does not allow a covalent union, it is necessary to use other ways to allow a fractionation of the content of the incubation solution for analysis. Based on the greater molecular weight of the aptamer-target complex, it is possible to use, for example, molecular exclusion columns, nitrocellulose filters, gradient centrifugation is sucrose or lithium and/or nondenaturant electrophoresis, to separate them from the unbound oligonucleotides [27].

4.4 Amplification and Characterization of Binding Sequences

Once the nucleotide sequences with high affinity for the target have been purified, it is necessary to amplify them for their characterization and construction of a new and more refined oligonucleotide library. For this step, methods well known in molecular biology such as PCR and/or in vitro cloning are used. From the newly refined and characterized oligonucleotide library, the process starts again, varying the conditions to increase the selection pressure [31].

5 SELEX FOR WHOLE CELLS

When focusing on the applications of aptamer technology in biomedicine, especially in those applications where the use of aptamers is proposed as blocking molecules with biological activity, it is common to concentrate on the development of these to bind proteins. In this sense, it is usual to use purified proteins as a target for SELEX; however, when the target proteins are part of an active complex, they have different structures associated with their physiological state (such as posttranscriptional modifications) or the binding domain in the case of transmembrane proteins. Thus it is likely that aptamers selected using said purified protein may not be able to interact with it in the physiological context [35]. A well-established example of this

phenomenon is the study published by Liu in 2009, where aptamers against the purified EGFRvIII ectodomain were produced and overexpressed in *Escherichia coli* (without posttranslational modifications). The obtained aptamers had a high affinity against the target protein sequence (K_d of 33×10^{-9} M); however, when challenged against the full-length EGFRvIII protein expressed on the surface of eukaryotic cells (a glycosylated protein), these aptamers were not able to bind [36]. These limitations led to the need to perform SELEX using whole cells [37,38]. This new methodology would allow the development of aptamers that would bind to the native conformation of the target protein, including possible posttranslational modifications [36].

Cell SELEX shows a couple of variations in comparison to a typical SELEX. For example: (1) When making SELEX using cell cultures as targets, the incubation process is carried out in culture plates with agitation, time, and temperature conditions determined by the best environment in which the aptamers should interact with the cells. (2) The use of cell lines that allow negative selection, for example, if it is desired for the aptamer to perform against tumor cells, normal cells of the same tissue should be used for negative selection [35,37,38].

6 APTAMER AS AN ANTIBIOTIC THERAPY

Among therapeutic agents, oligonucleotides are considered as a rival to antibodies or antibiotics (Table 13.1). Aptamers have high selectivity for their target and high tissue permeability, they are not immunogenic, they do not degrade rapidly in vivo and are easily stored and transported, and it is possible to modify them chemically to prevent their degradation [39]. This last characteristic is favorable for the following: (1) conjugate them with other molecules, influencing its pharmacokinetic behavior; (2) increasing the cellular uptake or decrease its clearance (phosphorothioate analogs); (3) increase its stability, preventing its degradation by nucleases [40].

Aptamers have been largely reported as excellent bioreceptors; this combined with their ability to easily perform chemical modifications on them would allow them to be used as versatile tools in antibiotic therapies without them having, by themselves, antimicrobial effects [20,28,39,44]. However, independently of the success of the in vitro antibacterial activity of aptamers, the increasing antibiotic resistance is still a challenge. Thus it is important to ask the question, how could antibiotic resistance mechanisms affect aptamers. There is still no

TABLE 13.1

Comparison Between Aptamer Properties, Antibodies, and Antibiotics [28,41–43].

Characteristic	Aptamers	Antibodies	Antibiotics
Affinity	ρM–μM	ρM–μM	ηM–μM
Specificity	High	High	Low (broad spectrum)
Production	In vitro	In vivo	In vitro
Batches variations	Low	High	Low
Chemical modification	Easy	Limited	Limited
Desaturation	Reversible	Irreversible	Irreversible

answer, but based on the known antibiotic resistance mechanisms, could it be possible to have an idea (Table 13.2).

Table 13.2 only represents a hypothetical interpretation. However, it is important to emphasize that, to advance in the development of aptamer-based antimicrobial therapies, studies oriented to determine and understand the effect of typical antimicrobial mechanisms are of great importance.

7 APTAMERS WITH ANTIBIOTICS ACTIVITY

The main studies that showed the development of aptamers with antimicrobial activity are described in the following.

7.1 Aptamers Against Mycobacterium tuberculosis

This corresponds to the first study that shows an aptamer with antimicrobial therapeutic activity. In 2007, Chen et al. developed by whole-cell SELEX an aptamer (named NK2) that binds to virulent strain *M. tuberculosis* (H37Rv). In this study, in vitro (interaction with whole and lysates cells) and in vivo assays (mice infected) were performed to determine the antibacterial activity of the NK2 aptamer using an attenuated strain of *M. bovis*, called bacillus Calmette–Guérin (BCG), as control. This study showed remarkable results like the following: (1) The aptamer NK2 binds to the surface of H37Rv with high affinity (K_d in the nanomolar range), but not to BCG, indicating that NK2 binds to some components of the surface of H37Rv that is not existing or existing in a lesser degree on BCG. (2) In in vitro assays, NK2 did not inhibit the growth of H37Rv. However, in an in vivo assay, mice treated with NK2 aptamer showed an increase in their interferon-γ (IFN-γ) production by

CD4+ T cells, a decrease in the interaction of H37Rv with the alveolar membrane and a decrease in the inflammatory reaction. (3) The treatment with NK2 decreased the number of live H37Rv in lung and spleen and prolonged the survival rate of mice. Overall, this study showed, for the first time, an aptamer with antibacterial activity [49].

Between 2012 and 2013, the interaction of H37Rv and macrophages in the presence of aptamers NK2 was investigated in more detail. Chen et al. published two studies, which reported that the NK2 aptamer and a selected pool of aptamers (10th pool of SELEX) could be obtained by whole cells SELEX. The studies showed that the 10th pool had a higher affinity to H37Rv with a K_d similar to NK2. These aptamers (NK2 and 10th pool) inhibit macrophage invasion by H37Rv and significantly stimulated the production of IFN-γ, interleukin-15, and interleukin-17 secreted from macrophages (in vitro analysis) [50]. Additionally, it was observed that these aptamers may play an active role against H37Rv, related to an inhibition of H37Rv adhesion/invasion of murine peritoneal macrophages and CD8+ T-cells. Interestingly, this last study has shown that the 10th aptamer-pool was more effective than the NK2 aptamer alone as an antibacterial agent, evidenced by a decrease of the presence of H37Rv in lung and spleen tissue, and an increase in the survival rate of mice with the treatment using the 10th aptamer [51].

As a general conclusion of these studies, it is possible to note that whole-cell SELEX allows to obtain aptamers that bind to the surface of bacteria (in this case *Mycobacterium*) inhibiting their ability to interact with their target tissue, and therefore, the ability to infect. In addition, the use of an aptamer pool can increase the antibacterial effects probably because the 10th pool can bind at various and diverse sites present on the surface of the bacterial cell.

TABLE 13.2
Antibiotic Resistance Mechanisms and Their Hypothetical Effect on the Activity of Aptamers Used as Antimicrobials.

Resistance Mechanism	Description	Could Affect the Aptamers Antimicrobial Activity?
Enzymatic inactivation	Bacterial enzymes that degrade or chemically modify antibiotics. For example, β-lactamases (β-lactamic ring hydrolysis) or bacterial acetyltransferase (aminoglycoside acetylation) [45]	Because of their chemical nature, aptamers could not be recognized by these enzymes. Probably, they could not affect the antimicrobial activity of aptamers
Alteration of antibacterial target	Through mutations, bacteria can change its proteins; among them, antibiotic-interaction sites, modifying or bypassing the antibiotic's target, that is, DNA gyrase modifications (for quinolone resistance) or peptidoglycan-precursor modification (vancomycin resistance) [6]	May affect the ability of the aptamer to interact with its target, thus, affecting its activity
Change in the membrane permeability	Gram-negative bacteria can reduce the permeability of their outer membrane by decreasing their number of purines. This phenomenon inhibits the internalization of antibiotics by diffusion [46]	May affect the ability of the aptamer to interact with its target, depending on its location, that is, could affect the aptamer internalization in gram-negatives, but not in gram-positives, where the target could be a teichoic acid
Efflux pumps	Efflux pumps are transport proteins involved in the extrusion of toxic substrates (as antibiotics, toxic metals, etc.) from the inner cell, and are found in prokaryotic and eukaryotic cells. These can also export virulence factors that are necessary for pathogenesis [47,48]	Efflux pumps may extrude the aptamers, whose target is inside the cell

In 2014, Pan et al. (from the same research group) published the development of the ZXL1 aptamer against mannose-capped lipoarabinomannan (ManLAM). ManLAM is predominant on the surface lipoglycan of *Mycobacterium tuberculosis* and it was purified from the strain H37Rv to be used as a SELEX target. This study showed that ZXL1 aptamer can bind to H37Rv's surface with high affinity (greater than NK2) and did not bind to other gram-negative bacilli like BCG, *M. smegmatis*, *M. intracellulare*, *M. avium*, *P. aeruginosa*, *S. aureus*, *E. coli*, *S. albus*, *M. luteus*, and *S. typhimurium*. This aptamer showed several advantages as a possible antibiotic therapy compared with the NK2 aptamer although they have similar binding affinities to the target in the H37Rv strain: (1) A significant inhibition of the ManLAM-induced immunosuppression of CD11c + dendritic cells. (2) Enhanced antigen–presenting activity of dendritic cells for CD4+ Th1-cell activation. (3) A single-dose aptamer ZXL1 significantly reduced the progression of H37Rv infections and live bacteria in the lungs of mice (as an antibiotic therapy) and rhesus monkeys without toxicological effects on lymphocytes. Finally, this study proposed two important ideas: (1) For the first time, a putative model of aptamer therapy and its effects were proposed; in this case, the ZXL1-mediated blocking of H37Rv ManLAM-induced immunosuppression. (4) Raised the need to delve into studies on toxicological and therapeutic effects of aptamers [52].

7.2 Aptamers Against Lipopolysaccharide

In 2008, Bruno et al. developed an aptamer against lipopolysaccharide (LPS) from *E. coli* O111: B4 and K2 strain. These aptamers were coupled to human C1qrs protein (the first component of the classical complement immune system that kills bacteria by the formation of pores), by a two system, 5′-biotinylated aptamer bound to streptavidin-C1qrs complex (indirect) and using sulfo-EGS as a cross-linker. These multivalent aptamers were able to decrease the quantity of bacteria in human serum samples, used as culture media. An important concept that the authors offers in this study was the idea that *"bacteria will not be able to easily evolve mechanisms to resist this surface attack mechanism"* because "in general, complement represents an ancient natural system of bacterial destruction that works quite well"[53]. Interestingly, in 2009, Bruno et al. showed that this kind of multivalent aptamer (aptamer-C1qrs) can be used as a therapy for other diseases when conjugating aptamers against cancerous cells (previously published by Ferreira et al. [54,55]) with C1qrs, and demonstrated that these aptamers had anticancer activity in vitro. These studies were the base of a patent granted in 2011 to Bruno et al. [56]. In this patent, the LPS aptamer, aptamer-C1qrs, and the 3′-conjugation of aptamers that increase its serum half-life by protection against nuclease activity (from human serum or bacteria) were protected.

Additionally, in 2009, Wen et al. published the development of a novel aptamer that binds to lipopolysaccharide and protects mice against endotoxemia. This study showed seven different aptamers that bind LPS, obtained by SELEX and used against commercial LPS as a target. Just one (N°19) of them showed the ability to protect C57BL/6 mice from damage caused by LPS (commercial). These aptamers significantly decreased NF-κβ activation of monocytes challenged by LPS and decreased to the IL-1 and TNF-α levels in the media of LPS-challenged monocytes, both in a dose-dependent manner. Although this study does not work directly with live microorganisms or showed an effect over bacterial cells, it has an important protective effect, showing an increase in the survival of mice from 0% to 60% when these mice were injected [57] and can have potential to be used as sepsis therapy.

7.3 Peptidoglycan-Specific Aptamers

In 2014, Ferreira et al. reported the development of aptamers antibac1 and antibac2. These aptamers were developed by SELEX to bind purified peptidoglycan with high affinity, oriented to the gram-positive bacteria *Staphylococcus aureus* and the gram-negative bacteria

E. coli TOP10, but not bind to fungal and human cells (*Candida albicans* and human fibroblasts). The authors proposed these aptamers for the development of biosensor aptamer-bases [58]. In 2017, this same group demonstrated that these aptamers can bind to several sepsis-producing agents as the gram-positive bacteria *S. aureus*, *Streptococcus pneumoniae*, *Listeria monocytogenes* and *Enterococcus, faecium*, and the gram-negative bacteria *Acinetobacter baumannii*, *Citrobacter freundii*, *Klebsiella pneumoniae*, *Morganella moraganii*, *Pseudomonas aeruginosa*, *Proteus mirabilis*, and *E. coli*. Interestingly, Graziani et al. showed that antibac1 and antibac2 increased the affinity for whole bacterial cells (K_d to ηM), different from purified peptidoglycan (Kd to μM) [59].

Later, in 2018, Ferreira et al. developed a study to determine the biomedical applications of the antibac1 and antibac2 aptamers, determining the stability and biodistribution of these in two different experimental infection models, bacterial-infected mice (*S. aureus*) and fungal infected mice (*C. albicans*), using radiolabeled aptamers with Technetium-99m. Stability was evaluated in saline, plasma, and cysteine solution, and its degradation by plasma nucleases was also assessed. For the biodistribution determination of these aptamers, the mice (immunosuppressed) were infected intramuscularly with *S. aureus* or *C. albicans* and the next day, the radiolabeled aptamer solution or the radiolabeled oligonucleotide (control) were injected by the tail vein in each animal. Finally, 3 h post aptamers injections, the mice were euthanized and tissue samples (blood, liver, spleen, stomach, heart, lung, kidney, infected muscle, and noninfected muscle) were used for radiolabeled determination by a gamma counter. This study demonstrates that the aptamers antibac1 and antibac2 were not degraded by plasma nucleases and were stable in the presence of saline solution and mice blood plasma. Some trans chelation was observed in the presence of cysteine. About biodistribution, the aptamers antibac1 and antibac2 and controls show a high accumulation in kidneys and liver, but just antibac1 and antibac2 show a major accumulation in the bacterial infections' focus. In fungal-infected mice, the biodistribution was similar to bacterial infected mice but did not show accumulations in the fungal infections' focus [60].

7.4 Aptamer Against Metallo-β-Lactamase

In 2009, Kim et al. development a aptamers that specific binding to the *Bacillus cereus* 5/B/6 metallo-β-lactamase (10 residues). Interestingly, this aptamer was able to inhibit the enzymatic activity effectively in a rapid, reversible, and noncompetitive manner [61]. The aptamers were selected by SELEX using the purified metallo-β-lactamase of *Bacillus cereus* 5/B/6 that was overexpressed, then purified, in *E. coli* MZ1. In this study, the B. *cereus* metallo-β-lactamase + and *E. coli* metallo-β-lactamase + were growth in the presence of cephalexin (an β-lactam antibiotic) and with or without aptamers (0.3 mM). These analyses reveal that the inhibitor aptamer can produce cell death in liquid cultures of both models. Kim et al. first showed an aptamer that could be used as adjuvant antibiotic therapy.

For 2019 Khan et al. studied the mechanism of cefuroxime hydrolysis of the aptamer reported by Kim et al. This study suggesting an allosteric mechanism for the enzyme inhibition by the aptamer, where the aptamer binding could alter conformational changes or local dynamics around the active site required for catalysis. These data are important for the evaluation of aptamer as an adjuvant antibiotic therapy, because, as Khan indicate in this study, "specific allosteric modulators can be found for a unique site on the target protein that consequently avoids adverse effects to the human host" [62].

Currently, there are several aptamers with an affinity for bacterial structures: some with antimicrobial activity and others used as a bioreceptor for the detection of bacteria (Table 13.3). The latter would be possible to use as antimicrobial therapy when coupled to molecules that possess antibiotic activity, in the same way as Bruno et al. [53,63]. As shown in Chapter 1 of this book, there are nanoparticles that possess antibiotic activity per se, or that can be activated. For example, the study published by Millenbaugh et al., in 2015, showed a bactericide effect in *S. aureus* using antibody-targeted gold nanoparticles and photoactivation of these gold nanoparticles. In this case, it produces a pulsed laser-induced photothermal damage by gold nanoparticles and the bioreceptor (antibody), causing the bioaccumulation of nanoparticle in the infection zone and later bacterial death [64].

The summarized studies of other aptamers that present the activity or potential to be used in the war against bacteria are provided in Table 13.3.

8 RIBOSWITCHES AND THEIR USE AS ANTIMICROBIAL TARGETS

As has been mentioned earlier, the need to develop novel antimicrobials for infectious agents has risen as a consequence of the increasing antibiotic resistance [82], and even the inclusion of new antibacterial drugs is projected soon to be ineffective [83]. Therefore, not only the development of new drugs but also the search for new antimicrobial targets is necessary for facing the problem of bacterial resistance.

In the look for new antibacterial targets, the capacity of antibiotics to interact with bacterial mRNAs and use these as a ligand has been used as a focus in the current literature; even more, it has been suggested that antibiotic mechanisms that were unknown at some point had these mRNAs as targets [84]. These nucleic acids are usually called riboswitches.

Riboswitches are gene-regulatory mRNA domains that respond to the intracellular concentration of a variety of metabolites, vitamins, coenzymes, etc. and control essential genes in many bacterial species, which can regulate several pathways from metabolite biosynthesis to membrane transport [85].

From a structural point of view, these gene-regulation motifs are commonly found in the untranslated regions of the mRNAs [86–88]. Thus, when the aptamer domain of the riboswitch is bound to the ligand, it triggers changes in the folding pattern of the expression platform, which can alter gene expression by several mechanisms [89].

The study of said mechanisms leads to the discovery of riboswitches in bacteria. The mechanisms for transport and synthesis of several organic molecules (and the set of proteins involved in these processes) are usually well known in bacteria. These usually include a negative feedback of the finished product at the end of an enzymatic pathway, sensing by other proteins and subsequent interaction with DNA/RNA to control the production of enzymes [87]. Thus, when studying the inhibition of B-complex vitamin biosynthesis genes by their final products thiamine, riboflavin and cobalamin, the hypothetical protein repressors that were supposed to be responsible for this process were not found [90,91]. This leads to the discovery of the regulatory role of mRNA sequences and suggested the participation of these nucleic acids in the sense and modulation of vitamin synthesis pathways [90,92]. For example, studies of the vitamin derivates, thiamine pyrophosphate (TPP) [93–95], flavin mononucleotide (FMN) [93], and adenosylcobalamin [94], have demonstrated the direct interaction of these with their respective mRNA to control the vitamin (B1, B2, and B12) operons. These studies report that metabolite binding can stabilize the conformation of a conserved RNA sensor (aptamer region) and can induce folding of other nonconserved RNA regions (expression region) and forms a structure that affects transcription termination or translation initiation. Thus, this metabolite-RNA interaction can cause "riboswitching" between alternative mRNA conformations that can alter gene expression [87].

Based on the capacity of these aptamers to regulate vital bacterial pathways, riboswitches can be very

TABLE 13.3
Aptamer With for Antibacterial Therapies and Biosensing.

Target	Description	Activity/Use	Reference
Salmonella enteritidis	RNA aptamer-based ligand that binds to the outer membrane protein C and specifically recognizes *S. enteritidis* without any cross-reactivity to other Salmonella serovars	Proposed as bioreceptor for detection of *S. enteritidis*	[65]
Salmonella typhimurium	RNA aptamers that bind to the pili structural protein of *S. typhimurium*	Inhibition of the invasion of *S. typhimurium* into human monocytes	[66]
Salmonella enteritidis *Salmonella typhimurium.*	DNA aptamer obtained by whole-cell SELEX, that bind to antibiotics resistant *Salmonella*	Inhibition of growth of *S. enteritidis* and *S. typhimurium* in bacterial cultures and a decrease in their membrane potential	[67]
Listeria spp.	DNA aptamers that specify binding to *Listeria* genus. This aptamer showed a cell capture performance equivalent to or better than a specific Listeria antibody	Development of a prototype of capture and detection assay of *Listeria* spp.	[68]
Listeria monocytogenes	DNA aptamers that recognize *L. monocytogenes* with high affinity and specificity obtained by whole-cell SELEX	Quantitative determination of *L. monocytogenes* by fluorescent bioassay	[69]
Campylobacter jejuni	DNA aptamers that bind to live *C. jejuni* cells with high affinity and specificity obtained by whole-cell SELEX	Development of a detection assay of *C. jejuni*	[70]
Enterotoxigenic *Escherichia coli* (ETEC) K88	DNA aptamers with high specificity to enterotoxigenic *E. coli* (ETEC) K88 obtained by SELEX against K88 fimbriae protein	Quantitative determination of *E. coli* by fluorescent bioassay	[71]
Escherichia coli	DNA aptamers that bind strongly to *E. coli* cells (ATCC 25922) obtained by whole-cell SELEX. Some of the aptamers described bind to meningitis/sepsis-associated *E. coli* (MNEC) clinical isolates	Determination of *E. coli* and MNEC by fluorescent bioassay	[72]
Poly-gamma-D-glutamic acid (PDGA), the major component of the *Bacillus anthracis* capsule.	DNA aptamers that bind to PDGA. These aptamers were coupled to Fc fragments of murine IgG to act as artificial antibodies	The construct can act as an opsonin. With potential use as future anthrax therapy	[73]
Bacillus anthracis spore (nonpathogenic sterne strain)	DNA aptamers development by SELEX using autoclaved anthrax spores	Detection of anthrax spores by electrochemiluminescence assay. Proposed as a future anthrax therapy	[74]

Continued

TABLE 13.3
Aptamer With for Antibacterial Therapies and Biosensing.—cont'd

Target	Description	Activity/Use	Reference
α-Toxin of *Staphylococcus aureus*	DNA aptamers that specifically bind to α toxin of *Staphylococcus aureus* obtained by SELEX	Inhibit the cytotoxic activity of α-toxin. Proposed as a functionally effective aptamer against α-toxin for treatment of *S. aureus* infections	[75]
Staphylococcus aureus	DNA aptamers that bind specify to *S. aureus* obtained by whole-cell SELEX	Rapid and specific detection of *S. aureus* by gold nanoparticle biosensor	[76]
Pathogenic species of *Streptococcus pyogenes*	DNA aptamers that bind specify to *S. pyogenes* obtained by whole cell SELEX	Development of different diagnostic tools such as biosensors	[77]
N-acyl homoserine lactone (HSL), of *Pseudomonas aeruginosa*	DNA aptamers that bind specify to HSL, a signal molecule of the quorum sensing system, obtained by SELEX. Aptamers Showed strong inhibitory activity on biofilm formation, decrease the protease, elastase, and pyocyanin secretion by *P. aeruginosa*	Proposed as potential drug candidates for the treatment of *P. aeruginosa* infection	[78]
Pseudomonas aeruginosa	DNA aptamer that strongly binds to *P. aeruginosa* cultured in biofilm and planktonic. Aptamers obtained by whole-cell SELEX.	Proposed as aptamer-drug conjugates and in biosensors	[79]
Murin lipopolysaccharide binding protein (mLBP)	RNA aptamer (2'-fluoro-pyrimidine-modified) that strongly binds to mLBP, an important protein in the activation of the pathophysiology of sepsis. Aptamers were obtained by SELEX using purified mLPB.	Proposed as diagnostic or therapeutic applications in human healthcare	[80]
Neisseria meningitidis	DNA aptamer that strongly binds to *Neisseria meningitidis* serogroup B obtained by whole-cell SELEX	Development of different diagnostic tools such as fluorescent biosensors	[81]

attractive targets for new antimicrobials (and face the growing bacterial resistance) for several reasons. First, riboswitches are designed to recognize small ligands; thus, the recognition of their targets is their physiological function, in opposition to other interactions between nucleic acids and small molecules developed in vitro [96]. Additionally, riboswitches are developed mainly by bacteria, not eukaryotes (with the notable exception of the TPP riboswitch), and this could minimize the cross-reactivity of bacterial riboswitch-targeted ligands. Finally, riboswitches can respond to ubiquitous and essential metabolites and are usually associated with mRNAs encoding proteins that are essential for survival or pathogenesis; therefore, they are often associated with highly conserved genes, a crucial characteristic to fight antimicrobial resistance [84,85,96–98].

To understand the attractiveness of riboswitches as a new antimicrobial aim, it is necessary to understand the chemical versatility of RNAs, which makes them a promising drug-target. Similar to proteins, RNA can fold into 3D structures with pockets that have the potential to bind ligands with high specificity [84,99] and undergo structural rearrangement and posttranscriptional modifications [100,101]. This can be observed in antibiotics that target the ribosome, which are aimed toward the rRNA rather than ribosomal proteins [102].

As rRNA currently is used as an antibacterial target [103−105], the exploration has expanded to other "nonriboswitch" RNA motifs to be used as drug targets. However, although most of these can associate with ligands, this interaction is usually fortuitous, making the design of ligands to target this motif a problematic task, due to lack of sensitivity [106,107]. Here is where riboswitches are fundamentally different; this is because they have evolved to bind ligands and show structural selectivity and a complex interphase with their ligands [96]. Next, a small description of the most studied riboswitches and their potential as antibacterial drugtargets are developed, and the most common ones are summarized in Table 13.4.

One example of the selectivity of these aptamers is the guanine-binding riboswitch from *Bacillus subtilis*, which can form a three-dimensional structure where the purine is completely enveloped, forming four hydrogen-bonding interactions, which can recognize each polar functional group of guanines, while the purine intercalates between two aromatic base triads. This closeness enables the riboswitch to discriminate against other purine analogs [108,109]. Thus, the interruption of this riboswitch, could disrupt the guanine-involved pathways and potentially, bacterial survival.

Several known antibacterial compounds have been described to function by targeting riboswitches. One prime example of this is pyrithiamine, an analog of thiamine that inhibits the growth of several bacterial and fungal species [110,111]. Similar to thiamine, pyrithiamine is readily phosphorylated inside cells to pyrithiamine pyrophosphate (PTPP) [112], which differs from thiamine pyrophosphate only in that the central thiazole ring is replaced by a pyridinium ring. In this case, PTPP can bind to several TPP riboswitches (in vitro) with comparable affinity to TPP, repressing the expression of a reporter gene fused to a TPP riboswitch inside bacteria [113]. In its phosphorylated

TABLE 13.4
Summary of Most Studied Riboswitches in Bacteria and Fungi.

Riboswitch	Effect	Reported Organism(s)	Reference
Cobalamin	Regulate cobalamin biosynthesis and transport of cobalamin and similar metabolites, and other genes	*E. coli*	[126]
FMN	Binds flavin mononucleotide (FMN) to regulate riboflavin biosynthesis and transport	*B. subtilis, E. coli, Salmonella enterica*	[122,123]
Glutamine	Binds glutamine to regulate genes involved in glutamine and nitrogen metabolism	*Prochlorococcus* sp., *Synechocystis* sp.	[124,125]
Glycine	Binds glycine to regulate glycine metabolism genes, including the use of glycine as an energy source	*Streptococcus pyogenes, B. subtilis*	[125,126]
Lysine	Binds lysine to regulate lysine biosynthesis, catabolism, and transport	*B. subtilis*	[116,117]
Purine (adenine, guanine)	Binds purines to regulate purine metabolism and transport. Different forms of the purine riboswitch bind guanine or adenine.	*B. subtilis*	[108,109, 127]
TPP	Binds thiamine pyrophosphate (TPP) to regulate thiamin biosynthesis and transport, as well as transport of similar metabolites. It is the only riboswitch found so far in eukaryotes	*B. subtilis, E. coli, A. oryzae*	[113]

form, pyrithiamine inhibits bacterial growth by repressing one or more TPP riboswitch-regulated genes [96]. For example, several strains of *B. subtilis*, *Escherichia coli* and the fungal specie *Aspergillus oryzae* that were cultured to resist the effect of pyrithiamine have a mutation in a conserved region of a TPP riboswitch [113]. In each case, the mutated riboswitch would normally regulate the expression of thiamine biosynthesis genes, suggesting its participation in bacterial survival.

A similar case happens for two analogs of the amino acid lysine, whose can inhibit bacterial growth. L-Aminoethylcysteine (AEC) and DL-4-oxalysine [114,115], both of these compounds bind to the lysine riboswitch from the *lysC* gene of *B. subtilis*, and both can repress the lysine riboswitch-regulated reporter gene in *B. subtilis* [116]. Furthermore, a mutation in the *lysC* riboswitch can grant resistance to several strains of *B. subtilis*, by disruption of the receptor formation and ligand binding in vitro, which prevents the riboswitch from regulating reporter gene expression. This suggests that AEC and DL-4-oxalysine inhibit bacterial growth at least in part by targeting a lysine riboswitch [116,117], making this aptamer an attractive target for further studies.

Another example is the case of the antibacterial riboflavin-analog roseoflavin, which can also target a riboswitch. Roseoflavin inhibits the growth of several gram-positive bacterial species through a mechanism that is reported to involve repression of riboflavin biosynthesis. Usually, all of the genes involved in riboflavin biosynthesis are expressed in a single operon, regulated by a single FMN-binding riboswitch [118–121]. According to current studies, roseoflavin can directly bind to the FMN riboswitch aptamers and downregulate the expression of an FMN riboswitch-*lacZ* reporter gene in *B. subtilis*. The role for this riboswitch in the antimicrobial mechanism of roseoflavin is supported by the researchers' observations that previously identified that roseoflavin-resistant strains have mutations within the FMN riboswitch [122]. This research line was able to be extended and advanced to the discovery, characterization, and production of the first riboswitch-targeted antimicrobial agent, *ribocil*, which can bind the FMN-riboswitch that mediates the *ribB* gene expression in gram-negative bacteria, causing the inhibition of growth of several species, including *E. coli* [123].

9 CONCLUSIONS AND PERSPECTIVES

Overall, aptamers and riboswitches have been shown to be not only promising but also useful and applicable targets for novel antimicrobial drugs. Their versatility, ease of synthesis, and safety present them with the above current antimicrobial therapies. However, the research of these aptamers and their biotechnological use is currently being developed and should still continue to generate knowledge for use in healthcare.

In the case of riboswitches, a focus in designing new highly active compounds to target riboswitches but not other cellular RNAs or proteins is also key and a challenge to be faced [96]. In reason to this, the existence of a variety of riboswitch classes in many bacterial species offers hope to develop new antimicrobials to be designed against the broad variety of bacterial pathogens and their increasing antimicrobial resistance.

ACKNOWLEDGMENTS
This study was supported by CONICYT Postdoctoral fellowship 3190734. The author also appreciates the personal support of the Fernanda Díaz and Diaz-García family (VDG).

REFERENCES

[1] G.L. French, The continuing crisis in antibiotic resistance, International Journal of Antimicrobial Agents 36 (2010) S3–S7.

[2] H.C. Neu, The crisis in antibiotic resistance, Science 257 (1992) 1064–1073.

[3] S. Heta, I. Robo, The side effects of the most commonly used group of antibiotics in periodontal treatments, Medical Sciences 6 (2018) 6.

[4] P. Lagacé-Wiens, E. Rubinstein, Adverse reactions to β-lactam antimicrobials, Expert Opinion on Drug Safety 11 (2012) 381–399.

[5] J.F. Westphal, D. Vetter, J.M. Brogard, Hepatic side-effects of antibiotics, Journal of Antimicrobial Chemotherapy 33 (1994) 387–401.

[6] P.A. Lambert, Bacterial resistance to antibiotics: modified target sites, Advanced Drug Delivery Reviews 57 (2005) 1471–1485.

[7] M. Frieri, K. Kumar, A. Boutin, Antibiotic resistance, Journal of Infection and Public Health 10 (2017) 369–378.

[8] A. Petchiappan, D. Chatterji, Antibiotic resistance: current perspectives, ACS Omega 2 (2017) 7400–7409.

[9] I.A. Rather, B.-C. Kim, V.K. Bajpai, Y.-H. Park, Self-medication and antibiotic resistance: crisis, current challenges, and prevention, Saudi Journal of Biological Sciences 24 (2017) 808–812.

[10] V.M. D'Costa, K.M. McGrann, D.W. Hughes, G.D. Wright, Sampling the antibiotic resistome, Science 311 (2006) 374–377.

[11] G.D. Wright, Solving the antibiotic crisis, ACS Infectious Diseases 1 (2015) 80–84.

[12] E. Holaskova, P. Galuszka, I. Frebort, M.T. Oz, Antimicrobial peptide production and plant-based expression systems for medical and agricultural biotechnology, Biotechnology Advances 33 (2015) 1005–1023.

[13] T. Weber, P. Charusanti, E.M. Musiol-Kroll, X. Jiang, Y. Tong, H.U. Kim, S.Y. Lee, Metabolic engineering of antibiotic factories: new tools for antibiotic production in actinomycetes, Trends in Biotechnology 33 (2015) 15–26.

[14] N.M.J. Vermeulen, D.E. Schwartz, Combinations and Methods for Reducing Antimicrobial Resistance, 1999.

[15] J. Clardy, M.A. Fischbach, C.T. Walsh, New antibiotics from bacterial natural products, Nature Biotechnology 24 (2006) 1541–1550.

[16] M.J. Hajipour, K.M. Fromm, A. Akbar Ashkarran, D. Jimenez de Aberasturi, I.R. de Larramendi, T. Rojo, V. Serpooshan, W.J. Parak, M. Mahmoudi, Antibacterial properties of nanoparticles, Trends in Biotechnology 30 (2012) 499–511.

[17] H.-Z. Lai, W.-Y. Chen, C.-Y. Wu, Y.-C. Chen, Potent antibacterial nanoparticles for pathogenic bacteria, ACS Applied Materials and Interfaces 7 (2015) 2046–2054.

[18] M.J. Smanski, H. Zhou, J. Claesen, B. Shen, M.A. Fischbach, C.A. Voigt, Synthetic biology to access and expand nature's chemical diversity, Nature Reviews Microbiology 14 (2016) 135–149.

[19] P. Colas, B. Cohen, T. Jessen, I. Grishina, J. McCoy, R. Brent, Genetic selection of peptide aptamers that recognize and inhibit cyclin-dependent kinase 2, Nature 380 (1996) 548–550.

[20] F. Pfeiffer, G. Mayer, Selection and biosensor application of aptamers for small molecules, Frontiers in Chemistry 4 (2016).

[21] D.L. Robertson, G.F. Joyce, Selection in vitro of an RNA enzyme that specifically cleaves single-stranded DNA, Nature 344 (1990) 467–468.

[22] C. Tuerk, L. Gold, Systematic evolution of ligands by exponential enrichment: RNA ligands to bacteriophage T4 DNA polymerase, Science 249 (1990) 505–510.

[23] A.D. Ellington, J.W. Szostak, In vitro selection of RNA molecules that bind specific ligands, Nature 346 (1990) 818–822.

[24] R. Stoltenburg, C. Reinemann, B. Strehlitz, SELEX—a (r) evolutionary method to generate high-affinity nucleic acid ligands, Biomolecular Engineering 24 (2007) 381–403.

[25] B.M. Bachtiar, C. Srisawat, E.W. Bachtiar, RNA aptamers selected against yeast cells inhibit Candida albicans biofilm formation in vitro, MicrobiologyOpen 8 (2019) e00812.

[26] H. Ulrich, C.A. Trujillo, A.A. Nery, J.M. Alves, P. Majumder, R.R. Resende, A.H. Martins, DNA and RNA aptamers: from tools for basic research towards therapeutic applications, Combinatorial Chemistry and High Throughput Screening 9 (8) (2006). Available online: https://www.ingentaconnect.com/content/ben/cchts/2006/00000009/00000008/art00006.

[27] Z. Zhuo, Y. Yu, M. Wang, J. Li, Z. Zhang, J. Liu, X. Wu, A. Lu, G. Zhang, B. Zhang, Recent advances in SELEX technology and aptamer applications in biomedicine, International Journal of Molecular Sciences 18 (2017) 2142.

[28] S. Song, L. Wang, J. Li, C. Fan, J. Zhao, Aptamer-based biosensors, TrAC Trends in Analytical Chemistry 27 (2008) 108–117.

[29] K.-M. Song, E. Jeong, W. Jeon, M. Cho, C. Ban, Aptasensor for ampicillin using gold nanoparticle based dual fluorescence–colorimetric methods, Analytical and Bioanalytical Chemistry 402 (2012) 2153–2161.

[30] J.F. Lee, J.R. Hesselberth, L.A. Meyers, A.D. Ellington, Aptamer database, Nucleic Acids Research 32 (2004) D95–D100.

[31] A.D. Ellington, J.W. Szostak, Selection in vitro of single-stranded DNA molecules that fold into specific ligand-binding structures, Nature 355 (1992) 850–852.

[32] G. Mayer, M.-S.L. Ahmed, A. Dolf, E. Endl, P.A. Knolle, M. Famulok, Fluorescence-activated cell sorting for aptamer SELEX with cell mixtures, Nature Protocols 5 (2010) 1993–2004.

[33] P. Bayat, R. Nosrati, M. Alibolandi, H. Rafatpanah, K. Abnous, M. Khedri, M. Ramezani, SELEX methods on the road to protein targeting with nucleic acid aptamers, Biochimie 154 (2018) 132–155.

[34] R. Stoltenburg, C. Reinemann, B. Strehlitz, FluMag-SELEX as an advantageous method for DNA aptamer selection, Analytical and Bioanalytical Chemistry 383 (2005) 83–91.

[35] M. Chen, Y. Yu, F. Jiang, J. Zhou, Y. Li, C. Liang, L. Dang, A. Lu, G. Zhang, Development of cell-SELEX technology and its application in cancer diagnosis and therapy, International Journal of Molecular Sciences 17 (2016) 2079.

[36] Y. Liu, C.-T. Kuan, J. Mi, X. Zhang, B.M. Clary, D.D. Bigner, B.A. Sullenger, Aptamers selected against the unglycosylated EGFRvIII ectodomain and delivered intracellularly reduce membrane-bound EGFRvIII and induce apoptosis, Biological Chemistry 390 (2009) 137–144.

[37] D. Shangguan, T. Bing, N. Zhang, Cell-SELEX: aptamer selection against whole cells, in: W. Tan, X. Fang (Eds.), Aptamers Selected by Cell-SELEX for Theranostics, Springer, Berlin, Heidelberg, 2015, pp. 13–33. ISBN 978-3-662-46226-3.

[38] K. Sefah, D. Shangguan, X. Xiong, M.B. O'Donoghue, W. Tan, Development of DNA aptamers using Cell-SELEX, Nature Protocols 5 (2010) 1169–1185.

[39] Zimbres, F.M.; Tárnok, A.; Ulrich, H.; Wrenger, C. Aptamers: Novel Molecules as Diagnostic Markers in Bacterial and Viral Infections?. BioMed Research International. Available online: https://www.hindawi.com/journals/bmri/2013/731516/abs/(accessed on Jan 9, 2020). 2013. 731516.

[40] J. Winkler, Oligonucleotide conjugates for therapeutic applications, Therapeutic Delivery 4 (2013) 791–809.

[41] J.J. Trausch, M. Shank-Retzlaff, T. Verch, Replacing antibodies with modified DNA aptamers in vaccine potency assays, Vaccine 35 (2017) 5495–5502.

[42] S. Yoshizawa, D. Fourmy, J.D. Puglisi, Structural origins of gentamicin antibiotic action, The EMBO Journal 17 (1998) 6437–6448.

[43] A.A. Peterson, S.W. Fesik, E.J. McGroarty, Decreased binding of antibiotics to lipopolysaccharides from polymyxin-resistant strains of *Escherichia coli* and *Salmonella typhimurium*, Antimicrobial Agents and Chemotherapy 31 (1987) 230–237.

[44] M. Majdinasab, A. Hayat, J.L. Marty, Aptamer-based assays and aptasensors for detection of pathogenic bacteria in food samples, TrAC Trends in Analytical Chemistry 107 (2018) 60–77.

[45] M.W. Vetting, S. Magnet, E. Nieves, S.L. Roderick, J.S. Blanchard, A bacterial acetyltransferase capable of regioselective N-acetylation of antibiotics and histones, Chemistry and Biology 11 (2004) 565–573.

[46] Resistance to antibiotics caused by decrease of the permeability in gram-negative bacteria, Presse Medicale 23 (11) (1994). Abstract – Europe PMC Available online: https://europepmc.org/article/med/8022741.

[47] M.A. Webber, L.J.V. Piddock, The importance of efflux pumps in bacterial antibiotic resistance, Journal of Antimicrobial Chemotherapy 51 (2003) 9–11.

[48] L.J.V. Piddock, Multidrug-resistance efflux pumps ? not just for resistance, Nature Reviews Microbiology 4 (2006) 629–636.

[49] F. Chen, J. Zhou, F. Luo, A.-B. Mohammed, X.-L. Zhang, Aptamer from whole-bacterium SELEX as new therapeutic reagent against virulent *Mycobacterium tuberculosis*, Biochemical and Biophysical Research Communications 357 (2007) 743–748.

[50] F. Chen, X. Zhang, J. Zhou, S. Liu, J. Liu, Aptamer inhibits *Mycobacterium tuberculosis* (H37Rv) invasion of macrophage, Molecular Biology Reports 39 (2012) 2157–2162.

[51] F. Chen, J. Zhou, Y.-H. Huang, F.-Y. Huang, Q. Liu, Z. Fang, S. Yang, M. Xiong, Y.-Y. Lin, G.-H. Tan, Function of ssDNA aptamer and aptamer pool against *Mycobacterium tuberculosis* in a mouse model, Molecular Medicine Reports 7 (2013) 669–673.

[52] Q. Pan, Q. Wang, X. Sun, X. Xia, S. Wu, F. Luo, X.-L. Zhang, Aptamer against mannose-capped lipoarabinomannan inhibits virulent *Mycobacterium tuberculosis* infection in mice and rhesus monkeys, Molecular Therapy 22 (2014) 940–951.

[53] J.G. Bruno, M.P. Carrillo, T. Phillips, In Vitro antibacterial effects of antilipopolysaccharide DNA aptamer-C1qrs complexes, Folia Microbiologica 53 (2008) 295–302.

[54] C.S.M. Ferreira, C.S. Matthews, S. Missailidis, DNA aptamers that bind to MUC1 tumour marker: design and characterization of MUC1-binding single-stranded DNA aptamers, Tumor Biology 27 (2006) 289–301.

[55] C.S.M. Ferreira, K. Papamichael, G. Guilbault, T. Schwarzacher, J. Gariepy, S. Missailidis, DNA aptamers against the MUC1 tumour marker: design of aptamer–antibody sandwich ELISA for the early diagnosis of epithelial tumours, Analytical and Bioanalytical Chemistry 390 (2008) 1039–1050.

[56] J.G. Bruno, J.C. Miner, Therapeutic Nucleic Acid-3′-conjugates, 2011.

[57] A. Wen, Q. Yang, J. Li, F. Lv, Q. Zhong, C. Chen, A novel lipopolysaccharide-antagonizing aptamer protects mice against endotoxemia, Biochemical and Biophysical Research Communications 382 (2009) 140–144.

[58] I.M. Ferreira, C.M. de Souza Lacerda, L.S. de Faria, C.R. Corrêa, A.S.R. de Andrade, Selection of peptidoglycan-specific aptamers for bacterial cells identification, Applied Biochemistry and Biotechnology 174 (2014) 2548–2556.

[59] A.C. Graziani, M.I. Stets, A.L. Lopes, P.H.C. Schluga, S. Marton, I.F. Mendes, A.S.R. Andrade, M.A. Krieger, J. Cardoso, High efficiency binding aptamers for a wide range of bacterial sepsis agents, Journal of Microbiology and Biotechnology 27 (2017) 838–843.

[60] I. Mendes Ferreira, C.M. de Sousa Lacerda, S. Roberta dos Santos, A.L. Branco de Barros, S. Odília Fernandes, V.N. Cardoso, A.S. Ribeiro de Andrade, Peptidoglycan aptamers biodistribution in infection-bearing mice, Nuclear Medicine Biomedical Imaging 3 (2018).

[61] S.-K. Kim, C.L. Sims, S.E. Wozniak, S.H. Drude, D. Whitson, R.W. Shaw, Antibiotic resistance in bacteria: novel metalloenzyme inhibitors, Chemical Biology and Drug Design 74 (2009) 343–348.

[62] N.H. Khan, A.A. Bui, Y. Xiao, R.B. Sutton, R.W. Shaw, B.J. Wylie, M.P. Latham, A DNA aptamer reveals an allosteric site for inhibition in metallo-β-lactamases, PloS One 14 (2019) e0214440.

[63] J.G.A. Bruno, Review of therapeutic aptamer conjugates with emphasis on new approaches, Pharmaceuticals 6 (2013) 340–357.

[64] D.G. Meeker, S.V. Jenkins, E.K. Miller, K.E. Beenken, A.J. Loughran, A. Powless, T.J. Muldoon, E.I. Galanzha, V.P. Zharov, M.S. Smeltzer, et al., Synergistic photothermal and antibiotic killing of biofilm-associated *Staphylococcus aureus* using targeted antibiotic-loaded gold nanoconstructs, ACS Infectious Diseases 2 (2016) 241–250.

[65] J.-Y. Hyeon, J.-W. Chon, I.-S. Choi, C. Park, D.-E. Kim, K.-H. Seo, Development of RNA aptamers for detection of Salmonella Enteritidis, Journal of Microbiological Methods 89 (2012) 79–82.

[66] Q. Pan, X.-L. Zhang, H.-Y. Wu, P.-W. He, F. Wang, M.-S. Zhang, J.-M. Hu, B. Xia, J. Wu, Aptamers that preferentially bind type IVB pili and inhibit human monocytic-cell invasion by *Salmonella enterica* serovar typhi, Antimicrobial Agents and Chemotherapy 49 (2005) 4052–4060.

[67] O.S. Kolovskaya, A.G. Savitskaya, T.N. Zamay, I.T. Reshetneva, G.S. Zamay, E.N. Erkaev, X. Wang, M. Wehbe, A.B. Salmina, O.V. Perianova, et al., Development of bacteriostatic DNA aptamers for Salmonella, Journal of Medicinal Chemistry 56 (2013) 1564–1572.

[68] S.H. Suh, L.-A. Jaykus, Nucleic acid aptamers for capture and detection of Listeria spp, Journal of Biotechnology 167 (2013) 454–461.

[69] N. Duan, X. Ding, L. He, S. Wu, Y. Wei, Z. Wang, Selection, identification and application of a DNA aptamer against Listeria monocytogenes, Food Control 33 (2013) 239–243.

[70] H.P. Dwivedi, R.D. Smiley, L.-A. Jaykus, Selection and characterization of DNA aptamers with binding selectivity to Campylobacter jejuni using whole-cell SELEX, Applied Microbiology and Biotechnology 87 (2010) 2323–2334.

[71] H. Li, X. Ding, Z. Peng, L. Deng, D. Wang, H. Chen, Q. He, Aptamer selection for the detection of Escherichia coli K88, Canadian Journal of Microbiology 57 (2011) 453–459.

[72] S. Marton, F. Cleto, M.A. Krieger, J. Cardoso, Isolation of an aptamer that binds specifically to E. coli, PloS One 11 (2016) e0153637.

[73] J.G. Bruno, M.P. Carrillo, R. Crowell, Preliminary development of DNA aptamer-Fc conjugate opsonins, Journal of Biomedical Materials Research Part A 90A (2009) 1152–1161.

[74] J.G. Bruno, J.L. Kiel, In vitro selection of DNA aptamers to anthrax spores with electrochemiluminescence detection, Biosensors and Bioelectronics 14 (1999) 457–464.

[75] J. Vivekananda, C. Salgado, N.J. Millenbaugh, DNA aptamers as a novel approach to neutralize Staphylococcus aureus α-toxin, Biochemical and Biophysical Research Communications 444 (2014) 433–438.

[76] Y.-C. Chang, C.-Y. Yang, R.-L. Sun, Y.-F. Cheng, W.-C. Kao, P.-C. Yang, Rapid single cell detection of Staphylococcus aureus by aptamer-conjugated gold nanoparticles, Scientific Reports 3 (2013) 1–7.

[77] DNA Aptamers Binding to Multiple Prevalent M-Types of Streptococcus pyogenes, Analytical Chemistry 83 (10) (2011). Available online: https://pubs.acs.org/doi/abs/10.1021/ac200575e.

[78] Z.G. Zhao, Y.M. Yu, B.Y. Xu, S.S. Yan, J.F. Xu, F. Liu, G.M. Li, Y.L. Ding, S.Q. Wu, Screening and antivirulent study of N-acyl homoserine lactones DNA aptamers against Pseudomonas aeruginosa quorum sensing, Biotechnology and Bioprocess Engineering 18 (2013) 406–412.

[79] J. Soundy, D. Day, Selection of DNA aptamers specific for live Pseudomonas aeruginosa, PloS One 12 (2017) e0185385.

[80] J. Aldag, T. Persson, R.K. Hartmann, 2′-Fluoro-Pyrimidine-Modified RNA aptamers specific for lipopolysaccharide binding protein (LBP), International Journal of Molecular Sciences 19 (2018) 3883.

[81] K. Mirzakhani, S.L.M. Gargari, I. Rasooli, S. Rasoulinejad, Development of a DNA aptamer for screening Neisseria meningitidis serogroup B by cell SELEX, Iranian Biomedical Journal 22 (2018) 193–201.

[82] R. Lewis, The rise of antibiotic-resistant infections, FDA Consumer 29 (1995) 11–15.

[83] K.E. Deigan, A.R. Ferré-D'Amaré, Riboswitches: discovery of drugs that target bacterial gene-regulatory RNAs, Accounts of Chemical Research 44 (2011) 1329–1338, https://doi.org/10.1021/ar200039b.

[84] W.C. Winkler, R.R. Breaker, Regulation of bacterial gene expression by riboswitches, Annual Reviews in Microbiology 59 (2005) 487–517, https://doi.org/10.1146/annurev.micro.59.030804.121336.

[85] Serganov, E. Nudler, A decade of riboswitches, Cell 152 (2013) 17–24, https://doi.org/10.1016/j.cell.2012.12.024.

[86] A.V. Sherwood, T.M. Henkin, Riboswitch-mediated gene regulation: novel RNA architectures dictate gene expression responses, Annual Reviews in Microbiology 70 (2016) 361–374, https://doi.org/10.1146/annurev-micro-091014-104306.

[87] R.R. Breaker, Riboswitches and translation control, Cold Spring Harbor Perspectives in Biology 10 (2018), https://doi.org/10.1101/cshperspect.a032797.

[88] J. Miranda-Ríos, M. Navarro, M. Soberón, A conserved RNA structure (thi box) is involved in regulation of thiamin biosynthetic gene expression in bacteria, Proceedings of the National Academy of Sciences of the United States of America 98 (2001) 9736–9741, https://doi.org/10.1073/pnas.161168098.

[89] X. Nou, R.J. Kadner, Coupled changes in translation and transcription during cobalamin-dependent regulation of btuB expression in Escherichia coli, Journal of Bacteriology 180 (1998) 6719–6728.

[90] J.B. Perkins, J. Pero, Biosynthesis of riboflavin, biotin, folic acid, and cobalamin, in: A.L. Sonenshein, J.A. Hoch, R. Losick (Eds.), Bacillus Subtillis and its Closest Relatives: From Genes to Cells, ASM Press, Washington, 2002, pp. 271–286.

[91] A.S. Mironov, I. Gusarov I, R. Rafikov, L.E. Lopez, K. Shatalin, R.A. Kreneva, D.A. Perumov, E. Nudler, Sensing small molecules by nascent RNA: a mechanism to control transcription in bacteria, Cell 111 (2002) 747–756, https://doi.org/10.1016/s0092-8674(02)01134-0.

[92] Nahvi, N. Sudarsan, M.S. Ebert, X. Zou, K.L. Brown, R.R. Breaker, Genetic control by a metabolite binding mRNA, Chemical Biology 9 (2002) 1043–1049, https://doi.org/10.1016/s1074-5521(02)00224-7.

[93] W. Winkler, A. Nahvi, R.R. Breaker, Thiamine derivatives bind messenger RNAs directly to regulate bacterial gene expression, Nature 419 (2002) 952–956, https://doi.org/10.1038/nature01145.

[94] K.F. Blount, R.R. Breaker, Riboswitches as antibacterial drug targets, Nature Biotechnology 24 (2006) 1558–1564, https://doi.org/10.1038/nbt1268.

[95] J.E. Barrick, R.R. Breaker, The distributions, mechanisms, and structures of metabolite-binding riboswitches, Genome Biology 8 (2007) 8, https://doi.org/10.1186/gb-2007-8-11-r239.

[96] E. Mehdizadeh Aghdam, M.S. Hejazi, A. Barzegar, Riboswitches: from living biosensors to novel targets of antibiotics, Gene 592 (2016) 244–259, https://doi.org/10.1016/j.gene.2016.07.035.

[97] J.R. Thomas, P.J. Hergenrother, Targeting RNA with small molecules, Chemical Reviews 108 (2008) 1171–1224, https://doi.org/10.1021/cr0681546.

[98] N.J. Baird, A.R. Ferré-D'Amaré, Idiosyncratically tuned switching behavior of riboswitch aptamer domains revealed by comparative small-angle X-ray scattering analysis, RNA 16 (2010) 598–609, https://doi.org/10.1261/rna.1852310.

[99] A.R. Ferré-D'Amaré, RNA-modifying enzymes, Current Opinion in Structural Biology 13 (2003) 49−55, https://doi.org/10.1016/s0959-440x(02)00002-7.

[100] J.A. Sutcliffe, Improving on nature: antibiotics that target the ribosome, Current Opinion in Microbiology 8 (2005) 534−542, https://doi.org/10.1016/j.mib.2005.08.004.

[101] R.L. Monaghan, J.F. Barrett, Antibacterial drug discovery-then, now and the genomics future, Biochemical Pharmacology 71 (2006) 901−909, https://doi.org/10.1016/j.bcp.2005.11.023.

[102] T.A. Steitz, On the structural basis of peptide-bond formation and antibiotic resistance from atomic structures of the large ribosomal subunit, FEBS Letters 579 (2005) 955−958, https://doi.org/10.1016/j.febslet.2004.11.053.

[103] D.J. Knowles, N. Foloppe, N.B. Matassova, A.I. Murchie, The bacterial ribosome, a promising focus for structure-based drug design, Current Opinion in Pharmacology 2 (2002) 501−506, https://doi.org/10.1016/s1471-4892(02)00205-9.

[104] G.J.R. Zaman, P.J.A. Michiels, in: P. McNamara (Ed.), Targeting RNA with Small Molecule Drugs in Trends in RNA Research, Nova Science Publishers, Inc., Hauppauge, NY, 2006, pp. 1−21.

[105] T. Hermann, Y. Tor, RNA as a target for small-molecule therapeutics, Expert Opinion on Therapeutic Patents 15 (2005) 49−62.

[106] J. Mulhbacher, D.A. Lafontaine, Ligand recognition determinants of guanine riboswitches, Nucleic Acids Research 35 (2007) 5568−5580, https://doi.org/10.1093/nar/gkm572.

[107] M. Mandal, B. Boese, J.E. Barrick, W.C. Winkler, R.R. Breaker, Riboswitches control fundamental biochemical pathways in Bacillus subtilis and other bacteria, Cell 113 (2003) 577−586, https://doi.org/10.1016/s0092-8674(03)00391-x.

[108] W.J. Robbins, The pyridine analog of thiamine and the growth of fungi, Proceedings of the National Academy of Sciences of the United States of America 27 (1941) 419−422, https://doi.org/10.1073/pnas.27.9.419.

[109] D.W. Woolley, A.C.G. White, Selective reversible inhibition of microbial growth with pyrithiamine, Journal of Experimental Medicine 78 (1943) 489−497, https://doi.org/10.1084/jem.78.6.489.

[110] Iwashima, Y. Wakabayashi, Y. Nose, Formation of thiamine pyrophosphate in brain tissue, Journal of Biochemistry 79 (1979) 845−847, https://doi.org/10.1093/oxfordjournals.jbchem.a131138.

[111] N. Sudarsan, S. Cohen-Chalamish, S. Nakamura, G.M. Emilsson, R.R. Breaker, Thiamine pyrophosphate riboswitches are targets for the antimicrobial compound pyrithiamine, Chemical Biology 12 (2005) 1325−1335, https://doi.org/10.1016/j.chembiol.2005.10.007.

[112] T. Shiota, J.E. Folk, F. Tietze, Inhibition of lysine utilization in bacteria by S-(betaaminoethyl) cysteine and its reversal by lysine peptides, Archives of Biochemistry and Biophysics 77 (1958) 372−377, https://doi.org/10.1016/0003-9861(58)90084-5.

[113] T. McCord, J. Ravel, C. Skinner, W. Shive, DL-4-oxalysine, an inhibitory analog of lysine, Journal of the American Chemical Society 79 (1957) 5693−5696, https://doi.org/10.1021/ja01578a029.

[114] K.F. Blount, X.J. Wang, J. Lim, N. Sudarsan, R.R. Breaker, Antibacterial compounds that target lysine riboswitches, Nature Chemical Biology 1 (2007) 44−49, https://doi.org/10.1038/nchembio842.

[115] N. Sudarsan, J.K. Wickiser, S. Nakamura, M.S. Ebert, R.R. Breaker, An mRNA structure in bacteria that controls gene expression by binding lysine, Genes and Development 17 (2003) 2688−2697, https://doi.org/10.1101/gad.1140003.

[116] K. Matsui, H.C. Wang, T. Hirota, H. Matsukawa, S. Kasai, K. Shinagawa, S. Otani, Riboflavin production by roseoflavin-resistant strains of some bacteria, Agricultural and Biological Chemistry 46 (1982) 2003−2008, https://doi.org/10.1080/00021369.1982.10865390.

[117] V.M. Berezovskii, A.I. Stepanov, N.A. Polyakova, L.S. Tulchinskaya, A.Y. Kukanova, Studies of a group of allo- and isoallxazine. XLVI.Synthesis and biological specificity of amino analogs, Bioorganicheskaia Khimiia 3 (1977) 521−524.

[118] M.S. Gelfand, A.A. Mironov, J. Jomantas, Y.I. Kozlov, D.A. Perumov, A conserved RNA structure element involved in the regulation of bacterial riboflavin synthesis genes, Trends in Genetics 15 (1999) 439−442, https://doi.org/10.1016/s0168-9525(99)01856-9.

[119] W.C. Winkler, S. Cohen-Chalamish, R.R. Breaker, An mRNA structure that controls gene expression by binding FMN, Proceedings of the National Academy of Sciences 99 (2002) 15908−15913, https://doi.org/10.1073/pnas.212628899.

[120] E.R. Lee, K.F. Blount, R.R. Breaker, Roseoflavin is a natural antibacterial compound that binds to FMN riboswitches and regulates gene expression, RNA Biology 6 (2009) 187−194, https://doi.org/10.4161/rna.6.2.7727.

[121] J.A. Howe, H. Wang, T.O. Fischmann, C.J. Balibar, L. Xiao, A.M. Galgoci, J.C. Malinverni, T. Mayhood, A. Villafania, A. Nahvi, N. Murgolo, C.M. Barbieri, P.A. Mann, D. Carr, E. Xia, P. Zuck, D. Riley, R.E. Painter, S.S. Walker, B. Sherborne, R. de Jesus, W. Pan, M.A. Plotkin, J. Wu, D. Rindgen, J. Cummings, C.G. Garlisi, R. Zhang, P.R. Sheth, C.J. Gill, H.T. Roemer, Selective small-molecule inhibition of an RNA structural element, Nature 526 (2015) 672−677, https://doi.org/10.1038/nature15542.

[122] S. Klähn, P. Bolay, P.R. Wright, R.M. Atilho, K.I. Brewer, M. Hagemann, R.R. Breaker, W.R. Hess, A glutamine riboswitch is a key element for the regulation of glutamine synthetase in cyanobacteria, Nucleic Acids Research 46 (2018) 10082−10094, https://doi.org/10.1093/nar/gky709.

[123] L. Huang, J. Wang, A.M. Watkins, R. Das, D.M. J Lilley, Structure and ligand binding of the glutamine-II riboswitch, Nucleic Acids Research 47 (2019) 7666−7675, https://doi.org/10.1093/nar/gkz539.

[124] J.T. Polaski, S.M. Webster, J.E. Johnson, R.T. Batey, Cobalamin riboswitches exhibit a broad range of ability to discriminate between methylcobalamin and adenosylcobalamin, Journal of Biological Chemistry 292 (2017) 11650–11658, https://doi.org/10.1074/jbc.m117.787176.

[125] A.M. Babina, N.E. Lea, M.M. Meyer, In vivo behavior of the tandem Glycine riboswitch in *Bacillus subtilis*, mBio 8 (2017), https://doi.org/10.1128/mbio.01602-17.

[126] Khani, N. Popp, B. Kreikemeyer, N. Patenge, A Glycine riboswitch in *Streptococcus pyogenes* controls expression of a sodium:alanine symporter family protein gene, Frontiers in Microbiology 9 (2018) 200, https://doi.org/10.3389/fmicb.2018.00200.

[127] R.R. Breaker, Prospects for riboswitch discovery and analysis, Molecular Cell 43 (2011) 867–879, https://doi.org/10.1016/j.molcel.2011.08.024.

CHAPTER 14

The Chemistry and Pharmacology of Antibiotics Used in the Treatment of Multidrug-Resistant Tuberculosis

MOTSHABI ALINAH SIBEKO • VICTOR CHIKE AGBAKOBA • JETHRO NKOMO • SHANGANYANE PERCY HLANGOTHI

1 INTRODUCTION

Bacteria are microscopic single-celled organisms without nucleus and specialized organelles. These organisms live in diverse environments, for example, in soil, water, and inside the human body [1,2]. They can be distinguished based on the nature of their cell walls, their shape, or their genetic makeup. Bacteria play various roles; some assist animals with digestion process [3,4] and help trees to grow [5,6], while others are useful in the production or fermented foods such as yoghurt [7]. The relationship between bacteria and humans is very complex; for instance, some bacteria are known to provide protective and stabilizing effect on the human body as resident microbes. Conversely, several other pathogenic bacteria invade and grow within the tissues of humans. Once they are in the body, they cause various diseases that are damaging to the body and sometimes leading to death. Examples of pathogenic bacteria include *pneumococcal pneumonia* responsible for pneumonia, *Vibrio cholera* responsible for cholera, and *Mycobacterium tuberculosis* responsible for tuberculosis (TB) [8].

TB is an airborne infectious disease that is caused by a bacterium known as *Mycobacterium tuberculosis* [9]. TB has been recognized by the World Health Organization (WHO) as one of the top 10 leading causes of death worldwide. In 2018, a report by the WHO on the global TB stated that 10 million people fell ill with TB; among the infected, 1.6 million people died from the disease in 2017 [10]. TB infections can be categorized into latent and active TB. Latent tuberculosis infection is the state in which a person infected does not show any symptoms; however, when left untreated, 10% of these infec-tions can develop into active TB. Active TB can be transmitted from person to person, and the bacteria can cause various symptoms. Some of the common symptoms include chronic coughing with blood con-taining sputum, fever, chest pains, night sweats, and weight loss [9,11,12]. The global incidences of TB differ from country to country, as can be seen in Fig. 14.1. Sur-prisingly, one third of the world's population is living with the pathogen as latent TB infection [13].

TB can affect most parts of the body, but in most cases, it affects the lungs, which is known as pulmonary TB [10]. As TB is an airborne disease, a patient infected with pulmonary TB can spread the infectious disease into the air by either coughing, spitting, speaking, or sneezing. It has been reported that 40,000 aerosol drop-lets can be released from a single cough, and each droplet can transmit the disease [10,14]. Presently, anti-biotics present the most effective means of treating TB. Antibiotics, also known as antibacterials, are drugs that can be used to either completely kill bacteria or inhibit their growth [8]. As a result, several antibiotics have been developed and successfully used in the treatment of TB. In the 1940s, Streptomycin was discovered as the first antibiotic, which was effective against *Mycobac-terium tuberculosis*; however, the bacteria rapidly devel-oped resistance to the monotherapy [15]. To resolve this problem, the administration of two drugs namely isoniazid and *para*-amino-salicylic acid was approved in the early 1950s to treat TB. Unfortunately, these drugs had to be taken for 12—24 months to prevent relapse. In the early 1970s, rifampicin, which was a more effective short-course chemotherapy, given for less than 12 months, was introduced [16].

Antibiotic Materials in Healthcare. https://doi.org/10.1016/B978-0-12-820054-4.00014-8

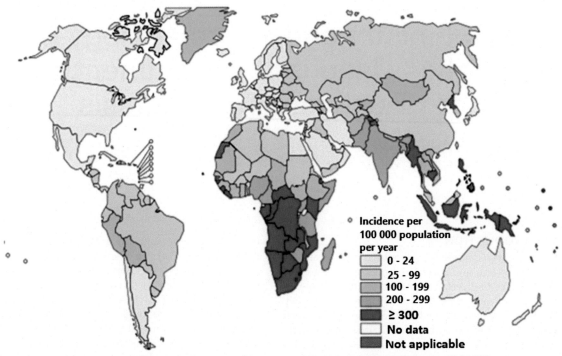

FIG. 14.1 Estimated TB incidence rates, 2016. WHO Global tuberculosis report, 2017 [13].

Over the past decades, rifampicin and isoniazid have been used as the first-line TB drug regimen. These antibiotics are effective inactive, drug-susceptible TB if the patient is compliant and completes the treatment [17]. In some cases, patients do not comply owing to the number of drugs one has to take, cost of drugs, adverse effects, and the long duration for full treatment [18]. The lack of compliance is a major contributing factor leading to the appearance of multidrug-resistant (MDR) TB strains, which are resistant to the two most potent first-line antituberculosis drugs. MDR-TB has been identified as the cause of the highest mortality rates, and its emergence has made the control of TB more challenging [17–19].

The WHO has recommended a second-line drug regimens for the treatment of MDR-TB, which includes fluoroquinolones, aminoglycosides, capreomycin, cycloserine, and thioamides [20]. In this chapter, we will discuss the epidemiology and socioeconomic impact of tuberculosis, as well as the chemistry, pharmacology, and the pharmacokinetics of various groups of antibiotics used to treat MDR-TB. Several reports have demonstrated that drug-resistant tuberculosis can be treated with the right combination and rational use of available antituberculosis drugs. Lastly, we will review some of the studies that have been done to evaluate the efficacy of different antibiotics on MDR-TB.

2 EPIDEMIOLOGY

Epidemiology is a Greek word that is derived from *epi* that means "upon, among," *demos* that means "people, district," and *logos* that means "study" [21]. In 1978, Lilienfeld defined epidemiology as "a method of reasoning about the disease that deals with biological inference derived from observations of disease phenomena in population groups" [22]. The definition of epidemiology has been continuously modified over the year, as a result of changes in society and the emergence of new diseases. In 2018, Mathilde et al. [23] compiled a literature review on the definitions of epidemiology that had been written between 1978 and 2017. Wasserthell-Smoller et al. [24] defined epidemiology "as the study of the health and disease in groups of people and the study of the factors that influence this distribution." Epidemiologists have the responsivity to investigate disease outbreaks within a population, by establishing the cause of the disease, identify people who are sick, their symptoms, when they contract the disease and where could they have been exposed, and most importantly how the disease develops and spreads through the population [25]. The data that are acquired are then analyzed systematically to establish strategies to prevent or control the disease and improve public health [26].

2.1 Epidemiology of TB

Since 1997, the WHO has published a global, annual report that is aimed at providing updated and comprehensive statistics about TB epidemic. The report outlines the measures taken in terms of prevention, diagnosis, and treatment of TB at global, regional, and country level [27]. The latest WHO report regarding the TB epidemic at the time of compiling this chapter was published in 2018. On a global scale, it was estimated that 10.0 million people developed TB in 2017; among them, 5.8 million were men, 3.2 million were women, and 1.0 million were children [10]. In 2016, out of an estimated 10.4 million new TB cases reported globally, approximately 1.7 people died from TB. It is clear that there is a drop in the number of dearth, as a 4% drop was reported in 2015 compared to 2016 [28]. The epidemiology of TB varies from region to region, and up to date, the WHO has identified about 30 countries as high burden countries (HBCs) of TB. Some of the countries that fall under HBCs include India, Indonesia, China, Nigeria, Pakistan, Bangladesh, Philippines, and South Africa [28,29]. These HBCs are responsible for 87% of TB cases worldwide, while low burden countries that include most countries in Western Europe contribute only 6% [10]. There are several contributing factors to the prevalence of TB in developing countries such as human immunodeficiency virus (HIV) and socioeconomic factors.

2.1.1 HIV and epidemiology of TB

On a global scale, *Mycobacterium tuberculosis* (MTb) infection remains one of the leading causes of morbidity and mortality. However, the risk of getting active TB is much greater in persons suffering from illnesses that impair the immune system. Among the illnesses, HIV has been identified as the most contributing factor in the TB epidemic. In countries with a generalized HIV epidemic, HIV positive people are 20−30 times more likely to develop active TB compared to people without HIV [30,31]. In 2007, 9.3 million new cases of TB were reported worldwide; out of that, 1.7 million were coinfected with HIV. The highest cases of TB-HIV coinfections

are mostly from Sub-Saharan Africa followed by South-East Asia [32]. Carbett et al. [33] reported on the burden of HIV and tuberculosis infection in Africa; as can be seen in Fig. 14.2, Africa has the highest rates of people coinfected with HIV-TB. From the figure, a person contributes 5% to the global total, and the red color represents African people, while the blue color is for the rest of the world. An estimated 8.8 million new cases of tuberculosis were reported in 2003, and from that 1.7 million people died. Moreover, 31% of these deaths were reported in Africa.

2.1.2 Socioeconomic factors associated with TB

Socioeconomics can be defined as a study that looks at how societies progress or regress as a result of the local, regional, or even the global economy. The prevalence of TB especially in developing countries has been widely associated with socioeconomic factors, for example, malnutrition, poverty, lack of knowledge, and poor socioeconomic status. Some of the studies that evaluated the relationship between TB and socioeconomic factors have highlighted that lower socioeconomic status is associated with increased risk of TB in most developing areas [34,35]. In the western countries, particularly United States and Europe, there was a dramatic decrease in TB incidences even before the introduction of first-line TB treatment and effective chemotherapy. The decline in TB incidences in United States from 113 per 100,000 persons per year in 1920 to less than 10 per 100,000 in the 1950s was attributed to improvements in housing, sanitation, and general socioeconomic status [36]. Several socioeconomic factors still play a significant role in TB epidemic in developing countries.

There are six socioeconomic status indicators that have been used to evaluate the impact of socioeconomic factors on the TB epidemic, namely overcrowding, income, poverty, public assistance, unemployment, and education [37]. As highlighted in the introduction, TB is an airborne disease, and aerosol droplets containing tubercle bacilli are released into the atmosphere when an infected person coughs or sneezes. The droplets

FIG. 14.2 Disproportionate burden of HIV, HIV-related tuberculosis, and *M. tuberculosis* coinfections in Africa [33].

remain suspended in the air; consequently, in the case of overcrowding, usually a decrease in fresh air increases the exposure to *M. tuberculosis* [37,38]. In most developing countries, the increase in the level of poverty results in poor nutrition and low body weight, this ultimately weakens the immune system and makes the body more vulnerable to invading organisms [39]. Poor housing infrastructure that is more exposed to flooding is also a major concern. Studies have shown that bacillus that is dried out or exposed to sunlight often is phenotypically too weak to start an infection. Poor housing with its attendant poor ventilation and increased dampness increase the risk of transmission and development of disease.

3 ANTIBIOTICS THAT CAN BE USED TO TREAT MDR-TB

The drugs that constitute the MDR-TB regimen are classified into three groups, following an extensive assessment of each drug's benefits and risks. Table 14.1 shows the groupings of drugs used to treat MDR-TB and their recommended inclusion criteria.

The MDR-TB treatment can be administered over a total period of 6–20 months, where the duration depends on the response of the patient to the specific drug combination. In South Africa, two regimens are traditionally used. The short-term treatment regimen

TABLE 14.1
Groups of Drugs Used in the Treatment of the MDR-TB [40].

Group	Inclusion criteria	Drug
Group A	Include all three medicines	Levofloxacin or Moxifloxacin Bedaquiline Linezolid
Group B	Add one or both medicines	Clofazimine Cycloserine or Terizidone
Group C	To complete the regimen and when medicines from Groups A and B cannot be used	Ethambutol Delamanid Pyrazinamide Imipenem—cilastatin or Meropenem Amikacin or streptomycin Ethionamide or Prothionamide p-aminosalicylic acid

Adapted: from WHO consolidated guidelines on drug-resistant tuberculosis treatment (2019).

is the most common and is given over a total duration of 9–11 months, where the intensive phase of the treatment is at least 4 months. It can be extended to 6 months in response to the effectiveness of the therapy in particular patients. The therapy usually consists of the following drugs: Linezolid, high dose isoniazid, bedaquiline, levofloxacin, clofazimine, pyrazinamide, and ethambutol. Linezolid and bedaquiline are administered over 2 and 6 months, respectively, irrespective of the duration of the intensive phase of the therapy. Table 14.2 illustrates the general overview of the duration of each drug in the short-term regimen.

The long-term treatment is given for a duration of 18–20 months. The intensive phase is 6 months and, depending on the response of the patient to the treatment, maybe elongated to 8 months. Linezolid, bedaquiline, levofloxacin, clofazimine, and terizidone form the core of the treatment. Linezolid is, in the long-term treatment regime, given throughout the duration of the therapy. Levofloxacin is many times substituted for moxifloxacin in necessary cases.

3.1 A Closer Look at the Chemistry, Pharmacology, and Pharmacokinetics of Some MDR-TB Drugs

The various nature and functions of the drugs used in the treatment of multidrug-resistant strains of MTb differ in many ways. However, they can be viewed by their chemistry (chemical structure), pharmacology, and pharmacokinetics. First, the chemistry of these drugs involves the various atoms, molecules, and chemical species that make up the bulk structure of the active compound. Second, the pharmacology of the drugs gives an insight into the effects of these chemical species in the body. Lastly, pharmacokinetics highlights the rate at which the drugs are adsorbed, distributed, metabolized, and eliminated from the body.

3.1.1 Isoniazid

This heterocyclic compound also called as isonicotinic acid hydrazide (INH) comprises a single pyridine ring with a hydrazide substitution, as shown in Fig. 14.3. Generally, INH is a prodrug, which means that it is not effective against MDR-TB as it is [41]. Nevertheless, the activity of INH in the human body depends on its activation by the KatG gene [42]. INH is a bactericidal drug that is well absorbed orally or intramuscularly and affects the synthesis of cell walls of rapidly dividing bacterial cells [43]. The minimum inhibitory concentration (MIC) is the lowest concentration of INH needed to prevent visible growth of bacterium. In the case of rapidly growing MTb, the MIC is 0.2 µM, but its activity is greatly lowered in the case of a slow growing MTb [44].

TABLE 14.2
General Overview of the Short-Term MDR-TB Treatment.

Drug	2 Months	4 Months	6 Months	9 Months
Linezolid	▓			
Isoniazid (High dose)	▓	▓		
Bedaquiline	▓	▓	▓	
Levofloxacin	▓	▓	▓	▓
Clofazimine	▓	▓	▓	▓
Pyrazinamide	▓	▓	▓	▓
Ethambutol	▓	▓	▓	

Adapted from: Interim clinical guidance on implementation of the modified short and long treatment regimens for people MDR-TB in South Africa (2018).

On the other hand, resistance to INH does occurs when the KatG gene is mutated [45]. Although INH is generally safe in the human body, exception does arise in the case of hepatic enzyme abnormalities. The side effects resulting from these abnormalities are known to result in hepatitis, especially in older patients [41]. Lastly, INH

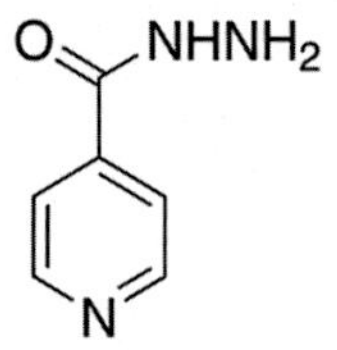

FIG. 14.3 Chemical structure of Isoniazid [41].

together with rifampicin, pyrazinamide, and ethambutol constitute the first-line oral antituberculosis drugs [19].

3.1.2 Rifamycins

The first rifamycin compound to be developed is rifampicin, and this was followed by two other derivatives, rifabutin and rifapentine [19,41,46]. The chemical structures of these antituberculosis drugs are very complex. Nevertheless, they all contain an aromatic naphthalene center that is then traversed by a 19-atom aliphatic polyketide bridge. Rifamycins target and bind themselves onto the enzyme RNA polymerase, thereby disrupting the transcription of segments of the bacterial DNA onto RNA [47]. Fig. 14.4 shows the clinically used rifamycins.

R = CH₃; Rifamycin (RIF)
R = cyclopentane; Rifabutin

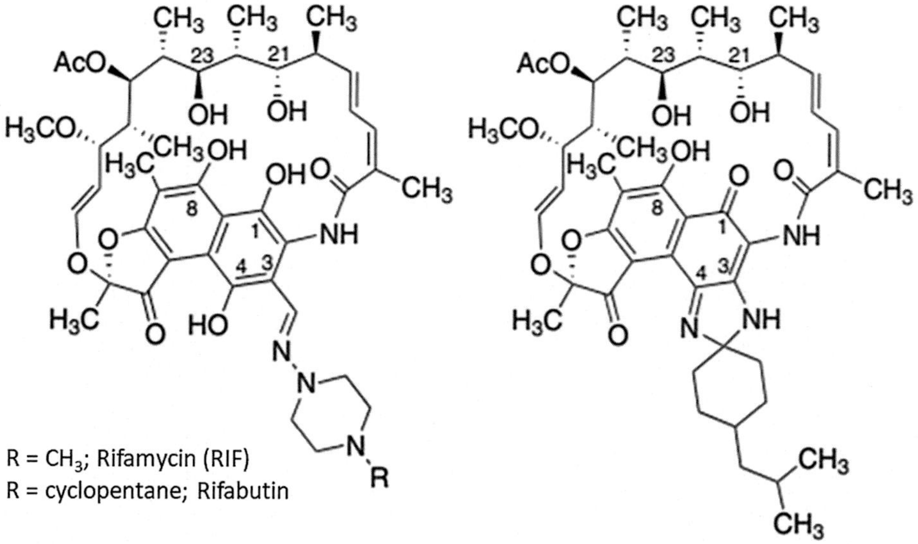

FIG. 14.4 Clinically used rifamycins—rifamycin and rifabutin (left) and rifapentine (right) [41].

Nevertheless, MTb can develop resistance to all three rifamycins as a result of the mutation of the RNA polymerase β-subunit gene (rpoB) [19,48]. Alternatively, resistance to rifamycin could arise via the Adenosine diphosphate-ribosylation of the alcohol functional group on carbon number 21 [41]. The usage of rifamycins in the treatment of MDR-TB is greatly restricted due to its high cost, limited availability of the drug, and inadequate susceptibility testing in developing countries [19]. Several adverse effects relate to the usage of rifamycins, some of which includes mild influenza symptoms, hepatoxicity, and altered liver function [41]. Furthermore, bodily fluids of patients taking rifamycins may turn orange-red color as a result of the furanonapthoquinone chromophore present in the rifamycin moiety. Lastly, rifamycins are known to have potential adverse interactions when coadministered with antiretroviral drugs [41].

3.1.3 Fluoroquinolones

There are several compounds that make up this group, such as moxifloxacin, levofloxacin, ofloxacin, and ciprofloxacin [19]. As shown in Fig. 14.5, their basic chemical structures consist of the heterocyclic aromatic compound "quinolone" with at least one fluoride substituent [41]. Fluoroquinolones have been used as the main drug in the treatments of MDR-TB as they have been found to be very effective [19,49].

Although fluoroquinolones share the same genetic target—the gyrA gene, they, however, are not all equally effective in the same way [49]. For instance, ciprofloxacin is known to be less effective compared to ofloxacin, levofloxacin, and moxifloxacin [19].

In addition, high dosages of levofloxacin are known to be very effective in the treatment of MTb strains that are otherwise resistant to ofloxacin. However, moxifloxacin is readily used due to its impressive sterilization activity compared to other fluoroquinolones [19]. The MIC does differ from one to the other. For instance, the MIC of moxifloxacin is 0.16 μM compared to 1.25 μM of levofloxacin [50]. Fluoroquinolones have several useful pharmacokinetic features. For instance, many fluoroquinolones have good oral bioavailability and can permeate into the cells. Cell permeability is particularly important when treating intracellular bacteria [51]. In addition, most fluoroquinolones are cleared from the body via the kidneys, with some others also able to be eliminated by liver metabolism and or other combined routes. The most common side effects of fluoroquinolones include skin reactions, disturbances of the central nervous system, and gastrointestinal upset [52,53]. Nonetheless, these side effects are usually mild and tend to be self-limiting. However, major side effects such as tendonitis (tendon damage due to collagen damage) and photosensitivity are not uncommon [52].

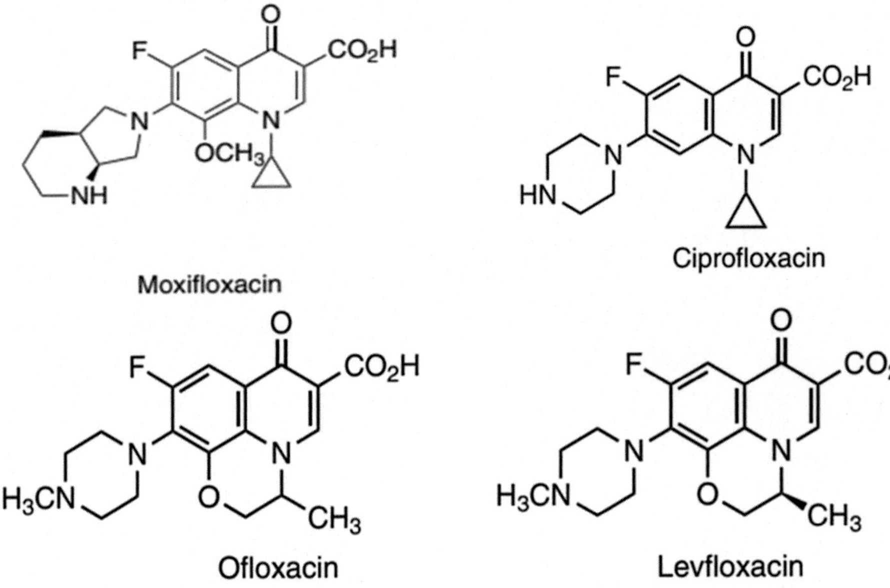

FIG. 14.5 Chemical structures of some fluoroquinolones [41].

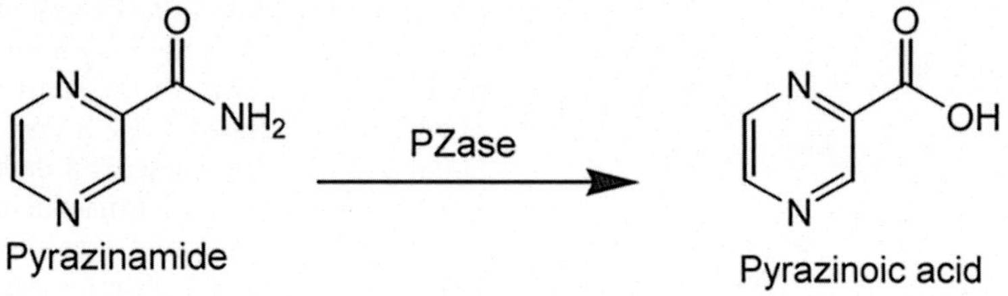

FIG. 14.6 Conversion of pyrazinamide into the active pyrazinoic acid by PZase [56].

3.1.4 Pyrazinamide

The rudimentary chemical structure of this drug consists of single pyrimidine aromatic heterocyclic with a carboxy amide functional group [54]. It is generally agreed that pyrazinamide (PZA) destroys MTb via intracellular acidification, which results from the hydrolysis of PZA by MTb nicotinamidase/pyrazinamidase [54,55]. PZA enters the mycobacteria as a prodrug via passive diffusion and active transport. Afterward, it is then converted by pyrazinamidase (PZase) into pyrazinoic acid, the active antibacterial moiety, as can be seen in Fig. 14.6 [54]. The activated PZA now inside the MTb proceeds in disrupting the membrane energetics and inhibits the membrane transport function [54,56].

Thus, PZA was widely used between 1950 and 1970 as a treatment for MTb; this was however before the advent of fluoroquinolones and rifampicin [19]. The low toxicity and low-cost nature of PZA is the reason why it is currently used during the entire duration of MDR-TB treatment [19]. Nevertheless, the effectiveness of PZA is usually accessed individually from one patient to another. The reason being that it could lead to hepatoxicity in elder people and persons living with alcohol addiction [19], especially, when PZA is used in combination with rifamycins and isoniazid [19]. Additionally, PZA is known to increase the concentration of uric acid in serum, which results to joint pains (nongouty arthralgia) [41]. The elimination of PZA from the body is dependent on whether the patient has a normal or abnormal renal and hepatic function [56,57]. In plasma, PZA has a half-life of 9–10 h, and about 70% are excreted in urine within 24 h [56,58].

3.1.5 Ethambutol

Ethambutol (EMB) is a low-cost first-line drug, which was discovered after the structural modification of N,N'-diisopropylethylenediamine, and their chemical structures are shown in Fig. 14.7 [59–61]. The potency of EMB is modest and highly dependent on the size and nature of alkyl group situated on the ethylenediamine nitrogens. In addition, the minimum pharmacophore needed for antitubercular activity is the ethylenediamine unit [41,62].

Despite the uncertainty with regards to the true mode of action, it is commonly agreed that EMB interferes with the construction of the arabinogalactan layer of the mycobacterial cell wall [63]. Generally, EMB is known to be well tolerated due to its low toxicity; however, it has been reported to induce ocular toxicity following the depletion of zinc and copper levels [64]. Approximately 50% of the initial dose of EMB is excreted unchanged in urine within the 24-h period of oral injection. During metabolism of EMB, the main path appears to involve an initial oxidation of the alcohol into an aldehydic intermediate, then followed by conversion to a dicarboxylic acid [65].

3.1.6 Clofazimine

This second-line is used in combination with other drugs for the treatment of MDR-TB. The chemical structure of this compound is quite complex but is characterized by the following chemical species, p-chlorophenyl, p-chloroaniline, and isopropylaminophenazine, as shown in Fig. 14.8. Clofazimine (CLOF) is a low-cost drug, and its activity is both intracellular and extracellular [66]. In addition, CLOF is known to have oral

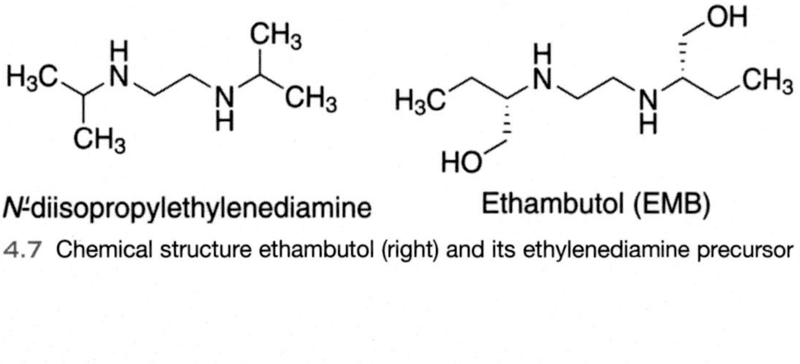

N,N'-diisopropylethylenediamine   Ethambutol (EMB)

FIG. 14.7 Chemical structure ethambutol (right) and its ethylenediamine precursor (left).

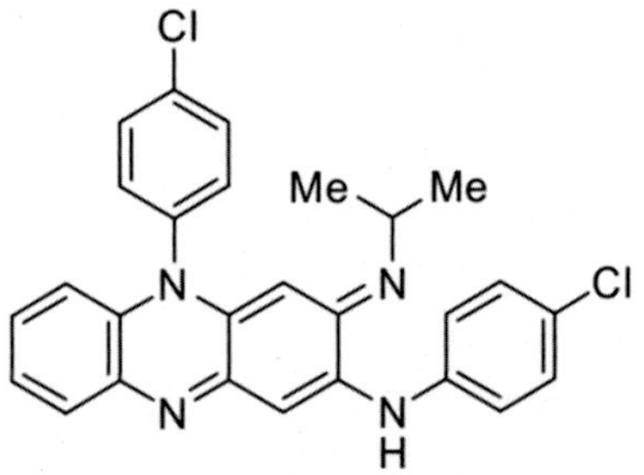

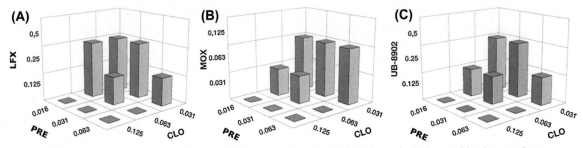

FIG. 14.8 Chemical structure of clofazimine.

absorption rate in the range of 45%–62% in the human body [67]. The highly lipophilic CLOF means that it tends to deposit predominantly in fatty tissues and cells of reticuloendothelial cells. CLOF is metabolized in the liver, and upon repeated oral dosages, has a half-life of about 70 days [68].

Despite its known usefulness, CLOF has serious adverse effects in humans; some of which include (1) gastrointestinal toxicity—about 40%–50% of patients experience abdominal pain, diarrhea, nausea, vomiting ,and even death (i.e., in cases of severe abdominal symptoms). (2) Discolouration of urine, feces, sweat and sputum, (3) eye pigmentation resulting from deposition of CLOF crystals, (4) headache, dizziness, drowsiness, fatigue, and taste disorder.

3.2 Studies Conducted to Study the Efficacy of Antibiotics on the Treatment of MDR-TB

Multidrug resistance has become a global crisis in the management of tuberculosis; as a result, it has become imperative to invest in the research of new drugs as well as the development of efficacious drug combinations and regimens. In 2016, WHO was tasked to create a priority list of other antibiotic-resistant bacteria that will assist and enable research and development of new and effective drugs [69]. Richeldi et al. [70] studied the clinical use of levofloxacin (LFX), which represents

one of the few second-line drugs in the therapeutic regimens of MDR-TB as the long-term treatment of drug-resistant tuberculosis. In the study, four patients with MDR-TB were treated twice a day for 9 months with 500 mg of LFX. The higher LFX daily dosages were targeted at address the pharmacodynamic properties of the drug. It was reported that the administration of LFX 500 mg for 9–24 months was tolerable and safe because the patients showed no significant alteration of either liver function tests, blood tests, or any other described side effect of the fluoroquinolone class. Furthermore, three patients with pulmonary MDR-TB showed radiologic and clinical improvement.

For over a decade now, a combination of two or more antibiotics have been studied to treat MDR-TB [71–74]. Gavin et al. [75] conducted a study to assess and compare the efficacy of three antituberculous combinations against MDR and drug-susceptible clinical isolates of *Mycobacterium tuberculosis* by an in vitro adaptation of the checkerboard assay. The first combination of antibiotics was clofazimine/pretomanid/levofloxacin, the second combination was clofazimine/pretomanid/moxifloxacin, and the third combination was clofazimine/pretomanid/UB-8902. The study was conducted using 7 MDR and 11 drug-susceptible clinical *M. tuberculosis* isolates from the Hospital Clinic de Barcelona. The interaction of the drugs was evaluated using the mean fractional inhibitory concentration index (FICI), which was by den Hollander et al. [76]. The FICI is regarded as the sum of the fractional inhibitory concentrations of each antibiotic present in the combination. The FIC was calculated using Eq. (14.1) [76].

$$FICI = FIC_A + FIC_B + FIC_C = \left(\frac{CIA}{MIC_A}\right) + \left(\frac{CIC_B}{MIC_B}\right) + \left(\frac{CIC_C}{MIC_C}\right) \qquad [14.1]$$

As can be seen from Fig. 14.9, all antibiotic combinations showed additive activity (0.75 < FICI < 4) against

FIG. 14.9 Three-dimensional representation of the growth pattern of each combination against *Mycobacterium tuberculosis* isolate 18S.

MDR and drug-susceptible isolates. Furthermore, no significant differences were reported between MDR and drug-susceptible isolates with the use of any of the combinations. Most importantly, MDR and drug-susceptible strains were susceptible to all of the drugs studied, and the similarity in the results obtained indicates that no factors associated with multidrug resistance, such as cross-resistance, decreased fitness, or impaired metabolic pathways, affected the efficacy of the different combinations against the isolates. Other combinations of fluoroquinolones, amikacin, and linezolid were also studied and found to be equivalent to those obtained with a standard combination of isoniazid, ethambutol, and rifampicin in in vitro conditions [72,77].

4 CONCLUDING REMARKS

The world health organization took a decision and declared TB as a global health emergency, and since 1997 there has been a report on the global status of the TB epidemic. It is very alarming to note that even in 2018, millions of people are still getting infected with the deadly disease and every year millions of people still die of the disease. This chapter highlighted the epidemic of TB and its severe impact on developing countries and highlighted on the current treatment of MDR-TB. It is clear that more efforts still need to be done in terms of eliminating this deadly disease.

ACKNOWLEDGMENTS

The National Research Foundation (NRF) of South Africa and Recycling and Economic Development Initiative of South Africa (REDISA), together with the Department of Science and Technology (DST)—Council for Scientific and Industrial Research (CSIR), Biorefinery Industry Development Facility (BIDF), are acknowledged for providing the bursary funding during the period of study.

REFERENCES

[1] M. Vos, et al., Micro-scale determinants of bacterial diversity in soil, FEMS Microbiology Reviews 37 (6) (2013) 936–954.

[2] J.P. Cabral, Water microbiology. Bacterial pathogens and water, International Journal of Environmental Research and Public Health 7 (10) (2010) 3657–3703.

[3] J.G. Lundgren, R.M. Lehman, Bacterial gut symbionts contribute to seed digestion in an omnivorous beetle, PloS One 5 (5) (2010) e10831.

[4] G. Tellez, et al., Digestive physiology and the role of microorganisms, Journal of Applied Poultry Research 15 (1) (2006) 136–144.

[5] S. Lladó, R. López-Mondéjar, P. Baldrian, Forest soil bacteria: diversity, involvement in ecosystem processes, and response to global change, Microbiology and Molecular Biology Reviews 81 (2) (2017) e00063-16.

[6] R. Utkhede, T. Vrain, J. Yorston, Effects of nematodes, fungi and bacteria on the growth of young apple trees grown in apple replant disease soil, Plant and Soil 139 (1) (1992) 1–6.

[7] S. Rezac, et al., Fermented foods as a dietary source of live organisms, Frontiers in Microbiology 9 (2018).

[8] E. Etebu, I. Arikekpar, Antibiotics: classification and mechanisms of action with emphasis on molecular perspectives, International Journal of Applied Microbiology and Biotechnology Research 4 (2016) 90–101.

[9] World Health Organization, Tuberculosis and Air Travel: Guidelines for Prevention and Control, World Health Organization, 2008.

[10] World Health Organization, Global Tuberculosis Report 2018, World Health Organization, 2018.

[11] S.H. Lee, Tuberculosis infection and latent tuberculosis, Tuberculosis and Respiratory Diseases 79 (4) (2016) 201–206.

[12] . W.W. Jensen, et al., Tuberculosis Infection Control in the Era of Expanding HIV Care and Treatment.

[13] Z.S. Bhat, et al., Drug targets exploited in *Mycobacterium tuberculosis*: Pitfalls and promises on the horizon, Biomedicine and Pharmacotherapy 103 (2018) 1733–1747.

[14] E.C. Cole, C.E. Cook, Characterization of infectious aerosols in health care facilities: an aid to effective engineering controls and preventive strategies, American Journal of Infection Control 26 (4) (1998) 453–464.

[15] J.F. Murray, D.E. Schraufnagel, P.C. Hopewell, Treatment of tuberculosis. A historical perspective, Annals of the American Thoracic Society 12 (12) (2015) 1749–1759.

[16] D.A. Mitchison, Role of individual drugs in the chemotherapy of tuberculosis, International Journal of Tuberculosis and Lung Disease 4 (9) (2000) 796–806.

[17] E.C. Rivers, R.L. Mancera, New anti-tuberculosis drugs in clinical trials with novel mechanisms of action, Drug Discovery Today 13 (23–24) (2008) 1090–1098.

[18] B. Jimmy, J. Jose, Patient medication adherence: measures in daily practice, Oman Medical Journal 26 (3) (2011) 155.

[19] J.A. Caminero, et al., Best drug treatment for multidrug-resistant and extensively drug- resistant tuberculosis, The Lancet Infectious Diseases 10 (9) (2010) 621–629.

[20] B. Villemagne, et al., Tuberculosis: the drug development pipeline at a glance, European Journal of Medicinal Chemistry 51 (2012) 1–16.

[21] P. Bartlett, L. Judge, The role of epidemiology in public health, Revue Scientifique et Technique-Office International des Epizooties 16 (2) (1997) 331–336.

[22] D.E. Lilienfeld, Definitions of epidemiology, American Journal of Epidemiology 107 (2) (1978) 87–90.

[23] M. Frérot, et al., What is epidemiology? Changing definitions of epidemiology 1978–2017, PloS One 13 (12) (2018) e0208442.

[24] S. Wassertheil-Smoller, J. Smoller, Biostatistics and Epidemiology: A Primer for Health and Biomedical Professionals, Springer, 2015.

[25] B. Waning, M. Montagne, W.W. McCloskey, Pharmacoepidemiology: Principles and Practice, 2001.

[26] R.S. Bhopal, Concepts of Epidemiology: Integrating the Ideas, Theories, Principles, and Methods of Epidemiology, Oxford University Press, 2016.

[27] World Health Organization, Global Tuberculosis Report 2013, World Health Organization, 2013.

[28] World Health Organization, Global Tuberculosis Report 2017, World Health Organization, 2017.

[29] B. Marais, A. Outhred, A. Zumla, Engaging high and low burden countries in the "TB end game", International Journal of Infectious Diseases 45 (2016) 100–102.

[30] C.K. Kwan, J.D. Ernst, HIV and tuberculosis: a deadly human syndemic, Clinical Microbiology Reviews 24 (2) (2011) 351–376.

[31] K. Naidoo, et al., HIV-associated tuberculosis, Clinical and Developmental Immunology 2011 (2011) 585919.

[32] World Health Organization, Global Tuberculosis Control: Epidemiology, Strategy, Financing: WHO Report 2009, World Health Organization, 2009.

[33] E.L. Corbett, et al., Tuberculosis in sub-Saharan Africa: opportunities, challenges, and change in the era of antiretroviral treatment, The Lancet 367 (9514) (2006) 926–937.

[34] A. Jiamsakul, et al., Socio-economic status and risk of tuberculosis: a case-control study of HIV-infected patients in Asia, International Journal of Tuberculosis and Lung Disease 22 (2) (2018) 179–186.

[35] S.M. Gelaw, Socioeconomic factors associated with knowledge on tuberculosis among adults in Ethiopia, Tuberculosis Research and Treatment 2016 (2016).

[36] L.G. Wilson, The historical decline of tuberculosis in Europe and America: its causes and significance, Journal of the History of Medicine and Allied Sciences 45 (3) (1990) 366–396.

[37] P. Mangtani, et al., Socioeconomic deprivation and notification rates for tuberculosis in London during 1982–91, BMJ 310 (6985) (1995) 963–966.

[38] E. Drucker, et al., Childhood tuberculosis in the Bronx, New York, The Lancet 343 (8911) (1994) 1482–1485.

[39] J.P. Cegielski, L. Arab, J. Cornoni-Huntley, Nutritional risk factors for tuberculosis among adults in the United States, 1971–1992, American Journal of Epidemiology 176 (5) (2012) 409–422.

[40] World Health Organization, WHO Consolidated Guidelines on Drug-Resistant Tuberculosis Treatment, 2019.

[41] G.A. Marriner, et al., The medicinal chemistry of tuberculosis chemotherapy, in: Third World Diseases, Springer, 2011, pp. 47–124.

[42] A. Argyrou, et al., Proteome-wide profiling of isoniazid targets in Mycobacterium tuberculosis, Biochemistry 45 (47) (2006) 13947–13953.

[43] World Health Organization, Companion Handbook to the WHO Guidelines for the Programmatic Management of Drug-Resistant Tuberculosis, World Health Organization, 2014.

[44] L.G. Wayne, H.A. Sramek, Metronidazole is bactericidal to dormant cells of Mycobacterium tuberculosis, Antimicrobial Agents and Chemotherapy 38 (9) (1994) 2054–2058.

[45] S. Gagneux, et al., Impact of bacterial genetics on the transmission of isoniazid-resistant Mycobacterium tuberculosis, PLoS Pathogens 2 (6) (2006) e61.

[46] Z. Ma, et al., Global tuberculosis drug development pipeline: the need and the reality, The Lancet 375 (9731) (2010) 2100–2109.

[47] D. Perlman, Structure-Activity Relationships Among the Semisynthetic Antibiotics, Academic Press, 1977.

[48] J.E. Bennett, R. Dolin, M.J. Blaser, Mandell, Douglas, and Bennett's Principles and Practice of Infectious Diseases: 2-Volume Set, vol. 1, Elsevier Health Sciences, 2014.

[49] A.S. Ginsburg, J.H. Grosset, W.R. Bishai, Fluoroquinolones, tuberculosis, and resistance, The Lancet Infectious Diseases 3 (7) (2003) 432–442.

[50] K.E. Lougheed, et al., New anti-tuberculosis agents amongst known drugs, Tuberculosis 89 (5) (2009) 364–370.

[51] S.E. Berning, The role of fluoroquinolones in tuberculosis today, Drugs 61 (1) (2001) 9–18.

[52] H.H. Liu, Safety profile of the fluoroquinolones, Drug Safety 33 (5) (2010) 353–369.

[53] L.A. Mitscher, Bacterial topoisomerase inhibitors: quinolone and pyridone antibacterial agents, Chemical Reviews 105 (2) (2005) 559–592.

[54] Y. Zhang, et al., Mode of action of pyrazinamide: disruption of Mycobacterium tuberculosis membrane transport and energetics by pyrazinoic acid, Journal of Antimicrobial Chemotherapy 52 (5) (2003) 790–795.

[55] A. Scorpio, Y. Zhang, Mutations in pncA, a gene encoding pyrazinamidase/nicotinamidase, cause resistance to the antituberculous drug pyrazinamide in tubercle bacillus, Nature Medicine 2 (6) (1996) 662.

[56] G. Momekov, et al., Pyrazinamide-pharmaceutical, biochemical and pharmacological properties and reappraisal of its role in the chemotherapy of tuberculosis, Pharmacia 61 (1) (2014) 38–67.

[57] C.A. Peloquin, Therapeutic drug monitoring in the treatment of tuberculosis, Drugs 62 (15) (2002) 2169–2183.

[58] S. Bareggi, et al., Clinical pharmacokinetics and metabolism of pyrazinamide in healthy volunteers, Arzneimittel Forschung 37 (7) (1987) 849–854.

[59] R. Shepherd, et al., Structure-activity studies leading to ethambutol, a new type of antituberculous compound, Annals of the New York Academy of Sciences 135 (2) (1966) 686–710.

[60] R. Shepherd, R. Wilkinson, Antituberculous agents. II. 1 N, N'- diisopropylethylenediamine and analogs, Journal of Medicinal Chemistry 5 (4) (1962) 823–835.

[61] J. Thomas, et al., A new synthetic compound with antituberculous activity in mice: ethambutol (dextro-2, 2'-(ethylenediimino)-di-1-butanol), American Review of Respiratory Disease 83 (6) (1961) 891–893.

[62] H. Häusler, et al., Ethambutol analogues as potential antimycobacterial agents, Bioorganic and Medicinal Chemistry Letters 11 (13) (2001) 1679–1681.

[63] A.E. Belanger, et al., The embAB genes of *Mycobacterium avium* encode an arabinosyl transferase involved in cell wall arabinan biosynthesis that is the target for the antimycobacterial drug ethambutol, Proceedings of the National Academy of Sciences 93 (21) (1996) 11919–11924.

[64] C.D. Mitnick, B. McGee, C.A. Peloquin, Tuberculosis pharmacotherapy: strategies to optimize patient care, Expert Opinion on Pharmacotherapy 10 (3) (2009) 381–401.

[65] L. Willis, D. Hayes, H.M. Mansour, Therapeutic liposomal dry powder inhalation aerosols for targeted lung delivery, Lung 190 (3) (2012) 251–262.

[66] C. Jagannath, et al., Chemotherapeutic activity of clofazimine and its analogues against *Mycobacterium tuberculosis*. In vitro, intracellular, and in vivo studies, American Journal of Respiratory and Critical Care Medicine 151 (4) (1995) 1083–1086.

[67] J.A. Seddon, et al., Paediatric use of second-line antituberculosis agents: a review, Tuberculosis 92 (1) (2012) 9–17.

[68] M.C. Cholo, et al., Clofazimine: current status and future prospects, Journal of Antimicrobial Chemotherapy 67 (2) (2011) 290–298.

[69] E. Tacconelli, et al., Discovery, research, and development of new antibiotics: the WHO priority list of antibiotic-resistant bacteria and tuberculosis, The Lancet Infectious Diseases 18 (3) (2018) 318–327.

[70] L. Richeldi, et al., Clinical use of levofloxacin in the long term treatment of drug resistant tuberculosis, Monaldi Archives for Chest Disease 57 (1) (2002) 39–43.

[71] E. Nuermberger, et al., Powerful bactericidal and sterilizing activity of a regimen containing PA-824, moxifloxacin, and pyrazinamide in a murine model of tuberculosis, Antimicrobial Agents and Chemotherapy 52 (4) (2008) 1522–1524.

[72] E. Rey-Jurado, et al., Activity and interactions of levofloxacin, linezolid, ethambutol and amikacin in three-drug combinations against *Mycobacterium tuberculosis* isolates in a human macrophage model, International Journal of Antimicrobial Agents 42 (6) (2013) 524–530.

[73] E. Rey-Jurado, et al., In vitro effect of three-drug combinations of antituberculous agents against multidrug-resistant *Mycobacterium tuberculosis* isolates, International Journal of Antimicrobial Agents 41 (3) (2013) 278–280.

[74] A. López-Gavín, et al., In vitro time-kill curves study of three antituberculous combinations against *Mycobacterium tuberculosis* clinical isolates, International Journal of Antimicrobial Agents 47 (1) (2016) 97–100.

[75] A. López-Gavín, et al., In vitro activity against *Mycobacterium tuberculosis* of levofloxacin, moxifloxacin and UB-8902 in combination with clofazimine and pretomanid, International Journal of Antimicrobial Agents 46 (5) (2015) 582–585.

[76] J.G. Den Hollander, J.W. Mouton, H.A. Verbrugh, Use of pharmacodynamic parameters to predict efficacy of combination therapy by using fractional inhibitory concentration kinetics, Antimicrobial Agents and Chemotherapy 42 (4) (1998) 744–748.

[77] E. Rey-Jurado, et al., Synergistic effect of two combinations of antituberculous drugs against *Mycobacterium tuberculosis*, Tuberculosis 92 (3) (2012) 260–263.

CHAPTER 15

Metal Oxide Nanoparticles: A Welcome Development for Targeting Bacteria

VICTORIA OLUWASEUN FASIKU* • SHESAN JOHN OWONUBI* •
NYEMAGA MASANJE MALIMA • DANIEL HASSAN • NEERISH REVAPRASADU

1 INTRODUCTION

There is an increasing rise in the number of reports on pathogenic organisms resistant to several antibiotic agents. Currently, it has been projected that antibacterial resistance will account for 10 million deaths in 2050 [1,2]. At present, over 70% of bacterial infections treated with known antibiotics have become almost impossible due to antibiotic resistance, and this has become a serious issue in the health sector [3]. Therefore, it is of great necessity that newer and more effective antibiotics or antibiotic delivery methods be developed. Resistance in bacteria is achieved primarily by the production of enzymes (by the bacteria) capable of modifying, degrading, and inactivating antibiotics [4]. For instance, the β-lactamase enzyme can be produced by bacteria to cleave the β-lactam ring of penicillin thus neutralizing treatment by penicillin [5].

In recent years, nanotechnology has shown excellent potential in various fields of medicine; hence, it has been explored by researchers in the eradication of pathogenic organisms and the treatment of infectious diseases. Particularly, in pharmaceutical technology, nanotechnology has offered greater advantages over conventional techniques and has become a center of attraction for several researchers and research groups [6]. Various studies on the application of nanotechnology are ongoing and they have brought about notable findings and innovation with overall improved physiochemical properties and therapeutic efficacy [7,8].

Nanoparticles are nanometer-sized materials with different characteristics when compared to the parent material (same material of larger particle size). A decrease in the size of a nanomaterial provides an increased surface/volume ratio that is very critical for various applications in nanotechnology [9,10].

Nanoparticles available for biomedical exploration include carbon-based nanostructures, polymeric nanoparticles, and metallic nanoparticles among others. There are various types of nanoparticles that have been explored for the treatment of infectious diseases caused by pathogenic microorganisms; however, metal nanoparticles appear to stand out owing to their numerous advantages over other materials. Nanoparticles have been reported to allow for strong, targeted, and extended antibacterial activity at smaller dosages [11]. The nature of the material used for synthesizing the nanoparticle as well as the size has a great effect on its antibacterial activity [12,13].

Metals such as silver (Ag), copper (Cu), gold (Au), titanium (Ti), and zinc (Zn) have been discovered to possess antibacterial properties as far back as 1500 BP in Egypt [14–16]. More recently, their potencies, as well as a broad spectrum of activity against several strains of resistant bacteria, are now been explored [17]. Metals can differentiate between mammalian cells and bacterial cells as a result of the cells deviating metal transport systems and metalloproteins. This serves as an advantage of metals over conventional antibiotics; hence, they can be used as effective, long-term antibacterial materials [14]. In addition, the small size of metal nanoparticles when compared to bacteria size permits a very strong interaction between the bacterial and the metal nanoparticles. Furthermore, these metals can be combined with oxygen to form metal oxide nanoparticles (MONPs) [18]. In vitro investigations of MONPs in the treatment of microbial infections have shown that they inhibit several ranges of bacterial infections. From experimental reports thus far, in comparison to antibiotic treatments, MONPs exhibit their bactericidal effect via different mechanisms,

---

*Authors contributed equally to this write-up.

Antibiotic Materials in Healthcare. https://doi.org/10.1016/B978-0-12-820054-4.00015-X

which include alteration of cell components, disruption of the cell wall, and DNA replication [5,19]. The ability to synthesize MONP with specific size and shape makes them potential antibacterial agents. The antibacterial efficacy of MONPs has been tested against various strains of Gram-positive and Gram-negative bacteria such as *Staphylococcus aureus* and *Pseudomonas aeruginosa*, respectively [20]. To determine the morphology, chemical composition, lattice parameter, crystallinity, and crystal structure of MONPs, techniques such as scanning electron microscopy, transmission electron microscopy, X-ray diffraction, energy-dispersive X-ray spectroscopy, and UV-visible spectroscopy have been widely employed [21]. Various types of MONP have been explored and have thus far found application in at least one of the various applications shown in Fig. 15.1. However, the application of MONPs in the field of biomedicine is the focus of this chapter.

2 TYPES OF MONPS

MONPs are of different types and they have unique physical, chemical, and biological characteristics that can be easily altered or manipulated [22]. MONPs, when formed, can have metallic, semiconductor, and insulator characteristics depending on their electronic structures. For instance, the reaction of metals from groups 3 to 12 with oxygen will lead to the formation of metallic and semiconductor MONPs, while groups 1, 2, and 13—18 on the periodic table will form insulator MONPs [23]. The individual properties of each MONP contribute to its antibacterial activity upon interaction with biological systems [24]. Examples of some of the MONPs discussed in this book chapter include zinc oxide, copper oxide, titanium dioxide, and magnesium oxide nanoparticles [25,26].

2.1 Zinc Oxide

Zinc oxide (ZnO)is one of the minerals found in the earth's crust and has been found useful as an additive in different products; however, over the years, it has been implicated in biomedical applications [27—29]. Owing to its wide bandgap features such as the ability to maintain high breakdown voltages and the ability to run at high power, zinc oxide is considered a versatile material. In addition, its high excision binding energy, which is responsible for the stability of electron—hole pair recombination, provides it with good luminescence properties that have made it relevant in several applications [27,30,31]. Zinc oxide is also quite easy to synthesize at a low cost, relatively eco-friendly and it is very sensitive [32—37]. Fig. 15.2 shows some other properties of zinc oxide nanoparticles.

Zinc oxide is an inorganic, polar, nontoxic, and nonhygroscopic crystalline material that is readily available, quite cheap, and safe. Stimulating properties such as antiinflammatory and wound healing activities have drawn much attention to zinc oxide nanoparticles [38—43]. Zinc oxide nanoparticles have found applications in UV light emitters, catalysts, transparent electronics, personal care products, chemical sensors, spin electronics, and coating/paints [44,45]. In addition, it can serve as a photonic material, wave propagator, gas sensor, and antibacterial agent [46—49]. Depending on the technique of preparation, zinc oxide nanoparticle can have these morphologies, viz sphere [50], nanoflowers, nanorods [51,52], nanospheres [51], dumbbell-shaped, rice flakes, rings [53], mulberry-like [54], and hexagonal prismatic rods [55].

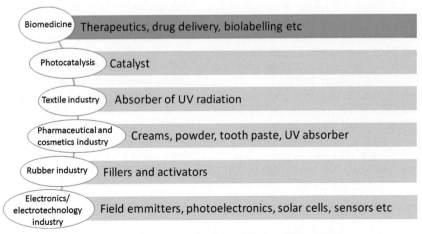

FIG. 15.1 Diverse applications of MONPs in different fields.

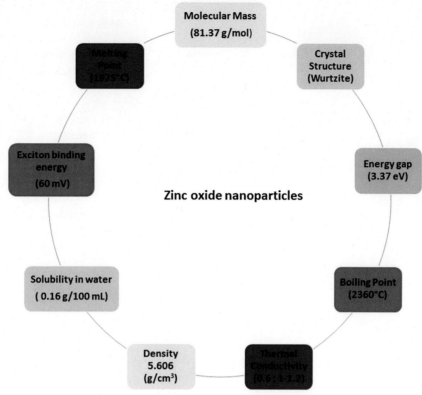

FIG. 15.2 Properties of zinc oxide nanoparticles.

2.2 Gold Oxide

Gold oxide (AuO) nanoparticles represent another type of MONPs with awesome properties. These physical and chemical properties allow for the use of gold oxide nanoparticles in various applications such as therapeutics, diagnostics, drug delivery [56,57], chemical and biological sensing, imaging [58], nonlinear optics, photovoltaics, and catalysis fields [59,60]. The prominent attributes that make them suitable for biotechnological application include large surface-to-volume ratio, low toxicity, shape-related optoelectronic properties, and excellent biocompatibility [61–63]. Biocompatibility and nontoxicity accounts for the wider application and the advantages of gold oxide nanoparticles over other MONPs [64]. Several chemical, physical, and biological "top-down" and "bottom-up" methods have been used for synthesizing gold oxide nanoparticles, and each method aims at controlling the size, shape, solubility, stability, and functionality [65]. Gold oxide nanoparticles with one-dimensional shapes usually have very complex physical properties and self-assembling characters when compared to their spherically shaped counterpart [66]. Gold oxide nanoparticles can exist in shapes such as nanosphere, nanoshell, nanocages, nanorods [67]. The ability to conjugate gold oxide nanoparticles with biological molecules, such as drugs and targeting agents, has made them promising biomaterials for applications in biological systems [68–71]. Other properties that make them outstandingly relevant in other fields are shown in Fig. 15.3.

2.3 Silver Oxide

The existence of silver has been almost since creation and its use since human civilization; however, its fabrication into nanoparticles and silver oxide (AgO) nanoparticles only received attention in recent years [72]. Silver oxide nanoparticles have become widely known and used in diverse applications such as oxidation catalysis [73], sensors [74], fuel cells [75], photovoltaic cells [76], plasmon photonic devices [77], and active cathode materials for zinc/silver oxide batteries [78,79]. Furthermore, silver oxide nanoparticles possess good fluorescence and optical properties that make it potentially useful in data storage devices and optical media for the enhancement of the readout signal [80,81]. Approximately, about 320 tons of silver

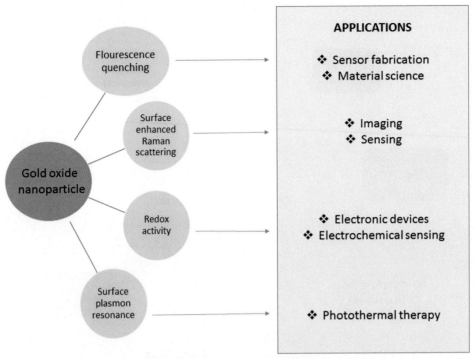

FIG. 15.3 Prominent properties of gold oxide nanoparticles and their applications.

oxide nanoparticles are produced yearly and are used in bioimaging and biosensing [82,83]. Silver oxide nanoparticles can be prepared via different techniques, and these include simple chemical methods [84,85], photosensitized reduction [86], and electrochemical synthesis [87]. In addition, a green synthesis that involves the use of microorganisms such as bacteria, fungi, yeast, actinomycetes, and plant extracts [88–91] has been discovered to be an excellent method for preparing silver oxide nanoparticles. The type of preparation method, temperature, nature of the solvent, and concentration often determines the size, morphology, and stability of the nanoparticles [88–90,92]. For example, by employing physical methods such as laser ablation or thermal decomposition, silver oxide nanoparticles with spherical or irregular shapes are produced [93]. Silver oxide nanoparticles have found application in electronics, catalysis, and in microorganism growth control in biological systems; thus, they are considered eco-friendly [88,94–96]. In addition, they have exhibited bactericidal activity against various microorganisms [90,94]. One major advantage of silver oxide nanoparticles over other nanoparticles is that, even in the solid state, it has the ability to act as an antibacterial agent [72].

2.4 Cerium Oxide

Cerium is a scarce metal with an atomic number of 58 that belongs to the lanthanide series in the periodic table. However, it is the most abundant rare earth metal that exists in two oxidation states (+3 and + 4) [97]. It can be transformed into nanoparticles to yield cerium oxide (CeO) nanoparticles (nanoceria). It can be prepared via traditional methods (hydrothermal, solvothermal, precipitation, etc.) and green synthesis (plant, fungus, polymer mediated). The preparation method is known to affect the shape of cerium oxide nanoparticle synthesized, and it can be spherical, octahedral, polyhedral, or cubic in shape [98–101]. The optical and electronic properties of cerium oxide nanoparticles are related to its valence electrons and thus increases its range of applications [102] as shown in Fig. 15.4.

Cerium oxide nanoparticles have a broad range of applications, and it can be used as a catalyst [103,104], ultraviolet radiation absorber [105,106], gas sensor, polishing agent, and especially a biomedical relevant material [107–109]. In recent times, their relevance in the biomedical field is beginning to grow due to their ability to act as protectors against radiation and toxicant-induced cell damage [110]. Although, poor solubility in water has limited its biological application,

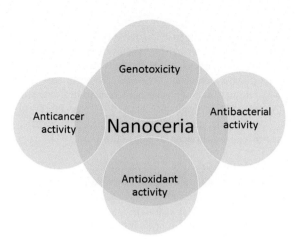

FIG. 15.4 Biomedical applications of cerium oxide nanoparticle (nanoceria).

coating with polymeric materials has been successful in encouraging enhanced stability, biocompatibility, and water solubility [111]. The biorelevant activities such as thermodynamic efficiency of redox cycling between 3+ and 4+ states on their surface [112] and their unique ability to absorb and release oxygen [113] have made them potential pharmacological agents [114], drug-delivery vehicles, bioanalysts, suitable biomedical materials [115,116], and bioscaffolds [117,118]. In addition, it has been discovered that they are able to mimic the activities of certain enzymes and free radicals involved in biochemical processes, such as superoxide dismutase (SOD) [112], catalase [119], peroxidase [120], oxidase [121], phosphatase [122], scavenging hydroxyl radicals [123], nitric oxide radicals [124], and peroxynitrite [125]. With the increasing use of nanoceria, concerns about exposure to humans and its eco-friendliness have been raised. Some reports document the toxicity of cerium oxide nanoparticles while others report low toxicity [122] and nonnanoceria-mediated cytotoxicity or inflammation [123,124].

2.5 Copper Oxide
Copper oxide (CuO) nanoparticles are formed from copper oxide, which is a compound made up of copper and oxygen [126]. It is a powder that is insoluble in water and slowly dissolves in alcohol but soluble in dilute acid, NH_4Cl, $(NH_4)_2CO_3$, and potassium cyanide solution. When in contact with carbon monoxide at high temperature, it reduces to metallic copper [127]. Nanoparticulate forms of copper oxide have different physical, chemical, optical, and magnetic properties, mechanical strengths, and electrical resistivity characteristics when compared to the bulk solid

material [128]. These are surface effect, superiority of the quantum size effect, volume effect, and macroscopic quantum tunneling effect in magnetic, optical absorption, chemical activity, and thermal resistance. An increasing interest is on copper oxide nanoparticles because of their good optical, catalytic, mechanical, and electrical properties [129,130] making them useful as catalysts, superconductors, as well as an electrode-active material. Copper oxide nanoparticles have shown potential applications in biomedicine, but a major drawback is the associated toxic effect that has been observed in mammalian cells [131–133]. Toxicity is as a result of an increase in the production of reactive oxygen species when cells are exposed to copper oxide nanoparticles that induce oxidative stress and ultimately damage the DNA and mitochondria [131,134–136]. Size, surface charge, and dissolution of copper oxide nanoparticles are major factors that influence its toxicity [137], and this toxicity has been reported to be further dependent on the exposure time [138,139]. Although the use of copper oxide nanoparticles as drug-delivery vehicles is still limited, it can be used in other applications, such as topic formulations, dressings, and coating textiles [128]. In the last decade, significant advancement has been achieved in developing methods for synthesizing copper oxide nanoparticles suitable for biomedical applications [140]. This is important because the method of preparation has a direct effect on the properties, size, and morphology of the nanoparticles; hence, it determines the application as well. Some of the methods used for synthesizing copper oxide nanoparticles are electrochemical methods, sonochemical synthesis, sol–gel techniques, microemulsion system, precipitation synthesis, and microwave irradiation.

2.6 Titanium Oxide
Of all the elements in the earth's crust, titanium is the ninth most abundant with an average concentration of 4400 mg/kg. Naturally, it does not occur in the metal state because of its high affinity for oxygen and other elements [141]. It exists in oxidation states of +4, but +3 and + 2 and can be converted into titanium oxide (TiO) nanoparticle. Titanium oxide is also known as titanium dioxide (titanium(IV) oxide), TiO_2; Titanium(II) oxide (titanium monoxide), TiO is white in color, very bright, and has a high refractive index with a yearly global consumption of four million tons [142]. Properties such as high stability, anticorrosive, and photocatalysis has led to their abundant production and wide use [143]. The physicochemical properties of titanium oxide nanoparticles are mainly affected by the shape, size, surface characteristics, and inner

structure. A decrease in the size of titanium oxide nano-particles has been reported to lead to harmful effects on human health hence raising considerable concerns [144,145]. However, by modifying its surface by coating, the cytotoxic effect can be reduced [146]. Other characteristics of titanium oxide nanoparticles are shown in Fig. 15.5 [147,148]. Approximately 70% of the total production volume of pigments in the world is based on titanium oxide nanoparticles [149] and it is one of the first five nanoparticles used in consumer products [150]. Common physicochemical methods used in preparing titanium oxide nanoparticles require high temperature, pressure, and toxic chemicals that limit their production and potential medical uses [151]. These methods include chemical vapor deposition, microemulsion, chemical precipitation, hydrothermal crystallization, and sol−gel methods [152,153]. Therefore, there is the need for an alternate method of synthesis such as green synthesis that is eco-friendlier and cost-effective [154]. Its application is found in coatings/paints, toothpaste, sunscreens pharmaceuticals and medicine [155−157]. In addition, other applications include self-cleaning and antifogging purposes such as self-cleaning tiles, self-cleaning windows, self-cleaning textiles, and antifogging car mirrors [158]. In nanomedicine, some potential application of

titanium oxide nanoparticle as imaging and nanotherapeutic tools has been reported [159]. Furthermore, nanopreparations involving titanium oxide nanoparticles are being explored for the potential treatment of acne vulgaris, recurrent condyloma acuminata and atopic dermatitis, hyperpigmented skin lesions, and other nondermatologic diseases [160]. In addition, under UV light irradiation, titanium oxide nanoparticle has exhibited antibacterial properties [159,161].

2.7 Iron Oxide

Iron oxide (FeO) nanoparticles remain one of the most common natural types of nanoparticles with eco-friendly properties. It exists as a crystalline structure with various forms of crystallinity, viz hematite (α-Fe_2O_3), maghemite (γ -Fe_2O_3), goethite FeOH(OH), and magnetite (Fe_3O_4) [162]. The hematite phase is cost-effective and nontoxic, and it has the most stable n-type semiconductor property with a bandgap of 2.1−2.2 eV [163]. They are known to be chemically inert materials, and as a result they find applications in imaging, targeting, drug delivery, and biosensors. The unique properties possessed by iron oxide nanoparticles are low or no toxicity, biocompatibility, potent magnetic and catalytic behavior, and superior role in multifunctional modalities [164]. Furthermore, with advancements in

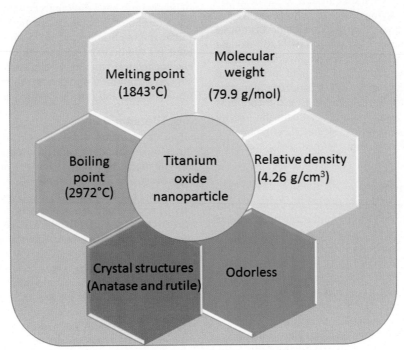

FIG. 15.5 Properties of titanium oxide nanoparticle.

the field of material science, tuning, and functionalizing, the surface of iron oxide nanoparticles is achievable, thus broadening its applications. Thermal decomposition, coprecipitation, sol—gel, microemulsion, hydrothermal, sonochemical, microwave, electrochemical, and biosynthesis are some of the methods used to synthesize iron oxide nanoparticles [165,166]. Considering, the method of synthesis has a relationship between size, shape, and magnetic properties of the iron oxide nanoparticles, and it is, therefore, important to employ methods suitable for the specific application [166]. Iron oxide nanoparticles are useful in applications such as magnetic, electrochemical, gas sensor, energy storage, cancer therapy, and magnetic storage, and biomedical treatments. Their nontoxic, magnetic, and semiconductor characteristics have made them to gain considerable attention in multifunctional biomedical applications in biological systems [167]. Some of the various biomedical applications of iron oxide nanoparticles include diagnosis by acting as a probe in magnetic resonance imaging [168,169] and biosensors for detecting biomolecules such as glucose, proteins, urea, and uric acid [170—172]. In addition, they have also found relevance in therapeutic nanomedicine such as cancer treatment and antibacterial activity [173,174] and in theranostics as nanocarriers, for enhancing drug activity in combination therapy (for example, IONPs and chemotherapeutic drugs) [175]. Recently, surface coating/alteration has been done on iron oxide nanoparticles with polymers, bioactive molecules, etc. in other to enhance their application in biomedicine [176].

To determine, study, and quantify various parameters such as crystallinity, morphology, surface charge, particle size, and magnetic properties of the previously discussed MONPs, various characterization techniques shown in Fig. 15.6 have been employed.

3 SYNTHESIS OF MONPS

As previously mentioned, several methods can be employed to synthesize MONPs and these preparation methods determine the chemical and physical properties, as well as the type of application suitable for the MONPs. For example, the chemical method of synthesis is preferred by most researchers for antibacterial characterization studies over time as a result of the efficacy observed, even though the actual mechanism behind its superior success is not known [18]. Some of the methods used for synthesizing MONPs include the following:

1. Sol—gel: It is a technique that involves condensing and hydroxylation of precursor molecules, it is often considered a wet route method. Factors such as temperature, pH of the gel, agitation, and the concentration of the precursors are responsible for the regulation of both (hydrolysis and condensation) reactions [177]. In addition, the structure of the MONP obtained can be affected by these factors. For instance, nanostructure with an amorphous or monodispersed phase can be formed [178,179]. An advantage of this method is the ability to control the homogeneity of the products obtained; TiO_2 and ZnO are examples of MONPs that have been synthesized via this method for bacterial efficacy [18].

2. Pyrolytic: In this process, an atomized solution is reacted with a soluble polymer in order for it to mix into the polymeric matrices; this is followed by heating the precursor molecules into the system. In some instances, the MONP serves as a catalyst during the process. An example of MONP that has been synthesized via this method is cobalt oxide (Co_3O_4) nanoparticles that have been employed for its bactericidal efficacy [180].

3. Chemical: This is the most preferred method of synthesizing MONPs. It is a reaction between high-boiling solvent, precursors, and surfactants at slow heating. This method involves the formation of monomers and their accumulation that, in turn, leads to the formation of highly monodispersed MONPs. This process allows for the control of the size and diameter of the nanoparticles during the growth and solvent boiling process, respectively. The antibacterial activities of MONPs such as cadmium oxide (CdO), CeO_2, iron oxide (Fe_2O_3 or Fe_3O_4), magnesium oxide (MgO), ZnO, vanadium pentoxide (V_2O_5), TiO_2, and graphene oxide that were synthesized via this method have been investigated [181].

4. Hydrothermal: Synthesis of MONPs using this method is often done in an aqueous medium and in autoclaves/reactors under 2000 psi pressure and temperature of 200°C to control the size and shape of the nanoparticles. Employing this method leads to the synthesizing of various types of nanoparticles because of the ease of controlling the hydrothermal temperature. Although it is difficult to dissolve aqueous solvents/mineralizers under ordinary condition however, they can be dissolved and recrystallized in heterogeneous reactions [182—184]. Common examples of MONPs with effective antibacterial potential that have been synthesized using this method are MgO nanoparticles, ZnO nanorods, and ZnO nanosheets.

5. Wet chemical: This method is easy, simple, and saves cost, and it involves mixing precursor

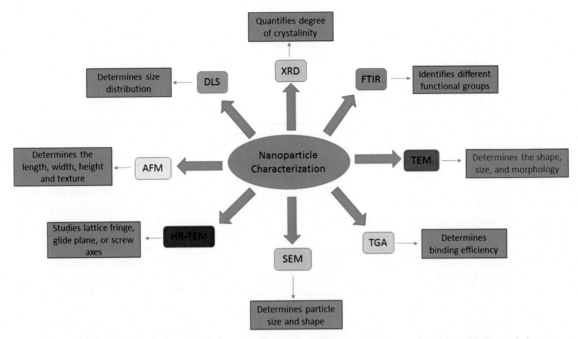

FIG. 15.6 Summary of the common techniques used to characterize the above-discussed MONPs and also the purpose of using the technique.

molecules in ultrapure water, then stirred for 30 min subsequently, heated for 45 min, and centrifuged. CuO nanorods, Fe_3O_4, and ZnO nanoparticles have been prepared using this method, and their efficacy against microorganisms have also been investigated [185].

6. Coprecipitation: Basically, the entire process involves brief nucleation and then the growth of nuclei on the surface of the crystal via diffusion of the solute. In this method, a metal hydroxide is formed by the conversion of salt precursor in the aqueous medium upon the addition of either sodium hydroxide or ammonium hydroxide. To obtain the needed MONPs, the chloride salts formed are washed and the hydroxide heated. Unfortunately, this method does not permit the control of size and size distribution; however, it can be adjusted by altering factors such as the ionic strength, pH, temperature, and salt used during the reaction process [186,187]. The antibacterial potential of metals oxide nanoparticles such as Fe_3O_4 and ZnO synthesized via this method has been investigated.

7. Sonochemical: In this method, metal salt and oxygen in solution are sonicated at high-intensity ultrasound sonication. Metal oxides are formed as a result of the reaction between reactive radicals, atoms, and metal ions in the solution. This method exerts changes in the size, distribution, composition, and morphologies of the nanoparticles due to the acceleration and collision of particles. In terms of incorporating nanoparticles into synthetic materials, this method is considered very effective. Examples of MONPs synthesized by this method is CuO and its bacterial activity has been evaluated [188,189].

8. Other methods employed in the synthesis of variousMONPs include biosynthesis/green synthesis [190], electrochemical [191,192], microwave [193], mycosynthesis [194], microemulsion [195,196], thermal decomposition [197], template/surface derived method [198], and solvothermal methods [199].

4 ANTIBACTERIAL EFFICACY OF MONPS

The bactericidal effect of MONPs is partially dependent on the size, concentration, and stability of the nanoparticle. The smaller the size of the MONPs, the better the chances of penetrating the cell membrane of the bacteria and exert its bactericidal capacity. For instance, zinc oxide nanoparticles have a bactericidal activity that is inversely proportional to its size but directly proportional to its concentration [200]. In addition, generally, the various properties of MONPs contribute to the antibacterial activity. For example, the alkalinity of the surface of calcium oxide (CaO) and magnesium oxide (MgO) nanoparticles impacts its antibacterial activity [201]. Their alkaline nature makes them more soluble in the medium when compared to other MONPs such as zinc oxide nanoparticles [202]. Furthermore, characteristics such as electrostaticity and photocatalysis of positively charged cerium oxide (CeO_2) and titanium dioxide (TiO_2), respectively, are known to contribute to their bactericidal action [203]. Additionally, the "gram" class of the bacteria has an effect on the antibacterial activity of MONPs. For instance, efficacy of MONPS is reportedly more pronounced against Gram-positive bacterial strains than Gram-negative bacterial strains [204]. Several studies have been done to investigate the antibacterial potency of different MONPs, herein is a summarized table of some of the findings (Table 15.1).

Apart from MONPs serving as direct antibacterial agent, they can also be employed as delivery vehicles for other therapeutic agents/molecules to treat bacterial infections [72,88]. Owing to their small size, they can penetrate cells and accumulate antibacterial agents at the site of target in some cases and in others, they can be drugs released over certain periods [227]. Other properties such as unique optical, physicochemical properties, biocompatibility, functional flexibility, tunable monolayers, controlled dispersity, high surface area for loading the density of drugs, stability, and non-toxicity have made them beneficial in drug-delivery application [228–230]. A typical example of MONPs that has been used to deliver different antibacterial agents is iron oxide nanoparticle. It has been found effective for the delivery of streptomycin [231,232], rifamycin, anthracycline, fluoroquinolone, tetracycline, and cephalosporin [233]. In addition, iron oxide nanoparticle has been directly coupled to amikacin, amoxicillin, bacitracin, cefotaxime, erythromycin, gentamicin, kanamycin, neomycin, penicillin, poly-myxin, streptomycin, and vancomycin to evaluate its ability as a drug-delivery vehicle in the treatment and management of bacterial infectious diseases [234].

5 MECHANISMS OF ACTION OF MONPS IN BACTERIA GROWTH INHIBITION

As the discovery of the antibacterial potency of MONPs, several researchers have investigated the molecular mechanism of their antibacterial action. This has become a great necessity because little information that elucidates the interaction of synthesized MONPs at the molecular level is available. The main mechanism of action that has been established so far lies in the generation of reactive oxygen species (ROS); however, other mechanisms of action include disturbance in

metal/metal ion homeostasis, photokilling, cell membranes damaged by electrostatic interaction, geno-toxicity, and protein/enzyme dysfunction. In most cases, for MONP to exhibit antibacterial activities, one or more of these mechanisms are combined [18]. These mechanisms of action are further discussed below and summarized in Fig. 15.7.

1. Protein and enzyme dysfunction: In this mechanism usually, the reaction between metal ions and thiol (−SH) groups in proteins and enzymes results in their loss of activity of bacteria. For example, silver reacts with soft bases of proteins and DNA such as sulfur and phosphorus, respectively, thus leading to DNA damage and subsequent cell death [235]. Metal ions act as a catalyst in the oxidation of amino acid chains that lead to the formation of protein-bound carbonyls. This, in turn, leads to protein degradation and loss of enzyme activity, and the degree of carbonylation is an indicator to the extent of protein damage [236,237]. Furthermore, MONPs have the ability to obstruct both catalytic and noncatalytic sites (by binding) leading to enzyme inactivity. Bacterial Fe−S dehydratases have been reportedly shown to be inactivated by MONPs [238,239].

2. Electrostatic interaction leading to cell membrane damage: Research has shown that there is an attraction between the electronegative chemical group of polymers present on the membrane of bacterial and metal ion. The surface area available for interaction determines the electrostatic force of attraction that will occur between the MONP and the bacteria; fortunately, MONP provides a larger surface area when compared to native particles with large sizes and this encourages efficacy. In other words, smaller particle size and larger surface area increase the bacterial inhibitory ability of MONPs; therefore, an attachment of MONPs leads to structural alteration and permeability of the cell membrane by the MONP. The charge difference is responsible for an electrostatic interaction that leads to the accumulation of MONP on the bacteria surface and subsequent penetration into the bacteria. This interaction causes bacterial toxicity and Gram-negative bacteria are known to be more susceptible to membrane damage because they contain more lipopolysaccharide in the outer lipid bilayer (with more charge) when compared to Gram-positive bacteria [240,241].

3. Metal/metal ion homeostasis disturbance: The survival of bacteria is largely dependent on metal ion homeostasis because it assists cofactors, catalysts, and coenzymes and fortunately a disturbance often occurs when they are exposed to MONPs [242].

TABLE 15.1
Some MONPs and Their Reported Antibacterial Potency.

Type of Metal Oxide Nanoparticle(s)	Type of Bacteria	Type of Investigation	Type of Technique	Results	References
Cerium oxide	*Pseudomonas aeruginosa* and *Streptococcus pneumoniae*	Zone of inhibition (ZOI)	Agar diffusion	0.0, ~3, and 4.67 mm at concentrations of 10, 50, and 100 mg/mL, respectively, for *Pseudomonas aeruginosa* while 0.0, ~3.60, and ~4.33 mm at the same concentrations, respectively, for *Streptococcus pneumoniae*	[205]
Iron oxide	*Bacillus cereus* and *Klebsiella pneumoniae*	Minimum inhibitory concentration (MIC)	Broth dilution	40%–50% loss in viable bacterial cells (both cells) at 5 and 40 μg/mL concentrations while 90%–99% loss in the cell viability for both cells was observed at the concentration of 80 μg/mL	[206]
Copper oxide	*Klebsiella pneumoniae, Salmonella typhimurium,* and *Enterobacter aerogenes*	Minimum inhibitory concentration (MIC)	Broth dilution	0.55, 0.15, and 0.30 μg/mL for *Klebsiella pneumoniae, Salmonella typhimurium,* and *Enterobacter aerogenes,* respectively	[140]
Zinc oxide, copper oxide, and iron oxide	*Bacillus subtilis* and *Escherichia coli*	Zone of inhibition (ZOI)	Agar diffusion	25, 21, and 15 mm, for zinc oxide, copper oxide, and iron oxide, respectively, for *Bacillus subtilis,* while *Escherichia coli* had ZOI of 19, 15, and 3 mm for ZnO, CuO, and Fe_2O_3, respectively	[20]
Magnesium oxide	*Bacillus subtilis* and *Escherichia coli*	Cell viability	CFU	53% survival of *Escherichia coli,* and only 21% of *Bacillus subtilis*	[207]
Cerium oxide	*Pseudomonas aeruginosa* and *Streptococcus pneumoniae*	Biofilm and minimum inhibitory concentration (MIC)	Broth dilution	*Pseudomonas aeruginosa* had values of 20 ± 5 and 70 ± 0.0 μg/mL for Planktonic culture and biofilm, respectively, while *Streptococcus pneumoniae* 110 ± 40 and 180 ± 80 μg/mL, respectively	[208]

Material	Bacteria	Assay	Method	Results	Ref.
Zinc oxide	Staphylococcus aureus and Escherichia coli	Minimum inhibitory concentration (MIC)	Broth dilution	1.5 and 3.1 mg mL⁻¹ for *Staphylococcus aureus and Escherichia coli*, respectively	[209]
Zinc oxide, copper oxide, and iron oxide	Escherichia coli, S. aureus, Pseudomonas aeruginosa, and Bacillus subtilis	Minimum bactericidal concentration (MBC)	Broth dilution	ZnO was 72%, 80%, 88%, and 84% more effective than Fe_2O_3, while 28%, 31%, 27%, 50%, and 40% more bactericidal than CuO against *E. coli, S. aureus, P. aeruginosa, and B. subtilis*, respectively	[20]
Cerium oxide	Bacillus subtilis and Streptococcus pneumonia	Zone of inhibition	Agar diffusion	*Bacillus subtilis* had ZOI of 0.0, 4.67 ± 0.33, and 10.33 ± 0.33 mm at concentrations of 1, 5, and 10 mg/mL, respectively, while 0.0, 3.33 ± 0.33, and 10.67 ± 0.33 mm at the same concentrations, respectively, for *Streptococcus pneumonia*	[194]
Zinc oxide	Staphylococcus aureus and Escherichia coli	Minimum inhibitory concentration (MIC)	Broth dilution	1.0 mg mL⁻¹ and 3.4 mg mL⁻¹ for *S. aureus and E. coli*, respectively	[210]
Silver oxide	Proteus vulgaris, Pseudomonas aeruginosa, Serratia marcescens, Salmonella typhi, Staphylococcus epidermidis, Methicillin-resistant Staphylococcus aureus, Bacillus subtilis, Streptococcus faecalis	Zone of inhibition (ZOI)	Disk diffusion	8, 8, 8, 7.5, 6, 6, and 6 mm, respectively	[211]
Cerium oxide	Pseudomonas aeruginosa	Zone of inhibition (ZOI)	Agar diffusion	3.33, 3.57, and 4.50 mm at concentrations of 10, 15, and 20 mg/mL, respectively	[212]
Silver oxide and zinc oxide	Staphylococcus aureus	Minimum inhibitory concentration (MIC) and MBC	Broth dilution	MIC value for silver was, on average, 4.86 ± 2.71 µg/mL, and the MBC was 6.25 µg/mL. For zinc, the average MIC was 500 ± 306.18 µg/mL and the MBC was 500 µg/mL, respectively	[213]

Continued

TABLE 15.1
Some MONPs and Their Reported Antibacterial Potency.—cont'd

Type of Metal Oxide Nanoparticle(s)	Type of Bacteria	Type of Investigation	Type of Technique	Results	References
Zinc oxide, copper oxide, and iron oxide	Escherichia coli, Pseudomonas aeruginosa, Staphylococcus aureus, and Bacillus subtilis	Zone of inhibition (ZOI)	Modified disk diffusion	ZnO, CuO, and Fe_2O_3 for Bacillus subtilis exhibited 25, 21, and 15 mm, respectively, for E. coli 19, 15, and 3 mm, respectively. Pseudomonas aeruginosa and Staphylococcus aureus exhibited similar patterns of zone inhibition; ZnO exhibited the greatest zone of inhibition followed by CuO and Fe_2O_3.	[20]
Zinc oxide	Campylobacter jejuni	Zone of inhibition and minimum inhibitory concentration (MIC)	Agar plates and broth dilution	Complete inhibition at ≥0.03 mg/mL of ZnO nanoparticles and MIC at concentration of 0.025 mg/mL	[214]
Cerium oxide	Pseudomonas aeruginosa	Zone of inhibition	Agar diffusion	Complete zone of inhibition at concentrations of 500, 750, and 1000 μg/L	[212]
Zinc oxide	Escherichia coli	Zone of inhibition and MIC/MBC	Disk diffusion and broth dilution	Particle sizes of 12 nm, 45 nm, and 2 μm exhibited zones of inhibition of 31, 27, and 22 mm, respectively	[215]
Cerium oxide	Bacillus subtilis	Minimum inhibitory concentration (MIC)	Broth microdilution	4 μg/mL	[216]
Copper oxide (annealed at different temperatures)	Escherichia coli, Pseudomonas aeruginosa, Bacillus subtilis, and Staphylococcus aureus	Minimum bactericidal concentration (MBC)	Broth dilution	Escherichia coli, Pseudomonas aeruginosa, Bacillus subtilis, and Staphylococcus aureus at 400°C: 30 ± 2, 35 ± 2, 45 ± 3, and 32 ± 5 μg/mL, respectively, 500°C: 45 ± 5, 50 ± 3, 62 ± 4, and 75 ± 2 μg/mL, respectively, 600°C: 60 ± 3, 75 ± 5, 85 ± 4, and 90 ± 5, μg/mL, respectively 700°C: 95 ± 4, 85 ± 5, 95 ± 5, and 100 ± 4 μg/mL, respectively	[217]

Nanoparticle	Test organism	Assay	Method	Findings	Reference
Zinc oxide, copper oxide, and zinc oxide–copper oxide	Streptococcus mutans	CFU on agar plates	CFU count	Complete bacterial eradication was observed during 2, 4, 6, and 24 h exposure time to zinc oxide–copper oxide, CFU for copper oxide nanoparticle at 2 h was 8.80, and complete bacterial eradication was observed in 4, 6, and 24 h treatment. The CFU for zinc oxide at 2, 4, 6, and 24 h was 96.78, 96.10, 55.89, and 219.32, respectively	[218]
Titanium dioxide	Methicillin-resistant Staphylococcus aureus	Zone of inhibition	Disk diffusion	The zone of inhibition of the titanium dioxide nanoparticle ranged from 16 to 44 mm, and it exhibited a better antibacterial activity compared to the antibiotics used in the study	[219]
Copper oxide	Escherichia coli, Pseudomonas aeruginosa, Klebsiella pneumonia, Enterococcus faecalis, Shigella flexneri, Salmonella typhimurium, Proteus vulgaris, Staphylococcus aureus	Minimum inhibitory concentration (MIC)	Broth dilution	MIC values for Escherichia coli, Pseudomonas aeruginosa, Klebsiella pneumonia, Enterococcus faecalis, Shigella flexneri, Salmonella typhimurium, Proteus vulgaris, and Staphylococcus aureus were 31.22, 125, 125, 250, 31.25, 125, 62.5, and 62.5, respectively	[220]
Magnesium oxide/germanium dioxide nanocomposite	Escherichia coli and Staphylococcus aureus	Minimum inhibitory concentration (MIC)	Broth dilution	MIC for Staphylococcus aureus was found to be 0.05 mg/mL whereas that of Escherichia coli was found to be 0.25 mg/mL	[221]
Magnesium oxide and titanium oxide	Bacillus subtilis and Staphylococcus aureus	Bactericidal efficacy	CFU count	Staphylococcus aureus showed bactericidal efficacy of 99.9% for MgO and TiO nanoparticles, while Bacillus subtilis was 94.5 and 69.0, respectively	[222]

Continued

TABLE 15.1
Some MONPs and Their Reported Antibacterial Potency.—cont'd

Type of Metal Oxide Nanoparticle(s)	Type of Bacteria	Type of Investigation	Type of Technique	Results	References
Zinc oxide, titanium oxide, and copper oxide	*Staphylococcus mutans* and *Streptococcus sanguinis*	MBC and minimum inhibitory concentration (MIC)	Broth dilution	MBC values of *Staphylococcus mutans* for ZnO, TiO, and CuO nanoparticles were 3.125, 0.195, and 25 µg/mL, respectively, while MIC values were 0.390, 0.097, and 12.5 µg/mL, respectively. MBC values of *Streptococcus sanguis* for ZnO, TiO, and CuO nanoparticles were 0.097, 0.097, and 0.1627 µg/mL, respectively, while MIC values were 0.0976, 0.0488, and 0.0976 µg/mL, respectively.	[223]
Cerium oxide	*Bacillus subtilis*	Cell viability	CFU count	~40% of survival and ~12% of survival at concentrations of 0.17 and 0.34 mg/mL, respectively	[224]
Zinc oxide	*Rothia dentocariosa* and *Rothia mucilaginosa*	MIC	Broth dilution	The MIC values of ZnO-NPs were determined to be 53 and 76 µg/mL for *Rothia dentocariosa* and *Rothia mucilaginosa*	[225]
Silver (Ag), cuprous oxide (Cu$_2$O), cupric oxide (CuO), zinc oxide (ZnO), titanium dioxide (TiO$_2$), tungsten oxide (WO$_3$), Ag + CuO, and Ag + ZnO composite	*Prevotella intermedia, Porphyromonas gingivalis, Fusobacterium nucleatum,* and *Aggregatibacter actinomycetemcomitans*	Minimum inhibitory concentration (MIC) and minimum bactericidal concentration (MBC)	Broth dilution	MIC and MBC values were in the range of <100 g/mL to 2500 g/mL and <100 g/mL to >2500 g/mL, respectively. The activity of the eight nanoparticles against the four species tested in descending order was Ag > Ag + CuO composite > Cu$_2$O > CuO > Ag + ZnO composite > ZnO > TiO$_2$ >WO$_3$. TiO$_2$ and WO$_3$ showed little or no antibacterial activity at the concentrations tested	[226]

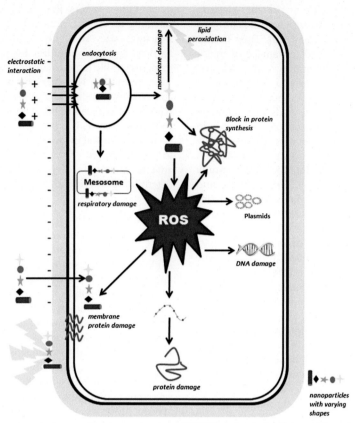

FIG. 15.7 Summary of some mechanisms of action of MONPs in bacteria. (Adapted from A. Raghunath and E. Perumal, Metal oxide nanoparticles as antimicrobial agents: a promise for the future, International Journal of Antimicrobial Agents 49 (2), 2017, 137–152.)

Metal ions from MONPs are slowly released through adsorption, dissolution, and hydrolysis [243] and positive charges possessed by the metal ions from the MONPs cause an electrostatic interaction with the DNA of the bacteria that leads to increased outer membrane permeability and disorganization. This, in turn, causes accumulation of MONPs that leads to delayed bacterial growth [18]. Additionally, a strong bond between MONPs and the outer membrane of the bacteria cell inhibits the transportation and activities of dehydrogenase and periplasmic enzymes [244]. Another way by which metal/metal ion homeostasis can be disturbed by MONPs is via the reduction of the metal ions generated by thiol (−SH) groups in enzymes and proteins of bacteria. This causes cell-wall disruption and inactivation of important metabolic proteins, obstructing respiration and further causes cell death [245,246].

4. Photokilling: This method of bacterial killing occurs upon exposing bacteria to MONPs in the presence of light; however, not all MONPs can be photo initiated

[247]. Electrons released during this process are trapped by MONPs and more ROS are generated [248]. Damage to proteins/DNA, diminution in superoxide SOD activity, and photochemical alteration of the cell membrane are the effect of exposing bacteria to transition MONPs in the presence of light [18,248].

5. Reactive oxidative species and oxidative stress production: Active redox cycling; prooxidant functional groups on the MONPs surface and cell−particle interactions are the key occurrences when reactive oxygen species are generated by MONPs upon exposure to bacteria [249]. The reactive site produced on the surface of the MONPs as a result of the change in electronic properties and particle size reduction is the point of interaction between molecular oxygen and electron donor/acceptor active sites. Some of the highly generated reactive oxygen species include superoxide anion, hydroxyl radicals, hydrogen peroxide, and organic hydroperoxides. For example, superoxide anion produced causes damage

to iron–sulfur (Fe–S) clusters in the electron transport chain and releases ferrous ions, leading to a decrease in ATP production [250]. They inhibit respiratory enzymes that lead to damage of major organic molecules such as amino acids, carbohydrates, lipids, nucleic acids, and proteins [250,251]. Furthermore, reactive oxygen species generated causes lipid peroxidation that leads to membrane architectural damage, conformational changes in membrane proteins, and alteration in membrane fluidity, integrity, and lateral organization [252]. A nonenzymatic antioxidant, glutathione, commonly found in plants, animals, fungi, some bacteria, and archaea is also affected (via oxidation) by metal ions released from MONPs. This leads to more production of ROS, and an imbalance in oxidants and antioxidants causes oxidative stress within the bacteria cell and ultimately induces cell death. The quantity of ROS generated by MONPs in bacteria is determined by the physicochemical properties of MONPs that include surface area, ability to diffuse, and electrophilic nature [251].

6. Inhibition of signal transduction and genotoxicity: This involves the suppression of bacterial cell division by MONPs, and this is achieved by interactions between the nucleic acids, genomic and plasmid DNA in particular, which further lead to disruption of the replication process of chromosomal and plasmid DNA [253]. The MONPs also affect signal transduction in bacteria by dephosphorylating the phosphotyrosine residues and thereby leading to signal transduction inhibition and subsequently inducing bacteria cell death [254,255].

6 LIMITATIONS

Despite several successful reports on the antibacterial efficacy of different MONPs, there are still concerns about their toxicity, thus causing limitations in the application and widespread commercialization. Several studies are ongoingly geared toward the investigation and understanding of the interaction of MONPs within the human body also because there is no clear understanding yet as to the mechanisms of toxicity in humans either [256]. In one study, researchers reportedly observed that the toxicity of MONPs is largely dependent on the bulkiness of the MONP. Hence, the researchers concluded that bulkier MONPs were less toxic, in comparison to their less bulky counterparts [257]. In addition, it is of importance to mention that the toxicity of certain MONPs depends on the dosage (concentration), duration of exposure, and the type of cell or tissue that is exposed

to it. Other factors that make the use of MONPs threating for in vivo antibacterial application includes structure, size, solubility, ability to cross different barriers, aggregation within the system, generation of oxidants, ability to percolate into the nucleus, and exposure time [199]. However, the major cause of toxicity concerns surrounding MONPs is due to their prooxidant nature that can lead to cell damage and damage to biomolecules. Hence, it is very crucial for extensive studies on the in vitro and in vivo toxicity assessment of MONPs to be conducted. The evaluation of the safety and adverse effect of MONPs will go a long way in determining its evident marketability.

7 CONCLUSION

The rise in antibiotic resistance and several strains of pathogenic microorganisms have led to numerous studies for newer strategies to improve the treatment and management of infectious diseases. Recently, the application of MONPs has shown to be a more effective alternative in comparison to conventional therapies. MONPs have been shown to exhibit antibacterial activities against various strains of bacteria as they are involved in several mechanisms that lead to bacterial membrane disruption/stability and ultimately cell death. These MONPs could be relevant in the pharmaceutical and biomedical industries as well as food preparation processes, water treatment, and coating-based applications as they have exhibited great antibacterial potential. MONPs may be used extensively in the nearest future for infectious disease treatment; however, toxicity findings still need to be carefully validated before in vivo applications are approved. Although there are hypotheses that toxicity in humans can be prevented or reduced by new techniques such as functionalization, ion doping, and conjugating with polymeric materials, validations to these findings need to be confirmed. Additionally, nanotechnologists, toxicologists, and microbiologists need a collaborative effort to determine the appropriate dosage against individual bacteria species as MONPs offer the promising antibacterial potential for diverse applications in nanomedicine.

ACKNOWLEDGMENTS

The financial assistance of the University of Zululand and the National Research Foundation, South Africa through the South African Research Chair Initiative (SARChI) is hereby acknowledged. SJO and NMM thank the National Research Foundation (NRF) for funding under the South African Research Chair for Nanotechnology.

REFERENCES

[1] R. Sugden, R. Kelly, S. Davies, Combatting antimicrobial resistance globally, Nature Microbiology 1 (2016) 16187.

[2] J. O'neill, Antimicrobial resistance: tackling a crisis for the health and wealth of nations, Review on Antimicrobial Resistance 1 (1) (2014) 1−16.

[3] A.M. Allahverdiyev, E.S. Abamor, M. Bagirova, M. Rafailovich, Antimicrobial effects of TiO_2 and Ag_2O nanoparticles against drug-resistant bacteria and leishmania parasites, Future Microbiology 6 (8) (2011) 933−940.

[4] A.P. Magiorakos, A. Srinivasan, R. Carey, Y. Carmeli, M. Falagas, C. Giske, S. Harbarth, J. Hindler, G. Kahlmeter, B. Olsson-Liljequist, Multidrug-resistant, extensively drug-resistant and pandrug-resistant bacteria: an international expert proposal for interim standard definitions for acquired resistance, Clinical Microbiology and Infection 18 (3) (2012) 268−281.

[5] J. Davies, D. Davies, Origins and evolution of antibiotic resistance, Microbiology and Molecular Biology Reviews 74 (3) (2010) 417−433.

[6] K. Adibkia, Y. Omidi, M.R. Siahi, A.R. Javadzadeh, M. Barzegar-Jalali, J. Barar, N. Maleki, G. Mohammadi, A. Nokhodchi, Inhibition of endotoxin-induced uveitis by methylprednisolone acetate nanosuspension in rabbits, Journal of Ocular Pharmacology and Therapeutics 23 (5) (2007) 421−432.

[7] K. Adibkia, Y. Javadzadeh, S. Dastmalchi, G. Mohammadi, F.K. Niri, M. Alaei-Beirami, Naproxen−Eudragit® RS100 nanoparticles: preparation and physicochemical characterization, Colloids and Surfaces B: Biointerfaces 83 (1) (2011) 155−159.

[8] A. Sabzevari, K. Adibkia, H. Hashemi, A. Hedayatfar, N. Mohsenzadeh, F. Atyabi, M.H. Ghahremani, R. Dinarvand, Polymeric triamcinolone acetonide nanoparticles as a new alternative in the treatment of uveitis: in vitro and in vivo studies, European Journal of Pharmaceutics and Biopharmaceutics 84 (1) (2013) 63−71.

[9] C. Buzea, I.I. Pacheco, K. Robbie, Nanomaterials and nanoparticles: sources and toxicity, Biointerphases 2 (4) (2007) MR17−MR71.

[10] K. Adibkia, M. Barzegar-Jalali, A. Nokhodchi, M. Siahi Shadbad, Y. Omidi, Y. Javadzadeh, G. Mohammadi, A review on the methods of preparation of pharmaceutical nanoparticles, Pharmaceutical Sciences 15 (4) (2009) 303−314.

[11] F. Martinez-Gutierrez, P.L. Olive, A. Banuelos, E. Orrantia, N. Nino, E.M. Sanchez, F. Ruiz, H. Bach, Y. Av-Gay, Synthesis, characterization, and evaluation of antimicrobial and cytotoxic effect of silver and titanium nanoparticles, Nanomedicine: Nanotechnology, Biology and Medicine 6 (5) (2010) 681−688.

[12] J.T. Seil, T.J. Webster, Antimicrobial applications of nanotechnology: methods and literature, International Journal of Nanomedicine 7 (2012) 2767.

[13] K. Adibkia, M. Alaei-Beirami, M. Barzegar-Jalali, G. Mohammadi, M.S. Ardestani, Evaluation and optimization of factors affecting novel diclofenac sodium-eudragit RS100 nanoparticles, African Journal of Pharmacy and Pharmacology 6 (12) (2012) 941−947.

[14] J.A. Lemire, J.J. Harrison, R.J. Turner, Antimicrobial activity of metals: mechanisms, molecular targets and applications, Nature Reviews Microbiology 11 (6) (2013) 371.

[15] J.W. Alexander, History of the medical use of silver, Surgical Infections 10 (3) (2009) 289−292.

[16] G. Borkow, J. Gabbay, Copper, an ancient remedy returning to fight microbial, fungal and viral infections, Current Chemical Biology 3 (3) (2009) 272−278.

[17] C. Malarkodi, S. Rajeshkumar, K. Paulkumar, M. Vanaja, G. Gnanajobitha, G. Annadurai, Biosynthesis and antimicrobial activity of semiconductor nanoparticles against oral pathogens, Bioinorganic Chemistry and Applications 2014 (2014).

[18] A. Raghunath, E. Perumal, Metal oxide nanoparticles as antimicrobial agents: a promise for the future, International Journal of Antimicrobial Agents 49 (2) (2017) 137−152.

[19] F. Cavalieri, M. Tortora, A. Stringaro, M. Colone, L. Baldassarri, Nanomedicines for antimicrobial interventions, Journal of Hospital Infection 88 (4) (2014) 183−190.

[20] A. Azam, A.S. Ahmed, M. Oves, M.S. Khan, S.S. Habib, A. Memic, Antimicrobial activity of metal oxide nanoparticles against Gram-positive and Gram-negative bacteria: a comparative study, International Journal of Nanomedicine 7 (2012) 6003.

[21] S. Mourdikoudis, R.M. Pallares, N.T. Thanh, Characterization techniques for nanoparticles: comparison and complementarity upon studying nanoparticle properties, Nanoscale 10 (27) (2018) 12871−12934.

[22] K. Lewis, A.M. Klibanov, Surpassing nature: rational design of sterile-surface materials, Trends in Biotechnology 23 (7) (2005) 343−348.

[23] J. Gangwar, B.K. Gupta, A.K. Srivastava, Prospects of emerging engineered oxide nanomaterials and their applications, Defence Science Journal 66 (4) (2016) 323−340.

[24] R. Singh, H.S. Nalwa, Medical applications of nanoparticles in biological imaging, cell labeling, antimicrobial agents, and anticancer nanodrugs, Journal of Biomedical Nanotechnology 7 (4) (2011) 489−503.

[25] H.H. Kung, Transition Metal Oxides: Surface Chemistry and Catalysis, Elsevier, 1989.

[26] C. Noguera, Physics and Chemistry at Oxide Surfaces, Cambridge University Press, 1996.

[27] C. Klingshirn, ZnO: material, physics and applications, ChemPhysChem 8 (6) (2007) 782−803.

[28] D.R. Lide, CRC Handbook of Chemistry and Physics, CRC Press, Boca Raton, FL, 2005. Internet Version 2005, http://www.hbcpnetbase.com. There is no corresponding record for this reference (2005).

[29] Z.L. Wang, Nanostructures of zinc oxide, Materials Today 7 (6) (2004) 26–33.

[30] Ü. Özgür, Y.I. Alivov, C. Liu, A. Teke, M. Reshchikov, S. Doğan, V. Avrutin, S.-J. Cho, H. Morkoç, A comprehensive review of ZnO materials and devices, Journal of Applied Physics 98 (4) (2005) 11.

[31] D. Reynolds, D.C. Look, B. Jogai, C. Litton, G. Cantwell, W. Harsch, Valence-band ordering in ZnO, Physical Review B 60 (4) (1999) 2340.

[32] N. Alvi, K. Ul Hasan, O. Nur, M. Willander, The origin of the red emission in n-ZnO nanotubes/p-GaN white light emitting diodes, Nanoscale Research Letters 6 (1) (2011) 130.

[33] A. Djurišić, Y. Leung, K. Tam, L. Ding, W. Ge, H. Chen, S. Gwo, Green, yellow, and orange defect emission from ZnO nanostructures: influence of excitation wavelength, Applied Physics Letters 88 (10) (2006) 103107.

[34] M. Willander, O. Nur, Q. Zhao, L. Yang, M. Lorenz, B. Cao, J.Z. Pérez, C. Czekalla, G. Zimmermann, M. Grundmann, Zinc oxide nanorod based photonic devices: recent progress in growth, light emitting diodes and lasers, Nanotechnology 20 (33) (2009) 332001.

[35] P. Batista, M. Mulato, ZnO extended-gate field-effect transistors as p H sensors, Applied Physics Letters 87 (14) (2005) 143508.

[36] B. Kang, F. Ren, Y. Heo, L. Tien, D. Norton, S. Pearton, p H measurements with single ZnO nanorods integrated with a microchannel, Applied Physics Letters 86 (11) (2005) 112105.

[37] Z.L. Wang, Towards self-powered nanosystems: from nanogenerators to nanopiezotronics, Advanced Functional Materials 18 (22) (2008) 3553–3567.

[38] H. Mirzaei, M. Darroudi, Zinc oxide nanoparticles: biological synthesis and biomedical applications, Ceramics International 43 (1) (2017) 907–914.

[39] C. Jayaseelan, A.A. Rahuman, A.V. Kirthi, S. Marimuthu, T. Santhoshkumar, A. Bagavan, K. Gaurav, L. Karthik, K.B. Rao, Novel microbial route to synthesize ZnO nanoparticles using Aeromonas hydrophila and their activity against pathogenic bacteria and fungi, Spectrochimica Acta Part A: Molecular and Biomolecular Spectroscopy 90 (2012) 78–84.

[40] G. Sangeetha, S. Rajeshwari, R. Venckatesh, Green synthesis of zinc oxide nanoparticles by aloe barbadensis miller leaf extract: structure and optical properties, Materials Research Bulletin 46 (12) (2011) 2560–2566.

[41] E. Sherly, J.J. Vijaya, N.C.S. Selvam, L.J. Kennedy, Microwave assisted combustion synthesis of coupled ZnO–ZrO2 nanoparticles and their role in the photocatalytic degradation of 2, 4-dichlorophenol, Ceramics International 40 (4) (2014) 5681–5691.

[42] K. Elumalai, S. Velmurugan, Green synthesis, characterization and antimicrobial activities of zinc oxide nanoparticles from the leaf extract of Azadirachta indica (L.), Applied Surface Science 345 (2015) 329–336.

[43] M. Stan, A. Popa, D. Toloman, A. Dehelean, I. Lung, G. Katona, Enhanced photocatalytic degradation properties of zinc oxide nanoparticles synthesized by using plant extracts, Materials Science in Semiconductor Processing 39 (2015) 23–29.

[44] M.S. Akhtar, S. Ameen, S.A. Ansari, O. Yang, Synthesis and characterization of ZnO nanorods and balls nanomaterials for dye sensitized solar cells, Journal of Nanoengineering and Nanomanufacturing 1 (1) (2011) 71–76.

[45] N.P. Sasidharan, P. Chandran, S.S. Khan, Interaction of colloidal zinc oxide nanoparticles with bovine serum albumin and its adsorption isotherms and kinetics, Colloids and Surfaces B: Biointerfaces 102 (2013) 195–201.

[46] R. Tayebee, F. Cheravi, M. Mirzaee, M.M. Amini, Commercial zinc oxide (Zn^{2+}) as an efficient and environmentally benign catalyst for homogeneous benzoylation of hydroxyl functional groups, Chinese Journal of Chemistry 28 (7) (2010) 1247–1252.

[47] L. Zhang, Y. Ding, M. Povey, D. York, ZnO nanofluids– A potential antibacterial agent, Progress in Natural Science 18 (8) (2008) 939–944.

[48] J. Xie, H. Deng, Z. Xu, Y. Li, J. Huang, Growth of ZnO photonic crystals by self-assembly, Journal of Crystal Growth 292 (2) (2006) 227–229.

[49] C. Liewhiran, S. Phanichphant, Improvement of flame-made ZnO nanoparticulate thick film morphology for ethanol sensing, Sensors 7 (5) (2007) 650–675.

[50] R. Jalal, E.K. Goharshadi, M. Abareshi, M. Moosavi, A. Yousefi, P. Nancarrow, ZnO nanofluids: green synthesis, characterization, and antibacterial activity, Materials Chemistry and Physics 121 (1–2) (2010) 198–201.

[51] N. Talebian, S.M. Amininezhad, M. Doudi, Controllable synthesis of ZnO nanoparticles and their morphology-dependent antibacterial and optical properties, Journal of Photochemistry and Photobiology B: Biology 120 (2013) 66–73.

[52] E.E. Hafez, H.S. Hassan, M. Elkady, E. Salama, Assessment of antibacterial activity for synthesized zinc oxide nanorods against plant pathogenic strains, International Journal of Scientific and Technology Research 3 (9) (2014) 318–324.

[53] M. Ramani, S. Ponnusamy, C. Muthamizhchelvan, From zinc oxide nanoparticles to microflowers: a study of growth kinetics and biocidal activity, Materials Science and Engineering: C 32 (8) (2012) 2381–2389.

[54] J. Ma, J. Liu, Y. Bao, Z. Zhu, X. Wang, J. Zhang, Synthesis of large-scale uniform mulberry-like ZnO particles with microwave hydrothermal method and its antibacterial property, Ceramics International 39 (3) (2013) 2803–2810.

[55] A. Stanković, S. Dimitrijević, D. Uskoković, Influence of size scale and morphology on antibacterial properties of ZnO powders hydrothemally synthesized using different surface stabilizing agents, Colloids and Surfaces B: Biointerfaces 102 (2013) 21–28.

[56] Y. Zhang, J. Qian, D. Wang, Y. Wang, S. He, Multifunctional gold nanorods with ultrahigh stability and tunability for in vivo fluorescence imaging, SERS detection, and photodynamic therapy, Angewandte Chemie International Edition 52 (4) (2013) 1148−1151.

[57] A.M. Youssef, M.S. Abdel-Aziz, S.M. El-Sayed, Chitosan nanocomposite films based on Ag-NP and Au-NP biosynthesis by Bacillus subtilis as packaging materials, International Journal of Biological Macromolecules 69 (2014) 185−191.

[58] D. Yin, X. Li, Y. Ma, Z. Liu, Targeted cancer imaging and photothermal therapy via monosaccharide-imprinted gold nanorods, Chemical Communications 53 (50) (2017) 6716−6719.

[59] Z. Zhang, J. Wang, X. Nie, T. Wen, Y. Ji, X. Wu, Y. Zhao, C. Chen, Near infrared laser-induced targeted cancer therapy using thermoresponsive polymer encapsulated gold nanorods, Journal of the American Chemical Society 136 (20) (2014) 7317−7326.

[60] E.L.L. Yeo, U. Joshua, J. Cheah, D.J.H. Neo, W.I. Goh, P. Kanchanawong, K.C. Soo, P.S.P. Thong, J.C.Y. Kah, Exploiting the protein corona around gold nanorods for low-dose combined photothermal and photodynamic therapy, Journal of Materials Chemistry B 5 (2) (2017) 254−268.

[61] T.K. Sau, A.L. Rogach, F. Jäckel, T.A. Klar, J. Feldmann, Properties and applications of colloidal nonspherical noble metal nanoparticles, Advanced Materials 22 (16) (2010) 1805−1825.

[62] M. Hu, J. Chen, Z.-Y. Li, L. Au, G.V. Hartland, X. Li, M. Marquez, Y. Xia, Gold nanostructures: engineering their plasmonic properties for biomedical applications, Chemical Society Reviews 35 (11) (2006) 1084−1094.

[63] N. Khlebtsov, L. Dykman, Biodistribution and toxicity of engineered gold nanoparticles: a review of in vitro and in vivo studies, Chemical Society Reviews 40 (3) (2011) 1647−1671.

[64] A. Tomar, G. Garg, Short review on application of gold nanoparticles, Global Journal of Pharmacology 7 (1) (2013) 34−38.

[65] Y.-C. Yeh, B. Creran, V.M. Rotello, Gold nanoparticles: preparation, properties, and applications in bionanotechnology, Nanoscale 4 (6) (2012) 1871−1880.

[66] L. Wang, Y. Liu, W. Li, X. Jiang, Y. Ji, X. Wu, L. Xu, Y. Qiu, K. Zhao, T. Wei, Selective targeting of gold nanorods at the mitochondria of cancer cells: implications for cancer therapy, Nano Letters 11 (2) (2010) 772−780.

[67] M. Das, K.H. Shim, S.S.A. An, D.K. Yi, Review on gold nanoparticles and their applications, Toxicology and Environmental Health Sciences 3 (4) (2011) 193−205.

[68] D.F. Moyano, V.M. Rotello, Nano meets biology: structure and function at the nanoparticle interface, Langmuir 27 (17) (2011) 10376−10385.

[69] J.A. Jamison, E.L. Bryant, S.B. Kadali, M.S. Wong, V.L. Colvin, K.S. Matthews, M.K. Calabretta, Altering protein surface charge with chemical modification modulates protein−gold nanoparticle aggregation, Journal of Nanoparticle Research 13 (2) (2011) 625−636.

[70] J.A. Khan, R.A. Kudgus, A. Szabolcs, S. Dutta, E. Wang, S. Cao, G.L. Curran, V. Shah, S. Curley, D. Mukhopadhyay, Designing nanoconjugates to effectively target pancreatic cancer cells in vitro and in vivo, PLoS One 6 (6) (2011) e20347.

[71] C.R. Patra, R. Bhattacharya, E. Wang, A. Katarya, J.S. Lau, S. Dutta, M. Muders, S. Wang, S.A. Buhrow, S.L. Safgren, Targeted delivery of gemcitabine to pancreatic adenocarcinoma using cetuximab as a targeting agent, Cancer Research 68 (6) (2008) 1970−1978.

[72] K.S. Siddiqi, A. Husen, R.A. Rao, A review on biosynthesis of silver nanoparticles and their biocidal properties, Journal of Nanobiotechnology 16 (1) (2018) 14.

[73] W. Wang, Q. Zhao, J. Dong, J. Li, A novel silver oxides oxygen evolving catalyst for water splitting, International Journal of Hydrogen Energy 36 (13) (2011) 7374−7380.

[74] V. Petrov, T. Nazarova, A. Korolev, N. Kopilova, Thin sol−gel SiO_2−SnO_x−AgO_y films for low temperature ammonia gas sensor, Sensors and Actuators B: Chemical 133 (1) (2008) 291−295.

[75] E. Sanli, B.Z. Uysal, M.L. Aksu, The oxidation of $NaBH_4$ on electrochemically treated silver electrodes, International Journal of Hydrogen Energy 33 (8) (2008) 2097−2104.

[76] Y. Ida, S. Watase, T. Shinagawa, M. Watanabe, M. Chigane, M. Inaba, A. Tasaka, M. Izaki, Direct electrodeposition of 1.46 eV bandgap silver (I) oxide semiconductor films by electrogenerated acid, Chemistry of Materials 20 (4) (2008) 1254−1256.

[77] J. Tominaga, The application of silver oxide thin films to plasmon photonic devices, Journal of Physics: Condensed Matter 15 (25) (2003) R1101.

[78] W. Parkhurst, S. Dallek, B. Larrick, Thermogravimetry-evolved gas analysis of silver oxide cathode material, Journal of the Electrochemical Society 131 (8) (1984) 1739−1742.

[79] S. Dallek, W. West, B. Larrick, Decomposition kinetics of AgO cathode material by thermogravimetry, Journal of the Electrochemical Society 133 (12) (1986) 2451−2454.

[80] Z.-J. Jiang, C.-Y. Liu, L.-W. Sun, Catalytic properties of silver nanoparticles supported on silica spheres, The Journal of Physical Chemistry B 109 (5) (2005) 1730−1735.

[81] A. Vaseashta, D. Dimova-Malinovska, Nanostructured and nanoscale devices, sensors and detectors, Science and Technology of Advanced Materials 6 (3−4) (2005) 312−318.

[82] M. Ahamed, M.S. AlSalhi, M. Siddiqui, Silver nanoparticle applications and human health, Clinica Chimica Acta 411 (23−24) (2010) 1841−1848.

[83] X. Chen, H.J. Schluesener, Nanosilver: a nanoproduct in medical application, Toxicology Letters 176 (1) (2008) 1−12.

[84] N.L. Yong, A. Ahmad, A.W. Mohammad, Synthesis and characterization of silver oxide nanoparticles by a novel method, International Journal of Scientific Engineering and Research 4 (2013) 155−158.

[85] Z. Zaheer, Silver nanoparticles to self-assembled films: green synthesis and characterization, Colloids and Surfaces B: Biointerfaces 90 (2012) 48−52.

[86] S. Jradi, L. Balan, X. Zeng, J. Plain, D.J. Lougnot, P. Royer, R. Bachelot, S. Akil, O. Soppera, L. Vidal, Spatially controlled synthesis of silver nanoparticles and nanowires by photosensitized reduction, Nanotechnology 21 (9) (2010) 095605.

[87] C. Fang, A.V. Ellis, N.H. Voelcker, Electrochemical synthesis of silver oxide nanowires, microplatelets and application as SERS substrate precursors, Electrochimica Acta 59 (2012) 346−353.

[88] A. Husen, K.S. Siddiqi, Phytosynthesis of nanoparticles: concept, controversy and application, Nanoscale Research Letters 9 (1) (2014) 229.

[89] K.S. Siddiqi, A. Husen, Fabrication of metal nanoparticles from fungi and metal salts: scope and application, Nanoscale Research Letters 11 (1) (2016) 98.

[90] K.S. Siddiqi, A. Husen, Fabrication of metal and metal oxide nanoparticles by algae and their toxic effects, Nanoscale Research Letters 11 (1) (2016) 363.

[91] A.R. Shahverdi, S. Minaeian, H.R. Shahverdi, H. Jamalifar, A.-A. Nohi, Rapid synthesis of silver nanoparticles using culture supernatants of Enterobacteria: a novel biological approach, Process Biochemistry 42 (5) (2007) 919−923.

[92] K.S. Siddiqi, A. Husen, Recent advances in plant-mediated engineered gold nanoparticles and their application in biological system, Journal of Trace Elements in Medicine and Biology 40 (2017) 10−23.

[93] A. Syafiuddin, M.R. Salim, A. Beng Hong Kueh, T. Hadibarata, H. Nur, A review of silver nanoparticles: research trends, global consumption, synthesis, properties, and future challenges, Journal of the Chinese Chemical Society 64 (7) (2017) 732−756.

[94] X.-F. Zhang, Z.-G. Liu, W. Shen, S. Gurunathan, Silver nanoparticles: synthesis, characterization, properties, applications, and therapeutic approaches, International Journal of Molecular Sciences 17 (9) (2016) 1534.

[95] H.H. Lara, E.N. Garza-Treviño, L. Ixtepan-Turrent, D.K. Singh, Silver nanoparticles are broad-spectrum bactericidal and virucidal compounds, Journal of Nanobiotechnology 9 (1) (2011) 30.

[96] S. Lokina, A. Stephen, V. Kaviyarasan, C. Arulvasu, V. Narayanan, Cytotoxicity and antimicrobial activities of green synthesized silver nanoparticles, European Journal of Medicinal Chemistry 76 (2014) 256−263.

[97] S. Rajeshkumar, P. Naik, Synthesis and biomedical applications of cerium oxide nanoparticles—a review, Biotechnology Reports 17 (2018) 1−5.

[98] D.A. Pelletier, A.K. Suresh, G.A. Holton, C.K. McKeown, W. Wang, B. Gu, N.P. Mortensen, D.P. Allison, D.C. Joy, M.R. Allison, Effects of engineered cerium oxide nanoparticles on bacterial growth and viability, Applied and Environmental Microbiology 76 (24) (2010) 7981−7989.

[99] S. Rojas, J.D. Gispert, S. Abad, M. Buaki-Sogo, V.M. Victor, H. Garcia, J.R.l. Herance, In vivo biodistribution of amino-functionalized ceria nanoparticles in rats using positron emission tomography, Molecular Pharmaceutics 9 (12) (2012) 3543−3550.

[100] H. Zhang, X. He, Z. Zhang, P. Zhang, Y. Li, Y. Ma, Y. Kuang, Y. Zhao, Z. Chai, Nano-CeO2 exhibits adverse effects at environmental relevant concentrations, Environmental Science and Technology 45 (8) (2011) 3725−3730.

[101] P. Demokritou, S. Gass, G. Pyrgiotakis, J.M. Cohen, W. Goldsmith, W. McKinney, D. Frazer, J. Ma, D. Schwegler-Berry, J. Brain, An in vivo and in vitro toxicological characterisation of realistic nanoscale CeO2 inhalation exposures, Nanotoxicology 7 (8) (2013) 1338−1350.

[102] I.A.P. Farias, C.C.L.d. Santos, F.C. Sampaio, Antimicrobial activity of cerium oxide nanoparticles on opportunistic microorganisms: a systematic review, Biomed Research International (2018) 2018.

[103] A. Trovarelli, Catalytic properties of ceria and CeO2-containing materials, Catalysis Reviews 38 (4) (1996) 439−520.

[104] C.-H. Wang, S.-S. Lin, Preparing an active cerium oxide catalyst for the catalytic incineration of aromatic hydrocarbons, Applied Catalysis A: General 268 (1−2) (2004) 227−233.

[105] N.N. Dao, M. Dai Luu, Q.K. Nguyen, B.S. Kim, UV absorption by cerium oxide nanoparticles/epoxy composite thin films, Advances in Natural Sciences: Nanoscience and Nanotechnology 2 (4) (2011) 045013.

[106] N. Zholobak, V. Ivanov, A. Shcherbakov, A. Shaporev, O. Polezhaeva, A.Y. Baranchikov, N.Y. Spivak, Y.D. Tretyakov, UV-shielding property, photocatalytic activity and photocytotoxicity of ceria colloid solutions, Journal of Photochemistry and Photobiology B: Biology 102 (1) (2011) 32−38.

[107] B. Courbiere, M. Auffan, R. Rollais, V. Tassistro, A. Bonnefoy, A. Botta, J. Rose, T. Orsière, J. Perrin, Ultrastructural interactions and genotoxicity assay of cerium dioxide nanoparticles on mouse oocytes, International Journal of Molecular Sciences 14 (11) (2013) 21613−21628.

[108] G. Pulido-Reyes, I. Rodea-Palomares, S. Das, T.S. Sakthivel, F. Leganes, R. Rosal, S. Seal, F. Fernández-Piñas, Untangling the biological effects of cerium oxide nanoparticles: the role of surface valence states, Scientific Reports 5 (2015) 15613.

[109] L. Peng, X. He, P. Zhang, J. Zhang, Y. Li, J. Zhang, Y. Ma, Y. Ding, Z. Wu, Z. Chai, Comparative pulmonary toxicity of two ceria nanoparticles with the same primary size, International Journal of Molecular Sciences 15 (4) (2014) 6072−6085.

[110] M. Culcasi, L. Benameur, A. Mercier, C. Lucchesi, H. Rahmouni, A. Asteian, G. Casano, A. Botta, H. Kovacic, S. Pietri, EPR spin trapping evaluation of ROS production in human fibroblasts exposed to

cerium oxide nanoparticles: evidence for NADPH oxidase and mitochondrial stimulation, Chemico-Biological Interactions 199 (3) (2012) 161–176.

[111] J.M. Perez, A. Asati, S. Nath, C. Kaittanis, Synthesis of biocompatible dextran-coated nanoceria with pH-dependent antioxidant properties, Small 4 (5) (2008) 552–556.

[112] C. Korsvik, S. Patil, S. Seal, W.T. Self, Superoxide dismutase mimetic properties exhibited by vacancy engineered ceria nanoparticles, Chemical Communications (10) (2007) 1056–1058.

[113] S. Deshpande, S. Patil, S.V. Kuchibhatla, S. Seal, Size dependency variation in lattice parameter and valency states in nanocrystalline cerium oxide, Applied Physics Letters 87 (13) (2005) 133113.

[114] I. Celardo, J.Z. Pedersen, E. Traversa, L. Ghibelli, Pharmacological potential of cerium oxide nanoparticles, Nanoscale 3 (4) (2011) 1411–1420.

[115] M. Li, P. Shi, C. Xu, J. Ren, X. Qu, Cerium oxide caged metal chelator: anti-aggregation and anti-oxidation integrated H_2O_2-responsive controlled drug release for potential Alzheimer's disease treatment, Chemical Science 4 (6) (2013) 2536–2542.

[116] J.A. Vassie, J.M. Whitelock, M.S. Lord, Targeted delivery and redox activity of folic acid-functionalized nanoceria in tumor cells, Molecular Pharmaceutics 15 (3) (2018) 994–1004.

[117] A.S. Karakoti, O. Tsigkou, S. Yue, P.D. Lee, M.M. Stevens, J.R. Jones, S. Seal, Rare earth oxides as nanoadditives in 3-D nanocomposite scaffolds for bone regeneration, Journal of Materials Chemistry 20 (40) (2010) 8912–8919.

[118] C. Mandoli, F. Pagliari, S. Pagliari, G. Forte, P. Di Nardo, S. Licoccia, E. Traversa, Stem cell aligned growth induced by CeO_2 nanoparticles in PLGA scaffolds with improved bioactivity for regenerative medicine, Advanced Functional Materials 20 (10) (2010) 1617–1624.

[119] B.A. Rzigalinski, K. Meehan, R.M. Davis, Y. Xu, W.C. Miles, C.A. Cohen, Radical Nanomedicine, 2006.

[120] X. Jiao, H. Song, H. Zhao, W. Bai, L. Zhang, Y. Lv, Well-redispersed ceria nanoparticles: promising peroxidase mimetics for H_2O_2 and glucose detection, Analytical Methods 4 (10) (2012) 3261–3267.

[121] A. Asati, S. Santra, C. Kaittanis, S. Nath, J.M. Perez, Oxidase-like activity of polymer-coated cerium oxide nanoparticles, Angewandte Chemie International Edition 48 (13) (2009) 2308–2312.

[122] J.M. Dowding, S. Das, A. Kumar, T. Dosani, R. McCormack, A. Gupta, T.X. Sayle, D.C. Sayle, L. von Kalm, S. Seal, Cellular interaction and toxicity depend on physicochemical properties and surface modification of redox-active nanomaterials, ACS Nano 7 (6) (2013) 4855–4868.

[123] M. Das, S. Patil, N. Bhargava, J.-F. Kang, L.M. Riedel, S. Seal, J.J. Hickman, Auto-catalytic ceria nanoparticles offer neuroprotection to adult rat spinal cord neurons, Biomaterials 28 (10) (2007) 1918–1925.

[124] J.M. Dowding, T. Dosani, A. Kumar, S. Seal, W.T. Self, Cerium oxide nanoparticles scavenge nitric oxide radical (NO), Chemical Communications 48 (40) (2012) 4896–4898.

[125] J.M. Dowding, S. Seal, W.T. Self, Cerium oxide nanoparticles accelerate the decay of peroxynitrite (ONOO−), Drug Delivery and Translational Research 3 (4) (2013) 375–379.

[126] M.J. Guajardo-Pacheco, J. Morales-Sánchez, J. González-Hernández, F. Ruiz, Synthesis of copper nanoparticles using soybeans as a chelant agent, Materials Letters 64 (12) (2010) 1361–1364.

[127] J. Singh, G. Kaur, M. Rawat, A brief review on synthesis and characterization of copper oxide nanoparticles and its applications, Journal of Bioelectronics and Nanotechnology 1 (9) (2016).

[128] M. Grigore, E. Biscu, A. Holban, M. Gestal, A. Grumezescu, Methods of synthesis, properties and biomedical applications of CuO nanoparticles, Pharmaceuticals 9 (4) (2016) 75.

[129] Y. Xi, C. Hu, P. Gao, R. Yang, X. He, X. Wang, B. Wan, Morphology and phase selective synthesis of CuxO (x= 1, 2) nanostructures and their catalytic degradation activity, Materials Science and Engineering: B 166 (1) (2010) 113–117.

[130] Y. He, A novel solid-stabilized emulsion approach to CuO nanostructured microspheres, Materials Research Bulletin 42 (1) (2007) 190–195.

[131] F. Perreault, S.P. Melegari, C.H. da Costa, A.L.d.O.F. Rossetto, R. Popovic, W.G. Matias, Genotoxic effects of copper oxide nanoparticles in Neuro 2A cell cultures, The Science of the Total Environment 441 (2012) 117–124.

[132] Y.-W. Baek, Y.-J. An, Microbial toxicity of metal oxide nanoparticles (CuO, NiO, ZnO, and Sb2O3) to *Escherichia coli, Bacillus subtilis*, and *Streptococcus aureus*, The Science of the Total Environment 409 (8) (2011) 1603–1608.

[133] T. Ostaszewska, M. Chojnacki, M. Kamaszewski, E. Sawosz-Chwalibóg, Histopathological effects of silver and copper nanoparticles on the epidermis, gills, and liver of Siberian sturgeon, Environmental Science and Pollution Research 23 (2) (2016) 1621–1633.

[134] G. Isani, M.L. Falcioni, G. Barucca, D. Sekar, G. Andreani, E. Carpenè, G. Falcioni, Comparative toxicity of CuO nanoparticles and $CuSO_4$ in rainbow trout, Ecotoxicology and Environmental Safety 97 (2013) 40–46.

[135] P. Ruiz, A. Katsumiti, J.A. Nieto, J. Bori, A. Jimeno-Romero, P. Reip, I. Arostegui, A. Orbea, M.P. Cajaraville, Short-term effects on antioxidant enzymes and long-term genotoxic and carcinogenic potential of CuO nanoparticles compared to bulk CuO and ionic copper in mussels *Mytilus galloprovincialis*, Marine Environmental Research 111 (2015) 107–120.

[136] R. Sankar, R. Maheswari, S. Karthik, K.S. Shivashangari, V. Ravikumar, Anticancer activity of Ficus religiosa engineered copper oxide nanoparticles, Materials Science and Engineering: C 44 (2014) 234–239.

[137] P.J.P. Espitia, N.d.F.F. Soares, J.S. dos Reis Coimbra, N.J. de Andrade, R.S. Cruz, E.A.A. Medeiros, Zinc oxide nanoparticles: synthesis, antimicrobial activity and food packaging applications, Food and Bioprocess Technology 5 (5) (2012) 1447–1464.

[138] M. Mortimer, K. Kasemets, A. Kahru, Toxicity of ZnO and CuO nanoparticles to ciliated protozoa Tetrahymena thermophila, Toxicology 269 (2–3) (2010) 182–189.

[139] M. Sajid, M. Ilyas, C. Basheer, M. Tariq, M. Daud, N. Baig, F. Shehzad, Impact of nanoparticles on human and environment: review of toxicity factors, exposures, control strategies, and future prospects, Environmental Science and Pollution Research 22 (6) (2015) 4122–4143.

[140] R. Rani, H. Kumar, R. Salar, S. Purewal, Antibacterial activity of copper oxide nanoparticles against gram negative bacterial strain synthesized by reverse micelle technique, International Journal of Pharmaceutical Research and Development 6 (2014) 72–78.

[141] H. Shi, R. Magaye, V. Castranova, J. Zhao, Titanium dioxide nanoparticles: a review of current toxicological data, Particle and Fibre Toxicology 10 (1) (2013) 15.

[142] M. Ortlieb, White giant or white dwarf?: particle size distribution measurements of TiO_2, GIT Laboratory Journal Europe 14 (9–10) (2010) 42–43.

[143] J. Riu, A. Maroto, F.X. Rius, Nanosensors in environmental analysis, Talanta 69 (2) (2006) 288–301.

[144] C. Wang, Y. Li, Interaction and nanotoxic effect of TiO_2 nanoparticle on fibrinogen by multi-spectroscopic method, The Science of the Total Environment 429 (2012) 156–160.

[145] P.O. Andersson, C. Lejon, B. Ekstrand-Hammarström, C. Akfur, L. Ahlinder, A. Bucht, L. Österlund, Polymorph-and size-dependent uptake and toxicity of TiO_2 nanoparticles in living lung epithelial cells, Small 7 (4) (2011) 514–523.

[146] R. Tedja, M. Lim, R. Amal, C. Marquis, Effects of serum adsorption on cellular uptake profile and consequent impact of titanium dioxide nanoparticles on human lung cell lines, ACS Nano 6 (5) (2012) 4083–4093.

[147] D.B. Warheit, T.R. Webb, K.L. Reed, S. Frerichs, C.M. Sayes, Pulmonary toxicity study in rats with three forms of ultrafine-TiO_2 particles: differential responses related to surface properties, Toxicology 230 (1) (2007) 90–104.

[148] C.M. Sayes, R. Wahi, P.A. Kurian, Y. Liu, J.L. West, K.D. Ausman, D.B. Warheit, V.L. Colvin, Correlating nanoscale titania structure with toxicity: a cytotoxicity and inflammatory response study with human dermal fibroblasts and human lung epithelial cells, Toxicological Sciences 92 (1) (2006) 174–185.

[149] R. Baan, K. Straif, Y. Grosse, B. Secretan, F. El Ghissassi, V. Cogliano, Carcinogenicity of carbon black, titanium dioxide, and talc, The Lancet Oncology 7 (4) (2006) 295–296.

[150] R.K. Shukla, V. Sharma, A.K. Pandey, S. Singh, S. Sultana, A. Dhawan, ROS-mediated genotoxicity induced by titanium dioxide nanoparticles in human epidermal cells, Toxicology in Vitro 25 (1) (2011) 231–241.

[151] Y.-F. Chen, H.-Y. Tsai, T.-S. Wu, Anti-inflammatory and analgesic activities from roots of Angelica pubescens, Planta Medica 61 (01) (1995) 2–8.

[152] N. Muhd Julkapli, S. Bagheri, S. Bee Abd Hamid, Recent advances in heterogeneous photocatalytic decolorization of synthetic dyes, The Scientific World Journal 2014 (2014).

[153] S. Valencia, X. Vargas, L. Rios, G. Restrepo, J.M. Marín, Sol–gel and low-temperature solvothermal synthesis of photoactive nano-titanium dioxide, Journal of Photochemistry and Photobiology A: Chemistry 251 (2013) 175–181.

[154] C. Jayaseelan, A.A. Rahuman, S.M. Roopan, A.V. Kirthi, J. Venkatesan, S.-K. Kim, M. Iyappan, C. Siva, Biological approach to synthesize TiO_2 nanoparticles using Aeromonas hydrophila and its antibacterial activity, Spectrochimica Acta Part A: Molecular and Biomolecular Spectroscopy 107 (2013) 82–89.

[155] T. Kaida, Optical characteristics of titanium oxide interference film and the film laminated with oxides and their applications for cosmetics, Journal of Cosmetic Science 55 (2004) 219–220.

[156] B. Trouiller, R. Reliene, A. Westbrook, P. Solaimani, R.H. Schiestl, Titanium dioxide nanoparticles induce DNA damage and genetic instability in vivo in mice, Cancer Research 69 (22) (2009) 8784–8789.

[157] R. Wolf, H. Matz, E. Orion, J. Lipozencic, Sunscreens—the ultimate cosmetic, Acta Dermatovenerologica Croatica 11 (3) (2003) 158–162.

[158] M. Montazer, S. Seifollahzadeh, Enhanced self-cleaning, antibacterial and UV protection properties of nano TiO_2 treated textile through enzymatic pretreatment, Photochemistry and Photobiology 87 (4) (2011) 877–883.

[159] Y. Yuan, J. Ding, J. Xu, J. Deng, J. Guo, TiO_2 nanoparticles co-doped with silver and nitrogen for antibacterial application, Journal of Nanoscience and Nanotechnology 10 (8) (2010) 4868–4874.

[160] A. Wiesenthal, L. Hunter, S. Wang, J. Wickliffe, M. Wilkerson, Nanoparticles: small and mighty, International Journal of Dermatology 50 (3) (2011) 247–254.

[161] M. Montazer, A. Behzadnia, E. Pakdel, M.K. Rahimi, M.B. Moghadam, Photo induced silver on nano titanium dioxide as an enhanced antimicrobial agent for wool, Journal of Photochemistry and Photobiology B: Biology 103 (3) (2011) 207–214.

[162] R.M. Cornell, U. Schwertmann, The Iron Oxides: Structure, Properties, Reactions, Occurrences and Uses, John Wiley & Sons, 2003.

[163] L. Vayssieres, C. Sathe, S.M. Butorin, D.K. Shuh, J. Nordgren, J. Guo, One-dimensional quantum-confinement effect in α-Fe_2O_3 ultrafine nanorod arrays, Advanced Materials 17 (19) (2005) 2320–2323.

[164] N.S. Vallabani, S. Singh, Recent advances and future prospects of iron oxide nanoparticles in biomedicine and diagnostics, 3 Biotech 8 (6) (2018) 279.

[165] D.L. Huber, Synthesis, properties, and applications of iron nanoparticles, Small 1 (5) (2005) 482–501.

[166] W. Wu, Z. Wu, T. Yu, C. Jiang, W.-S. Kim, Recent progress on magnetic iron oxide nanoparticles: synthesis, surface functional strategies and biomedical applications, Science and Technology of Advanced Materials 16 (2) (2015) 023501.

[167] P. Sangaiya, R. Jayaprakash, A review on iron oxide nanoparticles and their biomedical applications, Journal of Superconductivity and Novel Magnetism 31 (11) (2018) 3397–3413.

[168] Y. Ju, H. Zhang, J. Yu, S. Tong, N. Tian, Z. Wang, X. Wang, X. Su, X. Chu, J. Lin, Monodisperse Au–Fe$_2$C Janus nanoparticles: an attractive multifunctional material for triple-modal imaging-guided tumor photothermal therapy, ACS Nano 11 (9) (2017) 9239–9248.

[169] J. Xie, K. Chen, J. Huang, S. Lee, J. Wang, J. Gao, X. Li, X. Chen, PET/NIRF/MRI triple functional iron oxide nanoparticles, Biomaterials 31 (11) (2010) 3016–3022.

[170] Z. Chen, J.-J. Yin, Y.-T. Zhou, Y. Zhang, L. Song, M. Song, S. Hu, N. Gu, Dual enzyme-like activities of iron oxide nanoparticles and their implication for diminishing cytotoxicity, ACS Nano 6 (5) (2012) 4001–4012.

[171] Q. Wu, X. Wang, C. Liao, Q. Wei, Q. Wang, Microgel coating of magnetic nanoparticles via bienzyme-mediated free-radical polymerization for colorimetric detection of glucose, Nanoscale 7 (40) (2015) 16578–16582.

[172] F. Yu, Y. Huang, A.J. Cole, V.C. Yang, The artificial peroxidase activity of magnetic iron oxide nanoparticles and its application to glucose detection, Biomaterials 30 (27) (2009) 4716–4722.

[173] J.K. Patra, M.S. Ali, I.-G. Oh, K.-H. Baek, Proteasome inhibitory, antioxidant, and synergistic antibacterial and anticandidal activity of green biosynthesized magnetic Fe$_3$O$_4$ nanoparticles using the aqueous extract of corn (*Zea mays* L.) ear leaves, Artificial Cells, Nanomedicine, and Biotechnology 45 (2) (2017) 349–356.

[174] P. Nehra, R. Chauhan, N. Garg, K. Verma, Antibacterial and antifungal activity of chitosan coated iron oxide nanoparticles, British Journal of Biomedical Science 75 (1) (2018) 13–18.

[175] Y. Ren, H. Zhang, B. Chen, J. Cheng, X. Cai, R. Liu, G. Xia, W. Wu, S. Wang, J. Ding, Multifunctional magnetic Fe$_3$O$_4$ nanoparticles combined with chemotherapy and hyperthermia to overcome multidrug resistance, International Journal of Nanomedicine 7 (2012) 2261.

[176] A.K. Gupta, M. Gupta, Synthesis and surface engineering of iron oxide nanoparticles for biomedical applications, Biomaterials 26 (18) (2005) 3995–4021.

[177] G. Ennas, A. Musinu, G. Piccaluga, D. Zedda, D. Gatteschi, C. Sangregorio, J. Stanger, G. Concas, G. Spano, Characterization of iron oxide nanoparticles in an Fe$_2$O$_3$– SiO2 composite prepared by a sol– gel method, Chemistry of Materials 10 (2) (1998) 495–502.

[178] D. Niznansky, J.L. Rehspringer, M. Drillon, Preparation of magnetic nanoparticles (/spl gamma/-Fe/sub 2/O/sub 3/) in the silica matrix, IEEE Transactions on Magnetics 30 (2) (1994) 821–823.

[179] M.J. Hampden-Smith, L.V. Interrante, Chemistry of Advanced Materials: An Overview, Wiley-VCH, 1998.

[180] T. Ghosh, S.K. Dash, P. Chakraborty, A. Guha, K. Kawaguchi, S. Roy, T. Chattopadhyay, D. Das, Preparation of antiferromagnetic Co 3 O 4 nanoparticles from two different precursors by pyrolytic method: in vitro antimicrobial activity, RSC Advances 4 (29) (2014) 15022–15029.

[181] S.G. Kwon, Y. Piao, J. Park, S. Angappane, Y. Jo, N.-M. Hwang, J.-G. Park, T. Hyeon, Kinetics of monodisperse iron oxide nanocrystal formation by "heating-up" process, Journal of the American Chemical Society 129 (41) (2007) 12571–12584.

[182] K. Byrappa, M. Haber, Hydrothermal Technology for Crystal Growth, Noyes Publications [Imprint], 2001.

[183] D. Chen, R. Xu, Hydrothermal synthesis and characterization of nanocrystalline Fe$_3$O$_4$ powders, Materials Research Bulletin 33 (7) (1998) 1015–1021.

[184] C. Han, D. Zhao, C. Deng, K. Hu, A facile hydrothermal synthesis of porous magnetite microspheres, Materials Letters 70 (2012) 70–72.

[185] X. Wu, L. Zheng, D. Wu, Fabrication of superhydrophobic surfaces from microstructured ZnO-based surfaces via a wet-chemical route, Langmuir 21 (7) (2005) 2665–2667.

[186] T. Sugimoto, Formation of monodispersed nano-and micro-particles controlled in size, shape, and internal structure, Chemical Engineering and Technology: Industrial Chemistry-Plant Equipment-Process Engineering-Biotechnology 26 (3) (2003) 313–321.

[187] H.-C. Schwarzer, W. Peukert, Tailoring particle size through nanoparticle precipitation, Chemical Engineering Communications 191 (4) (2004) 580–606.

[188] E. Malka, I. Perelshtein, A. Lipovsky, Y. Shalom, L. Naparstek, N. Perkas, T. Patick, R. Lubart, Y. Nitzan, E. Banin, Eradication of multi-drug resistant bacteria by a novel Zn-doped CuO nanocomposite, Small 9 (23) (2013) 4069–4076.

[189] K.S. Suslick, D.J. Flannigan, Inside a collapsing bubble: sonoluminescence and the conditions during cavitation, Annual Review of Physical Chemistry 59 (2008) 659–683.

[190] Y. Abboud, T. Saffaj, A. Chagraoui, A. El Bouari, K. Brouzi, O. Tanane, B. Ihssane, Biosynthesis, characterization and antimicrobial activity of copper oxide nanoparticles (CONPs) produced using brown alga extract (*Bifurcaria bifurcata*), Applied Nanoscience 4 (5) (2014) 571–576.

[191] M.T. Reetz, W. Helbig, Size-selective synthesis of nanostructured transition metal clusters, Journal of the American Chemical Society 116 (16) (1994) 7401–7402.

[192] P. Pandey, S. Merwyn, G. Agarwal, B. Tripathi, S. Pant, Electrochemical synthesis of multi-armed CuO nanoparticles and their remarkable bactericidal potential

against waterborne bacteria, Journal of Nanoparticle Research 14 (1) (2012) 709.

[193] A. Roy, J. Bhattacharya, Microwave-assisted synthesis and characterization of CaO nanoparticles, International Journal of Nanoscience 10 (03) (2011) 413—418.

[194] K. Gopinath, V. Karthika, C. Sundaravadivelan, S. Gowri, A. Arumugam, Mycogenesis of cerium oxide nanoparticles using Aspergillus Niger culture filtrate and their applications for antibacterial and larvicidal activities, Journal of Nanostructure in Chemistry 5 (3) (2015) 295—303.

[195] J. Vidal-Vidal, J. Rivas, M. López-Quintela, Synthesis of monodisperse maghemite nanoparticles by the microemulsion method, Colloids and Surfaces A: Physicochemical and Engineering Aspects 288 (1—3) (2006) 44—51.

[196] V. Uskoković, M. Drofenik, Synthesis of materials within reverse micelles, Surface Review and Letters 12 (02) (2005) 239—277.

[197] S. Navaladian, B. Viswanathan, R. Viswanath, T. Varadarajan, Thermal decomposition as route for silver nanoparticles, Nanoscale Research Letters 2 (1) (2007) 44.

[198] L. D'Souza, R. Richards, Synthesis of Metal-Oxide Nanoparticles: Liquid—Solid Transformations, Synthesis, Properties, and Applications of Oxide Nanomaterials, John Wiley & Sons, Inc., Hoboken, New Jersey, 2007, pp. 81—117.

[199] M. Fernández-García, J.A. Rodriguez, Metal Oxide Nanoparticles, Encyclopedia of Inorganic and Bioinorganic Chemistry, 2011.

[200] K.R. Raghupathi, R.T. Koodali, A.C. Manna, Size-dependent bacterial growth inhibition and mechanism of antibacterial activity of zinc oxide nanoparticles, Langmuir 27 (7) (2011) 4020—4028.

[201] J. Sawai, K. Himizu, O. Yamamoto, Kinetics of bacterial death by heated dolomite powder slurry, Soil Biology and Biochemistry 37 (8) (2005) 1484—1489.

[202] L. Zhang, Y. Jiang, Y. Ding, M. Povey, D. York, Investigation into the antibacterial behaviour of suspensions of ZnO nanoparticles (ZnO nanofluids), Journal of Nanoparticle Research 9 (3) (2007) 479—489.

[203] I. Mohammed Sadiq, N. Chandrasekaran, A. Mukherjee, Studies on effect of TiO2 nanoparticles on growth and membrane permeability of Escherichia coli, Pseudomonas aeruginosa, and Bacillus subtilis, Current Nanoscience 6 (4) (2010) 381—387.

[204] M. Premanathan, K. Karthikeyan, K. Jeyasubramanian, G. Manivannan, Selective toxicity of ZnO nanoparticles toward Gram-positive bacteria and cancer cells by apoptosis through lipid peroxidation, Nanomedicine: Nanotechnology, Biology and Medicine 7 (2) (2011) 184—192.

[205] A. Arumugam, C. Karthikeyan, A.S.H. Hameed, K. Gopinath, S. Gowri, V. Karthika, Synthesis of cerium oxide nanoparticles using Gloriosa superba L. leaf extract and their structural, optical and antibacterial properties, Materials Science and Engineering: C 49 (2015) 408—415.

[206] S.A. Ansari, M. Oves, R. Satar, A. Khan, S.I. Ahmad, M.A. Jafri, S.K. Zaidi, M.H. Alqahtani, Antibacterial activity of iron oxide nanoparticles synthesized by co-precipitation technology against Bacillus cereus and Klebsiella pneumoniae, Polish Journal of Chemical Technology 19 (4) (2017) 110—115.

[207] J. Vidic, S. Stankic, F. Haque, D. Ciric, R. Le Goffic, A. Vidy, J. Jupille, B. Delmas, Selective antibacterial effects of mixed ZnMgO nanoparticles, Journal of Nanoparticle Research 15 (5) (2013) 1595.

[208] M.M. Masadeh, G.A. Karasneh, M.A. Al-Akhras, B.A. Albiss, K.M. Aljarah, S.I. Al-Azzam, K.H. Alzoubi, Cerium oxide and iron oxide nanoparticles abolish the antibacterial activity of ciprofloxacin against gram positive and gram negative biofilm bacteria, Cytotechnology 67 (3) (2015) 427—435.

[209] Z. Emami-Karvani, P. Chehrazi, Antibacterial activity of ZnO nanoparticle on gram-positive and gram-negative bacteria, African Journal of Microbiology Research 5 (12) (2011) 1368—1373.

[210] K.M. Reddy, K. Feris, J. Bell, D.G. Wingett, C. Hanley, A. Punnoose, Selective toxicity of zinc oxide nanoparticles to prokaryotic and eukaryotic systems, Applied Physics Letters 90 (21) (2007) 213902.

[211] R. Mie, M.W. Samsudin, L.B. Din, A. Ahmad, N. Ibrahim, S.N.A. Adnan, Synthesis of silver nanoparticles with antibacterial activity using the lichen Parmotrema praesorediosum, International Journal of Nanomedicine 9 (2014) 121.

[212] T.N. Ravishankar, T. Ramakrishnappa, G. Nagaraju, H. Rajanaika, Synthesis and characterization of CeO2 nanoparticles via solution combustion method for photocatalytic and antibacterial activity studies, ChemistryOpen 4 (2) (2015) 146—154.

[213] J.F. Hernández-Sierra, F. Ruiz, D.C.C. Pena, F. Martínez-Gutiérrez, A.E. Martínez, A.d.J.P. Guillén, H. Tapia-Pérez, G.M. Castañón, The antimicrobial sensitivity of Streptococcus mutans to nanoparticles of silver, zinc oxide, and gold, Nanomedicine: Nanotechnology, Biology and Medicine 4 (3) (2008) 237—240.

[214] Y. Xie, Y. He, P.L. Irwin, T. Jin, X. Shi, Antibacterial activity and mechanism of action of zinc oxide nanoparticles against Campylobacter jejuni, Applied and Environmental Microbiology 77 (7) (2011) 2325—2331.

[215] N. Padmavathy, R. Vijayaraghavan, Enhanced bioactivity of ZnO nanoparticles—an antimicrobial study, Science and Technology of Advanced Materials 9 (3) (2008) 035004.

[216] K. Krishnamoorthy, M. Veerapandian, L.-H. Zhang, K. Yun, S.J. Kim, Surface chemistry of cerium oxide nanocubes: toxicity against pathogenic bacteria and their mechanistic study, Journal of Industrial and Engineering Chemistry 20 (5) (2014) 3513—3517.

[217] A. Azam, A.S. Ahmed, M. Oves, M. Khan, A. Memic, Size-dependent antimicrobial properties of CuO nanoparticles against Gram-positive and-negative bacterial strains, International Journal of Nanomedicine 7 (2012) 3527.

[218] B. Ramazanzadeh, A. Jahanbin, M. Yaghoubi, N. Shahtahmassbi, K. Ghazvini, M. Shakeri, H. Shafaee, Comparison of antibacterial effects of ZnO and CuO nanoparticles coated brackets against Streptococcus mutans, Journal of Dentistry 16 (3) (2015) 200.

[219] A.S. Roy, A. Parveen, A.R. Koppalkar, M.A. Prasad, Effect of nano-titanium dioxide with different antibiotics against methicillin-resistant *Staphylococcus aureus*, Journal of Biomaterials and Nanobiotechnology 1 (1) (2010) 37.

[220] M. Ahamed, H.A. Alhadlaq, M. Khan, P. Karuppiah, N.A. Al-Dhabi, Synthesis, characterization, and antimicrobial activity of copper oxide nanoparticles, Journal of Nanomaterials 2014 (2014) 17.

[221] C. Avanzato, J. Follieri, I. Banerjee, K. Fath, Biomimetic synthesis and antibacterial characteristics of magnesium oxide—germanium dioxide nanocomposite powders, Journal of Composite Materials 43 (8) (2009) 897—910.

[222] S. Makhluf, R. Dror, Y. Nitzan, Y. Abramovich, R. Jelinek, A. Gedanken, Microwave-assisted synthesis of nanocrystalline MgO and its use as a bacteriocide, Advanced Functional Materials 15 (10) (2005) 1708—1715.

[223] F. Ahrari, N. Eslami, O. Rajabi, K. Ghazvini, S. Barati, The antimicrobial sensitivity of Streptococcus mutans and Streptococcus sangius to colloidal solutions of different nanoparticles applied as mouthwashes, Dental Research Journal 12 (1) (2015) 44.

[224] S.N. Patil, J.S. Paradeshi, P.B. Chaudhari, S.J. Mishra, B.L. Chaudhari, Bio-therapeutic potential and cytotoxicity assessment of pectin-mediated synthesized nanostructured cerium oxide, Applied Biochemistry and Biotechnology 180 (4) (2016) 638—654.

[225] S.T. Khan, M. Ahamed, J. Musarrat, A.A. Al-Khedhairy, Anti-biofilm and antibacterial activities of zinc oxide nanoparticles against the oral opportunistic pathogens Rothia dentocariosa and Rothia mucilaginosa, European Journal of Oral Sciences 122 (6) (2014) 397—403.

[226] M.A. Vargas-Reus, K. Memarzadeh, J. Huang, G.G. Ren, R.P. Allaker, Antimicrobial activity of nanoparticulate metal oxides against peri-implantitis pathogens, International Journal of Antimicrobial Agents 40 (2) (2012) 135—139.

[227] T. Buerki-Thurnherr, L. Xiao, L. Diener, O. Arslan, C. Hirsch, X. Maeder-Althaus, K. Grieder, B. Wampfler, S. Mathur, P. Wick, In vitro mechanistic study towards a better understanding of ZnO nanoparticle toxicity, Nanotoxicology 7 (4) (2013) 402—416.

[228] D.L. De Romana, K. Brown, J.X. Guinard, Sensory trial to assess the acceptability of zinc fortificants added to iron-fortified wheat products, Journal of Food Science 67 (1) (2002) 461—465.

[229] P. Patnaik, Handbook of Inorganic Chemicals, McGraw-Hill, 2002.

[230] S. Nair, A. Sasidharan, V.D. Rani, D. Menon, S. Nair, K. Manzoor, S. Raina, Role of size scale of ZnO nanoparticles and microparticles on toxicity toward bacteria and osteoblast cancer cells, Journal of Materials Science: Materials in Medicine 20 (1) (2009) 235.

[231] J. Hwang, E. Lee, J. Kim, Y. Seo, K.H. Lee, J.W. Hong, A.A. Gilad, H. Park, J. Choi, Effective delivery of immunosuppressive drug molecules by silica coated iron oxide nanoparticles, Colloids and Surfaces B: Biointerfaces 142 (2016) 290—296.

[232] M.E. El Zowalaty, S.H.H. Al Ali, M.I. Husseiny, B.M. Geilich, T.J. Webster, M.Z. Hussein, The ability of streptomycin-loaded chitosan-coated magnetic nanocomposites to possess antimicrobial and antituberculosis activities, International Journal of Nanomedicine 10 (2015) 3269.

[233] O. Ivashchenko, M. Lewandowski, B. Peplińska, M. Jarek, G. Nowaczyk, M. Wiesner, K. Załeski, T. Babutina, A. Warowicka, S. Jurga, Synthesis and characterization of magnetite/silver/antibiotic nanocomposites for targeted antimicrobial therapy, Materials Science and Engineering: C 55 (2015) 343—359.

[234] C.M. Istrate, A.M. Holban, A.M. Grumezescu, L. Mogoantă, G. Mogoşanu, T. Savopol, M. Moisescu, M. Iordache, B.S. Vasile, E. Kovacs, Iron oxide nanoparticles modulate the interaction of different antibiotics with cellular membranes, Romanian Journal of Morphology and Embryology 55 (3) (2014) 849—856.

[235] D.W. Hatchett, H.S. White, Electrochemistry of sulfur adlayers on the low-index faces of silver, The Journal of Physical Chemistry 100 (23) (1996) 9854—9859.

[236] I. Lynch, K.A. Dawson, Protein-nanoparticle interactions, Nano Today 3 (1—2) (2008) 40—47.

[237] P. Aggarwal, J.B. Hall, C.B. McLeland, M.A. Dobrovolskaia, S.E. McNeil, Nanoparticle interaction with plasma proteins as it relates to particle biodistribution, biocompatibility and therapeutic efficacy, Advanced Drug Delivery Reviews 61 (6) (2009) 428—437.

[238] L. Macomber, J.A. Imlay, The iron-sulfur clusters of dehydratases are primary intracellular targets of copper toxicity, Proceedings of the National Academy of Sciences 106 (20) (2009) 8344—8349.

[239] I.L. Calderón, A.O. Elías, E.L. Fuentes, G.A. Pradenas, M.E. Castro, F.A. Arenas, J.M. Pérez, C.C. Vasquez, Tellurite-mediated disabling of [4Fe—4S] clusters of *Escherichia coli* dehydratases, Microbiology 155 (6) (2009) 1840—1846.

[240] T.J. Beveridge, Structures of gram-negative cell walls and their derived membrane vesicles, Journal of Bacteriology 181 (16) (1999) 4725—4733.

[241] Y.-C. Chung, Y.P. Su, C.-C. Chen, G. Jia, H.L. Wang, J.G. Wu, J.G. Lin, Relationship between antibacterial activity of chitosan and surface characteristics of cell wall, Acta Pharmacologica Sinica 25 (7) (2004) 932—936.

[242] A. Gaballa, J.D. Helmann, Identification of a zinc-specific metalloregulatory protein, zur, controlling zinc transport operons in *Bacillus subtilis*, Journal of Bacteriology 180 (22) (1998) 5815—5821.

[243] D. Wang, Z. Lin, T. Wang, Z. Yao, M. Qin, S. Zheng, W. Lu, Where does the toxicity of metal oxide nanoparticles come from: the nanoparticles, the ions, or a combination of both? Journal of Hazardous Materials 308 (2016) 328—334.

[244] S. Rezaei-Zarchi, A. Javed, M. Javeed Ghani, S. Soufian, F. Barzegari Firouzabadi, A. Bayanduri Moghaddam, S.H. Mirjalili, Comparative study of antimicrobial activities of TiO2 and CdO nanoparticles against the pathogenic strain of *Escherichia coli*, Iranian Journal of Pathology 5 (2) (2010) 83–89.

[245] Y.H. Kim, D.K. Lee, H.G. Cha, C.W. Kim, Y.C. Kang, Y.S. Kang, Preparation and characterization of the antibacterial Cu nanoparticle formed on the surface of SiO2 nanoparticles, The Journal of Physical Chemistry B 110 (49) (2006) 24923–24928.

[246] A.B. Smetana, K.J. Klabunde, G.R. Marchin, C.M. Sorensen, Biocidal activity of nanocrystalline silver powders and particles, Langmuir 24 (14) (2008) 7457–7464.

[247] H. Chen, L. Wang, Nanostructure sensitization of transition metal oxides for visible-light photocatalysis, Beilstein Journal of Nanotechnology 5 (1) (2014) 696–710.

[248] W.-J. Chen, Y.-C. Chen, Fe$_3$O$_4$/TiO$_2$ core/shell magnetic nanoparticle-based photokilling of pathogenic bacteria, Nanomedicine 5 (10) (2010) 1585–1593.

[249] A. Nel, T. Xia, L. Mädler, N. Li, Toxic potential of materials at the nanolevel, Science 311 (5761) (2006) 622–627.

[250] V.V.T. Padil, M. Černík, Green synthesis of copper oxide nanoparticles using gum karaya as a biotemplate and their antibacterial application, International Journal of Nanomedicine 8 (2013) 889.

[251] M.J. Hajipour, K.M. Fromm, A.A. Ashkarran, D.J. de Aberasturi, I.R. de Larramendi, T. Rojo, V. Serpooshan, W.J. Parak, M. Mahmoudi, Antibacterial properties of nanoparticles, Trends in Biotechnology 30 (10) (2012) 499–511.

[252] T. Saito, T. Iwase, J. Horie, T. Morioka, Mode of photocatalytic bactericidal action of powdered semiconductor TiO$_2$ on mutans streptococci, Journal of Photochemistry and Photobiology B: Biology 14 (4) (1992) 369–379.

[253] K. Giannousi, K. Lafazanis, J. Arvanitidis, A. Pantazaki, C. Dendrinou-Samara, Hydrothermal synthesis of copper based nanoparticles: antimicrobial screening and interaction with DNA, Journal of Inorganic Biochemistry 133 (2014) 24–32.

[254] J. Kirstein, K. Turgay, A new tyrosine phosphorylation mechanism involved in signal transduction in Bacillus subtilis, Journal of Molecular Microbiology and Biotechnology 9 (3–4) (2005) 182–188.

[255] S. Shrivastava, T. Bera, A. Roy, G. Singh, P. Ramachandrarao, D. Dash, Characterization of enhanced antibacterial effects of novel silver nanoparticles, Nanotechnology 18 (22) (2007) 225103.

[256] E. Giamello, M.C. Paganini, D.M. Murphy, A.M. Ferrari, G. Pacchioni, A combined EPR and quantum chemical approach to the structure of surface Fs+ (H) centers on MgO, The Journal of Physical Chemistry B 101 (6) (1997) 971–982.

[257] A.M. Ferrari, G. Pacchioni, Metal deposition on oxide surfaces: a quantum-chemical study of the interaction of Rb, Pd, and Ag atoms with the surface vacancies of MgO, The Journal of Physical Chemistry 100 (21) (1996) 9032–9037.

CHAPTER 16

Metal Oxide–Based Nanocomposites as Antimicrobial and Biomedical Agents

SHESAN JOHN OWONUBI • NYEMAGA MASANJE MALIMA • NEERISH REVAPRASADU

1 INTRODUCTION

The spread of infectious diseases, the advent of multidrug-resistant microbes, and antibiotic-resistant diseases are threats to the global population, subject of considerable interest, and tremendous scientific challenge to the pharmaceutical and biomedical industry. The magnitude of antibiotic resistance may stir concern about the outgrowth and recurrence of microbial infections that are resistant to more than one drug [1]. It is noteworthy to mention that, in most cases, people infected with multidrug-resistant microbes do not recover easily as they are mostly confined to hospitals requiring several treatments with different types of antibiotics that are more expensive, less effective, and toxic [2]. Hence, great demand exists to develop or modify antimicrobial agents to improve their performance in this era of advanced science and technology [3]. Against the backdrop of many technologies available for the development or modification of antimicrobial agents, nanotechnology is highly attractive as it affords researchers of today a worthy platform to maneuver the vital properties of materials, for example, metal oxides, by transforming them into nanoparticles or related nanocomposites. The materials at nanoscale possess promising practical applications, and this has been proven to be useful in different treatment modalities including cell labeling diagnostics, biomarkers, targeted drug-delivery systems, nanodrugs for treatment of various diseases, antimicrobial agents, and contrast agents for biological imaging [4,5], tumor detection, and prognostic visual monitoring of therapy [6,7].

The search for sustainable methods to modernize medicine by the application of nanotechnology and the increase in demand for nanotechnology-derived products for improving human health has prompted researchers in the health industry to explore nanomaterials with superior flexibility, strength, performance, durability, and in some cases unique physicochemical properties. Recently, interest in the design of antimicrobial metal oxide–based nanocomposites (MOBNCs) has been on the rise. This is because the incorporation of metal oxide nanoparticles in composites has shown to yield advantages of being higher in performance, an unusual combination of properties as well as unique design possibilities [8]. MOBNCs have been shown to be beneficial in biomedicine; in gene therapy; DNA sequencing; diagnosis and screening of diseases; for delivery of drugs and imaging; tissue culturing and in cancer treatment [9,10].

The major highlight of this chapter is to present a general idea on MOBNCs, their synthetic and characterization protocols as well as their antimicrobial potential. The development of consumer and healthcare products that possess antimicrobial properties can be achieved through the fabrication of products that encompass these metal oxide nanocomposites with potential to inactivate microbes. Finally, we present the antimicrobial and biomedical applications of these MOBNCs.

2 METAL OXIDES

Inorganic metal oxides are among the many groups of nanomaterials that have been proven to be of great interest both scientifically and technologically [11]. Metal oxides can be insulators, semiconductors, or metallic, dependent on the electronic structure nature. The transformation of metal oxides from bulk into nanometre size significantly improves their hybrid properties compared to the original forms. Their peculiar features permit them to become multifaceted materials, possessing distinct electrical, catalytic, photoelectronic, optical, and magnetic properties covering practically all facets of materials science and solid-state physics, thus finding applications in superconductors, catalysis, gas sensing,

Antibiotic Materials in Healthcare. https://doi.org/10.1016/B978-0-12-820054-4.00016-1

electroceramics, and energy conversions [12]. Over the years, a number of metal oxides, zinc oxide (ZnO), titanium (IV) oxide (TiO_2), nickel oxide (NiO), copper (I) oxide (Cu_2O), copper (II) oxide (CuO), manganese (IV) oxide (MnO_2), Iron (III) oxide (Fe_2O_3), tungsten trioxide (WO_3), and cerium oxide (CeO_2) have been used in the architecture of nanomaterials and nanocomposites [8,13–15].

3 NANOCOMPOSITES

Nanocomposites refer to solid materials with two or more phase domains having distinguishable physical and chemical properties wherein at least one of these domains has a nanoscale structure that ranges in sizes from 1 to 100 nm [16,17]. The physicochemical properties of these nanomaterials are dependent on the interfacial characteristics and morphology of the constituent materials. In contrast to the individual materials, composites have extraordinarily good practical applications because of their improved toughness, high thermal insulation, low density, corrosion to resistance, stiffness, and strength [18]. The design of nanocomposites was encouraged by the quest for composite materials with enhanced properties and having filler sizes that are smaller.

Recently, the development of nanocomposites has received considerable attention making them important materials, both to academia and industry because they normally inherit the behavior of the constituent materials, and in some cases even yield multifunctional materials with surprisingly enhanced characteristics. Nanocomposites can be synthesized from a combination of three main precursor materials that are classified into metals, ceramics, and polymers. The nanocomposite could differ in properties from its parent component materials in optical, thermal, catalytic, electrical, electrochemical, and mechanical properties [16,17,19–21].

Nanocomposite materials can be in the form of mixed oxides wherein there is a solid solution made from mixed oxides, examples of such are titania (TiO_2)/tin oxide (SnO_2) nanocomposites; mixed oxides that form distinctive chemical compounds like zinc cobaltite ($ZnCo_2O_4$) and zinc stannate ($ZnSn_2O_4$); systems that do form solid solutions such as TiO_2/WO_3 nanocomposites; and those that form core/shell nanocomposite structures [22].

4 METAL OXIDE–BASED NANOCOMPOSITES

Some of the commonly reported metal oxide nanoparticles utilized in the fabrication of nanocomposites for biomedical applications overtime are magnetite (Fe_3O_4), hematite (Fe_2O_3), titania (TiO_2), alumina (Al_2O_3), and zirconia (ZrO_2) [23]. Metal oxide nanoparticles have demonstrated promising physical properties, which have encouraged antimicrobial, magnetic, and electrical conductivity properties [24]. These have led to their application as sensors, actuators, switchable electronic materials, scaffolds, drug-delivery agents, and imaging agents [23]. Of the numerous kinds, this chapter only focuses on three classes of MOBNCs, namely mixed metal oxide, metal–metal oxide, and polymer–metal oxide nanocomposites (Fig. 16.1).

4.1 Mixed Metal Oxide Nanocomposites

Interest in mixed metal oxides' applicability in diverse areas of research and technology has been on the rise. The combination of two or more different metals oxides can produce composite materials having unique properties (physical or chemical), which could lead to performances that are superior for various applications [25]. For a composite to be called as a "nanocomposite material," it should be noted that at least one of the materials should possess nanosized dimensions. Nanocomposites with two phases consisting of two metal oxides that coexist within the same matrix have been fabricated for varied applications including antibacterial activity and photocatalysis [26–30]. Mixed metal oxide nanocomposites such as TiO_2/WO_3 [29], ZnO/MgO [31] Co_3O_4/ZnO [32] TiO_2/Fe_2O_3 [33] TiO_2/Al_2O_3 [34,35] TiO_2/Fe_2TiO_5 [36], NiO/Co_3O_4 [37], Fe_2O_3/Co_3O_4 [37], MgO/Al_2O_3 [38], ZnO/SnO_2 [39], ZnO/CuO [40], TiO_2/MgO [41], MgO/CuO [42], CeO_2-MnOx [28], ZnO/Fe_3O_4 [43] $NdAlO_3$/Al_2O_3 [44], CuO/ZnO [45], Al_2O_3/ZrO_2 [46], and ZnO/NiO [30,47] have been synthesized and examined by employing various techniques. The results obtained from these studies showed that nanocomposites in comparison to single oxides possess higher photocarrier separation efficacy. Although the focus of their experiments was centered on oxides possessing two phases, those possessing double heterojunctions existing within more than two phases have not been explored. A few studies have been conducted on metal oxide nanocomposites with more than two phases such as ZnO/$ZnWO_3$/WO_3 [26], NiO/CeO_2/ZnO, CuO/NiO/ZnO [25], Mo/V/W mixed-oxide [48], oxide solid solution of La/Mo/V [49], pseudoquaternary system CaO/CoO/SiO_2 [50] and ZnO/CeO_2/TiO_2 [51], which have all resulted in materials with improved properties. For example, ZnO/$ZnWO_4$/WO_3 composites were shown to possess improved photocatalytic activities as a result of the existence of double heterojunctions by Wang et al. [26].

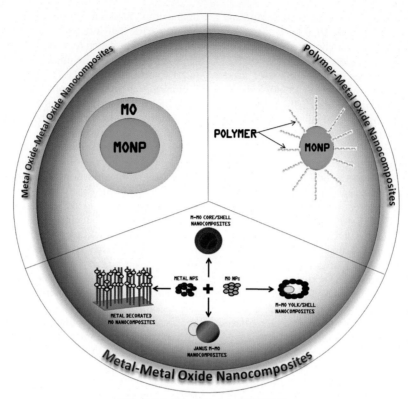

FIG. 16.1 Classification of MOBNCs.

4.2 Metal—Metal Oxide Nanocomposites

This class of nanocomposites is designed to comprise both a metal and a metal oxide, and they can be further categorized based on the geometric configuration of the nanohybrids into metal-decorated metal oxide nanoarrays; metal/metal oxide yolk/shell nanostructures; Janus noble metal—metal oxide nanostructures, and metal/metal oxide core/shell nanostructures.

Metal-decorated metal oxide nanoarrays are formed by loading noble metals on metal oxide nanoarrays. Here, different noble metals are loaded on different types of metal oxide nanoarrays [52—58] such as nanorods [59], nanotubes [60], nanowires [58,61,62], and nanosheets [63]. By employing photoreduction [59,64] and chemical reduction methods [59], noble metals are attached to the metal oxide nanoarrays. Photoreduction and chemical reduction techniques allow tuning of the level of metal coverage on the surface of metal oxide nanoarrays, which can be achieved by regulating the immersion time or by monitoring the amount of noble metal. Yu et al. recounted the synthesis of Au—Pd comodified TiO_2 nanotube films by photoreduction, highlighting the performance of Au—Pd—TiO_2 on the malathion elimination was higher

in this comodified form in comparison to that of the undecorated TiO_2 nanotube [65].

Metal—metal oxide yolk/shell nanostructures have a metal core that is solid and a hollow metal shell, with the core freely movable within the shell. Metal cores can be encapsulated easily within metal oxide shells resulting in the formation of yolk/shell nanohybrids by etching part of a metal core of preformed co/shell nanoparticles selectively [66—68]. When functional metal oxides like Fe_2O_3, ZnO, and NiO are used as shells, they contribute additional functionalities and properties to the entire architecture of the nanostructure [69,70]. However, due to corrosion of some metal oxides during the process of etching, it is more challenging to use them in preparation of yolk/shell nanostructures. Instead, their synthesis has successfully been achieved by incorporating SiO_2 as a sacrificial template. A number of yolk/shell nanomaterials viz: Au/ZrO_2 [71], Au/SnO_2 [72], and Au@r-GO/TiO_2 yolk/shell [73] have been synthesized by using the aforementioned technique. Researchers have also identified that, in addition to using SiO_2, a hard template such as carbon can be employed in fabricating metal/TiO_2 yolk/shell nanomaterials [74,75]. Some examples of nanostructures prepared using carbon as a template include Au/ZnO yolk/shell [76] and Au/NiO/yolk/shell [77].

Janus nanoparticles refer to characteristic types of materials whose surfaces possess two or more distinctive properties [78,79]. The unique surface of these nanoparticles encourages the particle to possess two different types of chemistry. Although there has been substantial effort put into the synthesis of different types of Janus nanomaterials considering this concept was described in a real-scientific context, only a few studies about Janus metal/metal oxide nanostructures are available [80−84]. Recently, Seh et al. [85] developed an effective method to synthesize Janus Au−TiO$_2$ nanorods, this was achieved by firstly preparing Au nanorods using seed-mediated growth, followed by growing TiO$_2$ on the surface of the nanorods anisotropically. In a similar feat, Pradhan and his coworkers achieved the fabrication of Janus Au−TiO$_2$ by first decorating the two surfaces of the particle with both hydrophilic and hydrophobic ligands and finally loading TiO$_2$ by surface sol−gel method on one side of the Au nanoparticles before attachment onto the surfaces [86]. Janus Au−ZnO has been successfully synthesized by Li et al. [87], wherein they employed zinc acetate dihydrate and Au nanocrystals as precursor materials. By the one-pot thermal exposure of a mixture of zinc precursors and Au, the preparation of Janus Au−ZnO hybrid nanopyramids, nanoflowers, and nanomultipods have been successfully designed and reported by Chen and coworkers [82]. Additionally, Wang et al. reportedly fabricated Janus Ag−Fe$_3$O$_4$, Au−Fe$_3$O$_4$, and AuAg−Fe$_3$O$_4$ nanohybrids using the widely known method of growing metal oxides over presynthesized noble metal seeds by pyrolysis of a metal carbonyl followed by oxidation in the air [88].

Metal−metal oxide core−shell nanostructures are formed when metal nanoparticles are enclosed using an apt method within a semiconductor shell so that deterioration and aggregation of metal nanoparticles caused by their contact with harsh chemicals can be reduced [13]. Several researchers have described the synthesis of Au/TiO$_2$ core−shell structures [89−94] and Pd/CeO$_2$ core−shell nanostructures [95−99].

4.3 Polymer−Metal Oxide Nanocomposite

The incorporation of metal oxide nanoparticles into polymers results in nanocomposites with tunable properties. These properties can be modified by appropriate combination with a choice of metal oxide nanoparticles, as a result of the high surface-to-volume ratio of the metal oxide nanoparticles. The properties change could be as a result of dispersion state, geometric shape and size of the nanoparticles. In most cases, there are weak

interactions between the polymer and metal-oxide nanoparticles that interfere with the properties of the composites, but dependent on the desire of researchers, these interactions can be tuned by modifying the nanoparticle surfaces [100,101]. This tunability can affect the biological, chemical or physical properties of the resulting nanocomposite significantly. Generally, metal oxide-polymer nanocomposites are prepared by blending or direct mixing of the metal oxide nanoparticles and the polymer. For example, in the presence of metal oxide nanoparticles, employing sol−gel processes and in situ polymerization of monomers [8,102]. This beneficial symbiosis existing between metal oxides and polymer, employed to design nanocomposites (metal oxide-polymer nanocomposites), have fascinated researchers worldwide and, as a result, they have been reportedly employed over decades for wide-ranging purposes. Demir et al. reportedly prepared PMAA/ZnO and PMAA/TiO$_2$ nanocomposites using an in situ polymerization process and identified that the particle dispersion homogeneity was heightened by the in situ copolymerization of methyl methacrylate (MMA) with a difunctional monomer [103]. One-step synthesis of polystyrene (PS)/TiO$_2$ nanocomposite was achieved by Wu et al. [104] using the miniemulsion polymerization technique. It was recognized that the development of the PS/TiO$_2$ nanocomposite was facilitated by the electrostatic interaction between Ti−OH and cetyl trimethyl ammonium bromide, and thereafter the TiO$_2$ particles were coated onto the PS to form the nanocomposite. Other metal oxide-polymer nanocomposites have been synthesized for use in various sectors of life such as poly(butylene succinate)/TiO$_2$, that was used in the decomposition of organic compounds [105], Chitosan/Fe$_3$O$_4$ that was used as a potentiometric urea biosensor [106], PVA/CuO used as nanofluid in heat transfer management system [107], PEG/iron oxide used in biomedicine as nanocarrier for doxorubicin delivery [108], Chitosan/ZnO used for dye degradation and as an antibacterial agent [109], PLA/TiO$_2$ used for biorecognition of anticancer drug daunorubicin [110], etc.

5 SYNTHESIS STRATEGIES OF METAL OXIDE−BASED NANOMATERIALS AND NANOCOMPOSITES

The fabrication of metal oxide-based nanomaterials and their nanocomposites with specific morphology and composition is one of the most tasking processes in their design. Owing to promising properties and potentials exhibited by synthesized metal oxide nanomaterials and their nanocomposites, huge interest in the synthesis strategies that ultimately define the properties of these nanomaterials or nanocomposites in making effective devices for numerous applications exist. Therefore, the judicious choice of synthetic method for the fabrication of these nanoscale materials is very crucial toward obtaining materials with good biological, optical, magnetic, and electrical properties that are dependent on their dimensions and size [111]. The choice of synthetic method is reliant on the targeted properties of the resultant nanoparticles/composites as the method of synthesis has been identified to influence the crystal structure, morphology, sizes, etc. [112]. The common synthesis methods for these nanomaterials are often grouped into three categories, namely gas phase, vapor-based, and solution-based synthesis methods [112]. Alternatively, the division of these synthesis approaches falls into two broad classes; one that encompasses physical methods called the top-down approach and another that includes wet methods called the bottom-up approach. It is imperative to mention that although the synthesis of monodispersed nanocomposite particles is a vital challenge of the physical method, the method is advantageous because it encourages large production of the synthesized nanocomposites. Wet chemical methods, on the other hand, encourage size uniformity of the nanocomposites easily, although varied shapes (e.g., nanorods, nanowires, nanotubes, etc.) can be achieved by altering the conditions of the reaction [113]. Despite several ways for the synthesis of metal oxide nanomaterials and nanocomposites exist; this chapter only covers physical vapor deposition, chemical vapor deposition, and the chemical methods (Fig. 16.2).

5.1 Physical Vapor Deposition

Physical vapor deposition (PVD) is a synthesis approach occurring in a closed chamber, wherein the source material is evaporated above its melting point allowing for evaporated particles to move onwards and get deposited on a substrate as a result of the free reaction path generated by a vacuum. By using this approach, nanomaterials can be deposited easily on the surfaces of metal oxide nanoarrays. The versatility of PVD was demonstrated by the deposition of Cu on zinc oxide to form Cu/ZnO composite system, which was then dispersed in ionic liquid for stabilization [114]. The huge demand for the detection and identification of explosives, narcotics, and other dangerous substances has compelled researchers to use PVD in synthesizing composites and thin films based on depositing oxides of Sn, Fe, and Ni on porous silicon substrates [115]. Bakhsheshi-Rad et al. synthesized a bilayer nano-TiO_2/fluoride doped hydroxyapatite composite coating on Mg—Zn—Ce alloy using PVD followed by electrochemical deposition method [116]. In this nanostructured assembly, the inner layer is TiO_2 while the top layer consists of FHA. An improvement in the corrosion resistance of the Mg alloy was noted after coating by TiO_2/FHA composite. Other researchers have reportedly employed this technique to formulate novel Pd/TiO_2 [117], Cu/ZnO nanocomposites [114], and ZnO nanowires [118].

5.1.1 Thermal evaporation

Thermal evaporation (TE) is among the simplest and oldest synthetic strategy used to fabricate metal oxides nanomaterials and their nanocomposites [119]. TE occurs in the presence of high temperature in a thermal furnace, which is essential for the source materials to be vaporized to facilitate the deposition of the nanomaterials at the nanoscale [120]. TE gives room for the control of important parameters such as optical, electrical, adhesion strength, gain structure, thickness, and

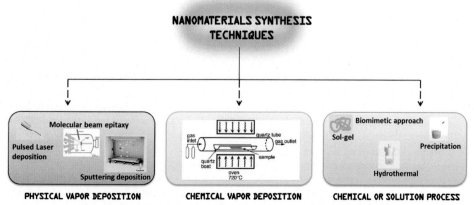

FIG. 16.2 Strategies of metal oxide based nanomaterials synthesis.

uniformity. [121]. For smooth operation using the TE technique, other components provide the capacity for more precise control of uniformity, components such as the substrates, process monitoring, auxiliary for vacuum pressure and source containers [122]. A simple thermal evaporation method was reportedly used to synthesize ZnO nanostructures on a cylindrical shape substrate [123,124]. The synthesized structures were considered good models for an examination of the size-dependent physicochemical properties of the materials. In another study, Chhikara and coworkers reported the fabrication of ZnO/MgO nanocomposite by the TE method [125]. The method was considered cost-effective and resulted in the formation of nanocomposites with good quality. It was also observed to be temperature-dependent regardless of the stoichiometric ratio of Zn and Mg precursors. Another benefit of TE is the ability to control the composition of the grown composite by regulating the amount of the precursors and/or by taking advantage of the melting point of the precursor materials [126].

5.1.2 Pulsed laser deposition

This method employs a narrow frequency bandwidth and high-power density laser as its source for vaporizing the desired material. The method is particularly employed when other synthetic techniques have failed to deposit the materials [127–129], and it has been successfully used in the synthesis of quantum dots [130], nanopowders [131], and nanotubes [132]. A matrix assisted-pulsed laser deposition (PLD) is commonly applied for polymers, biomaterials, coordinative and complex compounds, as well as hybrid metal organics [133]. The fabrication of titanium (TiO_2) nanoparticle-based films for gas sensing application was reported to have been achieved by PLD [134], and the sizes of the nanoparticles were found to be 10 nm in diameter [135], which encouraged the desired application. The synthesis of ZnO/MgZnO nanocomposites by PLD and the study on their spatial fluctuations of optical emissions have been investigated by Czekalla and colleagues [136]. Nanocomposite films of $Cu–Al_2O_3$ have been reportedly formed by embedding Cu nanoparticles in an amorphous Al_2O_3 using the PLD method [137]. During this process, it was observed that the nucleation and growth of the nanocomposite were dominated by processes occurring at the substrate surface rather than in the gas phase. The method has also been used to fabricate Pt/TiO_2 system from the sintered mixture targets of Pt and TiO_2 [138]. The atomic ratios between Pt and Ti present in the deposited films were found strongly dependent on the starting amounts of Pt in the target than on the influence of the laser. Furthermore, the study revealed that the energy levels formed by the interface between Pt and TiO_2 had an influence on the photoelectrochemical properties of Pt/TiO_2 nanocomposite electrodes. Aimon et al. demonstrated the synthesis of $BiFeO_3$/$CoFe_2O_4$ nanocomposites grown on $SrTiO_3$ by PLD by employing a combinatorial method through the ablation of $Bi_{1.2}FeO_3$ and $CoFe_2O_4$ targets [139].

5.1.3 Sputtering deposition

Sputtering deposition (SD) is a method that involves the ejection of metal atoms due to the bombardment of the solid target material by energetic particles [13]. To achieve this, the solid material is vaporized by a beam of inert gas ions through sputtering and the reaction occurs in a vacuum chamber by argon plasma. As sputtered atoms expelled to the gas phase are thermodynamically not in an equilibrium state, they can simply be deposited onto the substrate [13]. By the use of this synthetic approach, various metal oxide nanomaterials have been reportedly formed by metal targets magnetron sputtering [140], in which collimated beams of nanoparticles were formed and made to deposit nanostructured films on silicon substrates. The SD technique has been extensively utilized for the synthesis of a range of nanostructures like B, Si, W, ZnO, and carbon nanotubes [141–143]. By this technique, zinc oxide films (n-type and p-type) have been successfully fabricated by using plasma-assisted sputtering [144].

5.1.4 Molecular beam epitaxy

Molecular beam epitaxy (MBE) is used in fabricating insulators, metals, semiconductors, as well as metal oxide nanomaterials, as a result of its unique capability to regulate the monolayer level layering and also because it is compatible with surface-science techniques that encourage the ability to monitor the nanomaterial growth process. MBE is a vacuum deposition technique wherein thermal beam of atoms or molecules that are well defined react at a crystalline surface producing an epitaxial film that can be characterized in situ during growth. Epitaxial technologies have been employed to grow the oxides of superconductors despite higher technological costs, oxides like $YBa_2Cu_3O_{7-\delta}$ [145], Pr_2CuO_4, Ce_2CuO_4 [146], and $Bi_2Sr_2Cu_3Ca_{n-1}Cu_nO_{2n+4}$ for $n = 1–11$ [147,148]. This method has also been employed to grow ferroelectrics, for example, $BaTiO_3$ [149,150] and the incipient ferroelectric $SrTiO_3$ [151], and the ferrimagnet Fe_2O_3 [152].

To achieve specific intrinsic properties of sensitive materials, MBE has become adopted as the method of choice. An example is the achieving of the preparation of $EuTiO_3$ in as-grown film in its ground state, as it is known that $EuTiO_3$ is an antiferromagnet material on the brink of becoming ferromagnetic [153].

5.2 Chemical Vapor Deposition

Chemical vapor deposition (CVD) is an approach that entails the passing of precursor materials or gases toward heated substrates, encouraging chemical reaction to take place toward the surface of the substrate resulting in the formation of a solid deposit. The products formed in the chemical reaction are released through the exhaust chamber. Diverse energy sources such as hot filaments, plasmas, ions, electrons, and photons lasers are used in the reaction at a range of temperatures between 200 and 1200°C. Factors such as nucleation, growing structure, and crystallinity affect the physical properties of a deposited film [154]. The benefits of this method include better control of the reaction conditions, high flexibility, good conformal step coverage, high growth rate, and adaptability to large scale processing [155,156]. Notably, liquid injection-CVD and aerosol assisted-CVD are versatile techniques for the synthesis of MOBNCs and can, therefore, be adapted to the design of different metal-oxide nanostructures on various kinds of substrates.

ZnO-based nanocomposites are among the most studied metal-oxide nanostructures prepared by CVD-based approaches. In this view, Becker et al. performed the synthesis of Cu/ZnO nanocomposite by the CVD method in two steps, using Cu(II) diethylamino-2-propoxide and bis[bis(trimethylsilyl)amido]zinc as starting materials [157]. They also used the CVD approach to achieve the fabrication of Cu/Al_2O_3 and $Cu/ZnO/Al_2O_3$. Barreca et al. [158] achieved the synthesis of $ZnO-TiO_2$ nanocomposite using chemical vapor deposition by the initial growth of ZnO nanoplatelets and the subsequent dispersion of TiO_2 nanoparticles. An investigation of the gas sensing performances of the synthesized nanocomposite revealed better performances than the pristine ZnO systems. Because of the simplicity in operation, inexpensive equipment, universal precursor, and easy control of deposition kinetics, Wang and coworkers employed aerosol-assisted chemical vapor deposition (AA-CVD) for the fabrication of $Y_2O_3-CeO_2$ nanocomposite [159]. Another attempt toward the synthesis of metal oxide nanocomposites by AA-CVD was confirmed by Palgrave and Parkin, who achieved the fabrication of Au/TiO_2 and Au/WO_3 nanocomposites using $HAuCl_4$, $Ti(OiPr)_4$ and tungsten phenoxide as precursors materials for Au, Ti, and W, respectively [160]. Carbon-oxide nanocomposites add to the list of oxide-based systems that have been

reportedly synthesized by CVD. An interesting gas sensing investigation was performed on TiO_2-carbon nanotube material synthesized by CVD [161]. It is important to mention that research interest has also been channeled toward the synthesis of MnO_2/graphene [162], and IrO_2/CNT [163].

5.3 Chemical/Solution Process

Various chemical or solution techniques are generally used to make a number of metal oxide nanomaterials when the necessity of the process is required. A few of those techniques are presented later.

5.3.1 Coprecipitation method

This approach to synthesize metal oxide nanomaterials simply involves the dissolution of inorganic salts precursors in water or any other solvents to achieve a solution with homogenous distribution of ions. The metal salts are then precipitated as oxalates or hydroxides after attaining the critical concentration of species that precedes the growth and nucleation phases [112]. A number of experimental factors such as the concentration of the salt, solution pH, and temperature influence the size and shape of the resulting nanoparticles synthesized [164]. In addition, some other factors such as surface tension, viscosity, and stirring speed are also responsible for defining the properties of the final product. Once the salts are precipitated, the product is filtered, washed thoroughly and calcined to convert hydroxide into definite crystalline oxides [165]. Table 16.1 highlights different metal oxide nanocomposites synthesized by various techniques. Using this coprecipitation approach, NaOH, NH_3 or NH_4OH and Na_2CO_3 are commonly used as precipitating agents [166–168]. To avoid particle agglomeration that affects the size of the nanocomposites synthesized by this method, researchers have reportedly used surfactants [166]. The coprecipitation method has been shown to have advantages of being a water-based reaction, simple, low cost, flexible, encouragingly mild reaction conditions that enable nanomaterial size control [112]. However, a large amount of waste production and generally widespread low yield limit its use [169].

5.3.2 Sol—gel method

The synthesis of nanomaterials via sol—gel technique proceeds in a simple wet chemical reaction that involves hydrolysis and condensation processes to form a sol that, in turn, produces an integrated network as gel upon aging.

The sol—gel method involves the use of mild reaction conditions and composition of the materials

TABLE 16.1
Metal Oxide—Based Nanocomposites, Their Synthetic and Characterization Strategies, and Antimicrobial Applications.

MOBNC	Size (nm)	Synthesis approach	Characterization techniques	Target microorganism	Outcome	Ref.
Al_2O_3–TiO_2–Ag	–	CVD molecular layered and dry impregnation methods	AAS, TEM	E. coli	Remarkably high antibacterial capacity of the composite, where 99% of E. coli were permanently inactivated within a short time compared to the control in which bacteria were not inactivated	[242]
Zn/Fe_2O_3	<20	Hydrothermal	XRD, HRTEM, Mössbauer studies, ICP	E. coli and S. aureus	Antibacterial inhibition against S. aureus was better, in comparison to E. coli, and it was dependent on the stoichiometric weights between Zn and Fe	[243]
Cu/Cu_2O	7	Solvothermal	FTIR, XRD, TEM, TGA, Raman	S. cerevisiae	High antifungal activity of the composite with 3.73 µg/mL IC50 viability compared to the pristine Cu_2O nanoparticles	[244]
PA 6/ZnO	<100	melt spinning	SEM, TGA, DSC	S. aureus, and K. pneumoniae	The nanocomposites demonstrated good antimicrobial efficiency against both K. pneumoniae and S. aureus was observed to be dose-dependent	[245]
Ag (C, S)–TiO_2	<10	Sol–gel	XPS, BET, UV–Vis, FTIR, XRD, EDX	B. subtilis spores and E. coli cells	The prepared composite possessed strong antibacterial inhibition efficiency than P25–TiO_2	[246]
Chitosan-coated Ag/ZnO	30–65	Sol–gel	SEM, UV–Vis, FTIR, XRD	C. albicans	The composite exhibited a higher percentage of biofilm inhibition against C. albicans (92% at 50 mg/mL) compared to control.	[247]
Hybrid CH–α-Fe_2O_3	27–30	Self-assembly	UV-DRS, UV–Vis, SEM, FTIR, AFM, XRD	S. aureus and E. coli	The coated composite showed enhanced antibacterial inhibition than the uncoated one. There was a monotonical increase in the zone of bacteria inhibition with increasing α-Fe_2O_3 and CH-α-Fe_2O_3 concentration	[248]

Material	Size (nm)	Synthesis method	Characterization	Microorganism	Remarks	Reference
Zn–CuO	~30	Sonochemical irradiation	ESR, DSC, HR-SEM-FEI, SEM, XRD, ICP.	S. aureus MRSA, E. coli MDR and E. coli	Significantly higher antimicrobial activity of Zn–CuO compared to the pristine CuO and ZnO nanoparticles was noted within 10 min of exposure to bacteria	[249]
CuO/ZnO	—	Coprecipitation	SEM, XRD	A. trichoderma	The composite demonstrated superior inhibition of strain A. trichoderma compared to pure CuO and ZnO	[250]
ZnO–AgO	20–50	-	XRD, SEM-EDX	Lactobacillus sp. and Streptococcus mutans	Composites with silver and zinc oxide nanoparticles demonstrated higher antibacterial activity against S. mutans and Lactobacillus compared to the control group. In addition, ZnO had a greater effect on S. mutans than silver	[251]
Graphene/ZnO	30–40	Hydrothermal	XRD, TEM, Raman spectroscopy	S. typhi and E. coli	FLG/ZnO composite exhibited substantial antimicrobial activity against E. coli and S. typhi	[252]
Ag@Fe2O3	6–22	Biomimetic method	FTIR, UV–Vis, PXRD, and HRTEM	E. coli, S. aureus, C. albicans	The nanocomposite showed high antibacterial activity with a MIC value of 15.33 ± 4.62 and 12.67 ± 4.62 mg/mL, respectively, for S. aureus and E. coli. In addition, it demonstrated excellent antifungal activity against C. albicans fungus	[253]
PP/CuOP nanocomposite	40	Solvothermal	XRD, TEM, UV–Vis	E. coli	PP/CuOP displayed a higher release rate than PP/CuP in a short time, indicating efficient antimicrobial tendency. About 99% bacteria were eliminated	[254]

Continued

TABLE 16.1
Metal Oxide—Based Nanocomposites, Their Synthetic and Characterization Strategies, and Antimicrobial Applications.—cont'd

MOBNC	Size (nm)	Synthesis approach	Characterization techniques	Target microorganism	Outcome	Ref.
Mo—MoO$_3$/PPS	10—40	Solid-state reaction	XRD, FESEM, HRTEM	P. aeruginosa, S. aureus, K. pneumoniae, and A. fumigatus	The composite recorded high antibacterial and antifungal activities compared to the reported performance of Mo and MoO$_3$. The results of antifungal activities are also indicative of the possible anticancer properties of the nanocomposite	[255]
Chitosan/TiO$_2$	—	Chemical precipitation	FTIR, XRD, TEM, TGA	E. coli	The nanohybrid showed excellent antibacterial inhibition against E. coli (100%) within 24 h of exposure, compared to chitosan nanoparticles	[256]
Fe$_2$O$_3$/Au	<100	Coprecipitation	TEM	A. baumannii, A. baumannii, S. pyogenes, and S. saprophyticus	The prepared nanoeggs possessed superior antimicrobial activity against pathogenic bacteria under illumination with NIR	[257]
PAA/Fe$_3$O$_4$	14	Click chemistry	TEM, FTIR, XRD	B. subtilis, E. coli, and S. aureus	The composite was deemed effective against both gram-positive and gram-negative bacteria, in comparison with pure nanoparticles alone	[258]
CuO-cotton nanocomposite	10—15	Ultrasound irradiation	XRD, ICP, HR-SEM	E. coli and S. aureus	The coated composite demonstrated significant bactericidal activity, suggesting potential application in wound healing	[259]
Zn—SiO$_2$	<7	Deposition precipitation	XRD, FTIR, DLS, TEM, TGA, AFM, SEM	B. subtili, E. coli, C. parapsilosis, and A. niger	Enhanced antibacterial and antifungal activities against bacterial and fungal strains were observed under different solvents	[260]

CeO$_2$/CdO	27	Chemical precipitation and hydrothermal	XRD, HRTEM, FESEM, FTIR, TGA, EDX	*P. aeruginosa S. aureus S. pyogenes K. pneumoniae C. albicans F. oxysporum A. niger A. candidus*	The nanocomposite prepared in *n*-hexane showed higher antifungal activity while the one prepared in acetonitrile solvent showed higher antibacterial activity, indicating its potential use as an antimicrobial agent	[261]
Ag/CuO	7	Biosynthesis	XRD, XAFS, FESEM, EDS	*S. pneumoniae M. morganii*	Remarkably higher antibacterial property for Ag/CuO than Ag nanoparticles	[262]
CdO–NiO–ZnO	<37	Microwave assisted	XRD, SEM, EDS, and FTIR	*V. cholera, S. typhi, A. hydrophila, P. mirablis S. aureus, E. coli, P. aeruginosa, R. rhodochrous,* and *B. subtilis*	The composite showed improved antimicrobial activity against both gram-positive and gram-negative bacteria	[263]
NiO/CeO$_2$/ZnO	14—23	Coprecipitation	SEM, FTIR, EDS, PL, XRD, UV visible	*P. mirabilis, K. pneumonia, P. aeruginosa, E. coli, S. aureus*	The mixed metal oxide composite exhibited effective antibacterial activity against pathogenic bacteria, by completely damaging the functions of bacteria	[264]
Ag$_2$O/CeO$_2$/ZnO	50	Coprecipitation	XRD, SEM, EDS, and PL	*P. aeruginosa, S. aureus, P. mirabilis, E. coli, K. pneumoniae*	The composite was found to possess high antimicrobial activity against bacteria by damaging the cell wall and functions of bacteria	[265]
graphene/TiO$_2$/cotton	9.3	Simple dipping coating method	XRD, UV–Vis, FESEM, TEM, Raman, XPS	Bacteria (*S. aureus, E. coli*) and fungi (*C. albicans*)	The composite possessed excellent antibacterial and antifungal activities and was proved effective in killing the microorganisms compared to pure cotton and graphene oxide –cotton systems	[266]
CdS–ZnO	~94	Coprecipitation	XRD, EDX, FE-SEM, HRTEM	*E. coli, S. aureus,* and *K. pneumonia*	Strong antimicrobial activity against the tested pathogenic bacteria. Excellent antibacterial efficiency was obtained with the highest ZnO content in the samples	[267]

Continued

TABLE 16.1
Metal Oxide–Based Nanocomposites, Their Synthetic and Characterization Strategies, and Antimicrobial Applications.—cont'd

MOBNC	Size (nm)	Synthesis approach	Characterization techniques	Target microorganism	Outcome	Ref.
ZnO/Cu/graphene	~30	Wet chemical	XRD, UV–Vis, PL, SEM-EDX, HRTEM, FEM, XPS, FTIR	*E. coli* and *S. aureus*	The antibacterial efficiency was improved due to the combined effect of graphene and ZnO nanoparticles compared to the control	[268]
ZnO/Lanthanum	10–15	Coprecipitation	PL, EDX, HR-SEM, UV–Vis, FTIR, and XRD.	*B. subtilis, S. typhi, P. mirabilis,* and *S. aureus*	Substantial antibacterial growth inhibition *against S. typhi, P. mirabilis,* and utilized to treat typhoid fever and kidney stones	[269]
Ag−ZnO/Halloysite nanotubes	8	Hydrothermal	XRD, SEM, TEM, HRTEM, XPS	*E. coli*	The composite displayed superior inhibition on bacteria growth with an increase in its concentration than that on similar doses of pristine ZnO.	[270]
Cellulose@ZnO	10–30	Hydrothermal	XRD, SEM, TG, DTG, XPS	*S. aureus* and *E. coli*	Cellulose-coated ZnO exhibited an enhanced antibacterial activity than commercial ZnO	[271]
Polyolpolyester/CuO	24–33	microwave-induced synthesis	TEM, FTIR, ^{1}HNMR	*C. albicans, C. tropicalis, C. glabrata, C. krusei*	The synthesized compound was found to be effective against all candida species, signifying potential antifungal agents	[272]
ZnO/Polyurethane	24–71	Self-assembly	FESEM, HRTEM, EDX, UV–Vis	*S. aureus, K. pneumoniae*	Considerable growth inhibition of *K. pneumoniae* and *S. aureus* with over 98% reduction in both pathogenic bacteria	[273]
CuO/Polyurethane	~50	Electrospinning	SED-EDX, SEM	*E. coli, S. gallinarum*	All of the prepared composites with CuO concentration ranging from 7% to 12% displayed superior activity against the gram-negative strain	[274]
Ag@Fe$_3$O$_4$ γ-Fe$_2$O$_3$@Ag	20–40	Wet chemical	TEM, XRD	*P. aeruginosa, S. aureus E. coli, E. faecalis, C. albicans, C. tropicalis,* and *C. parapsilosis* strains	Both composites exhibited excellent antibacterial and antifungal activities, indicating their antimicrobial biomedical potential	[275]

Material	Size	Method	Characterization	Target	Activity	Ref.
CeO$_2$/Y$_2$O$_3$	10—20	Hydrothermal	EDX, SEM, FTIR, XRD, HRTEM, UV—Vis	*B. subtilis, P. aeruginosa, E. coli, and S. aureus.*	Exhibited substantial antibacterial activity gram-negative and gram-positive bacteria	[276]
ZnO—SiO$_2$	6.2	Deposition-precipitation method	TGA, XRD, SEM, TEM, DLS	*Bacteria (B. subtilis, E. coli) and fungi (C. parapsilosis and A. niger)*	The synthesized nanocomposite demonstrated better antibacterial activity than antifungal activity	[277]
Chitosan-Ag$_2$O	-	Solution casting	FTIR, SEM, UV—Vis, EDX, TGA	*E. coli, S. aureus, B. subtilis, and P. aeruginosa*	Excellent antibacterial ability for food packaging applications was observed	[278]
Maleimide polystyrene/SiO$_2$—Al$_2$O$_3$	>50	Sol—gel process	TGA, DSC. SEM, TEM, XRD, and FTIR	*B. cereus and E. coli*	Showed excellent antimicrobial activity against pathogenic bacteria	[279]
ZnO-alginate	16.7	microwave	XPS, XRD, TEM, UV—Vis	*E. coli and S. aureus*	Exhibits a remarkable antibacterial activity against the common pathogens	[280]
Polyindole/Ag—Co$_3$O$_4$	~20	Reflux condensation method	XRD, SEM-EDAX, FTIR, HRTEM	Fungi (*A. fumigatus, A. flavus, A. niger, C. albicans, A. terreus, C. tropicalis*), and bacteria (*B. subtilis, S. aureus, S. pneumoniae, E. coli, P. vulgaris, and K. pneumoniae*)	The nanocomposites exhibited improved antibacterial and antifungal activities against pathogenic bacteria and fungi. Hence, could be used as a potential candidate for the biomedical applications	[281]
TiO$_2$/GR sheets	6—12	Direct redox reaction	XPS, XRD, SAED, UV—vis, EDX, and TEM	*E. coli*	Under ambient visible light illumination, excellent photocatalytic and antibacterial activity were achieved	[282]
TiO$_2$/GO nanosheets	—	Sol—gel	XPS, UV—Vis	*E. coli*	Solar light irradiation enhanced graphene oxide/TiO$_2$ thin films antibacterial activity	[283]
GO-Fe$_3$O$_4$	30—36	Sonochemical	TEM, DLS, XRD, Raman	*P. viticola*	A remarkable powerful antifungal effect of GO-Fe$_3$O$_4$ was obtained at a very low concentration. GO-Fe$_3$O$_4$ showed the highest curative activity, followed by GO and Fe$_3$O$_4$	[281]

Continued

TABLE 16.1
Metal Oxide—Based Nanocomposites, Their Synthetic and Characterization Strategies, and Antimicrobial Applications.—cont'd

MOBNC	Size (nm)	Synthesis approach	Characterization techniques	Target microorganism	Outcome	Ref.
TiO_2/GR	10	Sol—gel	XPS, TEM, SEM, Raman, PL	C. elegans nematodes	Under irradiation, the composites revealed cytotoxic effects that were concentration-dependent on the tested nematodes	[284]
TiO_2/GO nanorods	3—5	Hydrothermal	XRD, XPS, TEM	E. coli	Higher antibacterial activity under simulated sunlight	[285]
ZnO/GR	22	Hydrothermal	XRD, XPS, TEM, UV—Vis, Raman.	E. coli	The hybrids displayed substantial antibacterial growth inhibition against E. coli	[286]
Polyaniline/$CoFe_2O_4$	~19.6	Polymerization	UV—Vis, XRD, SEM, EDAX, TEM, FTIR, TGA	C. albicans	The composite showed promising anticandidal activity in a dose-dependent manner, signifying potential applications in biomedicine	[287]
WO_3/GR	15	Solvothermal	XPS, TEM, AFM, Raman.	Bacteriophage, MS2 viruses	After 20 cycles of the RNA measurement, reduction in the RNA efflux exhibited was only <10%	[288]
TiO_2(Eu)/CuO	31—38	Coprecipitation	XRD, SEM, TEM, EDX, UV—Vis	E. faecalis and E. faecium	The inhibition of growth against E. faecium was lower than against E. faecalis	[289]
Au@TiO_2 Ag@TiO_2		Photochemical	BET, Raman, UV—Vis, XPs, TEM, SEM, XRD	E. coli	The presence of TiO_2 improved the antibacterial effect of Ag@TiO_2, and the growth of E. coli was completely inhibited at low concentration	[290]
Polyindole/Ni—ZnO	20—40	Coprecipitation and chemical oxidation	SEM, FTIR, XRD	Fungi (P. chrysogenum)	The nanocomposite showed higher antifungal activity than Ni—ZnO. On the other hand, polyindole, NiO, and ZnO showed no activity	[291]

Composite	Size (nm)	Synthesis method	Characterization	Microbe	Results	Ref.
Fe$_3$O$_4$/TiO$_2$ core—shell	-	Solvothermal	UV—Vis, TEM	*T. saprophyticus, S. pyogenes, S. aureus*	Composite demonstrated high capability to inhibit antibiotic-resistant strains growth as well as other several pathogenic bacteria	[292]
ZnO/Ag/Mn$_2$O$_3$	40—50	Thermal decomposition	UV—Vis, XPS, TEM, FE-SEM, XRD.	*E. coli, S. aureus*	The nanocomposite possessed high bactericidal activity compared to pure ZnO	[293]
PANI/V$_2$O$_5$	~11	In situ polymerization	FTIR, XRD	Bacteria (*S. aureus*) and fungi (*A. niger*)	The composite showed moderate antibacterial and antifungal activities	[294]
CS/PVA/ZnO	50—100	Solvothermal	XRD, SEM, TEM, FTIR	*S. aureus* and *E. coli*	Showed high antibacterial activity against *S. aureus* and *E. coli*	[295]
La$_2$CuO$_4$/CeO$_2$	26	Solvothermal	XRD, EDX, FESEM, HRTEM, UV—Vis DRS	*S. aureus* and *E. coli*	Enhanced antibacterial growth inhibition against both negative and positive bacterial strains	[296]
La$_2$CuO$_4$/CeO$_2$/rGO	29	Solvothermal	XRD, FESEM, HRTEM, EDX, UV—Vis DRS	*S. aureus* and *E. coli*	Demonstrated high antibacterial growth inhibition against both *S. aureus* and *E. coli*	[296]
CeO$_2$/CdO	10	Sol—gel method	XRD, FTIR, FESEM, EDX, HRTEM UV—Vis-DRS	*E. coli, P. vulgaris, S. aureus, S. pyogenes, C. albicans, P. aeruginosa,* and *K. pneumoniae*	The prepared nanoarrays demonstrated better antimicrobial activity against *P. aeruginosa* compared to other tested microbes	[297]
textile/Al$_2$O$_3$—TiO$_2$	50—80	Sol—gel method	XRD, FTIR, FESEM-EDX	*E. coli*	Remarkable antimicrobial activity of the composite was observed compared to the control, indicating its potential for antimicrobial and wound dressing applications	[298]

from molecular precursors that lead to disparity in materials properties, hence the approach has received considerable attention for the synthesis of nanomaterial [170]. Slow reaction rate also makes this method appealing for the synthesis of nanostructures consisting of many components, as it allows good engineering of the final product structure. The products fashioned by the use of sol—gel technique are either films [171,172] or colloidal powders [173—175]. This method is capable of producing both micro and nanostructured materials because the variation in reaction parameters greatly influences the size, shape, and structure of final synthesized nanomaterial [176,177]. In addition, this method is advantageous as the reaction usually occurs at room temperature, thus a rather cost-effective technique. The sol—gel method could also involve the use of inorganic precursors that undergo different chemical transformations to form 3D molecular networks. The hydrolysis and condensation of metal alkoxides are among the most common methods that form larger metal oxide molecules that polymerize to form the coating [178]. The sol—gel technique encourages the coating of substrates with complex shapes, which some commonly used coating procedures cannot achieve within the nanometer to micrometer scale. Some of the widely used substrates include nanotubes, fibers, or even organic/inorganic crystals and colloidal particles [178—180]. Successful employment of this method has been presented in the synthesis of iron-oxide silica nanocomposites using iron alkoxides as a precursor [181]. In addition, the fabrication of polyaniline/Cu-doped ZnO nanocomposite with enhanced antimicrobial activity was achieved by an inverse microemulsion technique [182].

5.3.3 Solvothermal/hydrothermal method

Solvothermal and hydrothermal routes are the two commonly used wet chemical approaches for the preparation of nanomaterials regarded as thermal methods. In solvothermal route, in the presence of an appropriate solvent, a precursor material is subjected to a chemical reaction under high temperature (higher than the boiling point of the solvent) in a closed chamber and pressure [13,183]. The solvent always plays a vital role in the solvothermal process, for example, in the synthesis of nanocrystalline TiO_2 using acetic acid as the solvent and tetrabutyl titanate as precursor material [184].

On the other hand, the hydrothermal route involves a heterogeneous chemical reaction in a closed system proceeding at a pressure greater than 1 atm, above room temperature and in an aqueous solvent [185]. It involves the mixing of precursors together with a base in water followed by thermal treatment in an autoclave in which the reaction temperature is set to be above the temperature of water [13]. The use of solvothermal/hydrothermal technique makes it easy to control the shape and the size of the nanomaterials by varying the chemical and thermodynamic parameters of the overall reaction [8].

For necessary improvement to the properties and size of the composites, capping agents, surfactants, and mineralizers have been reportedly employed [186—189]. Reported findings have presented approaches that involve a combination of the hydrothermal method with microwave [190] and sol—gel [191]. These approaches tend to provide the opportunity of varying the structural and physicochemical properties of the materials and encourage the development of single phased materials with improved stability [191]. Furthermore, researchers have shown that particle size, phase changes, morphology, and properties of the materials can be transformed considerably just by altering the pressure, reaction time, and temperature [189,192]. In most situations, calcination is required for eliminating the residuals and improving the crystallinity of the nanomaterials. Although solvothermal and hydrothermal processes are cost-effective and simple methods, the challenge of aggregation of metal nanoparticles is their major drawback [13,193—195]. A number of metal—MOBNCs such as Au/TiO_2 [74,196], ZnO/Au [197,198], $Pt/MnO_2/GS$ [199], and $CeO_2/rGO/Pt$ [200] have been successfully synthesized by hydrothermal method.

5.3.4 Biomimetic approach

In the quest for the greener fabrication of metal oxide nanocomposites, researchers have been focusing on alternative synthetic techniques to passivate both the rate at which wastes are generated during preparation and the cost involved in the synthesis and development of a rapid prototyping technique. The biomimetic process usually ensues at a relatively low temperature and it lessens the particle degradation in comparison to conventional techniques [8]. It involves the synthesis of nanomaterials that mimic mechanical, physicochemical, and biological properties of natural materials with the aim to employ these nanomaterials in biomedical devices or for tissue engineering as scaffolds. Considering this approach is a mild biological approach, researchers have reported it as an efficient method for the development of crystalline nanocomposite powders having controlled size, without the need of external additives such as sodium hydroxide or ammonium hydroxide. This inspiration to utilize natural or biological processes has led to significant achievement in the fabrication of highly ordered nanostructures [8].

This biomimetic approach encompasses alternate methods geared toward the formation of functional nanomaterials with biotechnological applications [201]. For example, the unparalleled properties associated with metal oxide synthesis in specific biological systems has inspired the synthesis of functionalized nanomaterial like manganese oxide (MnO_2) nanoflake reported by Oaki and Imai [202]. The researchers prepared MnO_2 nanoflake in aqueous solutions of poly(acrylic acid) and poly(ethyleneimine) containing mosaic interiors. This synthetic protocol allowed for the coordination of the organic polymers to the metal ions, which enabled interaction with the growing crystals. The results obtained in this study showed a tailored interaction between inorganic and organic moieties, which paved a way for the realization of a useful application of the biomimetic approach in the synthesis of functional nanomaterials.

In another study by Avanzato and Fath, MgO_2-GeO_2 nanocomposite powders were successfully prepared by the biomimetic approach and evaluation for antimicrobial properties [203]. The nanocomposite was prepared by cogelling the precursors in the presence of a biopolymer (poly-L-lysine), an acidic amino acid (aspartic acid) and a basic amino acid (histidine). The morphological characterization of the nanoparticles confirmed the formation of rhombic or cubic shapes. The results also showed the highest yield of the prepared nanocomposite within 2 h of reaction time. An examination of the antibacterial potential of the prepared nanocomposites against both gram-positive and gram-negative bacteria showed that they were more effective against *S. aureus* in comparison to *E. coli* at low concentration. This approach has also been reported to have been used in the synthesis of TiO_2–SiO_2–Ag nanocomposite [204], g-C_3N_4 (graphitic carbon nitride)/TiO2 nanosheets [205], Ag/SiO_2/organic-based biomimetic nanocomposite [206], etc.

6 TECHNIQUES FOR CHARACTERIZATION OF METAL OXIDE BASED NANOCOMPOSITES

Comprehensive characterization is one of the significant features in the understanding of the fashioned MOBNCs. Special characterization techniques are crucial to appreciate the structures at the nanoscale level, as well as at the interface of the component materials. These techniques are employed to get a better understanding of morphological and topographical features of the surfaces of the material composition, which aid in identifying including but not limited to the size,

crystallinity, phase, thermal, electrical, shape, magnetic and optical properties [15,207]. A very brief description of the most common techniques available for the characterization of nanomaterials and their nanocomposites is presented later [15,208].

6.1 X-ray Diffraction

X-ray diffraction (XRD) is a valuable technique that is versatile and nondestructive providing important information concerning the particle size, lattice constants, chemical composition and microstructure and crystal structure of a material [15]. By using this technique, the determination of the phase and unit cell information of the nanomaterial and nanocomposite understudy is made possible [209]. XRD functions as a result of the constructive interference of a beam of X-ray generated in a certain space of direction. The resultant diffractograms are gotten by the measurement of the angles of the diffracted X-ray beam by the crystalline phases in the nanostructured material. These diffractions patterns can also be used to ascertain by fingerprinting approach the nanomaterials using the International Center for Diffraction Data library. The details concerning the crystalline structures of diverse nanostructured materials and their nanocomposites can be obtained from diffractions and scattering analyses [210]. Interestingly, the diffraction analysis of nanocrystalline materials and nanocomposites can give the interlayer spacing that is important in determining the particle size of a nanomaterial. This interlayer spacing or *d*-spacing can be described as the distance between planes of atoms that give rise to diffraction peaks.

The *d*-spacing is acquired by using Bragg's equation:

$$n\lambda = 2d\,sin\theta \qquad 16.1$$

where n = order of the plane,
λ = X-ray wavelength
d = Interlayer spacing
θ = Glancing or diffraction angle.

In the determination of crystal size, the technique employs the Scherrer's equation (Eq. 16.1) [209,211,212].

$$D = K\lambda/_{\beta cos\theta} \qquad 16.2$$

where D (nm) = mean size of the crystalline domains
K = Dimensionless shape factor with a value close to unity
λ = X-ray wavelength
β = line broadening referring to the full width at half maximum
θ = Bragg angle

This formula is applied only for nanocomposites with a definite crystalline structure.

In many cases, the powder XRD (PXRD) technique is employed to determine the phase in a material, particularly when relating the synthesis method or effect of doping in the structure of the material [213−216]. For example, using chemical coprecipitation, talc/Fe_3O_4 magnetic nanocomposites were synthesized by Kalantari et al. By using PXRD as a tool, the surface binding of Fe^{3+} ions with talc in the nanocomposite was confirmed [167]. In addition, researchers have shown that when the PXRD technique is used together with transmission electron microscopy (TEM) it helps to determine the nanoscale dispersion within polymer matrix [217]. This was evident when Morgan and coworkers analyzed several polymer-layered silicate nanocomposites by both TEM and XRD, the discovered that when PXRD was used alone the diffraction peaks were not observed hence leading to negative results compared to positive results obtainable when both PXRD and TEM are used together. Studies have reported on the reduction in intensity of graphite oxide (GO) peaks in the nanocomposites when GO was exfoliated by inclusion of metal oxides [171,218], but by using XRD other researchers determined the existence and nonexistence of impurities from matrix structure whereby the fading of peaks in the purified spectra of multiwalled carbon nanotubes after acid washing confirms the removal of impurities [216].

6.2 Scanning Electron Microscopy

Scanning electron microscopy (SEM) utilizes high-energy electron rays to scan materials surface. By using SEM, the shape and morphology like roughness, homogeneity of dispersion of nanomaterial can be determined. In addition to morphological properties, more in-depth properties such as size, and topological information of materials can be obtained [219]. In studying the surface morphology of the nanomaterials and nanocomposites, SEM requires the surface to be electrically conductive, but if the material possesses a nonconducting surface, a thin layer of carbon or gold is used to coat the material [170]. When different metal oxides are incorporated in a nanocomposite through doping [216], or any other method, their effect on agglomeration [214], and shape [220], can be visible by SEM micrographs. It is important to shed some light on field emissive scanning electron microscopy (FESEM), which gives room for much higher magnification and resolution in comparison to SEM, this is due to its electron generation system [221]. Furthermore, high-quality images with minor electrical charging can be

obtained from FESEM as its detectors are optimized to work at high resolution and very low acceleration potential. FESEM also eliminates the prior deposition of conducting coatings onto the polymer nanocomposites, which is a requirement for the basic SEM [221].

6.3 Transmission Electron Microscopy

TEM as a microscopic technique works by the interaction between the material under analysis and a beam of electrons that is produced from an electron gun. This technique can be used to show morphological effects resulting from dispersion in the matrix, the effect of annealing, observation of defects and possible agglomeration [209,222].

It has received tremendous attention owing to its ability to present an image with a better resolution of 0.1−0.2 nm as well as additional information obtainable from the materials investigated. Information such as the core−shell structure of the nanoparticle dispersion [223], surface roughness [223], nanocomposites [220], gelling agents impact [224], particle size [177], layering in structure [209] and doping effects on morphology [216] can be deduced from TEM and recent advancement in the development of high-resolution transmission microscopy (HRTEM) has made additional information such as fringes, interplanar distance, dislocations, and defects be obtainable [170,219,225]. Furthermore, considerable information on the structural changes is obtainable by dark and bright-field images [209,226]; however, this technique suffers limitation during the preparation of sample because it requires low film thickness [170,219]. Additionally, selected area electron diffraction (SAED) is a related technique that makes available information regarding microdiffractions, crystal planes, and symmetry of the lattice of the nanomaterials [227].

6.4 Atomic Force Microscopy

Atomic force microscopy (AFM) refers to scanning probe microscopy that makes use of a small probe to scan on the surface of the sample material to obtain information about the surface properties of the material.

By measuring the forces of interaction between the AFM tip and the atom of the sample, AFM captures the surface images of the materials at the atomic resolution level [170]. The choice of mode of operation is particularly critical for the analysis because the technique works under various modes; tapping, contact, and conductive modes to collect surface images of both conducting and nonconducting materials [219,228]. The information concerning the distribution of the nanoparticles, the height of the matrix, and the height of nanoparticles dispersed within the matrix can as well be obtained from AFM [229]. Additionally, this technique can assist to determine the morphology, surface area, and roughness, shape, size, and distribution of the nanostructured materials [230].

6.5 Thermogravimetric Analysis

Thermogravimetric analysis (TGA) is a thermal characterization technique utilized to measure the weight of a material as a function of temperature at a constant heat rate. The technique usually involves the material being subjected to decomposition at high temperatures, with the results being presented in a plot of weight percent versus temperature under a controlled atmosphere.

Thermal analysis by TGA can serve as a key analysis to ascertain the thermal stability of a prepared nanocomposite in addition to other effects arising as a result of doping, curing or annealing [231]. It is important to mention that other thermal analysis techniques such as dynamic mechanical thermal analysis, thermal-mechanical analysis, and the more common differential scanning calorimetry (DSC) also exist. Common information obtainable from the thermogravimetric analysis is thermal decomposition steps of the nanocomposites, nature of the process of decomposition (endothermal or exothermic), moisture/solvent loss, stepwise weight loss, and final decomposed matter [232]. TGA was evidently important to Laachachi and coworkers in their investigation in the use of organoclays and metal oxide nanoparticles to improve fire retardancy and thermal stability of poly (methyl methacrylate) (PMMA), an attempt to study the nanocomposites thermal stability in comparison to PMMA doped by equal amounts of TiO_2 and Fe_2O_3. Their findings showed that TiO_2 induced better stability than Fe_2O_3 [233]. Using a similar approach, they investigated the best weight percent of metal oxide nanoparticles in the polymer matrix that resulted in a decrease in heat release [233]. In another study, the oxidation state of Mn in the nanocomposite was elucidated from the formula of the nanocomposite after determining the interlayer water by using TGA of MnO_2/CNT nanocomposite [234].

6.6 Ultraviolet–Visible Spectroscopy

Ultraviolet (UV)–visible spectroscopy is a spectrophotometric technique employed to measure photons of light within the UV–visible region, measuring the intensity of light before and after it passes through the material. It is a vital, although simple characterization tool utilized to characterize nanomaterials and their corresponding composites. Normally, metal and metal oxide nanomaterials exhibit characteristic surface plasmon resonance peaks in the UV–visible area because

of the interaction of incident UV radiation with the surface electrons of nanoparticles [235]. This surface plasmon resonance absorbance emanates from the surface electrons' vibrational energy's resonance with the interacting UV–visible light's wave energy. Considering the surface energy of nanomaterials or their surface electrons are very high, they tend to form stable dispersions in a liquid medium and this ensures that the surface plasmon resonance is characteristic of individual nanomaterials. Its peak is used to confirm the formation of the specific nanomaterial or in the detection of its existence in any polymer nanocomposite [221].

6.7 Fourier-Transform Infrared Spectroscopy

Fourier-transform infrared (FTIR) spectroscopy is a common and essential characterization technique utilized to get information on the existence of chemical functional groups in functionalized nanomaterials and the structure of a material [221]. FTIR identifies the stretching vibrations of molecules; hence, it can be used to identify the functional groups existent in any unknown inorganic and organic compound. FTIR is considered to be a valuable, noninvasive, cost-effective, and simple technique that can be used to deduce the existing chemical functional groups in a functionalized or stabilized molecule of nanocomposites and nanomaterials [236,237]. The absorption of electromagnetic radiation by the test material under analysis causes an increase in the frequencies of bond vibrations, which in turn creates an electronic transition between ground energy levels and several excited states. Considering that the vibrations are specific to the group of atoms involved in the vibration and the type of chemical bond, their frequencies signify the excitations of chemical bonds. This has encouraged the utilization of FTIR spectra to deduce the physicochemical interactions between nanomaterials and polymers.

6.8 Nuclear Magnetic Resonance Spectroscopy

Nuclear magnetic resonance (NMR) spectroscopy is among the more sophisticated techniques used to analyze and validate the structural characteristics of polymers, nanomaterials, and their nanocomposites. It relies on the magnetic nuclei populace in an exterior magnetic field to align the nuclei in a finite and anticipated number of orientations. Multi-nuclear NMR spectral analyses, particularly solid-state NMR investigations give information about the chemical environs for silicon, nitrogen, carbon, protons, etc. [221]. On one hand, ^{13}C NMR spectroscopy can be used to provide information about carbon-based nanomaterials and their polymer nanocomposites, on the other hand, both ^{1}H and ^{13}C solid-state NMR spectroscopic analysis assists in the validation of the surface chemistry and interactions of nanomaterials with the polymer in the nanocomposites [238].

6.9 Raman Spectroscopy

Raman spectroscopy is another analytical technique utilized for the investigation of possible flaws as well as phonon and electronic properties of the nanomaterials and their corresponding nanocomposites. The technique relies as a result of the molecular excitation, on the inelastic scattering of a high-frequency monochromatic source of light by molecules such as vibrational, translational, and rotational [221]. Raman spectroscopy identifies molecules' vibrational frequencies that are Raman active as they rely on the bond strength and mass of atoms. This enables the identification of important functional groups and linkages within nanocomposites and nanomaterials. Thus, surface-enhanced Raman spectroscopy, a more robust technique, combining good sensitivity with excellent chemical specificity is crucial. It allows for enhancement of signals of the analyzed molecules present on the surfaces of nanomaterials and nanocomposites and is mainly used for the characterization of carbon-based nanomaterials such as carbon nanotubes (CNT), graphene, reduced graphene oxide, carbon dots, and reduced carbon dots and also their polymer nanocomposites [221].

6.10 X-ray Photoelectron Spectroscopy

X-ray photoelectron spectroscopy (XPS) or electron spectroscopy for chemical analysis is utilized to estimate the surface chemical makeup and states of oxidation of the elements of nanomaterials and their corresponding polymer nanocomposite under high vacuum conditions [239]. Ideally, the emissions of electrons occur after irradiation by X-rays on sample nanostructured materials. The spectrum acquired from the XPS is as a result of measuring the kinetic energy that provides the binding energy and the number of electrons escaping from the surface of such materials [240]. It is of importance to mention that specific groups vary in binding energies and are reliant on the element's oxidation state, which allows for the easy detection from the XPS spectrum.

6.11 Photoluminescence Technique

Photoluminescence (PL) involves the rapid emission of light from the analyzed materials as a result of optical excitation. The nature of the optical excitation and the sample materials properties are the determining factors for the PL signals efficacy. Information about the bandgap, the composition of the sample material, evaluation of several diode materials and defects evaluation of light-emitting materials are easily obtainable using the PL [241].

7 ANTIMICROBIAL AND OTHER BIOMEDICAL POTENTIALS OF METAL OXIDE BASED NANOCOMPOSITES

MOBNCs have overtime established a wide range of applications in biomedicine (Fig. 16.3) as a result of the capacity to tailor their properties and the simplistic nature of their synthesis; they are routinely employed as biointerfaces for cellular and tissue engineering, vehicles for the controlled drug encapsulation and release, nanobiosensors for diagnosis, surgical tools, and dressings for wound healing [10,15,299].

7.1 Wound Dressing/Healing

The largest and outermost organ covering the body, the skin, is responsible for the protection of internal organs, ligaments, bones, and muscles, which are underneath from external physical, mechanical, chemical, and biological [300,301]. The skin's structure and its relative functions can be affected by surgical incisions, burns, cuts, and possibly illnesses, such as diabetes [302], and such occurrence requires the reestablishment of

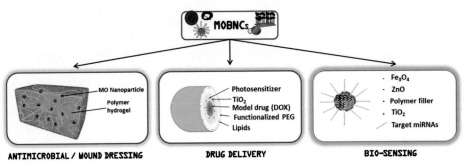

FIG. 16.3 Overview of biomedical application of MOBNCs.

its structure and functions as soon as possible to maintain homeostasis of the body. To achieve this reestablishment, after an injury to the skin, the process of healing is initiated almost immediately after the occurrence of injury to avoid the possible risk of contamination [303]. The most common infections to people are skin and soft tissue infections (SSTIs). The extent of SSTIs usually depends on the severity and etiology of the microbial invasion, with infections ranging from trivial superficial infections to life-threatening infections. During the initial stage of the process of infection, gram-positive organisms such as *Streptococcus pyogenes* (*S. pyogenes*) and *Staphylococcus aureus* (*S. aureus*) are the dominant organisms [304]. Later stages of the infection, for example, when chronic wounds are experienced, gram gram-negative organisms such as *E. coli* and *P. aeruginosa* species are more readily available [305]. The activation of the immune system in healthy human beings aids in eradicating pathogens, but in cases of the nonfunctional immune system, infection results causing the deterioration of extracellular matrix components (fibrin, elastin, collagen), growth factors and granulation tissue, thus compromising the regular wound healing process [306,307].

The natural healing of skin infections (wounds) is a common human biological process that is accomplished via precisely four greatly automated phases: hemostasis, inflammation, proliferation, and remodeling. These phases work together to promote speedy wound healing with minimal scarring and maximal function. This dynamic process encompasses coordinated, systematic and balanced activity of vascular, inflammatory, epithelial cells and connective tissues [308]. This makes the design of materials for application in wound healing very peculiar and care has to be taken in the choice of such materials. This is because such a material will need to demonstrate the capability to prevent the penetration by microorganisms into the wound. Accomplishing this requirement is what has led researchers in the design and synthesis of wound dressing materials that possess intrinsic biocidal activities, or by incorporation of antimicrobial agents within the designed materials to achieve such antimicrobial prowess [309].

Motivated by this desire to develop effective biomaterials for wound management, Lu and coworkers [310] incorporated Ag/ZnO nanoparticles into chitosan forming novel Ag/ZnO-chitosan nanocomposite for wound dressing. The results obtained from the evaluation of the possible antibacterial growth inhibition as well as the ability to heal wounds revealed high composite dressings showed enhanced blood-clotting capability, 13—14 days of moisture retention time, 21%—4% of swelling ratio and 81%—88% porosity that are beneficial

to accelerate and encourage faster wound healing. Furthermore, the assessment showed that the prepared nanocomposite possesses enhanced antibacterial activities against drug-sensitive and resistant pathogenic bacteria *P. aeruginosa*, *S. aureus*, and *E. coli* when compared with that of ZnO ointment gauze and pure chitosan.

Similarly, the investigation of the antimicrobial capacity of chitosan—PVP—TiO_2 nanocomposite as a prospective material for wound dressing in vivo has been presented by Archana et al. [311]. The prepared nanocomposite dressing was found to possess good biocompatibility against L929 and NIH3T3 fibroblast cells, as well as exceptional antimicrobial efficacy. When the prepared nanocomposite was compared to chitosan-treated groups, soframycin skin ointment and conventional gauze, it was found to accelerate the healing of albino rat model wounds with open excision-type wounds.

Considering the public health concern regarding skin infections, Cai et al. [312] successfully fabricated a uniform Fe_3O_4 NPs/CS/GE composite by electrospinning technique for potential wound dressing applications. The results revealed that the antibacterial activity of Fe_3O_4/CS/GE nanocomposite was heightened when magnetic Fe_3O_4 nanoparticles were incorporated in chitosan/gelatin nanofiber. The results further attested that the optimal antibacterial ability of Fe_3O_4/CS/GE nanocomposite against *S. aureus* and *E. coli* was achieved at filler loading ≥ 1 wt%, suggesting that the prepared nanocomposite hold potential applications as wound dressing material depending on how they are tailored.

In an effort to improve the radical scavenging capability and antimicrobial activity of a textile-based metal oxide nanocomposite, Parham et al. prepared textile/Al_2O_3—TiO_2 nanocomposite as a radical scavenger and antimicrobial wound dresser [298]. The assessment of the antimicrobial studies revealed a much higher antimicrobial prowess of textile/Al_2O_3—TiO_2 nanocomposite than those shown by textile/TiO_2 and textile/Al_2O_3/TiO_2 nanocomposites. They also reported that good interaction between the metal oxide nanoparticles and the textile is responsible for excellent antimicrobial and radical scavenging performance shown from *the* human skin fibroblast and *E. coli* cytotoxicity tests. Additionally, textile/Al_2O_3—TiO_2 nanocomposite showed the highest ability to scavenge radicals (38.2%), followed by textile/Al_2O_3/TiO_2 (35.5%) and textile/Al_2O_3 (35%). In view of the above, the researchers arranged the prepared nanocomposites in the following order of antimicrobial activity: textile/$Al_2O_3TiO_2$ > textile/Al_2O_3/TiO_2 > textile/TiO_2 > textile/Al_2O_3 > textile.

7.2 Drug Delivery

The development of advanced metal oxide nanocomposites for drug delivery has been of huge scientific interest to researchers involved in diverse fields from material science to nanomedicine. This has encouraged researchers to fabricate new systems employing nanocomposites made using metal oxide nanoparticles for drug delivery. A recent development on nanocomposite drug-delivery systems aims at the effective synthesis of drug delivery systems that can provide efficient encapsulation of therapeutics and discharge them upon a definitive stimulus. Researchers also aim at developing nanocomposites capable of encouraging combinational therapy by their ability to convert external signals to heat that generates highly oxidative species that are beneficial [313–315].

Researchers investigated the use of inorganic biomaterials; magnetic Fe_3O_4, SiO_2, and $CaSiO_4$ to successfully synthesize magnetic $Fe_3O_4/SiO_2/CaSiO_4$ nanocomposite for application in drug delivery [316]. Based on the superior drug loading capacity of $CaSiO_4$ employed in the synthesis of this nanocomposite, the researchers used the fabricated nanocomposite as the drug carrier using ibuprofen as a model drug. The nanocomposite registered a drug loading capability of 75 mg drug/g carrier and an estimated time of 60 h was enough for complete release of the loaded ibuprofen drug from the nanocomposite. The results presented findings showing that the combined advantages of both magnetic behavior of the prepared nanocomposite and sustained drug release make it a promising material for the prospective application in targeted drug delivery.

In another study, Yang and coworkers fabricated two-dimensional magnetic $WS_2@Fe_3O_4$ nanocomposite functionalized with polyethylene glycol (PEG) and coated with mesoporous silica (MS) for cancer treatment (model drug delivery and imaging-guided therapy) [317]. The researchers using an anticancer model drug, doxorubicin (DOX), investigated the drug loading ability of $WS_2-Fe_3O_4@MS-PEG$ nanocomposite. They recorded a significant increase in DOX loading with an increasing DOX amount that reached a maximum concentration of 0.5 mg/mL. In addition, they highlighted that the drug release behavior of the prepared $WS_2-IO@MS-PEG/DOX$ nanocomposite showed a drug release behavior that was pH-dependent, with the acceleration of the drug release under considerably acidic pH because of protonation of the amino group present in the DOX structure. Further observations presented revealed that stimulation of near-infrared light could initiate drug release from $WS_2-IO@MS-PEG/DOX$ nanocomposite. Their investigations revealed the fact that the combined therapeutic effect of chemotherapy and photothermal by $WS_2-IO@MS-PEG/DOX$ at the cellular level resulted in similar efficiency in cancer cell killing for both $WS_2-IO@MS-PEG/DOX$ without laser irradiation and free DOX. This infers that the combination of chemotherapy and photothermal could lead to enhanced therapeutic effect at the cellular level in comparison to single therapies. Bearing this in mind and the recent development of biomedical imaging worldwide, Yang and coworkers also investigated the behavior of $WS_2-IO@MS-PEG$ in vivo and they found that the incorporation of magnetic Fe_3O_4 nanoparticles within $WS_2-IO@MS-PEG$ nanocomposite makes it a promising magnetic resonance imaging T2 contrast agent. It is noteworthy to highlight that the integration of multiple imaging and therapy functionalities within a single nanoscale platform coupled with the inherent low cytotoxicity and good physiological stability of the as-synthesized $WS_2-Fe_3O_4@MS-PEG$ nanocomposite could make it an efficient imaging-guided therapy and potential drug carrier for drug delivery for cancer treatment.

A hierarchical three-dimensional Fe_3O_4/graphene nanosheet (GNS) nanocomposite synthesized by a simple in situ hydrothermal method was employed by Li and his colleagues for controlled drug delivery [318]. The as-prepared nanocomposites were tested for the ability to release model drug rhodamine B (RB). This investigation confirms effective model drug delivery, with exceptional loading of model drug RB. Li and coworkers concluded that the high efficiency of the nanocomposite in loading RB drug could be as a result of Fe_3O_4/GNS increased surface area wherein the hierarchical structure of Fe_3O_4 nanoflowers are highly encapsulated in the GNS matrix [319,320].

A study that investigated the potential application of dual surface-functionalized Janus nanocomposites for simultaneous stimulus-induced drug released and tumor cell targeting was recently performed by Wang et al. [321]. By utilizing a process of combined sol–gel reaction and miniemulsion, Wang and his coworkers fabricated Janus polystyrene (PS)/Fe_3O_4 @SiO_2 nanocomposite that was superparamagnetic. They decorated the surface of the designed nanocomposite with carboxyl groups to modify the chemical surface properties to allow for simultaneous stimulus-induced drug release and cell targeting without chemical or steric interference. Furthermore, tumor cell targeting was achieved by using a bis-amine linker to conjugate folic acid to the surface of PS. Stimulus-induced drug release was facilitated by immobilizing antitumor agent DOX via a pH-sensitive hydrazone bond to the silica shell. The study recorded that acidic conditions in the endosome caused the release of DOX from the Janus nanocomposite to be faster due to the acid-catalyzed hydrolysis of the hydrazone linker that simultaneously encouraged the increased amount of released DOX from these new Janus particles under acidic conditions. They recorded percentage DOX released in varied buffer conditions of 25.1%, 47.1%, and 82.6% (w/w) at pH 7.4, pH 6.0, and pH 5.0, respectively. Employing human MDA-MB-231 breast cell lines, Wang and his colleagues evaluated the capacity of these nanocomposites to effectively target tumors and kill cancerous cells. They recorded cytotoxicity profiles that were dose-dependent for control drug-free, DOX-loaded Janus and DOX particles with the results revealing that the Janus particles in the absence of drug were reasonably safe up to 3 mg/mL with cell viability with IC_{50} for the treatment estimated to be 3.3 ± 0.3 µg/mL. The Janus particles with covalently attached DOX (SJNCs-DOX) similarly caused cell viability reduction with 1030.2 ± 416.1 µg/mL estimated IC_{50}. These findings confirm that the suitability of the superparamagnetic Janus nanostructure with dual surface functionality for hyperthermia-induced sensitization of tumor cells and multimodal cell imaging, offering a multidimensional platform for cancer treatment.

7.3 Biosensing

Considerable progress has been made worldwide concerning the development of biosensors. Reflecting on the growing call for point-of-care, real-time, miniaturized, and efficient convenient instruments, medical devices and onsite environmental sampling instruments, biosensors have emerged as promising devices in biomedicine [322]. Recently, notable progress has been reported on the incorporation of nanostructured metal oxide nanocomposites in the architecture of high-performance biosensors [323–325]. Among the numerous kinds of nanomaterials that have been developed, metal oxide—based nanostructured materials have recently emerged as unique materials that offer an efficient surface for immobilizing biomolecules with desired orientation, high biological activity, and better conformation resulting in improving sensing behavior [326]. Additionally, the unique optical, electrical, and molecular properties of metal oxide nanomaterials coupled with the tenability by functionalization that allows for surface properties modification offer fascinating platforms for signal amplification by interfacing biorecognition elements with transducers. Owing to their unique varied morphologies at nanoscale and physicochemical properties,

biosensors designed using them have advantages of being sensitive, precise, specific, and stable, which extend their applications to both nonclinical and clinical applications [327].

Inspired by the pioneering study done by Dong and coworkers [328] in the area of classical detection techniques, Li et al. [329] successfully fabricated $Fe_3O_4@TiO_2$-DA(dopamine), which was then loaded with peptide nucleic acids, for detection of microMRNAs in the blood, in addition to identifying the mutant levels present. The results obtained from the optimization of electrochemical responses performed on the $Fe_3O_4@TiO_2$-DA dosages indicated an optimum dose of 1.0 mg mL^{-1} of $Fe_3O_4@TiO_2$-DA, a dose that any dose higher relates to a significant decrease in detection signals. Additionally, the electroanalytical method under optimized experimental conditions was employed to detect the mutated and wild miRNA samples, and they revealed the presence of a linear current correlation to the logarithm of the amount of miRNA obtained across the concentration range with the detection limit of 1.3 fM. Summarily, this study infers that the developed electroanalytical detection technique could be employed for probing miRNAs containing guanine present in blood with selectivity and high sensitivity and to identify their mutation levels. In this regard, the study anticipates the use of prepared nanocomposite in numerous biomedical applications including cancer diagnosis and early warning of cancer metastasis.

Li and coworkers demonstrated the synthesis of another mRNA sensing platform to probe free microRNAs in blood grounded on a "self-cleaning" functionalized microstructure [330]. They synthesized HDS-ZnO-APS sensing nanocomposite by using hydrophilic aminopropyltriethoxysilane (APS) and hexadecyltrimethoxysilane (HDS) by incorporating ZnO nanoparticles. The optimization of detection conditions revealed an increase in the colorimetric responses with increasing DNA probe concentrations up to 0.80 mM, of which the higher concentrations triggered a reduction in signals. Furthermore, identification of single-base mutant levels of miRNAs was detected using this developed colorimetric platform, and it revealed a linear relationship for the changes of product absorbances against the concentrations of single-base mutant miRNAs ranging from 1.0 to 200.0 pM ($R^2 = 0.9895$). This linearity suggests that the developed sensing system could be employed to efficiently in free miRNAs determination in blood with sensitivity and high selectivity, in addition to the ability to identify and quantify single-base mutant levels of miRNAs for profiling the pattern of gene expression.

Adopting the one-step codeposition technique, which has proven successful in the synthesis of biosensing systems, Xiao et al. synthesized PtAu—MnO_2 binary nanocomposites, using PtAu alloy and MnO_2, on graphene paper (GP) for flexible nonenzymatic glucose sensing [322]. They reported on the formation of coral-like nanocatalysts possessing large active surface areas when compared to other electrodes. The estimated active surface area of all electrodes was observed to increase in the order of PtAu—MnO_2/GP > PtAu/MnO_2/GP > PtAu/GP, suggesting PtAu—MnO_2/GP nanocomposite possessed the highest electrochemical activity of the as-prepared in acid solutions. The findings further revealed that the synthesized PtAu—MnO_2/GP binary nanocomposites showed superior catalytic glucose oxidation performance, having shown excellent reproducibility and stability, satisfactory selectivity, low detection limit (0.02 mM, $S/N = 3$), high sensitivity (58.54 $\mu A\,cm^{-2}\,mM^{-1}$), a wide linear range (0.1—30.0 mM), and tolerability to mechanical stress in comparison to low performance exhibited by MnO_2/GP, PtAu/GP, and PtAu/MnO_2/GP electrodes. This technique of metal and metal oxides cogrowth on freestanding carbon substrates has provided a scalable and convenient approach for the development of flexible electrochemical sensors with high-performance.

8 CONCLUSIONS, CHALLENGES, AND FUTURE PROSPECTS

MOBNCs have demonstrated themselves as versatile materials with numerous applications in different sectors. Owing to the simplicity in their synthesis and wide range of tunable properties, MOBNCs have found application as photocatalysts, antimicrobial agents, adsorbents, energy storage devices, sensors, drug delivery, medical devices, and surgical tools. Responding to a growing microbial resistance against common drugs and antibiotics, quite a number of investigations have been performed to design antimicrobial materials. It is worth noting that technological development in the synthesis of these nanocomposites has transformed the field of biomedicine, offering the possibility of bioimaging, early detection systems, diagnosis, and treatment of diseases caused by drug-resistant microorganisms.

These MOBNCs can be synthesized using solid, vapor, and solution phase techniques, wet chemical methods, and these techniques provide a simple and cost-effective route for their fabrication. However, the critical issues to be considered involve devising greener fabrication routes, as well as the ability to control and manipulate key properties of the nanostructured metal oxides to be incorporated within the composite. This capability would make it feasible to effectively harness the synergic effects of their constituents and consequently produce promising effects for possible antibacterial applications. It is well established that the type of synthesis protocol significantly affects the basic properties of the nanocomposite such as morphology, size, shape, dispersity, presence, and type of stress and defects in the crystal that in turn determines their interaction with bacterial and mammalian cells. In polymetallic oxide nanocomposites, the physicochemical parameters such as solubility and degree of agglomeration of the respective components are modified, in pursuit of attaining enhanced reactivity toward living organisms. Although a large number of existing fabrication strategies provide a rich foundation that may promote research devoted to biomedical applications, next-generation nanostructured composites could be developed using a low temperature in vitro biomimetic synthesis approach. In addition, for better exploration of MOBNCs, an in-depth study of their physicochemical and biological properties is of paramount importance.

Apart from the synthesis protocols, diverse characterization techniques including powder XRD, SEM, AFM, TGA, TEM, FTIR, UV—Vis, and Raman spectroscopy are available for the analysis of MOBNCs. These characterization tools discussed in this chapter provide insights into the surface properties, particle size, impurities, dopant level, thermal stability, and degradation pattern of the nanocomposite. In this chapter, we have reviewed MOBNCs as an emerging class of nanostructured materials suitable for antimicrobial and biomedical applications. We anticipate that the brief account of MOBNCs provided in this review provides a better understanding of the material and helps the readers with the design of nanocomposites for novel biomedical applications. In the future, such a rational design of MOBNCs will lead to the exploration of the basic understanding of material interactions as well as addressing the challenges of toxicity, long-term stability, and biocompatibility of MOBNCs.

ACKNOWLEDGMENTS

The financial assistance of the University of Zululand and the National Research Foundation, South Africa through the South African Research Chair Initiative (SARChI) is hereby acknowledged. SJO and NMM thank the National Research Foundation (NRF) for funding under the South African Research Chair for Nanotechnology.

REFERENCES

[1] F.C. Tenover, Mechanisms of antimicrobial resistance in bacteria, The American Journal of Medicine 119 (6) (2006) S3–S10.

[2] G.F. Webb, E.M. D'Agata, P. Magal, S. Ruan, A model of antibiotic-resistant bacterial epidemics in hospitals, Proceedings of the National Academy of Sciences 102 (37) (2005) 13343–13348.

[3] H.H. Lara, N.V. Ayala-Núñez, L.d.C.I. Turrent, C.R. Padilla, Bactericidal effect of silver nanoparticles against multidrug-resistant bacteria, World Journal of Microbiology and Biotechnology 26 (4) (2010) 615–621.

[4] P.D. Marcato, N. Durán, New aspects of nanopharmaceutical delivery systems, Journal of Nanoscience and Nanotechnology 8 (5) (2008) 2216–2229.

[5] R. Singh, H.S. Nalwa, Medical applications of nanoparticles in biological imaging, cell labeling, antimicrobial agents, and anticancer nanodrugs, Journal of Biomedical Nanotechnology 7 (4) (2011) 489–503.

[6] M. Jena, S. Mishra, S. Jena, S. Mishra, Nanotechnology-future prospect in recent medicine: a review, International Journal of Basic and Clinical Pharmacology 2 (4) (2013) 353–359.

[7] I.Y. Wong, S.N. Bhatia, M. Toner, Nanotechnology: emerging tools for biology and medicine, Genes and Development 27 (22) (2013) 2397–2408.

[8] S.S. Prasanna, K. Balaji, S. Pandey, S. Rana, Metal oxide based nanomaterials and their polymer nanocomposites, in: Nanomaterials and Polymer Nanocomposites, Elsevier, 2019, pp. 123–144.

[9] O.V. Salata, Applications of nanoparticles in biology and medicine, Journal of Nanobiotechnology 2 (1) (2004) 3.

[10] W. Qi, X. Zhang, H. Wang, Self-assembled polymer nanocomposites for biomedical application, Current Opinion in Colloid and Interface Science 35 (2018) 36–41.

[11] F. Wahid, C. Zhong, H.-S. Wang, X.-H. Hu, L.-Q. Chu, Recent advances in antimicrobial hydrogels containing metal ions and metals/metal oxide nanoparticles, Polymers 9 (12) (2017) 636.

[12] Y. Wu, H. Yan, M. Huang, B. Messer, J.H. Song, P. Yang, Inorganic semiconductor nanowires: rational growth, assembly, and novel properties, Chemistry–A European Journal 8 (6) (2002) 1260–1268.

[13] C. Ray, T. Pal, Recent advances of metal–metal oxide nanocomposites and their tailored nanostructures in numerous catalytic applications, Journal of Materials Chemistry A 5 (20) (2017) 9465–9487.

[14] A. Raghunath, E. Perumal, Metal oxide nanoparticles as antimicrobial agents: a promise for the future, International Journal of Antimicrobial Agents 49 (2) (2017) 137–152.

[15] A. Lateef, R. Nazir, Metal Nanocomposites: Synthesis, Characterization and Their Applications, Science and Applications of Tailored Nanostructures, first ed., One Central Press, Italy, 2017, pp. 239–240.

[16] S. Pina, J.M. Oliveira, R.L. Reis, Natural-based nanocomposites for bone tissue engineering and regenerative medicine: a review, Advanced Materials 27 (7) (2015) 1143–1169.

[17] M.A. Rafiee, J. Rafiee, Z. Wang, H. Song, Z.-Z. Yu, N. Koratkar, Enhanced mechanical properties of nanocomposites at low graphene content, ACS Nano 3 (12) (2009) 3884–3890.

[18] A.R. Horrocks, D. Price, D. Price, Fire Retardant Materials, Woodhead Publishing, 2001.

[19] T.E. Twardowski, Introduction to Nanocomposite Materials: Properties, Processing, Characterization, DEStech Publications, Inc., 2007.

[20] M. Mariano, N. El Kissi, A. Dufresne, Cellulose nanocrystals and related nanocomposites: review of some properties and challenges, Journal of Polymer Science Part B: Polymer Physics 52 (12) (2014) 791–806.

[21] H. Hu, L. Onyebueke, A. Abatan, Characterizing and modeling mechanical properties of nanocomposites-review and evaluation, Journal of Minerals and Materials Characterization and Engineering 9 (04) (2010) 275.

[22] M.R. Vaezi, Coupled semiconductor metal oxide nanocomposites: types, synthesis conditions and properties, in: Advances in Composite Materials for Medicine and Nanotechnology, IntechOpen, 2011.

[23] P. Schexnailder, G. Schmidt, Nanocomposite polymer hydrogels, Colloid and Polymer Science 287 (1) (2009) 1–11.

[24] A.K. Gaharwar, N.A. Peppas, A. Khademhosseini, Nanocomposite hydrogels for biomedical applications, Biotechnology and Bioengineering 111 (3) (2014) 441–453.

[25] A.O. Juma, E.A. Arbab, C.M. Muiva, L.M. Lepodise, G.T. Mola, Synthesis and characterization of CuO-NiO-ZnO mixed metal oxide nanocomposite, Journal of Alloys and Compounds 723 (2017) 866–872.

[26] Y. Wang, L. Cai, Y. Li, Y. Tang, C. Xie, Structural and photoelectrocatalytic characteristic of $ZnO/ZnWO_4/WO_3$ nanocomposites with double heterojunctions, Physica E: Low-dimensional Systems and Nanostructures 43 (1) (2010) 503–509.

[27] X. Jiang, X. Zhao, L. Duan, H. Shen, H. Liu, T. Hou, F. Wang, Enhanced photoluminescence and photocatalytic activity of $ZnO-ZnWO_4$ nanocomposites synthesized by a precipitation method, Ceramics International 42 (14) (2016) 15160–15165.

[28] Z. Wang, M. Yang, G. Shen, H. Liu, Y. Chen, Q. Wang, Catalytic removal of benzene over CeO_2-MnO_x composite oxides with rod-like morphology supporting PdO, Journal of Nanoparticle Research 16 (5) (2014) 2367.

[29] M.V. Dozzi, S. Marzorati, M. Longhi, M. Coduri, L. Artiglia, E. Selli, Photocatalytic activity of TiO_2-WO_3 mixed oxides in relation to electron transfer efficiency, Applied Catalysis B: Environmental 186 (2016) 157–165.

[30] P.P. Dorneanu, A. Airinei, N. Olaru, M. Homocianu, V. Nica, F. Doroftei, Preparation and characterization of NiO, ZnO and NiO–ZnO composite nanofibers by electrospinning method, Materials Chemistry and Physics 148 (3) (2014) 1029–1035.

[31] J. Vidic, S. Stankic, F. Haque, D. Ciric, R. Le Goffic, A. Vidy, J. Jupille, B. Delmas, Selective antibacterial effects of mixed ZnMgO nanoparticles, Journal of Nanoparticle Research 15 (5) (2013) 1595.

[32] R.K. Sharma, R. Ghose, Synthesis of Co_3O_4–ZnO mixed metal oxide nanoparticles by homogeneous precipitation method, Journal of Alloys and Compounds 686 (2016) 64–73.

[33] T. Kundu, M. Mukherjee, D. Chakravorty, T. Sinha, Growth of nano-α-Fe_2O_3 in a titania matrix by the sol–gel route, Journal of Materials Science 33 (7) (1998) 1759–1763.

[34] S. Sivakumar, C. Sibu, P. Mukundan, P.K. Pillai, K. Warrier, Nanoporous titania–alumina mixed oxides—an alkoxide free sol–gel synthesis, Materials Letters 58 (21) (2004) 2664–2669.

[35] W. Wunderlich, P. Padmaja, K. Warrier, TEM characterization of sol-gel-processed alumina–silica and alumina–titania nano-hybrid oxide catalysts, Journal of the European Ceramic Society 24 (2) (2004) 313–317.

[36] P.H. Camargo, G.G. Nunes, G.R. Friedermann, D.J. Evans, G.J. Leigh, G. Tremiliosi-Filho, E.L. de Sá, A.J. Zarbin, J.F. Soares, Titanium and iron oxides produced by sol–gel processing of [FeCl {Ti_2 (OPri) 9}]: structural, spectroscopic and morphological features, Materials Research Bulletin 38 (15) (2003) 1915–1928.

[37] M.M. Natile, A. Glisenti, New NiO/Co_3O_4 and Fe_2O_3/Co_3O_4 nanocomposite catalysts: synthesis and characterization, Chemistry of Materials 15 (13) (2003) 2502–2510.

[38] M. Nazari, R. Halladj, Adsorptive removal of fluoride ions from aqueous solution by using sonochemically synthesized nanomagnesia/alumina adsorbents: an experimental and modeling study, Journal of the Taiwan Institute of Chemical Engineers 45 (5) (2014) 2518–2525.

[39] A. Hamrouni, H. Lachheb, A. Houas, Synthesis, characterization and photocatalytic activity of ZnO-SnO_2 nanocomposites, Materials Science and Engineering: B 178 (20) (2013) 1371–1379.

[40] B. Li, Y. Wang, Facile synthesis and photocatalytic activity of ZnO–CuO nanocomposite, Superlattices and Microstructures 47 (5) (2010) 615–623.

[41] N. Bayal, P. Jeevanandam, Synthesis of TiO_2– MgO mixed metal oxide nanoparticles via a sol– gel method and studies on their optical properties, Ceramics International 40 (10) (2014) 15463–15477.

[42] K. Kaviyarasu, C.M. Magdalane, K. Anand, E. Manikandan, M. Maaza, Synthesis and characterization studies of MgO: CuO nanocrystals by wet-chemical method, Spectrochimica Acta Part A: Molecular and Biomolecular Spectroscopy 142 (2015) 405–409.

[43] M. Farrokhi, S.-C. Hosseini, J.-K. Yang, M. Shirzad-Siboni, Application of ZnO–Fe_3O_4 nanocomposite on the removal of azo dye from aqueous solutions: kinetics and equilibrium studies, Water, Air, and Soil Pollution 225 (9) (2014) 2113.

[44] S. Mathur, M. Veith, H. Shen, S. Hüfner, M.H. Jilavi, Structural and optical properties of $NdAlO_3$ nanocrystals embedded in an Al_2O_3 matrix, Chemistry of Materials 14 (2) (2002) 568–582.

[45] R.K. Sharma, R. Ghose, Synthesis of nanocrystalline CuO–ZnO mixed metal oxide powder by a homogeneous precipitation method, Ceramics International 40 (7) (2014) 10919–10926.

[46] Q. Ge, T. Lei, Y. Zhou, Microstructure and mechanical properties of hot pressed Al_2O_3–ZrO_2 ceramics prepared from ultrafine powders, Materials Science and Technology 7 (6) (1991) 490–494.

[47] R.K. Sharma, D. Kumar, R. Ghose, Synthesis of nanocrystalline ZnO–NiO mixed metal oxide powder by homogeneous precipitation method, Ceramics International 42 (3) (2016) 4090–4098.

[48] G. Schimanke, M. Martin, J. Kunert, H. Vogel, Characterization of Mo-V-W mixed oxide catalysts by ex situ and in situ X-ray absorption spectroscopy, Zeitschrift für Anorganische und Allgemeine Chemie 631 (6-7) (2005) 1289–1296.

[49] H. Arandiyan, M. Parvari, Studies on mixed metal oxides solid solutions as heterogeneous catalysts, Brazilian Journal of Chemical Engineering 26 (1) (2009) 63–74.

[50] S. Mukhopadhyay, K. Jacob, Phase equilibria in the system CaO-CoO-SiO_2 and Gibbs energies of formation of the quaternary oxides $CaCoSi_2O_6$, $Ca_2CoSi_2O_7$, and $CaCoSiO_4$, American Mineralogist 81 (1996) 963–972.

[51] X. Li, R. Zhao, H. Jiang, Y. Zhai, P. Ma, Preparation and catalytic properties of ZnO-CeO_2-TiO_2 composite, Synthesis and Reactivity in Inorganic Metal-Organic and Nano-Metal Chemistry 46 (5) (2016) 775–782.

[52] P. Roy, S. Berger, P. Schmuki, TiO_2 nanotubes: synthesis and applications, Angewandte Chemie International Edition 50 (13) (2011) 2904–2939.

[53] J.E. Yoo, K. Lee, M. Altomare, E. Selli, P. Schmuki, Self-organized arrays of single-metal catalyst particles in TiO_2 cavities: a highly efficient photocatalytic system, Angewandte Chemie International Edition 52 (29) (2013) 7514–7517.

[54] Z. Lian, W. Wang, S. Xiao, X. Li, Y. Cui, D. Zhang, G. Li, H. Li, Plasmonic silver quantum dots coupled with hierarchical TiO_2 nanotube arrays photoelectrodes for efficient visible-light photoelectrocatalytic hydrogen evolution, Scientific Reports 5 (2015) 10461.

[55] Z.-D. Gao, H.-F. Liu, C.-Y. Li, Y.-Y. Song, Biotemplated synthesis of Au nanoparticles–TiO_2 nanotube junctions for enhanced direct electrochemistry of heme proteins, Chemical Communications 49 (8) (2013) 774–776.

[56] H. Tang, G. Meng, Q. Huang, Z. Zhang, Z. Huang, C. Zhu, Arrays of cone-shaped ZnO nanorods decorated with Ag nanoparticles as 3D surface-enhanced Raman scattering substrates for rapid detection of trace polychlorinated biphenyls, Advanced Functional Materials 22 (1) (2012) 218–224.

[57] H.M. Chen, C.K. Chen, C.-J. Chen, L.-C. Cheng, P.C. Wu, B.H. Cheng, Y.Z. Ho, M.L. Tseng, Y.-Y. Hsu, T.-S. Chan, Plasmon inducing effects for enhanced photoelectrochemical water splitting: X-ray absorption approach to

electronic structures, ACS Nano 6 (8) (2012) 7362–7372.

[58] Y.-C. Pu, G. Wang, K.-D. Chang, Y. Ling, Y.-K. Lin, B.C. Fitzmorris, C.-M. Liu, X. Lu, Y. Tong, J.Z. Zhang, Au nanostructure-decorated TiO_2 nanowires exhibiting photoactivity across entire UV-visible region for photoelectrochemical water splitting, Nano Letters 13 (8) (2013) 3817–3823.

[59] Z. Chen, Y. Tang, C. Liu, Y. Leung, G. Yuan, L. Chen, Y. Wang, I. Bello, J. Zapien, W. Zhang, Vertically aligned ZnO nanorod arrays sentisized with gold nanoparticles for Schottky barrier photovoltaic cells, The Journal of Physical Chemistry C 113 (30) (2009) 13433–13437.

[60] L. Sun, J. Li, C. Wang, S. Li, Y. Lai, H. Chen, C. Lin, Ultrasound aided photochemical synthesis of Ag loaded TiO_2 nanotube arrays to enhance photocatalytic activity, Journal of Hazardous Materials 171 (1–3) (2009) 1045–1050.

[61] L. Chen, L. Luo, Z. Chen, M. Zhang, J.A. Zapien, C.S. Lee, S.T. Lee, ZnO/Au composite nanoarrays as substrates for surface-enhanced Raman scattering detection, The Journal of Physical Chemistry C 114 (1) (2009) 93–100.

[62] Z. Chen, Y. Tang, Y. Liu, G. Yuan, W. Zhang, J. Zapien, I. Bello, W. Zhang, C. Lee, S. Lee, ZnO nanowire arrays grown on Al: ZnO buffer layers and their enhanced electron field emission, Journal of Applied Physics 106 (6) (2009) 064303.

[63] J. Yang, Z. Li, W. Zhao, C. Zhao, Y. Wang, X. Liu, Controllable synthesis of Ag—CuO composite nanosheets with enhanced photocatalytic property, Materials Letters 120 (2014) 16–19.

[64] X. Zhao, B. Zhang, K. Ai, G. Zhang, L. Cao, X. Liu, H. Sun, H. Wang, L. Lu, Monitoring catalytic degradation of dye molecules on silver-coated ZnO nanowire arrays by surface-enhanced Raman spectroscopy, Journal of Materials Chemistry 19 (31) (2009) 5547–5553.

[65] H. Yu, X. Wang, H. Sun, M. Huo, Photocatalytic degradation of malathion in aqueous solution using an Au—Pd—TiO_2 nanotube film, Journal of Hazardous Materials 184 (1–3) (2010) 753–758.

[66] M. Kim, J.C. Park, A. Kim, K.H. Park, H. Song, Porosity Control of Pd@ SiO_2 Yolk—Shell nanocatalysts by the formation of nickel phyllosilicate and its influence on Suzuki coupling reactions, Langmuir 28 (15) (2012) 6441–6447.

[67] J. Lee, J.C. Park, H. Song, A nanoreactor framework of a Au@ SiO_2 yolk/shell structure for catalytic reduction of p-nitrophenol, Advanced Materials 20 (8) (2008) 1523–1528.

[68] J.C. Park, J.U. Bang, J. Lee, C.H. Ko, H. Song, Ni@ SiO_2 yolk-shell nanoreactor catalysts: high temperature stability and recyclability, Journal of Materials Chemistry 20 (7) (2010) 1239–1246.

[69] L. He, Y. Liu, J. Liu, Y. Xiong, J. Zheng, Y. Liu, Z. Tang, Core—shell noble-metal@ metal-organic-framework nanoparticles with highly selective sensing property, Angewandte Chemie International Edition 52 (13) (2013) 3741–3745.

[70] J. Li, X. Liang, J.B. Joo, I. Lee, Y. Yin, F. Zaera, Mass transport across the porous oxide shells of core—shell and yolk—shell nanostructures in liquid phase, The Journal of Physical Chemistry C 117 (39) (2013) 20043–20053.

[71] R. Güttel, M. Paul, F. Schüth, Ex-post size control of high-temperature-stable yolk—shell Au,@ ZrO_2 catalysts, Chemical Communications 46 (6) (2010) 895–897.

[72] L. Wang, H. Dou, Z. Lou, T. Zhang, Encapsuled nanoreactors (Au@ SnO_2): a new sensing material for chemical sensors, Nanoscale 5 (7) (2013) 2686–2691.

[73] M. Wang, J. Han, H. Xiong, R. Guo, Yolk@ shell nanoarchitecture of Au@ r-GO/TiO_2 hybrids as powerful visible light photocatalysts, Langmuir 31 (22) (2015) 6220–6228.

[74] W. Tu, Y. Zhou, H. Li, P. Li, Z. Zou, Au@ TiO_2 yolk—shell hollow spheres for plasmon-induced photocatalytic reduction of CO_2 to solar fuel via a local electromagnetic field, Nanoscale 7 (34) (2015) 14232–14236.

[75] I. Lee, J.B. Joo, Y. Yin, F. Zaera, A yolk@ shell nanoarchitecture for Au/TiO_2 catalysts, Angewandte Chemie International Edition 50 (43) (2011) 10208–10211.

[76] X. Li, X. Zhou, H. Guo, C. Wang, J. Liu, P. Sun, F. Liu, G. Lu, Design of Au@ ZnO yolk—shell nanospheres with enhanced gas sensing properties, ACS Applied Materials and Interfaces 6 (21) (2014) 18661–18667.

[77] P. Rai, J.-W. Yoon, H.-M. Jeong, S.-J. Hwang, C.-H. Kwak, J.-H. Lee, Design of highly sensitive and selective Au@ NiO yolk—shell nanoreactors for gas sensor applications, Nanoscale 6 (14) (2014) 8292–8299.

[78] F. Li, D.P. Josephson, A. Stein, Colloidal assembly: the road from particles to colloidal molecules and crystals, Angewandte Chemie International Edition 50 (2) (2011) 360–388.

[79] Z. Yang, A.H. Muller, C. Xu, P.S. Doyle, J.M. DeSimone, J. Lahann, F. Sciortino, S. Glotzer, L. Hong, D.A. Aarts, Janus Particle Synthesis, Self-Assembly and Applications, Royal Society of Chemistry, 2012.

[80] Z.W. Seh, S. Liu, M. Low, S.Y. Zhang, Z. Liu, A. Mlayah, M.Y. Han, Janus Au-TiO_2 photocatalysts with strong localization of plasmonic near-fields for efficient visible-light hydrogen generation, Advanced Materials 24 (17) (2012) 2310–2314.

[81] N.P. Herring, K. AbouZeid, M.B. Mohamed, J. Pinsk, M.S. El-Shall, Formation mechanisms of gold—zinc oxide hexagonal nanopyramids by heterogeneous nucleation using microwave synthesis, Langmuir 27 (24) (2011) 15146–15154.

[82] Y. Chen, D. Zeng, K. Zhang, A. Lu, L. Wang, D.-L. Peng, Au—ZnO hybrid nanoflowers, nanomultipods and nanopyramids: one-pot reaction synthesis and photocatalytic properties, Nanoscale 6 (2) (2014) 874–881.

[83] M.N. Tahir, F. Natalio, M.A. Cambaz, M. Panthöfer, R. Branscheid, U. Kolb, W. Tremel, Controlled synthesis of linear and branched Au@ ZnO hybrid nanocrystals and their photocatalytic properties, Nanoscale 5 (20) (2013) 9944–9949.

[84] K.X. Yao, X. Liu, L. Zhao, H.C. Zeng, Y. Han, Site-specific growth of Au particles on ZnO nanopyramids under ultraviolet illumination, Nanoscale 3 (10) (2011) 4195–4200.

[85] Z.W. Seh, S. Liu, S.Y. Zhang, M. Bharathi, H. Ramanarayan, M. Low, K.W. Shah, Y.W. Zhang, M.Y. Han, Anisotropic growth of titania onto various gold nanostructures: synthesis, theoretical understanding, and optimization for catalysis, Angewandte Chemie International Edition 50 (43) (2011) 10140–10143.

[86] S. Pradhan, D. Ghosh, S. Chen, Janus nanostructures based on Au– TiO_2 heterodimers and their photocatalytic activity in the oxidation of methanol, ACS Applied Materials and Interfaces 1 (9) (2009) 2060–2065.

[87] P. Li, Z. Wei, T. Wu, Q. Peng, Y. Li, Au– ZnO hybrid nanopyramids and their photocatalytic properties, Journal of the American Chemical Society 133 (15) (2011) 5660–5663.

[88] C. Wang, H. Yin, S. Dai, S. Sun, A general approach to noble metal– metal oxide dumbbell nanoparticles and their catalytic application for CO oxidation, Chemistry of Materials 22 (10) (2010) 3277–3282.

[89] Z. Bian, J. Zhu, F. Cao, Y. Lu, H. Li, In situ encapsulation of Au nanoparticles in mesoporous core–shell TiO_2 microspheres with enhanced activity and durability, Chemical Communications (25) (2009) 3789–3791.

[90] Q. Zhang, D.Q. Lima, I. Lee, F. Zaera, M. Chi, Y. Yin, A highly active titanium dioxide based visible-light photocatalyst with nonmetal doping and plasmonic metal decoration, Angewandte Chemie International Edition 50 (31) (2011) 7088–7092.

[91] X.-F. Wu, H.-Y. Song, J.-M. Yoon, Y.-T. Yu, Y.-F. Chen, Synthesis of core– shell Au@ TiO_2 nanoparticles with truncated wedge-shaped morphology and their photocatalytic properties, Langmuir 25 (11) (2009) 6438–6447.

[92] L. Han, C. Zhu, P. Hu, S. Dong, One-pot synthesis of a Au@ TiO_2 core–shell nanocomposite and its catalytic property, RSC Advances 3 (31) (2013) 12568–12570.

[93] Y. Yu, C.Y. Cao, Z. Chen, H. Liu, P. Li, Z.F. Dou, W.G. Song, Au nanoparticles embedded into the inner wall of TiO_2 hollow spheres as a nanoreactor with superb thermal stability, Chemical Communications 49 (30) (2013) 3116–3118.

[94] Z.W. Seh, S. Liu, S.-Y. Zhang, K.W. Shah, M.-Y. Han, Synthesis and multiple reuse of eccentric Au@ TiO_2 nanostructures as catalysts, Chemical Communications 47 (23) (2011) 6689–6691.

[95] N. Zhang, Y.-J. Xu, Aggregation-and leaching-resistant, reusable, and multifunctional Pd@ CeO_2 as a robust nanocatalyst achieved by a hollow core–shell strategy, Chemistry of Materials 25 (9) (2013) 1979–1988.

[96] M. Cargnello, N.L. Wieder, T. Montini, R.J. Gorte, P. Fornasiero, Synthesis of dispersible Pd@ CeO_2 core–shell nanostructures by self-assembly, Journal of the American Chemical Society 132 (4) (2009) 1402–1409.

[97] L. Adijanto, D.A. Bennett, C. Chen, A.S. Yu, M. Cargnello, P. Fornasiero, R.J. Gorte, J.M. Vohs, Exceptional thermal stability of Pd@ CeO_2 core–shell catalyst nanostructures grafted onto an oxide surface, Nano Letters 13 (5) (2013) 2252–2257.

[98] L. Adijanto, A. Sampath, A.S. Yu, M. Cargnello, P. Fornasiero, R.J. Gorte, J.M. Vohs, Synthesis and stability of Pd@ CeO_2 core–shell catalyst films in solid oxide fuel cell anodes, ACS Catalysis 3 (8) (2013) 1801–1809.

[99] A. Beltram, M. Melchionna, T. Montini, L. Nasi, R. Gorte, M. Prato, P. Fornasiero, Improved activity and stability of Pd@ CeO_2 core–shell catalysts hybridized with multi-walled carbon nanotubes in the water gas shift reaction, Catalysis Today 253 (2015) 142–148.

[100] A.C. Balazs, T. Emrick, T.P. Russell, Nanoparticle polymer composites: where two small worlds meet, Science 314 (5802) (2006) 1107–1110.

[101] F. Caruso, Nanoengineering of particle surfaces, Advanced Materials 13 (1) (2001) 11–22.

[102] Y. Haldorai, J.-J. Shim, Fabrication of metal oxide– polymer hybrid nanocomposites, in: Organic-Inorganic Hybrid Nanomaterials, Springer, 2014, pp. 249–281.

[103] M.M. Demir, P. Castignolles, Ü. Akbey, G. Wegner, In-situ bulk polymerization of dilute particle/MMA dispersions, Macromolecules 40 (12) (2007) 4190–4198.

[104] Y. Wu, Y. Zhang, J. Xu, M. Chen, L. Wu, One-step preparation of PS/TiO_2 nanocomposite particles via miniemulsion polymerization, Journal of Colloid and Interface Science 343 (1) (2010) 18–24.

[105] M. Miyauchi, Y. Li, H. Shimizu, Enhanced degradation in nanocomposites of TiO_2 and biodegradable polymer, Environmental Science and Technology 42 (12) (2008) 4551–4554.

[106] A. Ali, M. AlSalhi, M. Atif, A.A. Ansari, M.Q. Israr, J. Sadaf, E. Ahmed, O. Nur, M. Willander, Potentiometric urea biosensor utilizing nanobiocomposite of chitosan-iron oxide magnetic nanoparticles, Journal of Physics: Conference Series, IOP Publishing (2013) 2024.

[107] V. Pandey, G. Mishra, S. Verma, M. Wan, R. Yadav, Synthesis and ultrasonic investigations of CuO-PVA nanofluid, Materials Sciences and Applications 3 (9) (2012) 664.

[108] É. Allard-Vannier, S. Cohen-Jonathan, J. Gautier, K. Hervé-Aubert, E. Munnier, M. Soucé, P. Legras, C. Passirani, I. Chourpa, Pegylated magnetic nanocarriers for doxorubicin delivery: a quantitative determination of stealthiness in vitro and in vivo, European Journal of Pharmaceutics and Biopharmaceutics 81 (3) (2012) 498–505.

[109] Y. Haldorai, J.-J. Shim, Chitosan-zinc oxide hybrid composite for enhanced dye degradation and antibacterial activity, Composite Interfaces 20 (5) (2013) 365–377.

[110] M. Song, C. Pan, C. Chen, J. Li, X. Wang, Z. Gu, The application of new nanocomposites: enhancement effect of polylactide nanofibers/nano-TiO_2 blends on biorecognition of anticancer drug daunorubicin, Applied Surface Science 255 (2) (2008) 610–612.

[111] A.L. Rogach, D.V. Talapin, E.V. Shevchenko, A. Kornowski, M. Haase, H. Weller, Organization of matter on different size scales: monodisperse nanocrystals and their superstructures, Advanced Functional Materials 12 (10) (2002) 653—664.

[112] S. Stankic, S. Suman, F. Haque, J. Vidic, Pure and multi metal oxide nanoparticles: synthesis, antibacterial and cytotoxic properties, Journal of Nanobiotechnology 14 (1) (2016) 73.

[113] M. Khodaei, M. Enayati, F. Karimzadeh, Mechanochemically synthesized metallic-ceramic nanocomposite; mechanisms and properties, in: Advances in Nanocomposites-Synthesis, Characterization and Industrial Applications, IntechOpen, 2011.

[114] K. Richter, A. Birkner, A.V. Mudring, Stabilizer-free metal nanoparticles and metal—metal oxide nanocomposites with long-term stability prepared by physical vapor deposition into ionic liquids, Angewandte Chemie International Edition 49 (13) (2010) 2431—2435.

[115] V.A. Moshnikov, I. Gracheva, A.S. Lenshin, Y.M. Spivak, M.G. Anchkov, V.V. Kuznetsov, J.M. Olchowik, Porous silicon with embedded metal oxides for gas sensing applications, Journal of Non-crystalline Solids 358 (3) (2012) 590—595.

[116] H. Bakhsheshi-Rad, E. Hamzah, M. Daroonparvar, S.N. Saud, M. Abdul-Kadir, Bi-layer nano-TiO2/FHA composite coatings on Mg—Zn—Ce alloy prepared by combined physical vapour deposition and electrochemical deposition methods, Vacuum 110 (2014) 127—135.

[117] A. Honciuc, M. Laurin, S. Albu, M. Sobota, P. Schmuki, J. Libuda, Controlling the adsorption kinetics via nanostructuring: Pd nanoparticles on TiO2 nanotubes, Langmuir 26 (17) (2010) 14014—14023.

[118] Y. Kong, D. Yu, B. Zhang, W. Fang, S. Feng, Ultraviolet-emitting ZnO nanowires synthesized by a physical vapor deposition approach, Applied Physics Letters 78 (4) (2001) 407—409.

[119] M. Henini, Handbook of thin-film deposition processes and techniques, K.K. Schuegraph (ed.); Noyes, ISBN: 0-8155-1153-1, Microelectronics Journal 31 (3) (2000) 219.

[120] P. Savale, Physical vapor deposition (PVD) methods for synthesis of thin films: a comparative study, Archives of Applied Science Research 8 (5) (2016) 1—8.

[121] S.A. Campbell, The Science and Engineering of Microelectronic Fabrication (The Oxford Series in Electrical and Computer Engineering), 2001.

[122] D. Ortega, Q. Pankhurst, Nanoscience: Volume 1: Nanostructures through Chemistry, The Royal Society of Chemistry, 2013.

[123] Y. Zhang, L. Wang, X. Liu, Y. Yan, C. Chen, J. Zhu, Synthesis of nano/micro zinc oxide rods and arrays by thermal evaporation approach on cylindrical shape substrate, The Journal of Physical Chemistry B 109 (27) (2005) 13091—13093.

[124] H. Cheng, J. Cheng, Y. Zhang, Q.-M. Wang, Large-scale fabrication of ZnO micro-and nano-structures by microwave thermal evaporation deposition, Journal of Crystal Growth 299 (1) (2007) 34—40.

[125] D. Chhikara, K. Srivatsa, S.K. Muthusamy, On the synthesis and characterization of ZnO/MgO nanocomposite by thermal evaporation technique, Solid State Sciences 37 (2014) 108—113.

[126] M. Zhi, L. Zhu, Z. Ye, F. Wang, B. Zhao, Preparation and properties of ternary ZnMgO nanowires, The Journal of Physical Chemistry B 109 (50) (2005) 23930—23934.

[127] P. Willmott, J. Huber, Pulsed laser vaporization and deposition, Reviews of Modern Physics 72 (1) (2000) 315.

[128] M.N. Ashfold, F. Claeyssens, G.M. Fuge, S.J. Henley, Pulsed laser ablation and deposition of thin films, Chemical Society Reviews 33 (1) (2004) 23—31.

[129] S. Buzby, S. Franklin, S. Shah, Synthesis, Properties, and Applications of Oxide Nanomaterials, Wiley, Canada, 2007.

[130] T.J. Goodwin, V.J. Leppert, S.H. Risbud, I.M. Kennedy, H.W. Lee, Synthesis of gallium nitride quantum dots through reactive laser ablation, Applied Physics Letters 70 (23) (1997) 3122—3124.

[131] D. Geohegan, A. Puretzky, D. Rader, Gas-phase nanoparticle formation and transport during pulsed laser deposition of $Y_1 Ba_2 Cu_3 O_{7-d}$, Applied Physics Letters 74 (25) (1999) 3788—3790.

[132] Y. Zhang, H. Gu, S. Iijima, Single-wall carbon nanotubes synthesized by laser ablation in a nitrogen atmosphere, Applied Physics Letters 73 (26) (1998) 3827—3829.

[133] T. Lippert, D. Chrisey, A. Purice, C. Constantinescu, M. Filipescu, N. Scarisoreanu, M. Dinescu, Laser processing of soft materials, Romanian Reports in Physics 59 (2) (2007) 483.

[134] R. Rella, J. Spadavecchia, M. Manera, S. Capone, A. Taurino, M. Martino, A. Caricato, T. Tunno, Acetone and ethanol solid-state gas sensors based on TiO2 nanoparticles thin film deposited by matrix assisted pulsed laser evaporation, Sensors and Actuators B: Chemical 127 (2) (2007) 426—431.

[135] A. Caricato, M. Manera, M. Martino, R. Rella, F. Romano, J. Spadavecchia, T. Tunno, D. Valerini, Uniform thin films of TiO2 nanoparticles deposited by matrix-assisted pulsed laser evaporation, Applied Surface Science 253 (15) (2007) 6471—6475.

[136] C. Czekalla, J. Guinard, C. Hanisch, B. Cao, E. Kaidashev, N. Boukos, A. Travlos, J. Renard, B. Gayral, D.L.S. Dang, Spatial fluctuations of optical emission from single ZnO/MgZnO nanowire quantum wells, Nanotechnology 19 (11) (2008) 115202.

[137] C.N. Afonso, J. Gonzalo, R. Serna, J. De Sande, C. Ricolleau, C. Grigis, M. Gandais, D. Hole, P.D. Townsend, Vacuum versus gas environment for the synthesis of nanocomposite films by pulsed-laser deposition, Applied Physics A 69 (1) (1999) S201—S207.

[138] T. Sasaki, N. Koshizaki, J.-W. Yoon, K.M. Beck, Preparation of Pt/TiO2 nanocomposite thin films by pulsed laser deposition and their photoelectrochemical behaviors, Journal of Photochemistry and Photobiology A: Chemistry 145 (1—2) (2001) 11—16.

[139] N.M. Aimon, D. Hun Kim, H. Kyoon Choi, C. Ross, Deposition of epitaxial BiFeO₃/CoFe₂O₄ nanocomposites on (001) SrTiO₃ by combinatorial pulsed laser deposition, Applied Physics Letters 100 (9) (2012) 092901.

[140] F. Urban Iii, A. Hosseini-Tehrani, P. Griffiths, A. Khabari, Y.-W. Kim, I. Petrov, Nanophase films deposited from a high-rate, nanoparticle beam, Journal of Vacuum Science and Technology B: Microelectronics and Nanometer Structures Processing, Measurement, and Phenomena 20 (3) (2002) 995–999.

[141] L. Cao, K. Hahn, C. Scheu, M. Rühle, Y. Wang, Z. Zhang, C. Gao, Y. Li, X. Zhang, M. He, Template-catalyst-free growth of highly ordered boron nanowire arrays, Applied Physics Letters 80 (22) (2002) 4226–4228.

[142] L. Cao, Z. Zhang, L. Sun, C. Gao, M. He, Y. Wang, Y. Li, X. Zhang, G. Li, J. Zhang, Well-Aligned boron nanowire arrays, Advanced Materials 13 (22) (2001) 1701–1704.

[143] T. Karabacak, A. Mallikarjunan, J.P. Singh, D. Ye, G.-C. Wang, T.-M. Lu, β-phase tungsten nanorod formation by oblique-angle sputter deposition, Applied Physics Letters 83 (15) (2003) 3096–3098.

[144] V. Tvarozek, K. Shtereva, I. Novotny, J. Kovac, P. Sutta, R. Srnanek, A. Vincze, RF diode reactive sputtering of n-and p-type zinc oxide thin films, Vacuum 82 (2) (2007) 166–169.

[145] D. Berkley, B. Johnson, N. Anand, K. Beauchamp, L. Conroy, A. Goldman, J. Maps, K. Mauersberger, M. Mecartney, J. Morton, I nsitu formation of superconducting YBa₂Cu₃O₇₋ₓ thin films using pure ozone vapor oxidation, Applied Physics Letters 53 (20) (1988) 1973–1975.

[146] M. Naito, H. Sato, Stoichiometry control of atomic beam fluxes by precipitated impurity phase detection in growth of (Pr, Ce) 2CuO₄ and (La, Sr) 2CuO₄ films, Applied Physics Letters 67 (17) (1995) 2557–2559.

[147] M. Klausmeier-Brown, G. Virshup, I. Bozovic, J. Eckstein, K. Ralls, Engineering of ultrathin barriers in high Tc, trilayer Josephson junctions, Applied Physics Letters 60 (22) (1992) 2806–2808.

[148] D.G. Schlom, Perspective: oxide molecular-beam epitaxy rocks!, APL Materials 3 (6) (2015) 062403.

[149] R. McKee, F. Walker, J. Conner, E. Specht, D. Zelmon, Molecular beam epitaxy growth of epitaxial barium silicide, barium oxide, and barium titanate on silicon, Applied Physics Letters 59 (7) (1991) 782–784.

[150] R. McKee, F. Walker, E. Specht, G. Jellison Jr., L. Boatner, J. Harding, Interface stability and the growth of optical quality perovskites on MgO, Physical Review Letters 72 (17) (1994) 2741.

[151] I. Bozovic, J. Eckstein, G. Virshup, Superconducting oxide multilayers and superlattices: physics, chemistry, and nanoengineering, Physica C: Superconductivity 235 (1994) 178–181.

[152] S.A. Chambers, Epitaxial growth and properties of thin film oxides, Surface Science Reports 39 (5–6) (2000) 105–180.

[153] J.H. Lee, L. Fang, E. Vlahos, X. Ke, Y.W. Jung, L.F. Kourkoutis, J.-W. Kim, P.J. Ryan, T. Heeg, M. Roeckerath, A strong ferroelectric ferromagnet created by means of spin–lattice coupling, Nature 466 (7309) (2010) 954.

[154] S. Polarz, A. Roy, M. Merz, S. Halm, D. Schröder, L. Schneider, G. Bacher, F.E. Kruis, M. Driess, Chemical vapor synthesis of size-selected zinc oxide nanoparticles, Small 1 (5) (2005) 540–552.

[155] J. Hu, M. Ouyang, P. Yang, C.M. Lieber, Controlled growth and electrical properties of heterojunctions of carbon nanotubes and silicon nanowires, Nature 399 (6731) (1999) 48.

[156] U.K. Gautam, Y. Bando, P.M. Costa, X. Fang, B. Dierre, T. Sekiguchi, D. Golberg, Inorganically filled carbon nanotubes: synthesis and properties, Pure and Applied Chemistry 82 (11) (2010) 2097–2109.

[157] M. Becker, R.N. d'Alnoncourt, K. Kähler, J. Sekulic, R.A. Fischer, M. Muhler, The synthesis of highly loaded Cu/Al₂O₃ and Cu/ZnO/Al₂O₃ catalysts by the two-step CVD of CuIIdiethylamino-2-propoxide in a fluidized-bed reactor, Chemical Vapor Deposition 16 (1-3) (2010) 85–92.

[158] D. Barreca, E. Comini, A.P. Ferrucci, A. Gasparotto, C. Maccato, C. Maragno, G. Sberveglieri, E. Tondello, First example of ZnO– TiO₂ nanocomposites by chemical vapor deposition: structure, morphology, composition, and gas sensing performances, Chemistry of Materials 19 (23) (2007) 5642–5649.

[159] H. Wang, H. Song, C. Xia, D. Peng, G. Meng, Aerosol-assisted MOCVD deposition of YDC thin films on (NiO+ YDC) substrates, Materials Research Bulletin 35 (14–15) (2000) 2363–2370.

[160] R.G. Palgrave, I.P. Parkin, Aerosol assisted chemical vapor deposition using nanoparticle precursors: a route to nanocomposite thin films, Journal of the American Chemical Society 128 (5) (2006) 1587–1597.

[161] M. Khalilian, Y. Abdi, E. Arzi, Formation of well-packed TiO₂ nanoparticles on multiwall carbon nanotubes using CVD method to fabricate high sensitive gas sensors, Journal of Nanoparticle Research 13 (10) (2011) 5257.

[162] H. Lee, J. Kang, M.S. Cho, J.-B. Choi, Y. Lee, MnO₂/graphene composite electrodes for supercapacitors: the effect of graphene intercalation on capacitance, Journal of Materials Chemistry 21 (45) (2011) 18215–18219.

[163] Y. Chen, Y. Huang, K. Lee, D. Tsai, K. Tiong, Characterization of IrO₂/CNT nanocomposites, Journal of Materials Science: Materials in Electronics 22 (7) (2011) 890–894.

[164] B. Reddy, Advances in Nanocomposites-Synthesis, Characterization and Industrial Applications, 2011.

[165] A. La Rosa, M. Yan, R. Fernandez, X. Wang, E. Zegarra, Top-down and Bottom-Up Approaches to Nanotechnology.

[166] A.P. Jadhav, C.W. Kim, H.G. Cha, A.U. Pawar, N.A. Jadhav, U. Pal, Y.S. Kang, Effect of different surfactants on the size control and optical properties of Y₂O₃: Eu³⁺ nanoparticles prepared by coprecipitation method, The Journal of Physical Chemistry C 113 (31) (2009) 13600–13604.

[167] K. Kalantari, M.B. Ahmad, K. Shameli, R. Khandanlou, Synthesis of talc/Fe$_3$O$_4$ magnetic nanocomposites using chemical co-precipitation method, International Journal of Nanomedicine 8 (2013) 1817.

[168] H. Sadegh, R. Shahryari-ghoshekandi, M. Kazemi, Study in synthesis and characterization of carbon nanotubes decorated by magnetic iron oxide nanoparticles, International Nano Letters 4 (4) (2014) 129–135.

[169] A. Gaikwad, S. Navale, V. Samuel, A. Murugan, V. Ravi, A co-precipitation technique to prepare BiNbO$_4$, MgTiO$_3$ and Mg$_4$Ta$_2$O$_9$ powders, Materials Research Bulletin 41 (2) (2006) 347–353.

[170] M. Guglielmi, G. Kickelbick, A. Martucci, Sol-gel Nanocomposites, Springer, 2014.

[171] H.-J. Jeon, S.-C. Yi, S.-G. Oh, Preparation and antibacterial effects of Ag–SiO$_2$ thin films by sol–gel method, Biomaterials 24 (27) (2003) 4921–4928.

[172] J.-H. Lee, B.-O. Park, Transparent conducting ZnO: Al, in and Sn thin films deposited by the sol–gel method, Thin Solid Films 426 (1–2) (2003) 94–99.

[173] D.-H. Chen, X.-R. He, Synthesis of nickel ferrite nanoparticles by sol-gel method, Materials Research Bulletin 36 (7–8) (2001) 1369–1377.

[174] G. Ennas, A. Musinu, G. Piccaluga, D. Zedda, D. Gatteschi, C. Sangregorio, J. Stanger, G. Concas, G. Spano, Characterization of iron oxide nanoparticles in an Fe$_2$O$_3$– SiO$_2$ composite prepared by a sol– gel method, Chemistry of Materials 10 (2) (1998) 495–502.

[175] H. Zhang, G. Chen, Potent antibacterial activities of Ag/TiO$_2$ nanocomposite powders synthesized by a one-pot sol– gel method, Environmental Science and Technology 43 (8) (2009) 2905–2910.

[176] C.-L. Chiang, C.-C.M. Ma, Synthesis, characterization and thermal properties of novel epoxy containing silicon and phosphorus nanocomposites by sol–gel method, European Polymer Journal 38 (11) (2002) 2219–2224.

[177] R. Nazir, M. Mazhar, M.J. Akhtar, M.R. Shah, N.A. Khan, M. Nadeem, M. Siddique, M. Mehmood, N. Butt, Superparamagnetic bimetallic iron–palladium nanoalloy: synthesis and characterization, Nanotechnology 19 (18) (2008) 185608.

[178] Q. Chen, C. Boothroyd, A.M. Soutar, X.T. Zeng, Sol–gel nanocoating on commercial TiO$_2$ nanopowder using ultrasound, Journal of Sol-Gel Science and Technology 53 (1) (2010) 115–120.

[179] E.V. Benvenutti, C.C. Moro, T.M. Costa, M.R. Gallas, Silica based hybrid materials obtained by the sol-gel method, Química Nova 32 (7) (2009) 1926–1933.

[180] M. Krolow, C. Hartwig, G. Link, C. Raubach, J. Pereira, R. Picoloto, M. Gonçalves, N. Carreño, M. Mesko, Synthesis and characterisation of carbon nanocomposites, in: NanoCarbon 2011, Springer, 2013, pp. 33–47.

[181] L. Zhang, I. Djerdj, M. Cao, M. Antonietti, M. Niederberger, Nonaqueous sol–gel synthesis of a nanocrystalline InNbO$_4$ visible-light photocatalyst, Advanced Materials 19 (16) (2007) 2083–2086.

[182] X. Liang, M. Sun, L. Li, R. Qiao, K. Chen, Q. Xiao, F. Xu, Preparation and antibacterial activities of polyaniline/Cu$_{0.05}$Zn$_{0.95}$O nanocomposites, Dalton Transactions 41 (9) (2012) 2804–2811.

[183] G. Demazeau, Solvothermal reactions: an original route for the synthesis of novel materials, Journal of Materials Science 43 (7) (2008) 2104–2114.

[184] J. Ye, W. Liu, J. Cai, S. Chen, X. Zhao, H. Zhou, L. Qi, Nanoporous anatase TiO$_2$ mesocrystals: additive-free synthesis, remarkable crystalline-phase stability, and improved lithium insertion behavior, Journal of the American Chemical Society 133 (4) (2010) 933–940.

[185] K. Byrappa, M. Yoshimura, Handbook of Hydrothermal Technology, William Andrew, 2012.

[186] M.S. Akhtar, M.A. Khan, M.S. Jeon, O.-B. Yang, Controlled synthesis of various ZnO nanostructured materials by capping agents-assisted hydrothermal method for dye-sensitized solar cells, Electrochimica Acta 53 (27) (2008) 7869–7874.

[187] Y. Köseoğlu, M. Bay, M. Tan, A. Baykal, H. Sözeri, R. Topkaya, N. Akdoğan, Magnetic and dielectric properties of Mn$_{0.2}$Ni$_{0.8}$Fe$_2$O$_4$ nanoparticles synthesized by PEG-assisted hydrothermal method, Journal of Nanoparticle Research 13 (5) (2011) 2235–2244.

[188] Y. Wang, G. Xu, Z. Ren, X. Wei, W. Weng, P. Du, G. Shen, G. Han, Mineralizer-assisted hydrothermal synthesis and characterization of BiFeO$_3$ nanoparticles, Journal of the American Ceramic Society 90 (8) (2007) 2615–2617.

[189] W.W. Wang, Y.J. Zhu, L.X. Yang, ZnO–SnO$_2$ hollow spheres and hierarchical nanosheets: hydrothermal preparation, formation mechanism, and photocatalytic properties, Advanced Functional Materials 17 (1) (2007) 59–64.

[190] S. Verma, P. Joy, Y. Khollam, H. Potdar, S. Deshpande, Synthesis of nanosized MgFe$_2$O$_4$ powders by microwave hydrothermal method, Materials Letters 58 (6) (2004) 1092–1095.

[191] Z. Li, B. Hou, Y. Xu, D. Wu, Y. Sun, W. Hu, F. Deng, Comparative study of sol–gel-hydrothermal and sol–gel synthesis of titania–silica composite nanoparticles, Journal of Solid State Chemistry 178 (5) (2005) 1395–1405.

[192] H. Wang, Y. Ma, G. Yi, D. Chen, Synthesis of Mn-doped Zn$_2$SiO$_4$ rodlike nanoparticles through hydrothermal method, Materials Chemistry and Physics 82 (2) (2003) 414–418.

[193] Z. Hua, X.M. Wang, P. Xiao, J. Shi, Solvent effect on microstructure of yttria-stabilized zirconia (YSZ) particles in solvothermal synthesis, Journal of the European Ceramic Society 26 (12) (2006) 2257–2264.

[194] J. Wang, F. Ren, R. Yi, A. Yan, G. Qiu, X. Liu, Solvothermal synthesis and magnetic properties of size-controlled nickel ferrite nanoparticles, Journal of Alloys and Compounds 479 (1–2) (2009) 791–796.

[195] W.S. Nam, G.Y. Han, Characterization and photocatalytic performance of nanosize TiO$_2$ powders prepared by the solvothermal method, Korean Journal of Chemical Engineering 20 (6) (2003) 1149–1153.

[196] M. Hussain, M. Ahmad, A. Nisar, H. Sun, S. Karim, M. Khan, S.D. Khan, M. Iqbal, S.Z. Hussain, Enhanced photocatalytic and electrochemical properties of Au nanoparticles supported TiO$_2$ microspheres, New Journal of Chemistry 38 (4) (2014) 1424–1432.

[197] V.R. de Mendonça, C.J. Dalmaschio, E.R. Leite, M. Niederberger, C. Ribeiro, Heterostructure formation from hydrothermal annealing of preformed nanocrystals, Journal of Materials Chemistry A 3 (5) (2015) 2216–2225.

[198] C. Mondal, J. Pal, M. Ganguly, A.K. Sinha, T. Jana, T. Pal, A one pot synthesis of Au–ZnO nanocomposites for plasmon-enhanced sunlight driven photocatalytic activity, New Journal of Chemistry 38 (7) (2014) 2999–3005.

[199] H. Huang, Q. Chen, M. He, X. Sun, X. Wang, A ternary Pt/MnO$_2$/graphene nanohybrid with an ultrahigh electrocatalytic activity toward methanol oxidation, Journal of Power Sources 239 (2013) 189–195.

[200] X. Yu, L. Kuai, B. Geng, CeO$_2$/rGO/Pt sandwich nanostructure: rGO-enhanced electron transmission between metal oxide and metal nanoparticles for anodic methanol oxidation of direct methanol fuel cells, Nanoscale 4 (18) (2012) 5738–5743.

[201] O. Bello, K. Adegoke, R. Oyewole, Biomimetic materials in our world: a review, Journal of Applied Chemistry 5 (2013) 22–35.

[202] Y. Oaki, H. Imai, Biomimetic morphological design for manganese oxide and cobalt hydroxide nanoflakes with a mosaic interior, Journal of Materials Chemistry 17 (4) (2007) 316–321.

[203] C. Avanzato, J. Follieri, I. Banerjee, K. Fath, Biomimetic synthesis and antibacterial characteristics of magnesium oxide–germanium dioxide nanocomposite powders, Journal of Composite Materials 43 (8) (2009) 897–910.

[204] C. Liu, D. Yang, Y. Jiao, Y. Tian, Y. Wang, Z. Jiang, Biomimetic synthesis of TiO$_2$–SiO$_2$–Ag nanocomposites with enhanced visible-light photocatalytic activity, ACS Applied Materials and Interfaces 5 (9) (2013) 3824–3832.

[205] Z. Tong, D. Yang, T. Xiao, Y. Tian, Z. Jiang, Biomimetic fabrication of g-C$_3$N$_4$/TiO$_2$ nanosheets with enhanced photocatalytic activity toward organic pollutant degradation, Chemical Engineering Journal 260 (2015) 117–125.

[206] Y. Wu, M. Yan, J. Cui, Y. Yan, C. Li, A multiple-functional Ag/SiO$_2$/organic based biomimetic nanocomposite membrane for high-stability protein recognition and cell adhesion/detachment, Advanced Functional Materials 25 (36) (2015) 5823–5832.

[207] F. Opoku, E.M. Kiarii, P.P. Govender, M.A. Mamo, Metal Oxide Polymer Nanocomposites in Water Treatments, Descriptive Inorganic Chemistry Researches of Metal Compounds, IntechOpen, 2017.

[208] S. Laurent, D. Forge, M. Port, A. Roch, C. Robic, L. Vander Elst, R.N. Muller, Magnetic iron oxide nanoparticles: synthesis, stabilization, vectorization, physicochemical characterizations, and biological applications, Chemical Reviews 108 (6) (2008) 2064–2110.

[209] D. Wang, R. Kou, D. Choi, Z. Yang, Z. Nie, J. Li, L.V. Saraf, D. Hu, J. Zhang, G.L. Graff, Ternary self-assembly of ordered metal oxide– graphene nanocomposites for electrochemical energy storage, ACS Nano 4 (3) (2010) 1587–1595.

[210] M. Abareshi, S.M. Zebarjad, E. Goharshadi, Crystallinity behavior of MDPE-clay nanocomposites fabricated using ball milling method, Journal of Composite Materials 43 (23) (2009) 2821–2830.

[211] A. Azari, R.R. Kalantary, G. Ghanizadeh, B. Kakavandi, M. Farzadkia, E. Ahmadi, Iron–silver oxide nanoadsorbent synthesized by co-precipitation process for fluoride removal from aqueous solution and its adsorption mechanism, RSC Advances 5 (106) (2015) 87377–87391.

[212] L. Chai, Y. Wang, N. Zhao, W. Yang, X. You, Sulfate-doped Fe$_3$O$_4$/Al$_2$O$_3$ nanoparticles as a novel adsorbent for fluoride removal from drinking water, Water Research 47 (12) (2013) 4040–4049.

[213] M. Ikim, E.Y. Spiridonova, T. Belysheva, V. Gromov, G. Gerasimov, L. Trakhtenberg, Structural properties of metal oxide nanocomposites: effect of preparation method, Russian Journal of Physical Chemistry B 10 (3) (2016) 543–546.

[214] E. Flahaut, A. Peigney, C. Laurent, C. Marliere, F. Chastel, A. Rousset, Carbon nanotube–metal–oxide nanocomposites: microstructure, electrical conductivity and mechanical properties, Acta Materialia 48 (14) (2000) 3803–3812.

[215] L. Casas, A. Roig, E. Rodríguez, E. Molins, J. Tejada, J. Sort, Silica aerogel–iron oxide nanocomposites: structural and magnetic properties, Journal of Non-crystalline Solids 285 (1–3) (2001) 37–43.

[216] A.L.M. Reddy, S. Ramaprabhu, Nanocrystalline metal oxides dispersed multiwalled carbon nanotubes as supercapacitor electrodes, The Journal of Physical Chemistry C 111 (21) (2007) 7727–7734.

[217] A.B. Morgan, J.W. Gilman, Characterization of polymer-layered silicate (clay) nanocomposites by transmission electron microscopy and X-ray diffraction: a comparative study, Journal of Applied Polymer Science 87 (8) (2003) 1329–1338.

[218] H. Sun, L. Cao, L. Lu, Magnetite/reduced graphene oxide nanocomposites: one step solvothermal synthesis and use as a novel platform for removal of dye pollutants, Nano Research 4 (6) (2011) 550–562.

[219] V. Mittal, Polymer Nanotubes Nanocomposites: Synthesis, Properties and Applications, John Wiley & Sons, 2014.

[220] C.-L. Zhu, S.-W. Chou, S.-F. He, W.-N. Liao, C.-C. Chen, Synthesis of core/shell metal oxide/polyaniline nanocomposites and hollow polyaniline capsules, Nanotechnology 18 (27) (2007) 275604.

[221] N. Karak, Fundamentals of Nanomaterials and Polymer Nanocomposites, Nanomaterials and Polymer Nanocomposites, Elsevier, 2019, pp. 1–45.

[222] N. Karak, Chapter 1 - fundamentals of nanomaterials and polymer nanocomposites, in: N. Karak (Ed.), Nanomaterials and Polymer Nanocomposites, Elsevier, 2019, pp. 1—45.

[223] I.S. Chronakis, Novel nanocomposites and nanoceramics based on polymer nanofibers using electrospinning process—a review, Journal of Materials Processing Technology 167 (2—3) (2005) 283—293.

[224] B.J. Clapsaddle, A.E. Gash, J.H. Satcher Jr., R.L. Simpson, Silicon oxide in an iron (III) oxide matrix: the sol—gel synthesis and characterization of Fe—Si mixed oxide nanocomposites that contain iron oxide as the major phase, Journal of Non-crystalline Solids 331 (1—3) (2003) 190—201.

[225] H. Zou, S. Wu, J. Shen, Polymer/silica nanocomposites: preparation, characterization, properties, and applications, Chemical Reviews 108 (9) (2008) 3893—3957.

[226] S. Chen, J. Zhu, X. Wu, Q. Han, X. Wang, Graphene oxide— MnO_2 nanocomposites for supercapacitors, ACS Nano 4 (5) (2010) 2822—2830.

[227] R.L. Vander Wal, A.J. Tomasek, M.I. Pamphlet, C.D. Taylor, W.K. Thompson, Analysis of HRTEM images for carbon nanostructure quantification, Journal of Nanoparticle Research 6 (6) (2004) 555—568.

[228] P. Eaton, P. West, Atomic Force Microscopy, Oxford university press, 2010.

[229] H. Kim, D.-H. Seo, S.-W. Kim, J. Kim, K. Kang, Highly reversible Co_3O_4/graphene hybrid anode for lithium rechargeable batteries, Carbon 49 (1) (2011) 326—332.

[230] P. Daraei, S.S. Madaeni, N. Ghaemi, E. Salehi, M.A. Khadivi, R. Moradian, B. Astinchap, Novel polyethersulfone nanocomposite membrane prepared by PANI/Fe_3O_4 nanoparticles with enhanced performance for Cu (II) removal from water, Journal of Membrane Science 415 (2012) 250—259.

[231] S. Thomas, K. Joseph, S. Malhotra, K. Goda, M. Sreekala, Polymer Composites, Nanocomposites, Technology & Engineering, John Wiley & Sons, 2013.

[232] A. Jitianu, M. Crisan, A. Meghea, I. Rau, M. Zaharescu, Influence of the silica based matrix on the formation of iron oxide nanoparticles in the Fe_2O_3—SiO_2 system, obtained by sol—gel method, Journal of Materials Chemistry 12 (5) (2002) 1401—1407.

[233] A. Laachachi, E. Leroy, M. Cochez, M. Ferriol, J.L. Cuesta, Use of oxide nanoparticles and organoclays to improve thermal stability and fire retardancy of poly (methyl methacrylate), Polymer Degradation and Stability 89 (2) (2005) 344—352.

[234] H. Xia, M. Lai, L. Lu, Nanoflaky MnO_2/carbon nanotube nanocomposites as anode materials for lithium-ion batteries, Journal of Materials Chemistry 20 (33) (2010) 6896—6902.

[235] S. Eustis, M.A. El-Sayed, Why gold nanoparticles are more precious than pretty gold: noble metal surface plasmon resonance and its enhancement of the radiative and nonradiative properties of nanocrystals of different shapes, Chemical Society Reviews 35 (3) (2006) 209—217.

[236] S. Thakur, N. Karak, Green reduction of graphene oxide by aqueous phytoextracts, Carbon 50 (14) (2012) 5331—5339.

[237] R. Duarah, N. Karak, Facile and ultrafast green approach to synthesize biobased luminescent reduced carbon nanodot: an efficient photocatalyst, ACS Sustainable Chemistry & Engineering 5 (10) (2017) 9454—9466.

[238] G. Kickelbick, Introduction to Hybrid Materials, Hybrid Materials: Synthesis, Characterization, and Applications, 2007, pp. 1—48.

[239] J.-C. Dupin, D. Gonbeau, P. Vinatier, A. Levasseur, Systematic XPS studies of metal oxides, hydroxides and peroxides, Physical Chemistry Chemical Physics 2 (6) (2000) 1319—1324.

[240] J.F. Watts, J. Wolstenholme, An Introduction to Surface Analysis by XPS and AES, Wiley-VCH, May 2003, p. 224. ISBN 0-470-84713-1.

[241] M. Abbott, J. Cotter, F. Chen, T. Trupke, R. Bardos, K. Fisher, Application of photoluminescence characterization to the development and manufacturing of high-efficiency silicon solar cells, Journal of Applied Physics 100 (11) (2006) 114514.

[242] M.-A. Tartanson, L. Soussan, M. Rivallin, S. Pecastaings, C.V. Chis, D. Penaranda, C. Roques, C. Faur, Dynamic mechanisms of the bactericidal action of an Al_2O_3-TiO_2-Ag granular material on an *Escherichia coli* strain, Applied and Environmental Microbiology 81 (20) (2015) 7135—7142.

[243] T. Gordon, B. Perlstein, O. Houbara, I. Felner, E. Banin, S. Margel, Synthesis and characterization of zinc/iron oxide composite nanoparticles and their antibacterial properties, Colloids and Surfaces A: Physicochemical and Engineering Aspects 374 (1—3) (2011) 1—8.

[244] K. Giannousi, G. Sarafidis, S. Mourdikoudis, A. Pantazaki, C. Dendrinou-Samara, Selective synthesis of Cu_2O and Cu/Cu_2O NPs: antifungal activity to yeast *Saccharomyces cerevisiae* and DNA interaction, Inorganic Chemistry 53 (18) (2014) 9657—9666.

[245] A. Dural Erem, G. Ozcan, M. Skrifvars, Antibacterial activity of PA6/ZnO nanocomposite fibers, Textile Research Journal 81 (16) (2011) 1638—1646.

[246] D.B. Hamal, J.A. Haggstrom, G.L. Marchin, M.A. Ikenberry, K. Hohn, K.J. Klabunde, A multifunctional biocide/sporocide and photocatalyst based on titanium dioxide (TiO_2) codoped with silver, carbon, and sulfur, Langmuir 26 (4) (2009) 2805—2810.

[247] R. Thaya, B. Malaikozhundan, S. Vijayakumar, J. Sivakamavalli, R. Jeyasekar, S. Shanthi, B. Vaseeharan, P. Ramasamy, A. Sonawane, Chitosan coated Ag/ZnO nanocomposite and their antibiofilm, antifungal and cytotoxic effects on murine macrophages, Microbial Pathogenesis 100 (2016) 124—132.

[248] G.P. Halliah, K. Alagappan, A.B. Sairam, Synthesis, characterization of CH-α-Fe_2O_3 nanocomposite and coating on cotton, silk for antibacterial and UV spectral studies, Journal of Industrial Textiles 44 (2) (2014) 275—287.

[249] E. Malka, I. Perelshtein, A. Lipovsky, Y. Shalom, L. Naparstek, N. Perkas, T. Patick, R. Lubart, Y. Nitzan,

E. Banin, Eradication of multi-drug resistant bacteria by a novel Zn-doped CuO nanocomposite, Small 9 (23) (2013) 4069–4076.

[250] K. Phiwdang, M. Phensaijai, W. Pecharapa, Study of Antifungal Activities of CuO/ZnO Nanocomposites Synthesized by Co-precipitation Method, Advanced Materials Research, Trans Tech Publ, 2013, pp. 89–93.

[251] S. Kasraei, L. Sami, S. Hendi, M.-Y. AliKhani, L. Rezaei-Soufi, Z. Khamverdi, Antibacterial properties of composite resins incorporating silver and zinc oxide nanoparticles on Streptococcus mutans and Lactobacillus, Restorative Dentistry and Endodontics 39 (2) (2014) 109–114.

[252] S. Bykkam, S. Narsingam, M. Ahmadipour, T. Dayakar, K.V. Rao, C.S. Chakra, S. Kalakotla, Few layered graphene sheet decorated by ZnO nanoparticles for antibacterial application, Superlattices and Microstructures 83 (2015) 776–784.

[253] S. Kulkarni, M. Jadhav, P. Raikar, D.A. Barretto, S.K. Vootla, U. Raikar, Green synthesized multifunctional Ag@ Fe$_2$O$_3$ nanocomposites for effective antibacterial, antifungal and anticancer properties, New Journal of Chemistry 41 (17) (2017) 9513–9520.

[254] K. Delgado, R. Quijada, R. Palma, H. Palza, Polypropylene with embedded copper metal or copper oxide nanoparticles as a novel plastic antimicrobial agent, Letters in Applied Microbiology 53 (1) (2011) 50–54.

[255] N. Qureshi, R. Chaudhari, P. Mane, M. Shinde, S. Jadakar, S. Rane, B. Kale, A. Bhalerao, D. Amalnerkar, Nanoscale Mo-${\rm MoO} _ {3} $ entrapped in engineering thermoplastic: inorganic pathway to bactericidal and fungicidal action, IEEE Transactions on Nanobioscience 15 (3) (2016) 258–264.

[256] Y. Haldorai, J.J. Shim, Novel chitosan-TiO$_2$ nanohybrid: preparation, characterization, antibacterial, and photocatalytic properties, Polymer Composites 35 (2) (2014) 327–333.

[257] W.C. Huang, P.J. Tsai, Y.C. Chen, Multifunctional Fe$_3$O$_4$@ Au nanoeggs as photothermal agents for selective killing of nosocomial and antibiotic-resistant bacteria, Small 5 (1) (2009) 51–56.

[258] W. Zhang, X. Shi, J. Huang, Y. Zhang, Z. Wu, Y. Xian, Bacitracin-conjugated superparamagnetic iron oxide nanoparticles: synthesis, characterization and antibacterial activity, ChemPhysChem 13 (14) (2012) 3388–3396.

[259] I. Perelshtein, G. Applerot, N. Perkas, E. Wehrschuetz-Sigl, A. Hasmann, G. Gübitz, A. Gedanken, CuO–cotton nanocomposite: formation, morphology, and antibacterial activity, Surface and Coatings Technology 204 (1–2) (2009) 54–57.

[260] M. Arshad, A. Qayyum, G.A. Shar, G.A. Soomro, A. Nazir, B. Munir, M. Iqbal, Zn-doped SiO$_2$ nanoparticles preparation and characterization under the effect of various solvents: antibacterial, antifungal and photocatlytic performance evaluation, Journal of Photochemistry and Photobiology B: Biology 185 (2018) 176–183.

[261] C.M. Magdalane, K. Kaviyarasu, J.J. Vijaya, B. Siddhardha, B. Jeyaraj, Photocatalytic activity of binary metal oxide nanocomposites of CeO$_2$/CdO nanospheres: investigation of optical and antimicrobial activity, Journal of Photochemistry and Photobiology B: Biology 163 (2016) 77–86.

[262] N. Ghasemi, F. Jamali-Sheini, R. Zekavati, CuO and Ag/CuO nanoparticles: biosynthesis and antibacterial properties, Materials Letters 196 (2017) 78–82.

[263] K. Karthik, S. Dhanuskodi, C. Gobinath, S. Prabukumar, S. Sivaramakrishnan, Multifunctional properties of microwave assisted CdO–NiO–ZnO mixed metal oxide nanocomposite: enhanced photocatalytic and antibacterial activities, Journal of Materials Science: Materials in Electronics 29 (7) (2018) 5459–5471.

[264] M.A. Subhan, T. Ahmed, N. Uddin, A.K. Azad, K. Begum, Synthesis, characterization, PL properties, photocatalytic and antibacterial activities of nano multi-metal oxide NiO· CeO$_2$· ZnO, Spectrochimica Acta Part A: Molecular and Biomolecular Spectroscopy 136 (2015) 824–831.

[265] M.A. Subhan, N. Uddin, P. Sarker, H. Nakata, R. Makioka, Synthesis, characterization, low temperature solid state PL and photocatalytic activities of Ag$_2$O· CeO$_2$· ZnO nanocomposite, Spectrochimica Acta Part A: Molecular and Biomolecular Spectroscopy 151 (2015) 56–63.

[266] L. Karimi, M.E. Yazdanshenas, R. Khajavi, A. Rashidi, M. Mirjalili, Using graphene/TiO$_2$ nanocomposite as a new route for preparation of electroconductive, self-cleaning, antibacterial and antifungal cotton fabric without toxicity, Cellulose 21 (5) (2014) 3813–3827.

[267] T. Jana, S. Maji, A. Pal, R. Maiti, T. Dolai, K. Chatterjee, Photocatalytic and antibacterial activity of cadmium sulphide/zinc oxide nanocomposite with varied morphology, Journal of Colloid and Interface Science 480 (2016) 9–16.

[268] K. Ravichandran, N. Chidhambaram, S. Gobalakrishnan, Copper and Graphene activated ZnO nanopowders for enhanced photocatalytic and antibacterial activities, Journal of Physics and Chemistry of Solids 93 (2016) 82–90.

[269] A. Manikandan, E. Manikandan, B. Meenatchi, S. Vadivel, S. Jaganathan, R. Ladchumananandasivam, M. Henini, M. Maaza, J.S. Aanand, Rare earth element (REE) lanthanum doped zinc oxide (La: ZnO) nanomaterials: synthesis structural optical and antibacterial studies, Journal of Alloys and Compounds 723 (2017) 1155–1161.

[270] Z. Shu, Y. Zhang, Q. Yang, H. Yang, Halloysite nanotubes supported Ag and ZnO nanoparticles with synergistically enhanced antibacterial activity, Nanoscale Research Letters 12 (1) (2017) 135.

[271] S.-W. Zhao, M. Zheng, X.-H. Zou, Y. Guo, Q.-J. Pan, Self-assembly of hierarchically structured cellulose@ ZnO composite in solid–liquid homogeneous phase: synthesis, DFT calculations, and enhanced antibacterial activities, ACS Sustainable Chemistry & Engineering 5 (8) (2017) 6585–6596.

[272] E. Sharmin, S. Shreaz, F. Zafar, D. Akram, V. Raja, S. Ahmad, Linseed polyol-assisted, microwave-induced synthesis of nano CuO embedded in polyol-polyester matrix: antifungal behavior and coating properties, Progress in Organic Coatings 105 (2017) 200—211.

[273] S. Lee, Multifunctionality of layered fabric systems based on electrospun polyurethane/zinc oxide nanocomposite fibers, Journal of Applied Polymer Science 114 (6) (2009) 3652—3658.

[274] G. Ungur, J. Hrůza, Modified polyurethane nanofibers as antibacterial filters for air and water purification, RSC Advances 7 (78) (2017) 49177—49187.

[275] R. Prucek, J. Tuček, M. Kilianová, A. Panáček, L. Kvítek, J. Filip, M. Kolář, K. Tománková, R. Zbořil, The targeted antibacterial and antifungal properties of magnetic nanocomposite of iron oxide and silver nanoparticles, Biomaterials 32 (21) (2011) 4704—4713.

[276] C.M. Magdalane, K. Kaviyarasu, J.J. Vijaya, B. Siddhardha, B. Jeyaraj, Facile synthesis of heterostructured cerium oxide/yttrium oxide nanocomposite in UV light induced photocatalytic degradation and catalytic reduction: synergistic effect of antimicrobial studies, Journal of Photochemistry and Photobiology B: Biology 173 (2017) 23—34.

[277] A. Farrukh, M. Arshad, S. Haneef, N. Aslam, A. Afzaal, Antibacterial and antifungal activities of zinc-silicon oxides nanocomposite, Letters in Health and Biological Sciences 1 (1) (2016) 5—9.

[278] S. Tripathi, G. Mehrotra, P. Dutta, Chitosan—silver oxide nanocomposite film: preparation and antimicrobial activity, Bulletin of Materials Science 34 (1) (2011) 29—35.

[279] S. Ramesh, A. Sivasamy, K. Rhee, S. Park, D. Hui, Preparation and characterization of maleimide—polystyrene/SiO2—Al2O3 hybrid nanocomposites by an in situ sol—gel process and its antimicrobial activity, Composites Part B: Engineering 75 (2015) 167—175.

[280] L.V. Trandafilović, D.K. Božanić, S. Dimitrijević-Branković, A. Luyt, V. Djoković, Fabrication and antibacterial properties of ZnO—alginate nanocomposites, Carbohydrate Polymers 88 (1) (2012) 263—269.

[281] M. Elango, M. Deepa, R. Subramanian, G. Saraswathy, Synthesis, structural characterization and antimicrobial activities of polyindole stabilized Ag-Co3O4 nanocomposite by reflux condensation method, Materials Chemistry and Physics 216 (2018) 305—315.

[282] B. Cao, S. Cao, P. Dong, J. Gao, J. Wang, High antibacterial activity of ultrafine TiO2/graphene sheets nanocomposites under visible light irradiation, Materials Letters 93 (2013) 349—352.

[283] O. Akhavan, E. Ghaderi, Photocatalytic reduction of graphene oxide nanosheets on TiO2 thin film for photoinactivation of bacteria in solar light irradiation, Journal of Physical Chemistry C 113 (47) (2009) 20214—20220.

[284] O. Akhavan, E. Ghaderi, K. Rahimi, Adverse effects of graphene incorporated in TiO2 photocatalyst on minuscule animals under solar light irradiation, Journal of Materials Chemistry 22 (43) (2012) 23260—23266.

[285] K. Tam, C. Cheung, Y. Leung, A. Djurišić, C. Ling, C. Beling, S. Fung, W. Kwok, W. Chan, D. Phillips, Defects in ZnO nanorods prepared by a hydrothermal method, The Journal of Physical Chemistry B 110 (42) (2006) 20865—20871.

[286] T. Kavitha, A.I. Gopalan, K.-P. Lee, S.-Y. Park, Glucose sensing, photocatalytic and antibacterial properties of graphene—ZnO nanoparticle hybrids, Carbon 50 (8) (2012) 2994—3000.

[287] J.A. Khan, M. Qasim, B.R. Singh, W. Khan, D. Das, A.H. Naqvi, Polyaniline/CoFe2O4 nanocomposite inhibits the growth of *Candida albicans* 077 by ROS production, Comptes Rendus Chimie 17 (2) (2014) 91—102.

[288] O. Akhavan, M. Choobtashani, E. Ghaderi, Protein degradation and RNA efflux of viruses photocatalyzed by graphene—tungsten oxide composite under visible light irradiation, The Journal of Physical Chemistry C 116 (17) (2012) 9653—9659.

[289] R. Michal, E. Dworniczek, M. Caplovicova, O. Monfort, P. Lianos, L. Caplovic, G. Plesch, Photocatalytic properties and selective antimicrobial activity of TiO2 (Eu)/CuO nanocomposite, Applied Surface Science 371 (2016) 538—546.

[290] S.F. Chen, J.P. Li, K. Qian, W.P. Xu, Y. Lu, W.X. Huang, S.H. Yu, Large scale photochemical synthesis of M@TiO2 nanocomposites (M= Ag, Pd, Au, Pt) and their optical properties, CO oxidation performance, and antibacterial effect, Nano Research 3 (4) (2010) 244—255.

[291] D. Devadathan, R. Raveendran, Polyindole based nickel-zinc oxide nanocomposite-characterization and antifungal studies, International Journal of Chemical Engineering and Applications 5 (3) (2014) 240.

[292] W.J. Chen, P.J. Tsai, Y.C. Chen, Functional Fe3O4/TiO2 core/shell magnetic nanoparticles as photokilling agents for pathogenic bacteria, Small 4 (4) (2008) 485—491.

[293] R. Saravanan, M.M. Khan, V.K. Gupta, E. Mosquera, F. Gracia, V. Narayanan, A. Stephen, ZnO/Ag/Mn2O3 nanocomposite for visible light-induced industrial textile effluent degradation, uric acid and ascorbic acid sensing and antimicrobial activity, RSC Advances 5 (44) (2015) 34645—34651.

[294] C. Sridhar, N. Gunvanthrao Yernale, M. Prasad, Synthesis, spectral characterization, and antibacterial and antifungal studies of PANI/V2O5 nanocomposites, International Journal of Chemical Engineering 2016 (2016).

[295] Y. Gutha, J.L. Pathak, W. Zhang, Y. Zhang, X. Jiao, Antibacterial and wound healing properties of chitosan/poly(vinyl alcohol)/zinc oxide beads (CS/PVA/ZnO), International Journal of Biological Macromolecules 103 (2017) 234—241.

[296] S. Shanavas, A. Priyadharsan, V. Vasanthakumar, A. Arunkumar, P. Anbarasan, S. Bharathkumar, Mechanistic investigation of visible light driven novel La2CuO4/CeO2/rGO ternary hybrid nanocomposites for enhanced photocatalytic performance and antibacterial activity, Journal of Photochemistry and Photobiology A: Chemistry 340 (2017) 96—108.

[297] C.M. Magdalane, K. Kaviyarasu, J.J. Vijaya, C. Jayakumar, M. Maaza, B. Jeyaraj, Photocatalytic degradation effect of malachite green and catalytic hydrogenation by UV–illuminated CeO_2/CdO multilayered nanoplatelet arrays: investigation of antifungal and antimicrobial activities, Journal of Photochemistry and Photobiology B: Biology 169 (2017) 110–123.

[298] S. Parham, S. Chandren, D.H. Wicaksono, S. Bagherbaigi, S.L. Lee, L.S. Yuan, H. Nur, Textile/$Al_2O_3–TiO_2$ nanocomposite as an antimicrobial and radical scavenger wound dressing, RSC Advances 6 (10) (2016) 8188–8197.

[299] D.F. Emerich, C.G. Thanos, Nanotechnology and medicine, Expert Opinion on Biological Therapy 3 (4) (2003) 655–663.

[300] A.W.C. Chua, Y.C. Khoo, B.K. Tan, K.C. Tan, C.L. Foo, S.J. Chong, Skin tissue engineering advances in severe burns: review and therapeutic applications, Burns and Trauma 4 (1) (2016) 3.

[301] D. Sundaramurthi, U.M. Krishnan, S. Sethuraman, Electrospun nanofibers as scaffolds for skin tissue engineering, Polymer Reviews 54 (2) (2014) 348–376.

[302] M. Mühlstädt, C. Thomé, C. Kunte, Rapid wound healing of scalp wounds devoid of periosteum with milling of the outer table and split-thickness skin grafting, British Journal of Dermatology 167 (2) (2012) 343–347.

[303] F. Siedenbiedel, J.C. Tiller, Antimicrobial polymers in solution and on surfaces: overview and functional principles, Polymers 4 (1) (2012) 46–71.

[304] D. Simões, S.P. Miguel, M.P. Ribeiro, P. Coutinho, A.G. Mendonça, I.J. Correia, Recent advances on antimicrobial wound dressing: a review, European Journal of Pharmaceutics and Biopharmaceutics 127 (2018) 130–141.

[305] V. Ki, C. Rotstein, Bacterial skin and soft tissue infections in adults: a review of their epidemiology, pathogenesis, diagnosis, treatment and site of care, The Canadian Journal of Infectious Diseases and Medical Microbiology 19 (2) (2008) 173–184.

[306] G. Han, R. Ceilley, Chronic wound healing: a review of current management and treatments, Advances in Therapy 34 (3) (2017) 599–610.

[307] A.R. Siddiqui, J.M. Bernstein, Chronic wound infection: facts and controversies, Clinics in Dermatology 28 (5) (2010) 519–526.

[308] J.S. Boateng, K.H. Matthews, H.N. Stevens, G.M. Eccleston, Wound healing dressings and drug delivery systems: a review, Journal of Pharmaceutical Sciences 97 (8) (2008) 2892–2923.

[309] S. Dhivya, V.V. Padma, E. Santhini, Wound dressings—a review, Biomedicine 5 (4) (2015).

[310] Z. Lu, J. Gao, Q. He, J. Wu, D. Liang, H. Yang, R. Chen, Enhanced antibacterial and wound healing activities of microporous chitosan-Ag/ZnO composite dressing, Carbohydrate Polymers 156 (2017) 460–469.

[311] D. Archana, B.K. Singh, J. Dutta, P. Dutta, In vivo evaluation of chitosan–PVP–titanium dioxide nanocomposite as wound dressing material, Carbohydrate Polymers 95 (1) (2013) 530–539.

[312] N. Cai, C. Li, C. Han, X. Luo, L. Shen, Y. Xue, F. Yu, Tailoring mechanical and antibacterial properties of chitosan/gelatin nanofiber membranes with Fe_3O_4 nanoparticles for potential wound dressing application, Applied Surface Science 369 (2016) 492–500.

[313] M. Molina, M. Asadian-Birjand, J. Balach, J. Bergueiro, E. Miceli, M. Calderón, Stimuli-responsive nanogel composites and their application in nanomedicine, Chemical Society Reviews 44 (17) (2015) 6161–6186.

[314] R. Xing, K. Liu, T. Jiao, N. Zhang, K. Ma, R. Zhang, Q. Zou, G. Ma, X. Yan, An injectable self-assembling collagen–gold hybrid hydrogel for combinatorial antitumor photothermal/photodynamic therapy, Advanced Materials 28 (19) (2016) 3669–3676.

[315] K. Liu, R. Xing, Q. Zou, G. Ma, H. Möhwald, X. Yan, Simple peptide-tuned self-assembly of photosensitizers towards anticancer photodynamic therapy, Angewandte Chemie International Edition 55 (9) (2016) 3036–3039.

[316] B.-Q. Lu, Y.-J. Zhu, G.-F. Cheng, Y.-J. Ruan, Synthesis and application in drug delivery of hollow-core-double-shell magnetic iron oxide/silica/calcium silicate nanocomposites, Materials Letters 104 (2013) 53–56.

[317] G. Yang, H. Gong, T. Liu, X. Sun, L. Cheng, Z. Liu, Two-dimensional magnetic WS2@ Fe_3O_4 nanocomposite with mesoporous silica coating for drug delivery and imaging-guided therapy of cancer, Biomaterials 60 (2015) 62–71.

[318] X. Li, X. Huang, D. Liu, X. Wang, S. Song, L. Zhou, H. Zhang, Synthesis of 3D hierarchical Fe_3O_4/graphene composites with high lithium storage capacity and for controlled drug delivery, The Journal of Physical Chemistry C 115 (44) (2011) 21567–21573.

[319] Y. Chan, T. Wong, F. Byrne, M. Kavallaris, V. Bulmus, Acid-labile core cross-linked micelles for pH-triggered release of antitumor drugs, Biomacromolecules 9 (7) (2008) 1826–1836.

[320] F. Cavalieri, E. Chiessi, R. Villa, L. Viganò, N. Zaffaroni, M.F. Telling, G. Paradossi, Novel PVA-based hydrogel microparticles for doxorubicin delivery, Biomacromolecules 9 (7) (2008) 1967–1973.

[321] F. Wang, G.M. Pauletti, J. Wang, J. Zhang, R.C. Ewing, Y. Wang, D. Shi, Dual surface-functionalized janus nanocomposites of polystyrene/Fe_3O_4@ SiO_2 for simultaneous tumor cell targeting and stimulus-induced drug release, Advanced Materials 25 (25) (2013) 3485–3489.

[322] F. Xiao, Y. Li, H. Gao, S. Ge, H. Duan, Growth of coral-like PtAu–MnO_2 binary nanocomposites on free-standing graphene paper for flexible nonenzymatic glucose sensors, Biosensors and Bioelectronics 41 (2013) 417–423.

[323] B. Dakshayini, K.R. Reddy, A. Mishra, N.P. Shetti, S.J. Malode, S. Basu, S. Naveen, A.V. Raghu, Role of conducting polymer and metal oxide-based hybrids for applications in ampereometric sensors and biosensors, Microchemical Journal 147 (2019) 7–24.

[324] S. Kumar, M. Umar, A. Saifi, S. Kumar, S. Augustine, S. Srivastava, B.D. Malhotra, Electrochemical paper based cancer biosensor using iron oxide nanoparticles decorated PEDOT: PSS, Analytica Chimica Acta 1056 (2019) 135—145.

[325] N.P. Shetti, S.D. Bukkitgar, K.R. Reddy, C.V. Reddy, T.M. Aminabhavi, Nanostructured titanium oxide hybrids-based electrochemical biosensors for healthcare applications, Colloids and Surfaces B: Biointerfaces 178 (2019) 385—394.

[326] P.R. Solanki, A. Kaushik, V.V. Agrawal, B.D. Malhotra, Nanostructured metal oxide-based biosensors, NPG Asia Materials 3 (2011) 17.

[327] S.K. Vashist, J.H. Luong, Recent advances in electrochemical biosensing schemes using graphene and graphene-based nanocomposites, Carbon 84 (2015) 519—550.

[328] H. Dong, S. Jin, H. Ju, K. Hao, L.-P. Xu, H. Lu, X. Zhang, Trace and label-free MicroRNA detection using oligonucleotide encapsulated silver Nanoclusters as probes, Analytical Chemistry 84 (20) (2012) 8670—8674.

[329] R. Li, S. Li, M. Dong, L. Zhang, Y. Qiao, Y. Jiang, W. Qi, H. Wang, A highly specific and sensitive electroanalytical strategy for microRNAs based on amplified silver deposition by the synergic TiO_2 photocatalysis and guanine photoreduction using charge-neutral probes, Chemical Communications 51 (89) (2015) 16131—16134.

[330] S. Li, R. Li, M. Dong, L. Zhang, Y. Jiang, L. Chen, W. Qi, H. Wang, High-throughput, selective, and sensitive colorimetry for free microRNAs in blood via exonuclease I digestion and hemin-G-quadruplex catalysis reactions based on a "self-cleaning" functionalized microarray, Sensors and Actuators B: Chemical 222 (2016) 198—204.

Index

A

AA-CVD. *See* Aerosol-assisted chemical vapor deposition (AA-CVD)
ABS. *See* Acrylonitrile butadiene styrene (ABS)
ABX. *See* Antibiotics (ABX)
Acetobacter xylinum, 69
Acinetobacter baumannii, 18, 66, 124
2-(3-Acrylamidopropyldimethylammonio)ethanoate (APDMAE), 43
Acrylonitrile butadiene styrene (ABS), 143–144
Actinobacillus actinomycetemcomitans (Aa), 193–194
Actinobacillus succinogenes, 85–86
ACTs. *See* Artemisinin-based combination therapies (ACTs)
Acute wounds, 111
Addition polymerization, 35
Additive fabrication. *See* Three-dimensional printing (3D printing)
Additive layer manufacturing. *See* Three-dimensional printing (3D printing)
Advanced oxidation processes (AOPs), 223, 223f
AEC. *See* L-Aminoethylcysteine (AEC)
Aerosol-assisted chemical vapor deposition (AA-CVD), 293
AFM. *See* Atomic force microscopy (AFM)
Agarose, 77, 77f
Agarose nanoparticles (ANPs), 77
Acquired immunodeficiency syndrome (AIDS), 20–21
Alginate, 72–73, 72f
Aliphatic biobased polyesters, 84
Alphaherpesvirinae, 73
Alumina (Al$_2$O$_3$), 288
L-Aminoethylcysteine (AEC), 242
Aminopropyltriethoxysilane (APS), 309
Amoxicillin (AMX), 222, 222f, 227
AMPs. *See* Antibacterial peptides; Antimicrobial peptides (AMPs)
Amylopectin, 68f
Amylose, 68f
Anaerobiospirillum succiniciproducens, 85–86
ANPs. *See* Agarose nanoparticles (ANPs)

Antibacterial
  agents, 73
  efficacy of MONPs, 268–269
  hydrogels, 51
  materials in 3D printing, 147–154
    for biomedical applications and future trends, 154
Antibacterial peptides (AMPs), 148
Antibiotic nanomaterials
  engineered nanocomposites and applications, 4–7
    effect of organic and inorganic antibiotics, 5t–6t
  future perspectives, 7
  mechanism of nanomaterials, 3–4
  nanoparticles, 1–2
  physical and chemical properties nanomaterial influences, 2–3
Antibiotic polymer for biomedical applications
  antimicrobial polymers, 34
  applications, 42–46
    antimicrobial polymers for antimicrobial application, 44t–45t
  characterization, 35–41, 36t–38t
    cytotoxicity assessment, 41
    mechanical and thermal characterization, 40
    polymerization techniques, 39t
    surface and morphological characterization, 41
  mechanisms of action, 41–42
  properties, 35
  synthesis, 35
  types, 34
Antibiotic-resistant bacteria/bacterium (ARB), 65–66, 159. *See also* Multidrug-resistant bacteria (MDR)
  antibiotics in wastewater, 161–162
    treatment, 161, 162t
  bacteria inhibition, 163–164
  municipal wastewater treatment, 160–161
  nanotechnology in ARM, 163
  need for clean water, 160
  removal
    of antibacteria in wastewater, 162
    of antibiotics in wastewater, 162–163
Antibiotics (ABX), 1, 6–7, 14, 51, 65, 152–153, 221–222, 249. *See also* Therapeutic efficacy of antibiotics

Antibiotics (ABX) (*Continued*)
  chitosan-starch in antibiotic delivery, 57–58
  encapsulated scaffolds, 111–112, 114t
  scaffold wound dressings, 113–124
  wound healing mechanisms, 112–113
  polymer/metal nanocomposites for health applications
    biomedical postulations for antibiotic efficacy, 136
    future perspectives, 136–137
    polymer materials and architectures for antibiotic activity, 131–132
    structural components of metallic elements and metal oxides, 129–130
    syntheses of biocidal metal/metal oxide polymer nanocomposites, 135–136
  polymers, 66
  resistance mechanisms, 177–181
Anticancer drugs, 14
Antimicrobial peptides (AMPs), 66
Antimicrobial polymers, 34, 148–149
Antiretroviral therapy (ART), 18
Antiretroviral treatment (ARV treatment), 21
  challenges of, 21
Antiseptic surgery, 172
AOPs. *See* Advanced oxidation processes (AOPs)
APDMAE. *See* 2-(3-Acrylamidopropyldimethylammonio)ethanoate (APDMAE)
APS. *See* Aminopropyltriethoxysilane (APS)
Aptamers, 231–232
  as antibiotic therapy, 234–235
  with antibiotics activity, 235–238
  with for antibacterial therapies and biosensing, 239t–240t
  against lipopolysaccharide, 237
  against *M. tuberculosis*, 235–236
  against metallo-β-lactamase, 237–238
  riboswitches and use as antimicrobial targets, 238–242
  SELEX, 232–234

Note: Page numbers followed by "f" indicate figures and "t" indicate tables.

325